TOXIC CONTAMINANTS IN THE GREAT LAKES

Volume

14

in the Wiley Series in

Advances in Environmental Science and Technology

JEROME O. NRIAGU, Series Editor

TOXIC CONTAMINANTS IN THE GREAT LAKES

Edited by

Jerome O. Nriagu
National Water Research Institute
Burlington, Ontario, Canada

Milagros S. Simmons
The University of Michigan
Ann Arbor, Michigan

A WILEY-INTERSCIENCE PUBLICATION

JOHN WILEY & SONS

New York · Chichester · Brisbane · Toronto · Singapore

Library of Congress Cataloging in Publication Data:

Main entry under title:
Toxic Contaminants in the Great Lakes.

(Advances in environmental science and technology ; v. 14)
"A Wiley-Interscience publication."
Includes index.
1. Water—Pollution—Great Lakes Region. 2. Organic water pollution—Toxicology. I. Nriagu, Jerome O. II. Simmons, Milagros S. III. Series.

TD180.A38 vol. 14 628s [363.7′394] 83-16689
[TD223.3]

ISBN 0-471-89087-1

Printed in the United States of America

10 9 8 7 6 5 4 3 2 1

CONTRIBUTORS

ANDREN, ANDERS W., Water Chemistry Program, University of Wisconsin, Madison, Wisconsin

BORGMANN, UWE, Great Lakes Fisheries Research Branch, Canada Centre for Inland Waters, Burlington, Ontario

BRECHER, RONALD W., Canadian Wildlife Service, Ontario Region, National Wildlife Research Centre, Hull, Quebec

CAHILL, RICHARD A., Illinois State Geological Survey, Champaign, Illinois

CHAU, Y. K., National Water Research Institute, Canada Centre for Inland Waters, Burlington, Ontario

EADIE, BRIAN J., National Oceanic and Atmospheric Administration, Great Lakes Environmental Research Laboratory, Ann Arbor, Michigan

EDWARDS, C. J., International Joint Commission, Regional Office, Windsor, Ontario

ELLENTON, J. A., Canadian Wildlife Service, National Wildlife Research Centre, Ottawa, Ontario

EVANS, MARLENE S., Great Lakes Research Division, University of Michigan, Ann Arbor, Michigan

FINGLETON, DONALD, Energy and Environmental Systems Division, Argonne National Laboratory, Argonne, Illinois

FOX, G. A., Canadian Wildlife Service, Ottawa, Ontario

HALLETT, DOUGLAS J., Department of the Environment, Ontario Region, Toronto, Ontario

HODSON, PETER V., Great Lakes Fisheries Research Branch, Canada Centre for Inland Waters, Burlington, Ontario

MEYERS, PHILIP A., Oceanography Program, Department of Atmospheric and Oceanic Science, University of Michigan, Ann Arbor, Michigan

MINEAU, P., Canadian Wildlife Service, Canada Centre for Inland Waters, Burlington, Ontario

MURPHY, THOMAS J., Department of Chemistry, DePaul University, Chicago, Illinois

NORSTROM, R. J., Canadian Wildlife Service, National Wildlife Research Centre, Ottawa, Ontario

NRIAGU, JEROME O., National Water Research Institute, Burlington, Ontario

OWEN, ROBERT M., Oceanography Program, Department of Atmospheric and Oceanic Science, University of Michigan, Ann Arbor, Michigan

PARISO, MARY E., Bureau of Water Quality, Wisconsin Department of Natural Resources, Madison, Wisconsin

RICE, CLIFFORD P., Great Lakes Research Division, University of Michigan, Ann Arbor, Michigan

RYGWELSKI, KENNETH R., Cranbrook Institute of Science, Grosse Ile, Michigan

ST. AMANT, JAMES R., Bureau of Water Quality, Wisconsin Department of Natural Resources, Madison, Wisconsin

SCHMIDT, JILL A., Water Chemistry Program, University of Wisconsin, Madison, Wisconsin

SHEAR, H., Great Lakes Fisheries Research Branch, Canada Center for Inland Waters, Burlington, Ontario

SHEFFY, THOMAS B., Bureau of Water Quality, Wisconsin Department of Natural Resources, Madison, Wisconsin

SHIMP, N. F., Illinois State Geological Survey, Champaign, Illinois

SIMMONS, MILAGROS S., Department of Environmental and Industrial Health, The University of Michigan, Ann Arbor, Michigan

SONZOGNI, WILLIAM C., State Laboratory of Hygiene, University of Wisconsin, Madison, Wisconsin

STRACHAN, W. M. J., National Water Research Institute, Canada Centre for Inland Waters, Burlington, Ontario

SWAIN, WAYLAND R., Vakgroep Aquatische Oecologie, Amsterdam, The Netherlands

THOMAS, R. L., Great Lakes Research Branch, Canada Centre for Inland Waters, Burlington, Ontario

TISUE, THOMAS, Center for Great Lakes Studies, University of Wisconsin—Milwaukee, Milwaukee, Wisconsin

WESELOH, D. V., Canadian Wildlife Service, Canada Centre for Inland Waters, Burlington, Ontario

WHITTLE, D. MICHAEL, Great Lakes Fisheries Research Branch, Canada Centre for Inland Waters, Burlington, Ontario

WONG, PAUL T. S., Great Lakes Fisheries Research Branch, Canada Centre for Inland Waters, Burlington, Ontario

INTRODUCTION TO THE SERIES

The deterioration of environmental quality, which began when mankind first congregated into villages, has existed as a serious problem since the industrial revolution. In the second half of the twentieth century, under the ever-increasing impacts of exponentially growing population and of industrializing society, environmental contamination of the air, water, soil, and food has become a threat to the continued existence of many plant and animal communities of various ecosystems and may ultimately threaten the very survival of the human race. Understandably, many scientific, industrial, and governmental communities have recently committed large resources of money and human power to the problems of environmental pollution and pollution abatement by effective control measures.

Advances in Environmental Sciences and Technology deals with creative reviews and critical assessments of all studies pertaining to the quality of the environment and to the technology of its conservation. The volumes published in the series are expected to service several objectives: (1) stimulate interdisciplinary cooperation and understanding among the environmental scientists; (2) provide the scientists with a periodic overview of environmental developments that are of general concern or that are of relevance to their own work or interests; (3) provide the graduate student with a critical assessment of past accomplishment which may help stimulate him or her toward the career opportunities in this vital area; and (4) provide the research manager and the legislative or administrative official with an assured awareness of newly developing research work on the critical pollutants, and with the background information important to their responsibility.

As the skills and techniques of many scientific disciplines are brought to bear on the fundamental and applied aspects of the environmental issues, there is a heightened need to draw together the numerous threads and to present a coherent picture of the various research endeavors. This need and the recent tremendous growth in the field of environmental studies have clearly made some editorial adjustments necessary. Apart from the changes in style and format, each future volume in the series will focus on one particular theme or timely topic, starting with Volume 12. The author(s) of each pertinent section will be expected to critically review the literature and the most important recent developments in the particular field; to critically evaluate new concepts, methods, and data; and to focus attention on important unresolved or controversial questions and on probable future trends. Monographs embodying the results of unusually extensive and well-rounded investigations will also be published in the series. The net result of the new editorial policy should be more integrative and comprehensive volumes on key environmental issues and pollutants. Indeed, the development of realistic standards of environmental quality for many pollutants often entails such a holistic treatment.

JEROME O. NRIAGU, Series Editor

PREFACE

The Great Lakes basin encompasses an area of nearly 300,000 square miles and constitutes an intricate ecosystem unparalleled in the world. Its 40 million inhabitants include 30% of Canada's population and about 20% of the United States' population. The support of the huge population has entailed the deforestation, urbanization, and industrialization in the basin which have, expectably, created environmental problems specially in Lakes Erie, Ontario, and Michigan. The influx of wastes from industrial, municipal, and agricultural sources has adversely affected both the quality of the lakes and their fish resources.

The first comprehensive report on the pollution of the Great Lakes, which was issued in 1918, observed that industrial pollutants were not discharged in sufficient quantities to seriously affect the water use. However, the studies, conducted between 1913 and 1916, found localized bacterial pollution in nearshore waters which could constitute a direct threat to municipal water supplies. A reexamination of the pollution problems in a 1950 report noted that sewer services and the installation of treatment plants for domestic wastes had not kept pace with the population growth in the area. The rapid economic, industrial, and agricultural expansion which took place between 1913 and 1946 had resulted in major increases of sewage discharges and new wastes. Changes in manufacturing processes and commodity use were creating new and widespread problems and the urban and industrial complexes in the basins of the lower lakes were developed without adequate understanding of their environmental impacts on the lakes. Later studies in the 1960s showed that the pressure on the lower Great Lakes from human activities had reached the crisis stage. "Lake Erie is Dead" became a rallying cry for the effort to "save" the lakes. Understandably, the nutrient cycles dominated most of the early work on the pollution of the Great Lakes as demonstrated in the 1972 Great Lakes Water Quality Agreement which dealt primarily with water quality problems in relation to the eutrophication processes.

Since about 1970, there has been growing public concern about the occurrence of toxic chemicals in air, soil, water, and food chains in the Great Lakes basin. The concern was heightened by excessively high levels of mercury in fish, which forced the closing of commercial fishery in Lake St. Clair in 1970, and the other subsequent warnings issued periodically on the hazards of consuming mercury-contaminated fish from the Great Lakes. Nor was the concern diminished by the mirex residues in Lake Ontario fish and the PCB residues in Lake Michigan fish which attained concentrations high enough to warrant human health advisories against the consumption of fish from these lakes. At the present time, about 30,000 compounds of commercial and industrial significance are on the market and being used in the Great Lakes basin with some 2000 to 3000 new compounds being added to the list each year. About 500 toxic compounds have already been identified in the Great Lakes and many new contaminants are being discovered as the detection limits for hazardous substances are lowered by improvements in our analytical capability. The identification of toxic substances in the Great Lakes, the growing number of chemicals, and the lack of complete understanding about their behavior and effects continue to generate concern about the ability of chemical producers, users, and governments to ensure the safety and well-being of the lakes and the human beings living within the basin area. It is not surprising that toxic contaminants in the Great Lakes and the potential health effects of long-term exposure to them have recently emerged as a top priority research problem.

This volume deals primarily with the problem of toxic contaminants, their entry into, movement through, persistence in, and effects upon the Great Lakes. It provides comprehensive reviews of currently available information on contaminants of concern, notably chlorinated hydrocarbons, polynuclear aromatic hydrocarbons, toxaphene, mirex, as well as toxic metals and metalloids. The chapters have been selected to provide a perspective on the problems of hazardous substances in the Great Lakes with emphasis on how they vary in magnitude, effects, and persistence. The volume is clearly not intended to be an encyclopedic compilation of available data on every toxicant in the Great Lakes.

Toxic Contaminants in the Great Lakes should be fundamental reading for anyone interested in water pollution or the quality of our drinking waters. It should also be of interest to chemists and biologists concerned with the behavior and effects of contaminants in aquatic ecosystems.

Jerome O. Nriagu
Milagros S. Simmons

Ontario, Canada
Ann Arbor, Michigan
August 1983

CONTENTS

TOXIC CONTAMINANTS IN THE GREAT LAKES

1

PERSPECTIVES ON HUMAN HEALTH CONCERNS FROM GREAT LAKES CONTAMINANTS

William C. Sonzogni

Great Lakes Environmental Research Laboratory
National Oceanic and Atmospheric Administration
Ann Arbor, Michigan 48104

Wayland R. Swain

Professor
Vakgroep Aquatische Oecologie
Kruislaan 320, 1098 SM
Amsterdam, The Netherlands

Current address of Dr. Sonzogni: State Laboratory of Hygiene, 465 Henry Mall, University of Wisconsin, Madison, Wisconsin 53706.

---

1. INTRODUCTION

Although trace toxic chemical contaminants affect all levels of the Great Lakes ecosystem, they are considered a critical problem mainly because they pose a threat to human health. For example, certain chemical substances have accumulated in Great Lakes fish to the extent that the concentration in fish is several hundred thousand times higher than in the water. This has prompted advisories to be posted in many Great Lakes states warning that eating certain Great Lakes fish may be a health hazard. In some cases the fishery has even been closed. The purpose of this chapter is to review the basis for health concerns, particularly with regard to PCBs, and to evaluate the level of risk encountered.

PCBs are of scientific interest for a variety of reasons. First and foremost, very little data exist on the human health effects of other Great Lakes contaminants. Considerable attention has been given to PCBs because (1) the large size of the PCB reservoir and its poor degradability suggest the problem will persist for some time into the future, (2) PCBs bioaccumulate dramatically, (3) certain PCB compounds have been found to be carcinogenic in laboratory test animals, and (4) health effect studies have been undertaken throughout the world because PCBs are an ubiquitous environmental contaminant. Finally, PCBs are of interest because they may well serve as an indicator of a class of compounds for predicting the health effects of other contaminants based on varying structure–activity correlations.

2. CHARACTERISTICS OF PCBs

PCBs are a group of chlorinated hydrocarbons manufactured by the controlled chlorination of the biphenyl molecule. In the manufacturing process, chlorine replaces hydrogen at two or more sites (Fig. 1.1), so that more than 200 compounds are possible.

In the United States, PCBs were produced under the trade name Aroclor by the Monsanto Chemical Company. A wide variety of Aroclor products exist, for example, Aroclor 1248 and Aroclor 1254. The numerical designation refers to the specific mixture of PCBs contained in the formulation. The first two digits indicate the number of carbon atoms, and the latter two specify the percentage of chlorination by weight. Aroclors are complex mixtures, frequently containing as many as 60 different PCB compounds; they also have the potential for containing additional trace contaminants, such as polychlorinated dibenzofurans (PCDFs) and polychlorinated naphthalenes (PCNs) (Pomerantz et al., 1978). In 1977, all U.S. production of PCBs ceased, but prior to that date, more than 0.6×10^6 metric tons of PCB were produced in the United States alone (Zimmerman, 1982).

PCBs were first synthesized in the late 1800s, but were not used for industrial purposes in the United States until 1929. For nearly 50 years, U.S. industry manufactured and used PCBs because of their high dielectric constant, their chemical and thermal stability, their nonflammability, and their low cost. Their uses included insulating fluids in electrical transformers and capacitors in the power industry, heat transfer substances, cutting oils, hydraulic fluids, lubricating oils, and as plasticizers for making brittle plastic pliable. Application was also found for these compounds in paints, printing inks, carbonless copy paper, sealants, adhesives, and the like.

Regrettably, the characteristics which make the PCBs industrially desirable also make them persistent and allow them to accumulate in the environment. The chemical stability of PCB compounds is unique. They are

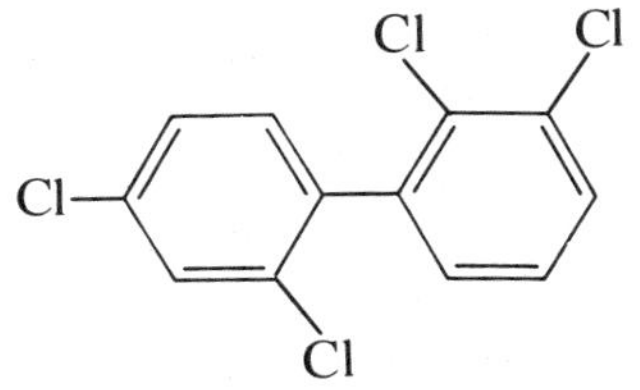

Figure 1.1 An example of a PCB molecule (2,2′,3,4′-tetrachlorobiphenyl).

resistant to acid–base reactions, hydrolysis, chemical oxidation, photodegradation, thermal changes, and most chemical agents. As a result of this stability, they are generally poorly metabolized by biologic systems. PCB solubility in water is very low, and appears to decrease with increasing chlorination (Hutzinger et al., 1974). They have relatively low vapor pressures, but in general, the lower the chlorine content, the more likely the compound is to volatilize (Pomerantz et al., 1978).

PCBs tend to bind tightly to particulate matter, notably soils and sediments. Thus, surface waters with low particulate loads may have barely detectable levels of PCB in the water mass, but high concentrations in the bottom muds. Of further concern is the effective half-life of these substances. Present estimates of PCB half-life fall into a range of 8–15 years in environmental samples. The persistence of these compounds measured in years or decades (IJC, 1975), combined with hydraulic detention times on the order of centuries for the sequentially coupled St. Lawrence Great Lakes (Rainey, 1967), yields the potential for bioaccumulation and bioconcentration virtually unparalleled elsewhere in North America.

Since PCBs are relatively insoluble in water and extremely soluble in oils and fats, they tend to partition out of the aquatic ecosystem into biologic tissue. Because of their persistence and the fact that they are poorly metabolized, these substances bioaccumulate in the food chain, often increasing several orders of magnitude in concentration at each succeeding trophic level.

3. TOXICOLOGY OF PCBs

A detailed study of the toxicology of a group of related substances as diverse as PCBs is beyond the scope of this chapter. However, a brief general review is presented to orient the reader for the discussion of effects which follows. A more detailed presentation is to be found in Kimbrough (1980).

The toxicological properties of PCBs vary among specific congeners and isomers. As a result, it is difficult to generalize regarding effects and toxicological properties of this wide group of substances. Evidence exists to suggest, however, that both persistence and toxicity of PCBs vary directly wth increasing chlorination (Safe Drinking Water Committee, 1980). However, it may not be presumed that a linear increase of chlorine content suggests a linearly increasing toxicity. Of critical importance are the degree, and the ring position, of the chlorine substitution on the biphenyl structure.

An example may be found in the biochemical induction of microsomal enzymes (hepatic mixed-function oxidases). The strength of this induction varies by both degree of chlorination and the relative positions of the substituted halogens. The structural requirements for aryl hydrocarbon hydroxylase (AHH) induction, for example, appear to involve the presence of at least two adjacent halogens in both of the lateral positions of the biphenyl molecule, and an absence of *ortho* position substitution (Goldstein, 1980). Halogenation of both *meta* and *para* positions seems to be a prerequisite. The 3,5,3′,5′-tetrachlorobiphenyl is inactive, despite the fact that its half-life is 10 times greater than 3,4,3′,4′-tetrachlorobiphenyl, a potent cytochrome P-448 inducer (Goldstein, 1980).

Matthews et al. (1978), in their exhaustive review of the biochemistry of PCBs, reported the enzyme induction threshold in nonhuman experiments ranged from 0.5 to 25 μg/g PCB. It is of interest that increases in these enzymes are frequently correlated with increased hepatic neoplasia (DHEW Subcommittee on Health Effects of PCBs and PBBs, 1978), and thus their etiology is a major health concern.

Further, microsomal enzyme induction may be due to metabolites of PCB, their congeners, or to non-PCB impurities in commercial mixtures. For example, trace amounts of polychlorinated dibenzofurans (PCDFs), an impurity commonly observed in some PCB formulations, may be a much more potent inducer of hepatic microsomal mixed-function oxidase enzymes (DHEW Subcommittee on Health Effects of PCBs and PBBs, 1978) than PCBs.

4. OBSERVED EFFECTS OF PCBs

4.1. Animal Studies

A variety of effects of PCB ingestion in animals has been reported, including (1) hepatic neoplasia, (2) mixed-function oxidase system alterations, (3) reproductive dysfunction, (4) hepatic porphyria, (5) liver hypertrophy, and (6) increased bile flow (Safe Drinking Water Committee, 1980). Kimbrough et al. (1978) have reviewed the acute and chronic toxicity of PCBs in animals and have summarized LD_{50} values. In general, the acute toxicity of PCBs appears to be low. The chronic toxicity is of greater concern, especially with regard to environmental exposure (Safe Drinking Water Committee, 1980). The observed chronic effects of PCBs in several nonhuman species, with special relevance to the Great Lakes contamination problem, are discussed below.

4.1.1. Mink

Effects of PCB toxicity in mink are of interest, because Great Lakes fish have been used as animal feed by commercial mink ranchers (Aulerich and Ringer, 1977). Owners who fed their animals coho salmon from Lake Michigan observed a high incidence of reproductive failure and kit mortality. The cause of these effects was apparently the high levels of PCBs (up to 20 mg/kg) in the salmon. It has been demonstrated that PCB levels as low as 2 mg/kg in fish prevent survival of newborn mink (Fetterolf, 1975).

In laboratory studies, mink exposed to PCBs at levels of 30 mg/kg suffered 100% mortality between the beginning of the breeding season and the end of the whelping period. The PCB mixtures designated as Aroclor 1242, 1248, and 1254 are all toxic to mink at dietary levels of 100 mg/kg, resulting in reproductive dysfunction and increased mortality (Task Force on PCB, 1976). Other observed effects of PCB toxicity in mink include anorexia, gastric ulcers, bloody stools, and hypertrophy of the liver, kidneys, and heart (Federal Register, 1977).

Although results of PCB toxicity studies in mink have significance for the Great Lakes, mink appear to be particularly susceptible to PCBs. Thus, their utility as an index of low-level PCB exposure to humans is in question. The ferret, the European relative to the North American mink, is not nearly as sensitive to PCBs (Ringer, 1982).

4.1.2. Primates

Rhesus monkeys suffered illness and reproductive dysfunction when exposed to PCBs at levels of 100–300 mg/kg (Kimbrough et al., 1978). Observed effects included facial swelling, edema, changes in menstrual cycles, an increased incidence of abortion, lower birth weight of infants, and an increase in susceptibility to infectious diseases (Allen and Barsotti, 1976; Federal Register, 1977).

At lower dietary levels of PCBs (5.0 and 2.0 mg/kg), rhesus monkeys continued to exhibit acute exposure symptomology, including facial swelling, chloracne, and loss of hair within one or two months (Barsotti et al., 1976). These levels of PCBs also had a marked effect on reproduction. After matings with control males, only one out of eight females of the 5.0 mg/kg PCB group, and five out of eight females of the 2.0 mg/kg PCB group, carried their pregnancies to term, although all of the control females exhibited normal term births. At the time of birth, the infants of the exposed groups were smaller, and by 2 months of age developed symptoms of acute PCB intoxication, including swollen eyelids and lips, loss of hair, and increased pigmentation of the skin. At even lower levels of exposure (1 and 0.5 mg/kg PCB), infant monkeys appeared to be more susceptible to infection and exhibited hyperlocomotor activity and learning retardation (Bowman et al., 1978). However, when tested at 44 months, these animals were hypoactive relative to controls (Bowman and Heironimus, 1981).

Allen and Barsotti (1976) attributed a rapid increase in PCB levels in tissues of nursing infant rhesus monkeys to consumption of PCB-contaminated milk from their mothers. Within 2 months following birth, the nursing offspring exhibited the usual dermal symptoms of acute exposure to PCB, and within 8 months, three of the six infants died. The three surviving infants showed marked improvements in their physical state when they were removed from dietary sources of PCBs.

In subsequent studies, adult female monkeys were removed from PCB diets for a period of one year prior to conception (Allen et al., 1980). Although the physical condition of these animals improved and PCB levels in adipose tissue declined sharply, four of 15 progeny were stillborn. Of the live-born infants, four died after weaning at 4 months of age. It is of interest that the birth weight of those infants which survived to term was markedly reduced when compared with nonexposed infants.

Despite the severity of effects on the rhesus monkey from relatively low PCB exposure, the significance is not clear, since this species may be particularly sensitive to PCBs. Kimbrough et al. (1978) point out that squirrel monkeys fed Kanechlor 400 did not show effects below a total dose of 300 mg/kg of body weight given over 20–32 weeks.

4.1.3. Birds

PCBs appear to be very toxic to avian species, causing changes in thyroid activity, increased susceptibility to infection, reduction in the size of the spleen, hepatic microsomal enzyme induction, adverse behavioral effects, reproductive dysfunction, and increased mortality (Kimbrough et al., 1978; Roberts et al., 1978).

In the early 1970s, the reproductive success of Great Lakes herring gulls in Lakes Erie and Ontario was reduced to near zero, presumably as a function of exposure to PCB and related organochlorine substances (Roberts et al., 1978; Nelson, 1979). Supporting this contention, it has been observed that PCB levels of 1–2 mg/kg in eggs have marked effects on reproduction in chickens. Levels of up to 180 mg/kg have been reported in gull eggs in the Great Lakes basin (Task Force on PCB, 1976). Since PCB levels in the eggs roughly reflect the levels in the livers of the adult birds, PCBs in the eggs yield an approximate indication of the bioaccumulation levels to be found in the adult gulls.

A marked decline in the PCB content of herring gull eggs has been noted in recent months. Apparently in response to these reductions, the number of adult gulls has increased, as have reproduction levels (Nelson, 1979).

4.1.4. Mice

When female mice were fed 32 mg/kg PCB 10–16 days after conception, progeny born to these mothers developed a neurological dysfunction termed *spinning syndrome* (Chou et al., 1979; Tilson et al., 1979). Other subtle neurobehavioral deficiencies were observed in prenatally exposed mice that did not develop frank neurotoxicological disease. In these asymptomatic mice, learning defects and altered neuromuscular reflexes were observed, even though PCBs were below detection levels in their bodies (Tilson et al., 1979). Sluggish responses and diminished habituation have been observed in mice perinatally exposed to levels of PCB well below that which produces neurotoxic or neuromotor dysfunction (Storm et al., 1981).

4.2. Human Exposure

4.2.1. Occupational Exposure

Reports of occupational exposures to PCBs date back to the 1930s. The skin condition known as chloracne was first reported by Schwartz (1936) for workers who inhaled chlorobiphenyls. Additional occupational exposures have been reported for marine electricians, capacitor and transformer manufacturer workers, and machinists. Symptoms included cutaneous eruptions and systemic manifestations. Cordle et al. (1978) report that some exposures were fatal, but that these exposures were often due to mixtures of chlorinated hydrocarbons, usually chlorinated naphthalenes and PCBs.

Table 1.1 provides PCB concentrations in three different worker groups from Finland who were exposed to various levels of PCBs. Although the amount of PCBs in capacitor factory workers exceeded the amounts in persons without special exposures to PCBs (control group) by about two orders of magnitude, no biological differences were noted. The exposed workers were given a number of clinical tests, including tests for induction of liver microsomal enzyme activity and steroid metabolism. No alterations were observed.

Cordle et al. (1978) reported that a study (Ouw et al., 1974) of Australian capacitor factory workers exposed to Aroclor 1242 found that workers complained of "burning" skin, persistent body odor caused by the PCB "fumes," and skin lesions. There was, however, a poor correlation between the severity of complaint and blood PCB concentrations. Those individuals with blood PCB concentrations less than 200 μg/kg showed no significant health effects. Cordle et al. (1978) also reviewed the Japanese literature regarding occupational exposure to PCBs. In an appraisal of Japanese carbonless copy paper factory workers two years after use of PCBs in processes ceased, Hasegawa et al. (1973) found no skin, liver function, urine, or blood abnormalities.

Thus, although there have been several instances of occupational exposure to PCBs, the effects on the exposed populations are frequently confounded by the co-occurrence of additional chemical substances, including

Table 1.1. Concentrations of PCBs in Blood (mg/kg fat) of Various Worker Groups in Finland[a]

Worker Group	Average	Range	Number of Individuals
Capacitor factory workers	313	100–700	11
Analytical laboratory workers handling PCBs	53	33–71	6
Controls (no special PCB exposure)	5.4	3.6–9.9	9

[a] Data from Karppanen and Kolho (1974) as reported by Cordle et al. (1978).

other chlorinated hydrocarbon compounds. Among occupationally exposed individuals, the majority of short-term observations do not suggest significant deviations from control populations. No evidence of an increased incidence in cancer attributed to occupational exposure to PCBs is available in the literature. This is not surprising, however, since long-term studies of occupationally exposed individuals have not been made.

4.2.2. Exposure from Consumption

A substantial portion of the population of North America has been exposed to PCBs, although frank illness among the exposed population has not generally been observed. To date, the bulk of the information on the effects of acute exposure to PCBs is derived from an incident in Japan in which rice oil accidentally contaminated with Kanechlor 400 (a commercial mixture of PCBs) was ingested. Individuals who were exposed in this manner developed symptoms and physical characteristics now associated with PCB intoxication. This set of symptoms, including chloracne, increased skin pigmentation, and swollen eyelids, was so characteristic that it was given the name *Yusho* (rice oil) disease (Harada, 1976; Higuchi, 1976; Wong and Hwang, 1981).

The extent of the contamination of the food ingested is unclear, but samples of the contaminated oil contained PCB concentrations from 2000 to 3000 mg/liter (Cordle et al., 1978).

According to the summary of Cordle et al. (1978) of dose–response epidemiologic studies of the *Yusho* incident, the average cumulative intake of PCBs causing overt symptoms was about 2000 mg, with the lowest approaching 500 mg. As a result of the uncertainty about the actual PCB content of the oil, and hence, about the amount consumed, it is possible that the ingested levels that produced an effect were slightly lower (e.g., 200 mg may be the lowest level producing an effect).

Infants exposed *in utero* were small for gestational age and tended to be born prematurely (Funatsu et al., 1972). Harada (1976) followed these infants and noted subsequent disturbances in responsiveness and neuromuscular functioning including an average IQ of 70; sluggish, clumsy, and jerky movements; apathy and hypotonia; autonomic dysfunctioning and growth retardation. In related studies, Harada (1976) reports diminished sensory and motor nerve conductance in exposed adults, consistent with the syndrome observed in exposed infants. The extent of the persistence of this substance is evidenced by the fact that infants born to exposed mothers several years after the incident showed elevated levels of blood PCBs, reduced birth weight, and brown pigmentation of the skin (Abe et al., 1975; Harada, 1976).

Interpretation of the health effects of PCBs from the Japanese *Yusho* incident are confounded by the discovery (Kuratsune et al., 1976) that the commercial PCB mixture which caused the contamination contained polychlorinated dibenzofurans (PCDFs) at levels up to 5 mg/liter. Polychlorinated dibenzofurans have been reported to be 200 to 500 times more toxic than

PCBs (DHEW Subcommittee on Health Effects of PCBs and PBBs, 1978). Kashimoto et al. (1981) cite animal studies that indicate PCDFs may be as much as 10,000 times more toxic than PCBs.

In 1979, an incident similar to the Japanese *Yusho* experience occurred in Taiwan. Patient symptoms included increased pigmentation of skin, swelling of gums and eyelids, and abdominal pains (Kashimoto et al., 1981). As was the case in the Japanese incident, PCDFs were found in the contaminated oil in a ratio of 1:500 PCDFs to PCBs. Kashimoto et al. (1981) suggest that the PCDFs were primarily responsible for the toxic effects in both exposures.

Unfortunately, it is not certain whether the PCDFs, the PCBs, or a combination of both, were the direct etiologic agent for *Yusho*. However, it can be concluded from *Yusho*-related studies that (1) lower-chlorinated PCB compounds are excreted more rapidly than higher-chlorinated compounds, and (2) PCBs may be transferred from mother to child in the fetal stage and through breast feeding.

5. POTENTIAL EFFECTS OF LONG-TERM EXPOSURE

The potential for long-term chronic effects of orally ingested PCBs is of significant concern because of the widespread dissemination of these compounds in environmental media. Summarizing market basket surveys of PCB levels in food, Calabrese and Sorenson (1977) estimate average daily ingestion of PCB to be in the range of 5–10 μg PCB/day, based on a level of 0.1 mg/kg PCB in 3.0% of food consumed. It is clear, however, that this estimated mean does not properly represent special populations at risk, particularly those individuals whose preferences and behaviors result in much higher intake levels.

An example of such a special risk population is to be found in the case of the Lake Michigan sport fishermen. Individuals who habitually consume large quantities of fish from Lake Michigan will have substantially higher intake of PCB than the general population. For example, an individual who eats 50 g of fish per day (1.8 oz/day) containing PCB at the currently applicable U.S. Federal Food and Drug action limit of 5.0 mg/kg will ingest 250 μg PCB/day. For individuals following diets high in fish protein, exposure levels may be similar, or even higher. One such popular diet recommends five fish meals per week totaling nearly 100 g (3.5 oz.) of fish consumed per day. Assuming a contamination range of 2–5 mg/kg PCB in the flesh of the fish, the conscientious subscriber to this diet would consume between 200 and 500 μg of PCB per day. The concern for chronic exposure among populations at special risk is therefore obvious, particularly in view of the fact that many of the PCB compounds are poorly metabolized and tend to accumulate over time.

Toxicity from chronic exposures is usually classified in four distinct types: (1) carcinogenicity, (2) mutagenicity, (3) teratogenicity, and (4) toxicity for all other reasons. Each of these categories is discussed briefly with particular reference to the role of PCBs.

5.1. Carcinogenic Effects

Any chemical that increases the incidence of malignant neoplasia to levels higher than those found in a comparable control group not similarly exposed is said to be carcinogenic (Safe Drinking Water Committee, 1977). Current theory provides two distinct mechanisms by which a chemical can serve as the etiologic agent of neoplastic disease. The first mechanism, termed *initiation,* is the process by which a substance provides the stimulus to induce neoplasia. The second mechanism, *promotion,* is the process whereby the previously initiated cells are stimulated to evolve into neoplastic disease. Zimmerman (1982) cites evidence that PCBs serve as promoters in test animals.

Results of laboratory animal experiments have been extensively reviewed by Kimbrough et al. (1978), PCB Risk Assessment Work Force (1979), Safe Drinking Water Committee (1980), Technical Committee from the Michigan Departments of Public Health, Agriculture and Natural Resources (1981), and Zimmerman (1982). The evidence for PCB carcinogenesis in humans is presumptive based on animal studies, although Humphrey (personal communication, Michigan Department of Public Health, Lansing, 1982) has observed a slight increase in cancer incidence among a population consuming large quantities of fish from Lake Michigan.

A complete discussion of PCB-induced animal carcinogenicity is beyond the scope of this volume. However, several pertinent studies are of interest. Selikoff (1972) notes that PCBs fed chronically to rats caused hepatic adenofibrosis and biliary epithelial hyperplasia. These lesions were morphologically indistinguishable from those induced by *p*-dimethylamino-azobenzene, a known carcinogen. Ito et al. (1974) report that many Kanechlor compounds, a Japanese commercial formulation of PCB, have tumorigenic action in the rat liver, but only Kanechlor 500 induced hepatocellular carcinomas in the livers of mice. Kimbrough and Linder (1974) report the induction of hepatomas, adenofibrosis, and increased liver weights in mice fed 300 mg/kg of Aroclor 1254 for eleven months. Kimbrough et al. (1975) report observing hepatocarcinomas in 26 of 184 rats fed 100 mg/kg Aroclor 1260 for 21 months. In contrast, only one of 173 control animals developed hepatocellular carcinoma.

In a National Cancer Institute (1978) bioassay, Fisher 344 rats were orally administered PCBs at levels of 25, 50, and 100 mg/kg, and compared to controls after a period of 104–105 weeks. Although it was concluded that

Aroclor 1254 was not carcinogenic to the rats under the conditions of the experiment, a high incidence of hepatocellular lesions was observed. A dose–response relationship between the PCBs and liver lesions was found. Moreover, because of the small numbers of animals and other limitations of the experiment, the possibility of carcinogenic effects of PCB exposure in rats could not be ruled out. Further, good agreement was noted between the National Cancer Institute (1978) data and Kimbrough et al. (1975) data in terms of the percent of rats with tumors at the 100 mg/kg level.

5.2. Mutagenic Effects

Chemicals capable of modifying genetic material such that a heritable change is produced are considered to be mutagenic. Mutagenic effects of environmental contaminants in general are summarized in Sutton and Harris (1972). More recently, the mutagenic potential of PCBs has been reviewed by the Safe Drinking Water Committee (1977) and the Committee on the Assessment of Human Health Effects of Great Lakes Water Quality (1981). Overall, the potential for PCB mutagenic effects appears to be quite small. As a result, PCBs are not normally considered to be mutagenic.

5.3. Teratogenic Effects

A chemical that acts to produce congenital abnormalities in a fetus is considered to be teratogenic. Studies of rhesus monkeys (Barsotti et al., 1976) indicate that surviving offspring of pregnant females fed 2.5 and 5.0 mg/kg PCB (Aroclor 1248) for at least 6 months were distinctively small. Low-dose fetal toxicity has also been observed in mink (Aulerich and Ringer, 1977). Placental passage and fetopathy have been demonstrated in experimental studies with mammals such as rabbits, mice, rats, and rhesus monkeys (Allen et al., 1980; Bowman et al., 1978; Chou et al., 1979; Grant et al., 1971; Masuda et al., 1979; and Tilson et al., 1979).

In the strictest definition of the term, PCBs have not yet been shown to be teratogenic in humans. However, an increasing amount of evidence relates PCBs to transplacental passage and human fetopathy (Polishuk et al., 1977; Masuda et al., 1978; Kodama and Ota, 1980; Kimbrough et al., 1978; and Jacobson et al., 1982).

5.4. Other Toxic Effects

Wasserman et al. (1973) have demonstrated morphological alteration of the zona fascinulata of the adrenal gland in rats administered 200 mg/kg of Aroclor 1221 for 6 weeks. These morphologic changes were accompanied by increased levels of circulated corticosterone.

Several studies have reported immunosuppression in a variety of experimental animals, including chickens, rabbits, and guinea pigs (Vos and Koeman, 1970; Vos and Beems, 1971; and Vos, 1972). Calabrese and Sorenson (1977) note that the immunosuppressive actions of PCBs may cause increased susceptibility of exposed fish to fungal disease, exposed ducks to viral hepatitis, and hepatic tumors in exposed rats.

An increasing body of evidence suggests that PCBs may act synergistically with other materials to potentiate toxicity. Grant et al. (1971) report that in rats fed carbon tetrachloride and Aroclor 1254, the PCB mixture potentiated the toxicity of carbon tetrachloride. Observed residues of Aroclor 1254 in rats with carbon tetrachloride-damaged livers were higher in the heart, blood, kidney, liver, and testes.

Dahlgren et al. (1972) reported on the effects of additions of organochlorine compounds to pheasants. These authors note that pheasants fed 50 mg PCB along with dieldrin yielded only additive, and not synergistic interactions. Heath (1970) reports similar findings related to the additive effects of DDE.

Zimmerman (1982) indicates that PCBs are potentially synergistic in humans. He further suggests that, because of the wide variety of contaminants contained in natural fish populations, the possibility for increased toxic effects of these substances exists, since the potential for synergism is relatively high. He concludes this line of reasoning by suggesting that the possibility for additive or synergistic effects should be considered in establishing new standards for PCBs in Great Lakes fish.

6. CONSUMPTION LIMITS

6.1. Methods for Estimating Safe Exposures to Toxic Chemicals

A primary problem in assessing the risks to human health from exposure to toxic chemicals is estimating the exposure that can be tolerated without undesirable consequences. Testing for chronic, low-level effects in man is difficult, because effects are often not manifested until long after the exposure. In some instances, data from accidental or occupational exposures can be utilized to establish exposure limits. However, in most cases, these data do not exist, and estimations have to be based on statistical extrapolation from high levels of exposure in animals. While high exposure levels in laboratory animals reduce both the time required for tumor development and the number of animals required to produce statistically valid results (Rall, 1979), extrapolating results of these studies to humans must be done with caution. Currently, scientific controversy exists over the validity of predicting human response to toxic chemicals from animal studies (Crouch and Wilson, 1979; Golberg, 1979; Rall, 1979; and Squire, 1981). The appropriateness of such extrapolation has been questioned because of the relative uncertainty of the response curve at increasingly lower dose levels (PCB Risk Assessment Work Force, 1979). Further, the possibility of varying biological repair mechanisms among species complicates such estimates. These and other inherent limitations, which have been well documented in the literature, must be considered when using the results of laboratory tests on animals to assess chronic effects. Neverthless, viable alternatives to animal laboratory studies to estimate long-term risks of low levels of contaminants do not presently exist.

There are two primary methods of estimating maximum allowable exposure to toxic chemicals from laboratory animal studies. These are the Acceptable Daily Intake approach and the Risk Modeling approach (Safe Drinking Water Committee, 1980).

6.1.1. Acceptable Daily Intake

The Acceptable Daily Intake (ADI) approach, which was first applied to additives in food, is based on the premise that there is a level of ingestion of a given substance which will pose no lifetime risk to human health if consumed daily. The assumption is that a threshold level exists below which no toxic effects will occur. The ADI approach relies heavily on the use of safety factors, whereby the experimental threshold dose is arbitrarily increased, often by several orders of magnitude, to provide a margin of safety. The ADI approach is currently not favored for setting criteria for environmental carcinogens (Safe Drinking Water Committee, 1980).

6.1.2. Risk Modeling

A second approach used to estimate exposure from toxic chemicals is the risk modeling approach. Modeling approaches are used primarily to assess the risks from carcinogens, because quantitative theories of carcinogenesis lend themselves well to mathematical simulation (Safe Drinking Water Committee, 1980). A number of different types of models exist to assess human risk from low exposures, including dichotomous response models, tolerance distribution models, logistic models, "hitless" models, time-to-tumor occurrence models, and simple linear models. For a more complete discussion of mathematical simulation tools used to estimate exposure from toxic chemicals, the reader is referred to the Safe Drinking Water Committee (1980) report.

6.2. U.S. FDA Standards for PCBs

The U.S. Food and Drug Administration (FDA) became actively involved in the control and regulation of PCBs in 1969 when the chemical was first discovered in food items. In 1973, temporary tolerance limits were established to protect the consumer from food products indirectly contaminated

Table 1.2. FDA 1973 Temporary Tolerances for PCBs in Food

Food	PCB Concentration (mg/kg)
Milk (fat basis)	2.5
Dairy products (fat basis)	2.5
Poultry (fat basis)	5.0
Fish (edible portion)	5.0
Eggs	0.5
Finished animal feed	0.2
Animal feed components	2.0
Infant and junior foods	0.2
Paper food-packaging material without PCB-impermeable barrier	10.0

with PCBs (Jelinek and Corneliussen, 1976). These tolerance levels are presented in Table 1.2.

The present 5 mg/kg guideline for fish established in 1973 by the FDA (Federal Register, 1973) was designed to prevent individuals from exceeding a PCB dose of 1 μg/kg of body weight per day. This guideline was based primarily on the Japanese *Yusho* incident, where the average total PCB dose that was associated with effects in man was 2000 mg. Applying a safety factor of 10:1, a total of 200 mg of PCBs was determined to be tolerable for a "protracted period of time" (Federal Register, 1973). This level permits ingestion of 4 μg/kg of body weight per day over the life span of a 70-year-old man. However, the lowest dose in the Japanese *Yusho* incident observed to produce an effect was 500 mg. Applying a similar kind of safety factor logic, it was reasoned that a daily dose of 1 μg/kg of body weight could be tolerated over a long period. Thus, a 5 ppm tolerance level in fish was designed to protect fish consumers from exceeding the daily PCB dose of 1 μg/kg of body weight.

Humphrey (1976) noted that the recommended dose is being exceeded by many individuals consuming Lake Michigan fish. He also reported that, if the 1974 average annual intake of PCBs by sport fishermen was continued, the 200 mg safety limit would be exceeded in slightly over 4 years, and the 500 mg level would be achieved in about 10.5 years.

Because exposure is a continuous phenomena, and since PCBs are not readily metabolized, the 1 μg/kg of body weight per day safety margin may not be low enough, at least for some population groups (Technical Committee from the Michigan Departments of Public Health, Agriculture and Natural Resources, 1981). This guideline may be particularly inappropriate for infants or nursing mothers.

As a result of increased concern over PCBs, the federal FDA lowered the temporary tolerance levels for several food categories in 1979. The tolerance level for milk and dairy products was lowered from 2.5 to 1.5 mg/kg, and the guideline for eggs was lowered from 0.5 to 0.3 mg/kg. The action level for poultry was reduced from 5 to 3 mg/kg (Nelson, 1979). The rationale for this action was partly based on the availability of several studies that indicated PCBs were potentially carcinogenic (Federal Register, 1977).

The tolerance level in fish and shellfish (5 mg/kg) was also scheduled to be lowered from 5 to 2 mg/kg, but objections filed by the National Fisheries Institute delayed this action. The contention was that the lowering of the tolerance level of PCBs in fish would have a significant impact on the economic livelihood of the freshwater commercial fisheries, especially the Great Lakes commercial fishery. In June of 1979, the federal FDA reduced the tolerance level in fish to 2 mg/kg, but implementation was postponed. Currently, the action level for PCBs in fish remains at 5 mg/kg in the United States, although there is considerable pressure to lower the standard to 2 mg/kg. Michigan's Toxic Substances Control Commission recommended a 2 ppm standard for skin-on fillets to be adopted for the State of Michigan (Zimmerman, 1982).

6.3. Standards for PCBs in Canada and Other Countries

In 1976, the Health Protection Branch of Canada's Department of National Health and Welfare established a temporary tolerance level of 2 mg/kg for the edible portion of fish. This guideline was developed to protect the public from the risk of buying and eating contaminated fish (Task Force on PCBs, 1976). The Canadian government also has established action levels for PCBs in various food items, but those guidelines are not widely publicized.

Few standards currently exist for PCBs in fish and edible foods in other countries. Nevertheless, current or new standards in other countries could have a major impact in the Great Lakes region as they relate to the future of the eel fishery in Lake Ontario. Lake Ontario eels, which often contain high levels of PCBs and other contaminants, are not sold for human consumption in the United States. However, Lake Ontario eels are exported to Europe where they are considered a delicacy. Recently, exports have been reduced as a result of the high levels of PCBs and other contaminants. Thus, the future of the eel industry will depend, at least in part, on PCB criteria developed outside of North America.

6.4. Other PCB Criteria

The 1978 Great Lakes Water Quality Agreement between the United States and Canada set an objective for PCBs in fish at 0.1 mg/kg (whole fish, calculated on a wet weight basis). This objective was established for the protection of birds and other animals that consume Great Lakes fish (IJC, 1978). Although PCB levels in Great Lakes fish appear to be declining, the 0.1

mg/kg objective is not likely to be met by most fish in the near future (Sonzogni and Simmons, 1981; Swain, 1982).

The U.S. Environmental Protection Agency has established a water quality criteria of 0.001 μg/liter for PCBs to protect freshwater and marine aquatic life (U.S. EPA, 1976). The tendency of PCBs to bioaccumulate in organisms, often by factors of 100,000 or more, was largely responsible for this criteria. Although this criteria was not developed to protect human health, since man is frequently the ultimate consumer in the aquatic food chain, the implications are obvious.

7. ESTIMATED RISK ASSOCIATED WITH CONSUMING CONTAMINATED FISH

7.1. Current Exposure/Dose Levels

The exposure to which an individual is subjected is a function of the concentration of the substance of concern over the course of duration of exposure. Dose, as contrasted with exposure, is calculated based upon the mass of the substance of concern per unit of body mass per unit of time. Because compounds, such as PCBs, tend to be cumulative, residual, and persistent, small repeated exposures can result in relatively high body burdens.

7.2. PCB Intake from Eating Fish

For the average North American citizen, oral exposure to PCBs is low because the mean consumption of PCB contaminated fish is low. However, certain individuals consume large quantities of fish, and these subpopulations may receive relatively larger PCB exposures.

Table 1.3, based on a U.S. FDA study (PCB Risk Assessment Work Force, 1979), presents the intake of PCBs from eating selected commercial fish that are known to contain PCBs. The annual exposure is given, assuming no-tolerance limits, and assuming tolerance levels of 5, 2, and 1 mg/kg. Except for the no-tolerance estimate, mean PCB concentrations were calculated by eliminating PCB values that exceeded the stated tolerance. Note that mean PCB levels used in the calculation were derived from a domestic survey that included non-Great Lakes fish that are comparatively low in PCBs. The mean PCB concentration of the fish considered in each of the categories in Table 1.3 is, in fact, less than 1 mg/kg.

The 50 and 90 percentile intake in Table 1.3 reflects average and high consumption of the exposed population. The average annual per capita consumption of fish in the United States is about 6.8 kg/yr (15 lb/yr), most of which are marine species. Approximately 15.2% of the U.S. population consumes the fish species from which Table 1.3 was determined.

Table 1.3. PCB Intake by U.S. Commercial Fish Eaters Based on FDA 1978–1979 Survey[a] (PCB Risk Assessment Work Force, 1979)

Consumption	Assuming No Tolerance	Assuming Tolerance of 5 mg/kg	Assuming Tolerance of 2 mg/kg	Assuming Tolerance of 1 mg/kg
Average fish consumption—50 percentile (μg/kg of body weight per day)	0.12	0.11	0.08	0.05
Fish high consumption—90 percentile (μg/kg of body weight per day)	0.32	0.29	0.21	0.13

[a] For assumed tolerances, PCB values exceeding the tolerance were eliminated from the data set used to estimate intake.

The consumption of Great Lakes sport fish, particularly Lake Michigan fish, provides a much greater PCB exposure to fish eaters. A study conducted by the Michigan Department of Public Health (Humphrey, 1976) determined that the 50 percentile intake from a population of individuals bordering Lake Michigan, who regularly eat sport fish, was 1.7 μg/kg·day, whereas the 90 percentile intake was 3.9 μg/kg·day (Table 1.4). These intakes are based on 1974 PCB levels in cooked fish, which range from 1.03 to 4.6, 0.48 to 5.38, and 0.36 to 2.06 μg/g for lake trout, salmon, and other fish, respectively.

PCBs in whole lake trout and coho salmon during this period averaged 22.91 and 10.45 μg/g, respectively, indicating that preparation and cooking

Table 1.4. Intake of PCBs by Fish Consumers (μg/kg of body weight per day)

Consumption	Average for Consumers of Selected U.S. Species with High PCB Levels[a]	Consumption of Lake Michigan Sport Species of Fish[b]			
		1974 Survey Levels	25% Reduction from 1974 Survey	50% Reduction from 1974 Survey	75% Reduction from 1974 Survey
50 Percentile consumption level	0.12	1.7	1.28	0.85	0.42
90 Percentile consumption level	0.32	3.9	2.92	1.95	0.98

[a] PCB Risk Assessment Work Force (1979); about 15.2% of total U.S. population consumes species of interest.

[b] 1974 Survey results as reported by Humphrey (1976); the survey was based on a sample from the approximately 382,000 sport fishermen in the 18 counties in the State of Michigan which border Lake Michigan.

reduce the amount of PCBs actually consumed. Note that the Lake Michigan sport fish consumers' PCB intake is more than an order of magnitude greater than the national average (no-tolerance) levels given in Table 1.3. PCB levels have decreased in Lake Michigan since 1974, however. Consequently, PCB intake reductions of 25, 50, and 75% below 1974 levels are given in Table 1.4. Although the intake reductions given in Table 1.4 are arbitrary, they are in accordance with observed reductions in the PCB content of whole fish in Lake Michigan. Swain (1982) reports that between 1974 and 1982, the average amount of PCBs found in Lake Michigan lake trout and bloater chubs decreased about 75%. Average PCB concentrations in both coho salmon and bloater chubs have also fallen below the FDA's standard of 5 mg/kg. In fact, in bloater chubs, the PCB concentration has fallen below the more stringent standard of 2 mg/kg that is proposed.

The 1974 PCB level in cooked Lake Michigan sport fish reported by Humphrey (1976) averaged 2.49 mg/kg. Reductions of 25, 50, and 75% would result in PCB levels in cooked fish of 1.87, 1.25, and 0.62 mg/kg, respectively. Although current data on the PCB content of cooked fish are not available, the decrease in raw, whole, Lake Michigan sport fish suggests that the levels in cooked fish may also have decreased from 50 to 75% of their 1974 level.

7.3. PCB Exposure Other Than from Eating Fish

Non-occupationally acquired exposures to PCB may also be obtained through drinking water, food consumption (in addition to fish), and respired ambient air. Infants may also receive PCBs through breast milk or by placental transfer.

Swain (1980, 1982) and Sonzogni and Swain (1980) have made estimates of annual PCB exposures to Great Lakes residents from drinking potable water and respiring ambient air. An individual consuming drinking water with a concentration of 4 ng/liter would ingest about 2.9 μg of PCBs/yr. The intake from respiring ambient air in nonurban Great Lakes areas, assuming an air concentration of 1.5 ng/m^3, yields an exposure of 7.9 μg/yr. Table 1.5 compares these intakes with consumption of domestic commercial fish and Great Lakes sports species of fish. Intakes from other sources are small relative to consumption of fish. Consumption of a single 0.45-kg (1-lb) serving of Great Lakes salmon or trout per year provides an exposure to PCB some three orders of magnitude higher than drinking water and respired ambient air considered together. Note that the intake of PCBs from Great Lakes fish given in Table 1.5 was derived from the intake value in μg/kg of body weight per year given by Humphrey (1976) by assuming an average body weight of 70 kg.

Infant populations represent yet another high-risk population. In addition to their exposure during gestation as a function of their mothers' circulating blood titer of PCB passed transplacentally to the infant, at parturition they

Table 1.5. Comparison of Annual PCB Intake Acquired from Drinking Water and Respired Ambient Air with Consumption of Fish

Source	Annual PCB Intake (μg)
Drinking water	2.9
Respiration (Nonurban air)	7.8
U.S. domestic commercial fish (PCB Risk Assessment, 1979)	
50 percentile consumption (No-tolerance level)	3,088
90 percentile consumption (No-tolerance level)	8,066
Lake Michigan sport species consumption (Humphrey, 1976)	
50 percentile consumption	43,435
90 percentile consumption	99,645

may be exposed to additional sources of PCB, and at levels substantially greater than those experienced *in utero*.

Breast milk constitutes one of these major additional sources of PCBs. Intake via this route is particularly significant, since the infant is exposed to PCBs at a very early and sensitive stage in development. In this regard, liver microsomal enzyme systems are often not fully developed at birth, so that newborns may have more difficulty detoxifying foreign chemicals such as chlorinated biphenyls (Swain, 1980). Based on 1971–1972 data, Nisbet (1976) warned that ". . . exposure of breast-fed infants is likely to be very much greater than that of any adult, even an adult who likes fish."

Wickizer et al. (1981) reported that, based on a survey of nursing mothers from 68 of the 83 counties in the State of Michigan, all milk samples contained PCBs. Over 70% of the mothers tested had PCB levels greater than 1 μg/g (on a fat weight basis) in their breast milk. The average PCB concentration in the fat content of breast milk was 1.5 μg/g.

Given a PCB concentration of 1.5 μg/g (fat basis) in mothers' milk, and assuming an infant consumes an average of 600 mL/day of breast milk with a fat content of 4 g/100 ml (Wickizer et al., 1981), a daily intake of 36 μg is estimated. The annual intake, assuming the infant continues to breast-feed throughout the first year, would amount to 13.1 mg. Given a typical birth weight of 4 kg, the daily dose during the child's early development would be about 9 μg/kg of body weight. This compares with the maximum daily PCB dose rate of 1 μg/kg of body weight per day recommended for adult intake by the U.S. Food and Drug Administration. While the health effects of feeding PCB-contaminated breast milk to infants are just being investigated, precautionary measures for nursing mothers have been recommended (Wickizer et al., 1981). Early efforts at understanding the effect of PCB exposure on infants have suggested a number of subtle behavioral deficits which may have implications for subsequent development (Jacobson et al., 1982; Fein et al., 1982).

7.4. Estimates of the Probabilities of Health Effects from Eating Great Lakes Fish

In order to estimate the risk to humans from consumption of PCB-contaminated fish, the PCB Risk Assessment Work Force (1979) utilized available carcinogenicity data from experimental animals as the basis of their assessment. This work was later reviewed by the Technical Committee from the Michigan Departments of Public Health, Agriculture, and Natural Resources (1981).

The PCB Risk Assessment Work Force (1979) used a linear, single-hit extrapolation model to estimate risk. Of the various models used for high to low dose risk extrapolation, the linear model is the most conservative and least likely to underestimate risk (PCB Risk Assessment Work Force, 1979; Safe Drinking Water Committee, 1980). The model is based on the assumption that risk is directly proportional to the environmental exposure.

Tables 1.6 and 1.7 present the estimated cancer risks of eating contaminated fish at the 50 and 90 percentile consumption levels reported in Table 1.3. Risks were determined based on upper (99%) confidence bounds in order to eliminate the effect of sample size (PCB Risk Assessment Work Force,

Table 1.6. Lifetime Risk Estimates of Consuming PCB-Contaminated Fish Based on Applicable Animal Studies. Risks Expressed as Risk of Additional Cancers Per 100,000 Persons Consuming Fish at the 50 Percentile Level over Their Lifetime[a]

Animal Study	Additonal Cancers per 100,000 Consumers of Selected U.S. Fish Species with Highest PCB Levels	Additional Cancers per 100,000 Lake Michigan Sport Fish Eaters			
		1974 Survey Levels	25% Reduction from 1974 Survey	50% Reduction from 1974 Survey	75% Reduction from 1974 Survey
Kimbrough—rat liver carcinoma	1.3	18.4	13.8	9.2	4.6
NCI bioassay—total malignancies for male or female	4.1	58.0	43.7	29.0	14.3
NCI bioassay—liver carcinoma and adenomas for male and female	0.9	12.8	9.6	6.4	3.2
NCI bioassay—hematopoietic malignancies for male and female	2.7	38.2	28.2	19.1	9.4

[a] Adapted from PCB Risk Assessment Work Force (1979).

Table 1.7. Lifetime Risk Estimates of Consuming PCB-Contaminated Fish Based on Applicable Animal Studies. Risks Expressed as Risk of Additional Cancers Per 100,000 Persons Consuming Fish at the 90 Percentile Level over Their Lifetime[a]

Animal Study	Additonal Cancers per 100,000 Consumers of Selected U.S. Fish Species with Highest PCB Levels	Additional Cancers per 100,000 Lake Michigan Sport Fish Eaters: 1974 Survey Levels	25% Reduction from 1974 Survey	50% Reduction from 1974 Survey	75% Reduction from 1974 Survey
Kimbrough—rat liver carcinoma	3.4	41.4	31.0	20.7	10.4
NCI bioassay—total malignancies for male or female	10.6	129.2	96.7	64.6	33.5
NCI bioassay—liver carcinoma and adenomas for male and female	2.5	30.5	22.8	15.2	7.7
NCI bioassay—hematopoietic malignancies for male and female	7.0	85.3	21.4	42.6	21.4

[a] Adapted from PCB Risk Assessment Work Force (1979).

1979). Interpreting the data in this way also contributed to the conservative nature of the estimate. Note that the annual cancer incidence given is in addition to that observed to be occurring within the population-at-large (at a rate of 25,000–100,000). Tables 1.6 and 1.7 show the reduction in risk that will result from incremental reduction in PCB intake. Because the model used to estimate risk is linear, the anticipated reductions in risk are directly proportional to intake reduction.

Importantly, the cancer risks estimated in Tables 1.6 and 1.7 carry a high degree of uncertainty. While an unavoidable lack of certainty exists, the data presented are the best currently available. The assumptions made in the model formulation tend to force the estimated risks toward the upper or conservative limits.

7.4.1. Probability of PCB-Induced Cancer Mortalities

In order to make comparisons to other hazards resulting in mortalities, it is of interest to estimate the cancer-related deaths that may result from consuming PCB-laden fish. Although not specified in Tables 1.6 and 1.7, it is assumed that all additional cancer incidence projected to result from eating PCB-contaminated fish will be hepatic neoplasms. As discussed earlier, the liver appears to be the main target for chronic PCB-related effects. Further, in laboratory studies, only heptocarcinogenic effects were observed.

The American Cancer Society (1980) projects approximately 9400 deaths per year will be due to liver cancer in the United States. Additionally, they project an incidence rate of 13,000 new liver cancers per year. Of the total annual deaths due to cancers of all types (386,686, or 168.4 per 100,000 individuals), about 2.2% are attributed to liver cancer (American Cancer Society, 1980).

Although the liver cancer case fatality rate is less than 100%, in keeping with the attempt to develop an upper bound to risk, it will be assumed for the sake of comparison that all cancers projected in Tables 1.6 and 1.7 will result in mortality. With the 100% case fatality rate assumption, an upper limit projection of the average number of deaths per year anticipated from eating PCB-laden fish is given in Table 1.8. The results clearly suggest an elevated risk beyond the irreducible to consumers of Lake Michigan sports species of fish.

Table 1.9 compares the risk of consuming PCB-laden fish with the risks of cancer deaths from other causes. Note that PCB-related cancer risks in Table 1.9 are considerably less than those for smoking or alcohol. However, the risk of the 90 percentile Great Lakes fish consumer is in the same ranges as diagnostic medical X-rays and daily diet soda (saccharin) consumption. Importantly, as discussed previously, the reported risks from PCB consumption are believed to be upper-bound estimates. Further, current PCB levels in fish have decreased from the 1974 levels used as the basis for lifetime PCB intake in Table 1.9. This would suggest that the risk of eating Great Lakes fish may be less than given in Table 1.9, and that the risk to average and infrequent Great Lakes fish eaters may be relatively small. On the other

Table 1.8. Projected Average Maximum Annual Mortalities per 100,000 Persons Eating PCB-Contaminated Fish[a]

Percentile	U.S. Nationwide Consumers of Selected (PCB-Contaminated) Fish Species	Lake Michigan Sport Fish Consumers (1974 Contaminant Levels)
50 Percentile consumers over 70-year life span	0.03	0.46
90 Percentile consumers over 70-year life span	0.08	1.02

[a] Based on the average of the cancer incidence rates given for the four animal studies in Tables 1.6 and 1.7, and assuming all cancers result in mortalities. Fish consumption PCB intake rates are given in Table 1.4.

Table 1.9. Average Annual Risk of Cancer-related Deaths from Various Causes[a]

Risk Type	Approximate Individual Risk per Year
Cosmic-ray induced cancer from living in Denver versus New York	1 in 100,000
Average U.S. diagnostic medical X-rays	1 in 100,000
One diet soda/day (saccharin)	1 in 100,000
Four tablespoons peanut butter/day (aflatoxin)	1 in 25,000
Miami or New Orleans drinking water	1 in 800,000
1/2 lb charcoal-broiled steak weekly	1 in 2,500,000
Alcohol (average of smokers and nonsmokers)	1 in 20,000
Smoking	1 in 800
Person in room with smoker	1 in 100,000
Regular use of contraceptive pills	1 in 50,000
50 percentile lifetime consumer of Great Lakes fish at 1974 PCB level[b]	1 in 200,000
90 percentile lifetime consumer of Great Lakes fish at 1974 PCB level[b]	1 in 100,000

[a] All risk estimates, except fish consumption risks, adapted from Hutt (1978); assumes all cancers result in mortalities.

[b] Conservative estimate; adapted from Table 1.8.

hand, uncertainty about the actual risk and the preliminary observation of increased frequency of reported cancers among consumers of sport species of fish from Lake Michigan (Humphrey, Michigan Department of Public Health, personal communication, 1982) attest to the need for additional efforts in cause-specific risk assessment.

8. SUMMARY AND FUTURE CONSIDERATIONS

Of the more than 400 potentially toxic compounds that have been found in the water, sediment, and biota in the Great Lakes, few have been studied with regard to the actual risk they pose to human health. Probably the most intensively investigated in this regard are polychlorinated biphenyls (PCBs), largely because these persistent chemicals have bioaccumulated dramatically in edible Great Lakes fish. Animal experiments and observations of humans accidentally exposed to PCBs suggest that environmental exposures are not an acute hazard, but are a possible long-term carcinogenic and teratogenic concern. Toxicological studies indicate that some PCB formulations may biochemically induce microsomal enzymes, which are frequently correlated with increased hepatic neoplasia. Both the degree of chlorination and the relative position of the substituted chlorines appear to affect the strength of this induction.

Individuals who habitually consume large quantities of fish from the Great Lakes, especially from Lake Michigan, will have substantially higher intakes of PCBs than the general population. Based on conservative extrapolations from animal studies, such individuals face an increased, although relatively small, risk of developing cancer. More information, such as may be provided by ongoing epidemiological studies of subpopulations exposed to high PCB levels, is needed to better define the associated risks. Further, more information is needed on the possible health effects of PCBs transferred from mother to infant, as well as on possible synergistic effects of PCBs.

Despite concern over potential health effects of contaminants found in the Great Lakes, the fact that levels of contaminants such as PCBs and DDT are decreasing is cause for guarded optimism. It would appear that control measures for these compounds are beginning to be effective. As environmental levels of contaminants decrease, the health risk associated with these contaminants should also decrease. Optimism aside, it is clear that our knowledge of the chronic human health effects of Great Lakes contaminants is minimal, and the need for careful and innovative health effects research will continue into the foreseeable future.

ACKNOWLEDGMENTS

The assistance of T. R. Crane in assembling information for this manuscript is appreciated. The authors also acknowledge the assistance of A. J. Davis in the preparation of this manuscript. Great Lakes Environmental Research Laboratory (GLERL) contribution number 330.

REFERENCES

Abe, S., Y. Inoue, and M. Takamatsu. (1975). Polychlorinated biphenyl residues in plasma of Yusho children born to mothers who had consumed oil contaminated by PCB. *Acta Medica Fukuoka* **66:** 605–609.

Allen, J. R. and D. A. Barsotti. (1976). The effects of transplacental and mammary movement of PCBs on infant Rhesus monkeys. *Toxicology* **6:** 331.

Allen, J. R., D. A. Barsotti, and L. A. Carstens. (1980). Residual effects of polychlorinated biphenyls on adult nonhuman primates and their offspring. *J. Toxicol. Environ. Health* **6:** 55–66.

American Cancer Society. (1980). *Cancer facts and figures 1981*. American Cancer Society, Inc., New York.

Aulerich, R. and R. Ringer. (1977). Current status of PCB toxicity to mink, and effect on their reproduction. *Environ. Contam. Toxicol.* **6:** 279–292.

Barsotti, D., R. Marlar, and J. Allen. (1976). Reproductive dysfunction in Rhesus monkeys exposed to low levels of polychlorinated biphenyls (Arochlor 1248). *Food Cosmet. Toxicol.* **14:** 99–103.

Bowman, R. E. and M. P. Heironimus. (1981). Hypoactivity in adolescent monkeys perinatally exposed to PCBs and hyperactivity as juveniles. *Neurobehavior. Toxicol. Teratol.* **3:** 15–18.

Bowman, R. E., M. P. Heironimus, and J. R. Allen. (1978). Correlation of PCB body burden with behavioral toxicology in monkeys. *Pharmacol. Biochem. Behavior* **9:** 49–56.

Calabrese, E. J. and A. J. Sorenson. (1977). The health effects of PCBs with particular emphasis on human high risk groups. *Rev. Environ. Health* **2:** 285–304.

Chou, S. M., T. Miike, W. M. Payne, and G. J. Davis. (1979). Neuropathology of "spinning synrome" induced by prenatal intoxication with a PCB in mice. *Ann. N.Y. Acad. Sci.* **320:** 373–395.

Committee on the Assessment of Human Health Effects of Great Lakes Water Quality. (1981). 1981 Annual Report. Report to the Great Lakes Water Quality Board/Great Lakes Science Advisory Board, International Joint Commission.

Cordle, F., P. Corneliussen, C. Jelinek, B. Hackley, R. Lehman, J. McLaughlin, R. Rhoden, and R. Shapiro. (1978). Human exposure to polychlorinated biphenyls and polybrominated biphenyls. *Environ. Health Persp.* **24:** 157–172.

Crouch, E. and R. Wilson. (1979). Interspecies comparison of carcinogenic potency. *J. Toxicol. Environ. Health* **5:** 1095–1118.

Dahlgren, R. B., R. L. Linder, and C. W. Carlson. (1972). Polychlorinated biphenyls: their effects on penned pheasants. *Environ. Health Persp.* **1:** 89.

Department of Health, Education, and Welfare Subcommittee on Health Effects of PCBs and PBBs. (1978). General summary and conclusions. *Environ. Health Persp.* **24:** 191–198.

Federal Register, Vol. 38, No. 129, July 6, 1973, p. 10896.

Federal Register, Vol. 42, No. 22, February 2, 1977.

Fein, G. G., S. W. Jacobson, P. M. Schwartz, and J. L. Jacobson. (1982). Intrauterine exposure to environmental toxins: Behavioral effects in the human newborn. Manuscript in preparation, University of Michigan, Ann Arbor.

Fetterolf, C. M. (1975). PCB body burdens deny full use of the Great Lakes fishery resource. Statement to the National Conference of Polychlorinated Biphenyls, November 19–21.

Funatsu, I., F. Yamashita, Y. Ito, S. Tseugama, T. Funatsu, T. Yoshikane, M. Hayashi, T. Kato, M. Yakushiji, G. Okamoto, S. Yamasaki, T. Arima, T. Kuno, H. Ide, and I. Ibe. (1972). Polychlorinated biphenyl (PCB)-induced fetopathy. *Kurume Med. J.* **19:** 43–51.

Goldberg, L. (1979). Implications for human health. *Environ. Health Persp.* **32:** 273–277.

Goldstein, J. A. (1980). Structure–activity relationships for the biochemical effects and their relationships to toxicity. In: R. D. Kimbrough, Ed., *Halogenated biphenyls, terphenyls, naphthalenes, dibenzodioxins and related products.* Elsevier/North-Holland Biomedical Press, Amsterdam, pp. 151–190.

Grant, D. L., D. C. Villeneuve, K. A. McCully, and W. Phillips. (1971). Placental transfer of polychlorinated biphenyls in the rabbit. *Environ. Physiol.* **1:** 61–66.

Harada, M. (1976). Intrauterine poisoning. Clinical and epidemiological studies and significance of the problem. Bulletin of the Institute of Constitutional Medicine, Kumamoto University, 25 pp.

Hasegawa, J., M. Sato, and H. Tsuruta. (1973). Report on survey of work areas environmental where PCB is handled and the health of workers handling PCB. Ministry of Labor, Bureau of Labor Standards and Institute of Industrial Health, Japan.

Heath, R. G. (1970). Effects of polychlorinated biphenyls on birds. Proceedings of the XV International Ornithological Congress, The Hague.

Higuchi, K., Ed. (1976). *PCB poisoning and pollution.* Academic Press, New York, p. 184.

Humphrey, H. E. B. (1982). Personal communication.

Humphrey, H. E. B. (1976). Evaluation of changes in the level of polychlorinated biphenyls (PCB) in human tissue. Michigan Dept. of Public Health, Lansing.

Hutt, P. B. (1978). *Food Drug Cosmetic Law J.* **33:** 558. In: N. J. McCormick (1981) *Reliability and risk analysis*. Academic Press, New York, p. 348.

Hutzinger, O., S. Safe, and V. Zitko. (1974). *The chemistry of PCBs*. CRC Press, Cleveland.

International Joint Commission. (1975). Water quality objectives subcommittee report: Persistent compounds. Water Quality Board Report (1974) to the International Joint Commission, Appendix A, pp. 123 and Appendices.

International Joint Commission. (1978). Great Lakes Water Quality Agreement of 1978. Great Lakes Regional Office, Windsor, Ontario.

Ito, N., N. Hiroshi, S. Makiura, and M. Arai. (1974). Histopathological studies on liver tumorigenesis in rats treated with polychlorinated biphenyls. *Gann* **65:** 545.

Jacobson, S. W., J. L. Jacobson, P. M. Schwartz, and G. G. Fein. (1982). *Intrauterine exposure of human newborns to PCBs: Measures of exposure*. In press, Ann Arbor Science Press, Ann Arbor.

Jelinek, C. F. and P. E. Corneliussen. (1976). Levels of PCB in the U.S. food supply. In: U.S. Environmental Protection Agency (1976) Conference Proceedings, National Conference on Polychlorinated Biphenyls (November 19–21, 1975, Chicago, Illinois), EPA-560/6-75-004, Washington, D.C., pp. 147–154.

Karppamen, E. and E. Kolho. (1974). The concentration of PCBs in human blood and adipose tissue in three different research groups. National Swedish Environmental Protection Board Publication 1974, 4 E (1974). Cited in: F. Cordel et al., *Environ. Health Persp.* **24:** 157.

Kashimoto, T. et al. (1981). Role of polychlorinated dibenzofuran in Yusho (PCB poisoning). *Arch. Environ. Health* **36:** 321.

Kimbrough, R. D. and R. E. Linder. (1974). Introduction of adenofibrosis and hepatomas of the liver in BALB/CJ mice by polychlorinated biphenyls (Aroclor 1254). *J. Natl. Cancer Inst.* **53:** 547–552.

Kimbrough, R. D., R. A. Squire, R. E. Linder, J. D. Strandberg, R. J. Montaii, and V. W. Burse. (1975). Induction of liver tumors in Sherman strain female rats by polychlorinated biphenyl Aroclor 1260. *J. Natl. Cancer Inst.* **55:** 1453–1459.

Kimbrough, R., J. Buckley, L. Fishbein, G. Flamm, L. Kasza, W. Marcus, S. Shibko, and R. Teske. (1978). Animal toxicology. *Environ. Health Persp.* **24:** 173–184.

Kimbrough, R. D., Ed. (1980). *Halogenated biphenyls, terphenyls, naphthalenes, dibenzodioxins and related products*. Elsevier/North-Holland Biomedical Press, Amsterdam, p. 406.

Kodama, H. and H. Ota. (1980). Transfer of polychlorinated biphenyls to infants from their mothers. *Arch. Environ. Health* **35:** 95–100.

Kuratsune, M., Y. Masuda, and J. Nagayama. (1976). Some of the recent findings concerning *Yusho,* National Conference on Polychlorinated Biphenyls, U.S. Environmental Protection Agency, Office of Toxic Substances, Washington, D.C., Report No. 560/6-75-004.

Masuda, Y., R. Kagawa, H. Koroki, M. Kuratsane, T. Yoshimura, I. Taki, M. Kusuda, F. Yamashita, and M. Hayashi. (1978). Transfer of polychlorinated biphenyls from mothers to foetuses and infants. *Bull. Environ. Contam. Toxicol.* **16:** 543–546.

Masuda, Y., R. Kagawa, B. Kuroki, S. Tokudome, and M. Kuratsune. (1979). Transfer of various polychlorinated biphenyls to the foetuses and offspring of mice. *Food Cosmet. Toxicol.* **17:** 623–627.

Matthews, H., G. Fries, A. Gardner, L. Garthoff, J. Goldstein, Y. Ku, and J. Moore. (1978).

Metabolism and biochemical toxicity of PCBs and PBBs. *Environ. Health Persp.* **24:** 147–155.

National Cancer Institute. (1978). Bioassay of Arochlor 1254 for possible carcinogenicity. Carcinogenesis Technical Report Series No. 38, CAS No. 2732-18-8, MCI-CG-TR-38; through PCB Risk Assessment Work Force, 1979.

Nelson, S. (1979). PCBs revisited. Environmental Midwest, U.S. Environ. Protection Agency, Region 5, Chicago, Illinois, October–November, p. 10–14.

Nisbet, I. C. T. (1976). Environmental transport and occurrence of PCB in 1975. National Conference on Polychlorinated Biphenyls, November 19–21, 1975, Chicago, Illinois. NTIS PB-253-248, pp. 254–256.

Ouw, K. H., G. R. Simpson, and D. S. Siyali. (1974). The use and health effects of Arochlor 1242, a polychlorinated biphenyl in an electrical industry in N.S.W. Australia. Report Div. of Occupational Health and Radiation Control, Health Commissions of New South Wales, Australia.

PCB Risk Assessment Work Force. (1979). An assessment of risk associated with the human consumption of some species of fish contaminated with polychlorinated biphenyls (PCBs). Report prepared for the U.S. Food and Drug Administration.

Polishuk, Z. W., D. Wasserman, M. Wasserman, S. Cucos, and M. Ron. (1977). Organochlorine compounds in mother and fetus during labor. *Environ. Res.* **13:** 278–284.

Pomerantz, I., J. Burke, D. Firestone, J. McKinney, J. Roach, and W. Trotter. (1978). Chemistry of PCBs and PBBs. *Environ. Health Persp.* **24:** 133–146.

Rainey, R. H. (1967). Natural displacement of pollution from the Great Lakes. *Science* **155:** 1242–1243.

Rall, D. P. (1979). Relevance of animal experiments to humans. *Environ. Health Persp.* **32:** 297–300.

Ringer, R. (1982). Toxicology of PCBs in small mammals. Paper presented at the International Symposium on PCBs in the Great Lakes, Lansing, Michigan, March 15–17, 1982.

Roberts, J. R., D. W. Rodgers, J. R. Bailey, and M. A. Rorke. (1978). Polychlorinated biphenyls: Biological criteria for an assessment of their effects on environmental quality. National Research Council of Canada, Report No. 16077, Ottawa, p. 172.

Safe Drinking Water Committee, National Research Council. (1977). *Drinking water and health,* Vol. 1, National Academy Press, Washington, D.C.

Safe Drinking Water Committee, National Research Council. (1980). *Drinking water and health,* Vol. 3. National Academy Press, Washington, D.C.

Schwartz, L. (1936). Dermatitis from synthetic resins and waxes. *Am. J. Public Health* **26:** 586.

Selikoff, I. J., (Ed. (1972). Polychlorinated biphenyl—environmental impact: A review of the panel on hazardous trace substances. *Environ. Res.* **5:** 249.

Sonzogni, W. C. and M. S. Simmons. (1981). Notes on Great Lakes trace metal and toxic organic contaminants. Great Lakes Environmental Planning Study Contribution No. 54.

Sonzogni, W. C. and W. R. Swain. (1980). Perspectives on U.S. Great Lakes chemical toxic substances research. *J. Great Lakes Res.* **6:** 265–274.

Squire, R. A. (1981). Ranking animal carcinogens: A proposed regulatory approach. *Science* **214:** 877–880.

Storm, J. E., J. L. Hart, and R. F. Smith. (1981). Behaviour of mice after pre- and postnatal exposure to Arochlor 1254. *Neurobehavior. Toxicol. Teratol.* **3:** 5–9.

Sutton, H. E. and M. I. Harris, Eds. (1972). *Mutagenic effects of environmental contaminants.* Academic Press, New York.

Swain, W. R. (1980). An ecosystem approach to the toxiology of residue forming xenobiotic organic substances in the Great Lakes. In: *Ecotoxicology working papers: Testing for the effects of chemicals on ecosystems*. National Research Council, National Academy of Sciences, pp. 194–257.

Swain, W. R. (1982). An overview of the scientific basis for concern with polychlorinated biphenyls in Great Lakes. In press, Ann Arbor Science Press, Ann Arbor.

Task Force on PCB. (1976). Background to the regulation of polychlorinated biphenyls (PCB) in Canada. Environment Canada and Health Welfare Canada, April 1, 1976.

Technical Committee from the Michigan Departments of Public Health, Agricultural and Natural Resources. (1981). PCBs in Great Lakes fish: An evaluation of the proposed FDA guidelines. Report to the Toxic Substances Control Commission, October 9, 1981.

Tilson, H. A., G. J. Davis, J. A. McLachlan, and G. W. Lucier. (1979). The effects of polychlorinated biphenyls given prenatally on the neurobehavioral development of mice. *Environ. Res.* **18:** 466–474.

U.S. Environmental Protection Agency. (1976). *Quality criteria for water*. U.S. Environmental Protection Agency, Washington, D.C.

Vos, J. B. (1972). Toxicology of PCBs for mammals and for birds. *Environ. Health Persp.* **1:** 105.

Vos, J. B. and R. B. Beems. (1971). Dermal toxicity studies of technical polychlorinated biphenyls and fractions thereof in rabbits. *Toxicol. Appl. Pharmacol.* **19:** 617.

Vos, J. B. and J. H. Koeman. (1970). Comparative toxicological study with polychlorinated biphenyls in chickens with special reference to prophyria, edema formation, liver necrosis and tissue residues. *Toxicol. Appl. Pharmacol.* **17:** 565.

Wasserman, D., M. Wasserman, S. Cucos, and M. Djaraherian. (1973). Functions of adrenal gland-zona fasciculata in rats receiving polychlorinated biphenyls. *Environ. Res.* **6:** 354.

Wickizer, T. M., L. B. Brilliant, R. Copeland, and R. Tilden. (1981). Polychlorinated biphenyl contamination of nursing mothers' milk in Michigan. *Am. J. Public Health* **71:** 132.

Wong, K. C. and M. W. Hwang. (1981). Children born to PCB-poisoned mothers. *Clin. Med.* (Taipei) **7:** 83–87.

Zimmerman, N. (1982). Polychlorinated biphenyls in Great Lakes fish: Toxicological justification for lowering the acceptable standard to 2 ppm. Paper presented at International Symposium on PCBs in the Great Lakes, March 15–17, 1982.

2

CONTAMINANTS RESEARCH AND SURVEILLANCE—A BIOLOGICAL APPROACH

Harvey Shear

Program Coordinator
Great Lakes Fisheries Research Branch
Canada Centre for Inland Waters
867 Lakeshore Rd.
Burlington, Ontario L7R4A6

1. INTRODUCTION

With the publication of Rachel Carson's *Silent Spring* in 1962, public attention was drawn for the first time to the dangers of man-made chemicals as they impacted the natural food chains. Although her work dealt mostly with DDT and its effects on birds, the attention that this book generated spurred a plethora of scientific activities in North America, in universities, government agencies, and private industry.

An area of particular interest to the scientific community was the Laurentian Great Lakes. The five Great Lakes (Ontario, Erie, Huron, Michigan, and Superior) represent a freshwater resource unparalleled in the world. Fully 20% of the world's fresh surface water is contained in these lakes (IJC, 1978). One-sixth of the population of the United States and Canada reside in the Great Lakes watershed, and a substantial portion of the industry and commerce of North America is directly attributable to this resource.

Canada and the United States have been concerned about the degradation of the quality of these lakes for some time. Indeed, one of the earliest references to the International Joint Commission (a Canada–U.S. body set up by treaty in 1912 to resolve boundary water conflicts) was an investigation into the pollution of the Niagara River and other boundary waters (IJC, 1918). Over the years, there were additional studies carried out on Great Lakes pollution, mainly addressing the question of excessive nutrient enrichment (IJC, 1951, 1967, 1969). These studies culminated in the signing of the Great Lakes Water Quality Agreement of 1972 (GLWQA, 1972). This agreement laid the foundation for a major cleanup of phorphorus entering the Lakes and also indicated an approach to be taken in managing contaminants. Water quality objectives were to be developed for metals and synthetic organic substances to protect the most sensitive use of the water. Considerable progress was made in this area, but it was recognized that the full significance of synthetic organic compounds and their impact on the entire ecosystem, including man, had not been fully appreciated. In 1978, a revised agreement was signed (GLWQA, 1978) emphasizing the ecosystem approach to management and the significance of materials such as PCBs, mirex, polychorinated styrenes, and so forth. This has resulted in an increased emphasis on holistic examinations of contaminants in the ecosystem; studies on synergism of contaminants; studies of the dynamics of contaminants within organisms; the development of new diagnostic tools to assess the effects of chronic, low-level exposure; the establishment of a tissue archive for retroactive contaminant analysis; and the development of predictive tools such as quantitative structure–activity correlations to deal with the multitude of organic compounds.

This review briefly outlines some of the major Canadian government programs dealing with contaminants in the Great Lakes. The emphasis is on biology and highlights activities in both surveillance and research.

2. SURVEILLANCE

With the signing of the 1972 Water Quality Agreement (GLWQA, 1972) and the subsequent 1978 agreement (GLWQA, 1978), a major surveillance program was initiated. Although the initial emphasis was largely on measuring phosphorus loadings and concentrations, the surveillance program is now very much involved in the assessment of the degree of contamination in the Lakes. The biological program consists of determining levels of metals and organic contaminants in fish, plankton, benthos, and herring gull eggs on a long-term basis to assess trends of contaminants through time, as well as the spatial distribution of contaminants within and between lakes. Another major activity is the development of a fish health program to investigate the well-being of certain fish species in the Great Lakes related to the effects of contaminants on these fish.

The contaminants in aquatic biota programs carried out in Canada consist of whole-body burden determinations of so-called routine, and nonroutine contaminants. The routine contaminants are:

Routine Contaminants Monitored in Biota

Metals	Organic Compounds
Arsenic	Dieldrin
Cadmium	Chlordane
Chromium	DDT and metabolites
Copper	Heptachlor epoxide
Lead	Mirex
Mercury	PCB
Nickel	
Selenium	
Zinc	

These materials are analyzed in all samples collected.

The nonroutine compounds are analyzed on selected samples each year. Compounds to be analyzed are chosen from the list of priority compounds published by the IJC (1977). The nonroutine compounds analyzed in 1981 were:

Some Nonroutine Contaminants Monitored in Biota

Polychlorinated styrenes
Tetrachlorodibenzo-*p*-dioxin
Lindane
Phthalate esters
Chlorinated phenols
Polybrominated biphenyls
Chlorinated benzenes
Mirex photodecomposition products

Some Nonroutine Contaminants Monitored in Biota (*Continued*)

Toxaphene
Polychlorinated dibenzofurans
Organolead
Chlorinated diphenyl ethers
Polychlorinated terphenyls
Polynuclear aromatic hydrocarbons

The surveillance of contaminants in biota has been in place since 1977. It is thus possible to present trends for up to 5 years on the routine contaminants. These trends are presented in Tables 2.1 and 2.2.

2.1. Metals

A few significant features emerge from these tables. Firstly, the levels of metals (total metal) in whole fish samples are generally extremely low, well below any guidelines established for human consumption of fish flesh. In particular, one should note the low levels of mercury in fish from Lake St. Clair and western Lake Erie. In 1970, the fishery in Lake St. Clair was closed because of excessive levels of mercury in the fish from the lake, especially walleye (OME, 1977; OME/OMNR, 1982). With the reduction in mercury output from the offending industry, levels of mercury in fish began to decline to those indicated in Table 2.1. Another interesting feature is that of mercury levels in Lake Superior as compared to the other lakes. It is apparent that Lake Superior fish are still showing the effects of mercury inputs from the pulp and paper industry, although this industry reduced its emissions in the early 1970s (J. Kelso, personal communication). This points up the rather unique nature of Lake Superior. It is such a cold, deep, oligotrophic lake that all natural processes are slowed down, including reductions in the body burden of mercury in fish.

As noted earlier, the concentrations of metals in whole fish flesh are generally well below any guidelines for human consumption of fish. It must be stressed, however, that the form of metal in fish is critical to any assessment of the degree of contamination. Recent work by Chau et al. (1980) has shown that significant levels of alkylated lead compounds occur in Great Lakes fish. For example, they found total lead levels of 30–185 ng/g in fish species from Lakes Ontario and St. Clair. However, upon analysis of the various organolead species (tetramethyl, trimethyl-ethyl, dimethyl-diethyl, methyl-triethyl, tetraethyl) significant concentrations were detected in many fish. Wong et al. (1978b) pointed out that organolead compounds are generally more toxic than inorganic lead compounds. Thus, it is clear that the measurement of total lead in fish flesh, and the establishment of guidelines for lead based on total lead are potentially in error.

Table 2.1. Levels of Trace Metals in Whole Fish, 1977–1979[a]

Year	As[b]	Cd	Cr	Cu	Pb	Hg	Ni	Se	Zn
				Lake Ontario (Lake Trout)					
1977	<0.10	0.20	0.3–0.6	1.4–2.1	<1.00	0.13–0.18	<0.50	0.44	14.5
1978	<0.05	<0.02	<0.2	0.94–1.08	<0.10	0.16–0.20	<0.05	0.30	8.5–10.9
1979	0.48	<0.02	<0.2	1.3–1.6	<0.10	0.14–0.22	<0.05–0.16	0.44	10.7
				Lake Ontario (Rainbow Smelt)					
1977	<0.10	<0.20	1.13	0.77–1.39	<1.0	0.04–0.10	<0.5	0.33–0.36	21.40–22.27
1978	0.50	0.03	<0.20	0.4–0.8	<0.1–0.28	0.04–0.08	<0.05–0.22	0.36	24.0
1979	0.60	<0.02–0.06	<0.2–0.3	0.3–0.5	<0.1–0.2	0.04–0.10	0.1–0.2	0.34	22.8–23.6
				Lake Huron (Splake)					
1979	0.15	<0.02	<0.2	0.62	<0.10	0.16	<0.05	0.7	11.8
1980	0.18	<0.02	<0.2	0.85	<0.10	0.18	<0.05	0.8	11.6
				Lake Superior (Lake Trout)					
1980	0.36	<0.02	<0.20	0.70	<0.10	0.32	<0.05	0.38	11.64
				Lake Erie (Walleye)					
1977	<0.10	<0.20	0.4	3.3	<1.0	0.20	<0.5	0.26	12.9
1978	0.24	<0.02	<0.2	0.5	<0.1	0.17	0.05	0.34	12.1
1979	0.21	0.02	<0.2	0.3	<0.1	0.04	<0.05	0.29	23.7
				Lake St. Clair (Walleye)—Edible Portions Only					
	1970	1971	1972	1973	1974	1975	1976	1977	1978	1979–81
Hg	2.1	1.8	1.3	1.1	0.98	0.81	0.93	0.96	0.70	0.5–1.0

[a] All data expressed as μg/g wet weight, whole fish basis, unless otherwise noted.

[b] National Health and Welfare Guidelines for Metals in Fish Products (edible portions) are: Hg, 0.5 μg/g; As, 5.0 μg/g; Cu, 100.0 μg/g; Zn, 100.0 μg/g; Fe, 25.0 μg/g.

Table 2.2. Levels of Selected Organic Compounds (μg/g) in Whole Fish from Lake Ontario, 1975–1981

Year	Organic Compound			
	ΣDDT	PCB	Dieldrin	Mirex
		Lake Trout		
1977	2.66	4.95	0.04	0.27
1978	1.16	7.10	0.18	0.21
1979	1.58	3.79	0.20	0.23
1980	0.62	4.79	0.10	0.18
1981	1.39	2.82	0.15	0.15
		Coho Salmon		
1977	1.43	3.03	0.07	0.16
1978	0.64	3.00	0.10	0.08
1979	0.81	1.21	0.10	0.05
1980	0.74	2.30	0.07	0.10
		Smelt		
1977	0.60	1.50	0.02	0.11
1978	0.44	1.82	0.05	0.06
1979	0.39	0.80	0.04	0.06
1980	0.25	1.12	0.04	0.08
		Spottail Shiners		
1975	0.24	0.69	—	—
1976	—	—	—	—
1977	0.16	0.65	—	0.013
1978	0.099	0.32	—	0.029
1979	0.026	0.15	—	0.001
1980	0.041	0.27	—	0.011

Lead is by no means the only metal capable of undergoing alkylation. Mercury, arsenic, tin, and selenium have all been shown to alkylate under the appropriate conditions (Baker et al., 1981). This is discussed more fully in Section 3 of this chapter, but it should be noted that the same word of caution regarding lead may also apply to these metals.

2.2. Synthetic Organic Compounds

The data on organic compounds presents a much clearer picture of both spatial distribution and trends in time of whole-body burdens. It is very obvious, for example, that DDT levels in fish have declined substantially

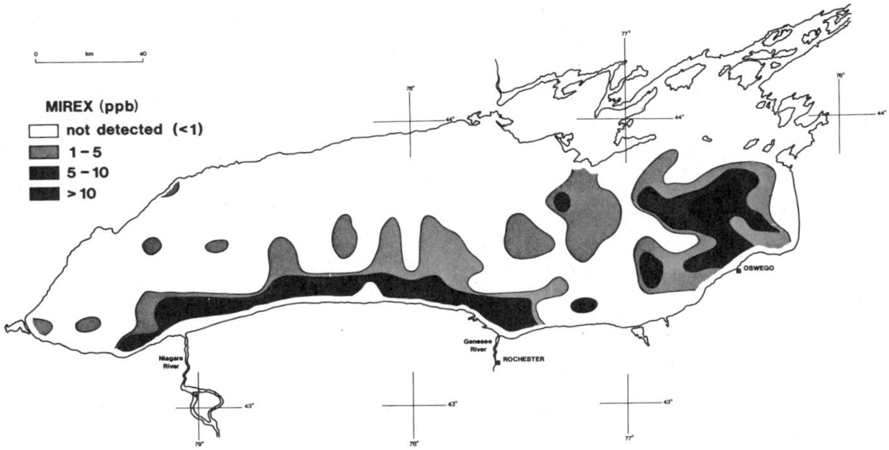

Figure 2.1 Distribution of mirex in the surficial 3 cm of sediment in Lake Ontario, 1968.

from 1977 to 1980. This would appear to be in direct response to the ban placed on DDT in 1970. The rise in ΣDDT in lake trout in 1981 is, as yet, unexplained. Additional data will be needed before a real upward trend can be assumed. Although PCBs have undergone a similar ban on manufacture and use, declines in levels of PCBs are not quite so apparent. This may be the result of the ubiquitous nature of PCBs and the continued slow input of material into the environment through leaching of poorly designed waste disposal sites, combustion of PCB-contaminated oils, and the input of PCBs through atmospheric deposition on a global basis.

Another compound of concern, particularly in Lake Ontario, is mirex. Mirex was first reported in Lake Ontario in 1974 by Kaiser (1974) in longnose gar and northern pike from the Bay of Quinte. Subsequent to this, Holdrinet et al. (1978) investigated mirex in the sediments of Lake Ontario and discovered two sources for this pesticide, based on mirex concentrations and the sediment distribution patterns of Lake Ontario (Fig. 2.1).

Mirex levels in aquatic biota have been measured since the inception of the contaminants surveillance program. Levels have generally remained constant or at most have shown a modest decline (Table 2.2). This pattern is reflected in the herring gull egg data (IJC, 1981a) for the years 1977 to 1980 (see Table 2.3). There was an obvious, dramatic decline in mirex from 1974 to 1977, and then a very marked shift to an almost constant level in eggs from two separate colonies. The reason for this change in the rate of decline is not apparent. It may be necessary to resurvey the sediments of the lake in search of new sources, or perhaps the biota of the lake have reached a quasi-equilibrium with the mirex in the sediments.

Of the nonroutine contaminants monitored, perhaps the most notorious

Table 2.3. Mirex Levels (μg/g) in Herring Gull Eggs from Lake Ontario Colonies[a]

Year	Muggs Island	Snake Island
1974	7.4	6.6
1975	3.4	6.0
1976	—	—
1977	2.1	3.0
1978	1.4	1.8
1979	1.8	2.0
1980	1.7	1.7

[a] Source: IJC (1981a).

is dioxin (tetrachlorodibenzo-*p*-dioxin) and, in particular, one isomer: 2,3,7,8 TCDD. This isomer has been shown to be extremely toxic to laboratory test animals (McConnell et al., 1978; Schwetz et al., 1973; Murray et al., 1979). Although a long-term data base does not exist to determine trends in dioxin levels in fish, one does exist for herring gulls. Concentrations of 2,3,7,8 TCDD have declined from a high of over 1000 ppt (ng/kg) in gull eggs in 1971 to a current (1980) level of about 70 ppt (IJC, 1981a). Data from a limited number of sport fish from Lakes Ontario and Superior, taken in 1981, indicate concentrations in whole fish of 14–74 ppt, and from nondetectable to 57 ppt in edible portions (Ontario Ministry of the Environment, unpublished). Canada's Departments of National Health and Welfare and Fisheries and Oceans have set a guideline of 20 ppt of 2,3,7,8 TCDD in edible portions of fish for the protection of human consumers of fish.

The source of dioxin in Lake Ontario has been traced to the Niagara River area and, in particular, to the chemical waste dumps along the U.S. shore of the river near Niagara Falls, New York. Work is continuing in Canada to monitor biota in the river and in the Lake. Studies at the Canada Department of Fisheries and Oceans have been initiated to determine what, if any, effect dioxin is having on the health of the fishery resource.

2.3. Kingston Basin Study

In an attempt to understand the dynamics of contaminants in a natural system, a 2-year study was undertaken in the Kingston Basin of Lake Ontario by Fisheries and Oceans (IJC, 1981a; Shear, 1981; M. Whittle, unpublished). This study examined the temporal and spatial variability in metals and organic compounds in a pelagic (free swimming) and demersal (bottom dwelling) food chain. Plankton and benthos account for a major portion of the food consumed by many fish species, and probably play a major role in the transfer of contaminants through the food chain.

The study revealed several interesting things. Highest levels of contaminants were observed in the spring, data being normalized for lipid content. Smaller organisms showed more of a range in concentrations than larger organisms. The higher concentrations in spring probably reflect land runoff of contaminants as a source, since bottom and surface plankton had similar contaminant levels, thus, ruling out bottom sediment resuspension as a source (see Figs. 2.2 and 2.3). Another interesting feature of this study was the difference in accumulation of contaminants in the food chain. Chlorinated hydrocarbons showed the expected pattern of increased concentration through the various trophic levels, whereas trace metals showed the opposite pattern (Figs. 2.2 and 2.3). This is probably a manifestation of surface adsorption phenomena rather than true uptake by the organisms. Cells (plankton) with a large surface area relative to their volumes would be expected to adsorb metals to a significant extent. The only exception to this was mercury. This is probably the result of organomercury compounds being absorbed in lipid and biomagnified, similar to the chlorinated hydrocarbons.

The study concluded that organisms living a demersal existence were exposed to higher levels of chlorinated hydrocarbons than were pelagic organisms.

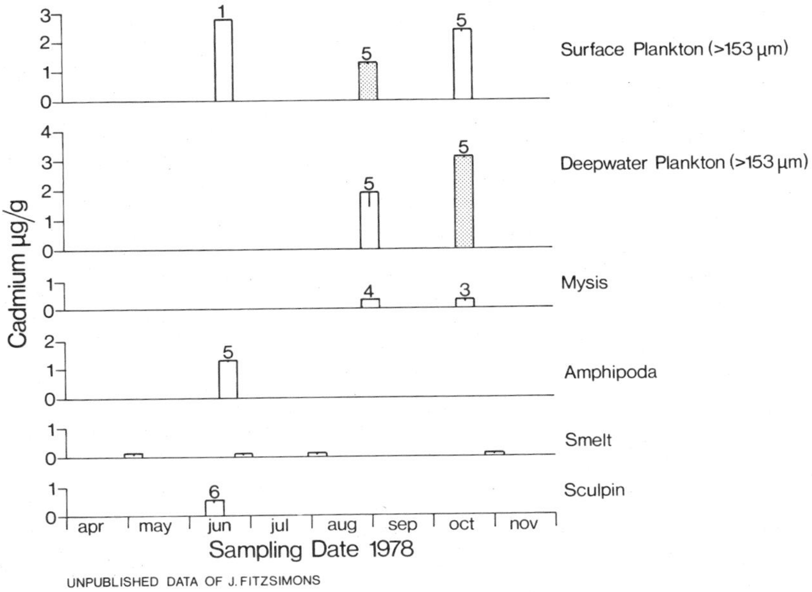

Figure 2.2 Concentrations of cadmium in biota from the Kingston Basin of Lake Ontario, 1978.

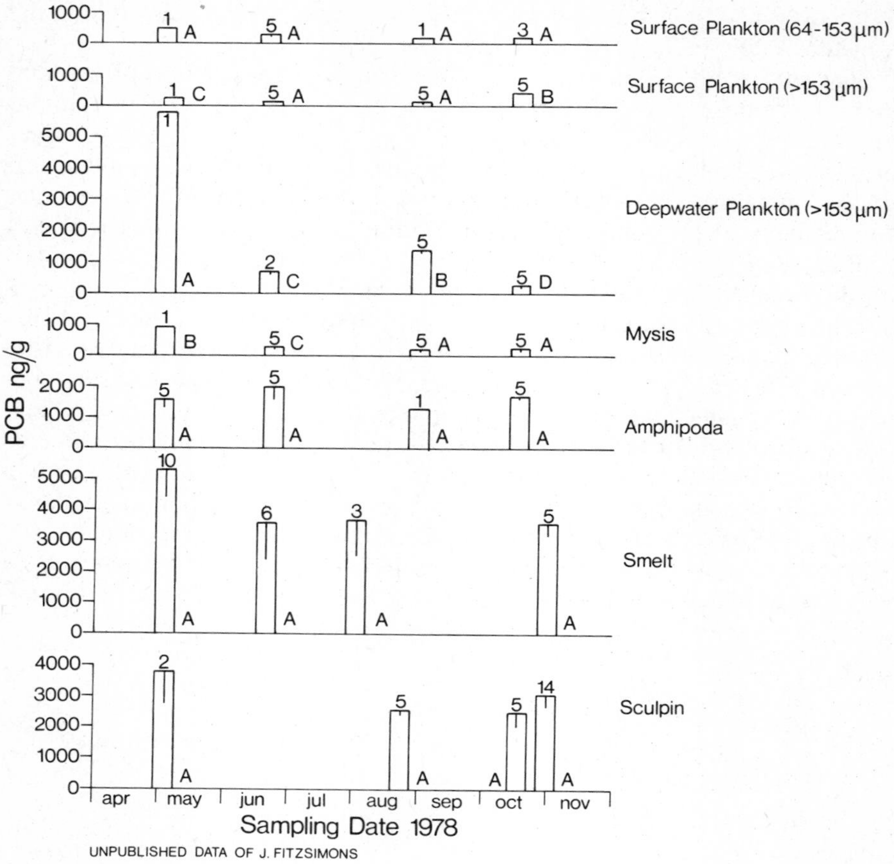

Figure 2.3 Concentrations of PCB in biota from the Kingston Basin of Lake Ontario, 1978.

2.4. Biological Tissue Archive

Annex 12 of the 1978 Great Lakes Water Quality Agreement (GLWQA, 1978) stipulates that a biological tissue archive should be established to permit retroactive contaminant analysis in the event of the discovery of a new compound of concern in the Great Lakes. The Canadian Wildlife Service established such an archive with herring gull eggs as part of a research program in 1970 (Hallett et al., 1981) and the Department of Fisheries and Oceans, Great Lakes Biolimnology Laboratory began to establish a fish tissue archive in 1979. The problem that had to be overcome in developing and maintaining this archive was that of proper sample preservation. Various methods exist for the preservation of fish: canning; freeze-drying; freezing at −20°C, −40°C, −80°C; freezing in liquid nitrogen; oven-drying. Several

Table 2.4. Losses of Organic Compounds from Biological Samples after Freeze-Drying, Oven-Drying, and Deep Freezing[a]

Treatment	Compound	Percentage Loss
Freeze-drying (lake trout)	HCB	70
	*pp'*DDE	45
	PCB	20–35
Oven-drying (60°C) (plankton)	HCB	33
	*pp'*DDE	31
	PCB	41
Deep freezing (−40°C) (lake trout)	*pp'*DDE	15
	HCB	0

[a] Source: M. Whittle, unpublished data.

of these methods were tested to determine their suitability in maintaining known concentrations of contaminants in samples previously analyzed. The results of these tests are summarized in Table 2.4. It can be seen that oven-drying and freeze-drying were totally unsuitable for the retention of chlorinated hydrocarbon compounds. Deep freezing appeared to be the most satisfactory method, and a long-term study was initiated to ascertain the best temperature for long-term storage of fish tissue homogenates, and the losses that were incurred by that storage. Table 2.5 presents some typical results of these studies. Deep freezing of lake trout tissue at −20°C results in significant losses of HCB, *pp'*DDE, and PCB over an 8- to 41-week period. This may be due to volatilization of compounds or separation of the homogenate caused by lipid migration (M. Whittle, unpublished data). It appears that freezing at lower temperatures retards loss of organic contami-

Table 2.5. Summary of Long-Term Storage Losses of Organic Contaminants in Biological Tissues

Treatment	Compound	Time (weeks)	Percentage Loss
Deep freezing (−20°C) (lake trout)	HCB	8	24
	*pp'*DDE	8	60
	PCB	41	32
Deep freezing (−40°C) prefrozen at −196°C	*pp'*DDT	52	6–9
	*pp'*DDD	52	4–16
	*pp'*DDE	52	15–21
	Oxychlordane	52	24
	Dieldrin	52	18–39
	Mirex	52	31–38
	γ Chlordane	52	71
	HCB	52	0
	Photo mirex	52	0

nants, but does not eliminate losses completely. Ultra deep freezing in liquid nitrogen (−196°C) would probably be the most successful method of maintaining sample integrity, but this is clearly an impractical solution, because very-long-term storage in liquid nitrogen presents some very difficult problems. The ultimate solution may be a quick freezing at −196°C followed by maintenance at −80°C (M. Whittle, unpublished), although even this method has its losses. Currently, a frozen acid extract is being tested as a possible long-term storage procedure.

2.5. Fish Health Assessment

The surveillance program described so far is largely one of measuring levels of materials in biota. The logical extension of this is effects monitoring, that is, investigating the well-being of Great Lakes fish in relation to the stresses imposed on them, particularly contaminants. To be useful as a surveillance tool, an indicator of stress on fish has to be readily observable by field biologists. It was decided to concentrate on external body tumors as an initial indicator of fish health. Considerable work has been done on fish tumors and their relation to contaminant stress (Stich et al., 1976; Sonstegard, 1977).

In the program initiated by the Department of Fisheries and Oceans, a major survey was conducted on several species of fish in a number of lo-

Table 2.6. Availability of Fish at Sampling Locations and the Frequency of Pathological Anomalies at These Sites[a,b]

Species	Pathological Condition	Number of Sites Species Captured	Number of Sites Condition Recorded
Brown bullhead	Papilloma	14	8
Carp	Gonadal tumor	12	3
Carp X goldfish hybrid	Gonadal tumor	6	6
Channel catfish	Papilloma	8	1
Coho	Thyroid hyperplasia	4	3
Drum	Epidermal hyperplasia	9	1
Gizzard shad	Epidermal hyperplasia	8	0
Golden and silver redhorse	Papilloma	9	2
Longnose sucker	Papilloma	2	1
Pike	Lymphosarcoma	9	0
Walleye	Lymphocystis	15	7
White sucker	**Papilloma**	**23**	**17**
Yellow perch	Gonadal tumor	16	8

[a] Source: V. Cairns, unpublished data.

[b] 28 sites visited in 1980.

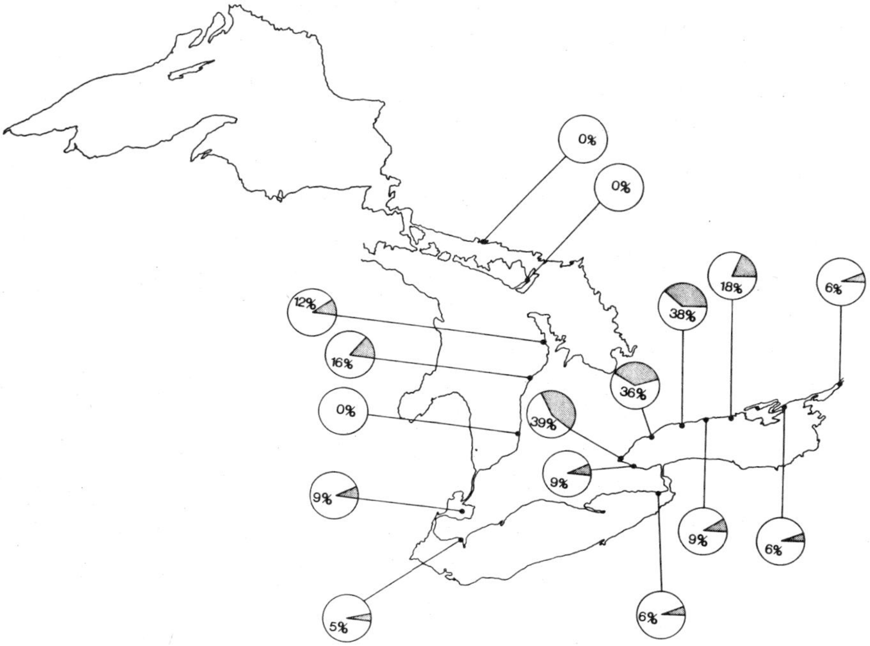

Figure 2.4 Prevalence (%) of epidermal papilloma on white suckers.

cations along the Canadian Great Lakes shoreline. The results of this survey, Table 2.6, indicate that the preferred species for monitoring tumors is the white sucker, *Catostomus commersoni*. The animal is widely distributed throughout the lakes, and tumor frequency (lip papilloma) is high. Results of a major survey of white suckers revealed some very interesting features (Fig. 2.4). Although the results are circumstantial at the moment because no data on contaminant burdens in these animals are available, it seems that there is a pattern of increased frequency of sucker lip papillomas when frequency is normalized for animal size, and that these are associated with areas of known industrial effluent discharge, for example, Lake Ontario.

Work is now in progress investigating the migratory behavior of one spawning population of suckers in Lake Ontario. The development or disappearance of tumors in this population over time is being examined. This ongoing work should elucidate the ontogeny of tumors.

3. RESEARCH

Pure research on contaminants, as defined earlier, is a topic so extensive that one could easily cover several chapters of this volume. I only attempt to highlight some of the more important work carried out over the past few years in the Great Lakes.

3.1. Alkylation of Metals

The occurrence of biological methylation was first observed early in the nineteenth century but was not thoroughly investigated until 1892 when Gosio (1892) observed a garlic odor emanating from mould-infected potatoes. This gas was later identified as trimethylarsine (Challenger, 1945). Thayer (1973) showed that biological methylation is a process that is common among living organisms.

There have been numerous studies on the environmental processes involved in methylation since the discovery of methylation of mercury by natural sediments (Jensen and Jernelöv, 1969). Wong et al. (1975) were one of the first groups of investigators to use a system consisting of natural water and sediment to demonstrate methylation of arsenic, mercury, and lead. In connection with the *in vitro* studies on methylation Chau et al. (1977) developed an *in situ* gas collector. They were able to detect *in situ* production of tetramethyllead from Hamilton Harbour sediment spiked with trimethyllead acetate. Similar results were found with sediment spiked with selenium, the gases produced being dimethyl selenide and dimethyl diselenide.

Alkylated lead compounds are discussed at length in Chapter 16 of this volume.

The factors that affect alkylation in nature have been investigated in selected locations. Baker et al. (1981) found that pH had a significant, although variable effect on *in situ* methylation. Tetramethyllead production (from trimethyllead acetate) increased with increasing pH in both biological and purely chemical systems. Mercury methylation ($HgCl_2 \rightarrow HgCH_3$) was found to occur at slightly acidic pH (5.5–6.5) and no $Hg(CH_3)_2$ was detected. Arsenic methylation was enhanced at low pH (below 5.5) but occurred over a wide range (3.5–7.5). With selenium, the reverse was true. Methylation was very low at pH 3.5 and increased with increasing pH.

The toxicity of organometals has been investigated on algae and bacteria. Algal species can tolerate quite high levels of inorganic arsenic (Wong et al., 1977). Only preliminary results can be reported on the effect of organoarsenicals on algae. A mixture of volatile arsines generated from sediments of the Moira River (Ontario) suppressed *Chlorella pyrenoidosa* primary production and cell growth by about 45% as compared to a control (Chau and Wong, 1981). Extract measurement of methylarsine was not possible, but arsenic levels in exposed algae were 10 times higher than in control cultures.

Much of the organotin found in environmental samples occurs as a result of the widespread useage of organotin in plastic stabilizers and biocides (Brinckman et al., 1981). Chau et al. (1981), however, studied the methylation of tin in the aquatic environment. They found that sediments enriched with either Sn(II) or Sn(IV) compounds would produce very different organotin compounds. For example, $Sn(CH_3)_4$ was seldom produced unless $(CH_3)_3SnCl$ was added.

Table 2.7. Toxicity of Several Tin Compounds to *Ankistrodesmus falcatus* Primary Production

Compound	Median Inhibitory Concentration (mg/liter)
Me_3Sn	5.50
Et_3Sn	0.20
Pr_3Sn	0.02
Bu_3Sn	0.02
ϕ_3Sn	0.01
Me_2Sn	21.00
Et_2SN	16.00
ϕ_2Sn	8.00
MeSn	23.00
ϕSn	19.00
Sn(IV)	12.00
Sn(II)	14.00

Wong et al. (1982a), in a study on the toxicity of organotin compounds on aquatic biota, found that the form of the organic portion of the organotin molecule (and the degree of organic substitution) governed the toxicity to algae. A few examples are given in Table 2.7.

3.2. Synergism

As noted earlier in this chapter, the concept of water quality objectives was developed by the International Joint Commission as a means of managing contaminant inputs to the Great Lakes (IJC, 1975). The objectives published in the 1978 agreement (GLWQA, 1978) were developed singly, each being designed to protect aquatic life. Wong et al. (1978a) showed that these ob-

Table 2.8. The Effect of a 10-Metal Mixture on Percentage Reduction in Primary Production in *Anabaena* Culture and Natural Lake Ontario Phytoplankton

	Algal Species	
Metal Mixture	Natural Population	*Anabaena*
Control (no metals added)	100%	100%
0.1X	86%	85%
0.5X	77%	48%
1.0X	59%	37%

Table 2.9. The Effect of a 10-Metal Mixture on Percentage Reduction in Acetylene Reduction by *Anabaena flos-aquae*

Metal Mixture	Acetylene Reduction (%)
Control	100%
1X	74%
10X	37%

jective levels of the agreement were indeed nontoxic to algae when present individually, but were highly toxic when present together. Wong et al. (1982b) have subsequently shown that the 10-metal mixture is not only inhibitory to primary production (reducing primary production from 41 to 69% of control values in four algal cultures and one natural phytoplankton population), but was also inhibitory to growth, acetylene reduction, and uptake of δ-amino isobutryic acid (a nonmetabolizable analog of alanine). Tables 2.8 to 2.10 illustrate selected data on the effects of metal mixtures on these physiological and biochemical parameters. The concentrations of metals used at the 1X level were: As, 50 μg/liter; Cd, 0.2 μg/liter; Cr, 50 μg/liter; Cu, 5 μg/liter; Fe, 300 μg/liter; Pb, 25 μg/liter; Hg, 0.2 μg/liter; Ni, 25 μg/liter; Se, 10 μg/liter; and Zn, 30μg/liter.

The exact mode of action of the metals in the mixture is now becoming clear. Some combinations may act antagonistically, some may act synergistically, and some additively. Borgmann (1980) found that of six metals tested (Cd, Hg, Zn, As, Cu, Pb) in binary combinations, only the Zn:As combination showed a large divergence in observed versus calculated growth times of freshwater copepods. Synergism in multimixtures of metals was cumulative. For all other metal combinations, toxicity was accounted for by the summation of noninteractive, single, metal effects.

Table 2.10. The Effect of a 10-Metal Mixture on Growth of *Ankistrodesmus falcatus*

Metal Mixture	Doubling Time (days)	Cell Numbers at Steady State (cells/ml)
Control	3.5	1.0×10^6
0.1X	4.0	9.3×10^5
0.5X	5.0	8.8×10^5
1.0X	8.5	5.7×10^5

3.3. Metal Toxicity Studies

Traditional studies on the toxicity of metals to plankton invertebrates and fish have been carried out for several years by the Department of Fisheries and Oceans in support of the development of Water Quality Objectives for the International Joint Commission. A major part of the work involved the development and subsequent refinement of the lead objective (see Chapter 16). Work on other metals, however, has been carried out. Borgmann (1981) was able to determine the actual free copper concentration of added copper in natural Lake Ontario and Hamilton Harbour waters using a copepod bioassay procedure. Copepod growth was affected by free copper concentrations as low as 10^{-10} to 10^{-9} mol/liter (0.064 μg/liter). This agrees very closely with the work of Van Den Berg et al. (1979) who found a free copper concentration of $10^{-10.3}$ mol/liter in extracellular products of *Anabaena cylindrica, Navicula pelliculosa,* and *Scenedesmus quadricauda* to which copper had been added. It would appear that natural complexing ligands (Van Den Berg et al., 1979) and artificially added ones (Borgmann, 1981) can achieve the same level of complexing.

Borgmann et al. (1980) determined the toxicity of several other metals on growth and mortality of natural assemblages of copepods using natural food and water. They found that, except for As, there was a seasonal cycle of toxicity to metals in natural waters. They investigated Cd, Cu, Hg, Pb, and As. The reason for this cyclical toxicity was unclear for Cd, Cu, and Hg, but appeared to be related to seston dry weight for Pb. This toxicity study should be compared to the *in situ* study carried out in the Kingston Basin of Lake Ontario (see Section 2.3). A definite seasonality was observed in metal concentrations in biota. Variability in metal toxicity could be associated with pulses of complexing ligands entering the lake from land runoff during periods of snow-melt or heavy rain. This seasonal pulse would also likely deliver increased loadings of metals to the lake. There is obviously additional research needed in this area to more clearly identify the seasonal patterns of contaminants' loading and toxicity.

3.4. Organic Contaminants

Each year there are several hundred new organic compounds synthesized. Of these, a substantial number find their way into the aquatic environment. I have already discussed, at length, the surveillance of these materials. Research on their toxicological properties is extensive in the literature, but is predominantly oriented toward mammalian or avian studies. For example, a good deal of the data used in the formulation of an ecosystem objective for mirex (IJC, 1981b), was derived from studies on rats and birds (Hallett et al., 1978; Norstrom et al., 1980; Sundaram et al., 1980; Villeneuve et al., 1979a; Villeneuve et al., 1979b, Villeneuve et al., 1979c; Weseloh et al.,

1979). The argument used in the objective (IJC, 1981b) was that insufficient data existed on long-term chronic toxicity of mirex to aquatic organisms to justify any modification of the existing objective.

Work is now being carried out on some selected organic compounds. For example, Niimi (1979, 1981) has investigated levels of hexachlorobenzene (HCB) in Lake Ontario salmonids and has studied the kinetics of HCB in salmonids from the same lake. He found mean values of 80, 62, and 36 ng/g HCB in whole fish homogenates of lake trout, rainbow trout, and coho salmon, respectively. It appears as though these three species have different rates of uptake of HCB throughout their life cycle.

In the follow-up study, Niimi and Cho (1981) investigated the kinetics of HCB in rainbow trout (*Salmo gairdneri*) and also examined kinetics in lake trout and coho salmon. They concluded that the half life of HCB in rainbow trout, under their laboratory conditions, was about 7 months, but could be as long as several years.

The results of this study were used to examine the kinetics of salmonids in Lake Ontario. There was good agreement between observed levels of HCB in rainbow trout in the lake and that predicted from laboratory studies. There was much poorer agreement between predicted and observed levels of HCB in coho salmon and lake trout. These differences were attributed to body weight, age, and feeding habits. Niimi and Cho (1981) concluded that observed levels of HCB in Lake Ontario salmonids were due primarily to the limited availability of HCB in the lake environment.

The problem in dealing with toxicity and contaminant dynamics of organic compounds is the overwhelming number of compounds to be tested. One way around this obstacle is the institution of studies on quantitative structure–activity relationships (QSAR). The QSAR will allow one to predict the behavior of a whole family of compounds based on the detailed study of just one member of the family. This approach is only now being adopted to the aquatic toxicology field. This clearly shows promise as a way of unplugging the data bottle neck that currently exists.

4. CONCLUSIONS

In this chapter, I have tried to give a mere sampling of the sort of work being carried out by the Canadian government on contaminants problems in the Great lakes. There are complementary programs in the United States (dealt with elsewhere in this volume), as well as at countless universities in both countries. The omission of their work was deliberate; the focus had to be restricted.

In the late 1960s and early 1970s we perceived cultural eutrophication as a major problem in the lakes. We developed a program which addressed that issue, resulting in large-scale restoration of the lakes. The contaminants problem is a more difficult one, but is, by no means, intractable. Coordinated effort by government, universities, and private industry can lead to a substantial reduction of contaminant inputs to the ecosystem.

REFERENCES

Baker, M. D., P. T. S. Wong, Y. K. Chau, C. I. Mayfield, and W. E. Innis. (1981). Methylation of Pb, Hg, As and Se in the acidic aquatic environment. Proceedings International Conference on Heavy Metals in the Environment, Amsterdam, Sept. 1981.

Borgmann, U. (1980). Interactive effects of metals in mixtures on biomass production kinetics of freshwater copepods. *Can. J. Fish. Aquat. Sci.* **37:** 1295–1302.

Borgmann, U. (1981). Determination of free metal ion concentrations using bioassays. *Can. J. Fish. Aquat. Sci.* **38:** 999–1002.

Borgmann, U., R. Cove, and C. Loveridge (1980). Effect of metals on the biomass production kinetics of freshwater copepods. *Can. J. Fish. Aquat. Sci.* **37:** 567–575.

Brinckman, F. E., J. A. Jackson, W. R. Blair, G. J. Olson, and W. P. Iverson. (1981). Ultratrace speciation and biogenesis of methyltin transport species in estuarine waters. In: *Trace Metals in Sea Water*. NATO Adv. Res. Inst. Erice, Sicily, Italy. Plenum Publishing Corp.

Carson, R. L. (1962). *Silent spring*. Fawcett, Greenwich, Connecticut, 304 pp.

Challenger, F. (1945) Biological Methylation. *Chem. Rev.* **36:** 315–361.

Chau, Y. K. and P. T. S. Wong. (1981). Some environmental aspects of organo-arsenic, lead and tin. In: F. E. Brinckman and R. H. Fish, Eds., *Environmental speciation and monitoring needs for trace metal-containing substances from energy-related processes*. Proceedings DOE/NBS Workshop, May 1981, pp. 65–80.

Chau, Y. K., W. J. Snodgrass, and P. T. S. Wong. (1977). A sampler for collecting evolved gases from sediment. *Water Res.* **11:**807–809.

Chau, Y. K., P. T. S. Wong, O. Kramar, G. A. Bengert, R. B. Cruz, J. O. Kinrade, J. Lye, and J. C. Vanloon. (1980). Occurrence of tetraalkyllead compounds in the aquatic environment. *Bull. Environ. Contam. Toxicol.* **24:** 265–269.

Chau, Y. K., P. T. S. Wong, O. Kramar, and G. A. Bengert. (1981). Methylation of tin in the aquatic environment. *Proceedings International Conference on Heavy Metals in the Environment*. Amsterdam, Sept. 1981.

GLWQA. (1972). Great Lakes Water Quality Agreement with Annexes and texts and terms of reference, between the United States of America and Canada. Ottawa, April 1972.

GLWQA. (1978). Great Lakes Water Quality Agreement of 1978. Agreement with Annexes and terms of reference, between the United Stated of America and Canada. Ottawa, April 1978.

Gosio, B. (1892). Action de quelques moissisures sur les composes fixes d'arsenic. *Arch. Ital. Biol.* **18:** 253–293.

Hallett, D. J., K. S. Khera, D. R. Stoltz, I. Chu, D. C. Villeneuve, and G. Trivett. (1978). Photomirex: Synthesis and assessment of acute toxicity, tissue distribution and mutagenicity. *J. Agric. Food Chem.* **26:** 388–391.

Hallett, D. J., H. Shear, D. V. Weseloh, and P. Mineau. (1981). Surveillance of wildlife contaminants on the Great Lakes. *Verh. Internat. Verein. Limnol.* **21:** 1734–1740.

Holdrinet, M. Van Hove, R. Frank, R. L. Thomas, and L. J. Hetling. (1978). Mirex in the sediments of Lake Ontario. *J. Great Lakes Res.* **4:** 69–74.

International Joint Commission. (1918). Final report of the International Joint Commission on the Pollution of Boundary Waters Reference. International Joint Commission, August 1918. Washington. Government Printing Office.

International Joint Commission. (1951). Report of the International Joint Commission on the Pollution of Boundary Waters. International Joint Commission, Washington and Ottawa.

International Joint Commission. (1967). Summary report on the Pollution of the Niagara River. International Joint Commission, October 1967. Washington and Ottawa.

International Joint Commission. (1969). Report to the International Joint Commission on the pollution of Lake Erie, Lake Ontario and the international section of the St. Lawrence River. International Joint Commission, 3 vols. Washington and Ottawa.

International Joint Commission. (1975). Annual report of the Water Quality Objectives Subcommittee. International Joint Commission, Windsor, Ontario, 239 pp.

International Joint Commission. (1977). Status report on the persistent toxic pollutants in the Lake Ontario basin. International Joint Commission, Windsor, Ontario, 95 pp.

International Joint Commission. (1978). Environmental management strategy for the Great Lakes system. Final report of the International Reference Group on Great Lakes Pollution from Land Use Activities, Windsor, Ontario, 115 pp.

International Joint Commission. (1981a). Great Lakes surveillance. Annual report of the Great Lakes Water Quality Board to the International Joint Commission, Windsor, Ontario, 174 pp.

International Joint Commission. (1981b). Report of the Aquatic Ecosystem Objectives Committee. International Joint Commission Science Advisory Board, Windsor, Ontario, 48 pp.

Jensen, S. and A. Jernelöv. (1969). Biological methylation of mercury in aquatic organisms. *Nature* **223:** 753–754.

Kaiser, K. L. E. (1974). Mirex: An unrecognized contaminant of fishes from Lake Ontario. *Science* **185:** 523–525.

McConnell, E. E., J. A. Moore, J. K. Haseman, and M. W. Harris. (1978). The comparative toxicity of chlorinated dibenzo-*p*-dioxins in mice and guinea pigs. *Toxicol. Appl. Pharmacol.* **44:** 335–356.

Murray, F. J., F. A. Smith, K. D. Nitschke, C. G. Humiston, R. J. Kochiba, and B. A. Schwetz. (1979). Three-generation reproductive study of rats given 2,3,7,8-tetrachlorodibenzo-*p*-dioxin (TCDD) in the diet. *Toxicol. Appl. Pharmacol.* **50:** 241–252.

Niimi, A. J. (1979). Hexachlorobenzene (HCB) levels in Lake Ontario salmonids. *Bull. Environ. Contam. Toxicol.* **23:** 20–24.

Niimi, A. J. (1981). Gross growth efficiency of fish (K_1) based on field observations of annual growth and kinetics of persistent environmental contaminants. *Can. J. Fish. Aquat. Sci.* **38:** 250–253.

Niimi, A. J., and C. Y. Cho. (1981). Elimination of hexachlorobenzene (HCB) by rainbow trout (*Salmo gairdneri*) and an examination of its kinetics in Lake Ontario salmonids. *Can. J. Fish. Aquat. Sci.* **38:** 1350–1356.

Norstrom, R. J., D. J. Hallett, F. I. Onuska, and M. Comba. (1980). Mirex and its degradation products in Great Lakes herring gulls. *Environ. Sci. Technol.* **14:** 860–868.

Ontario Ministry of the Environment. The decline in mercury concentration in fish from Lake St. Clair, 1970–1976. Ontario Ministry of the Environment Report Number AQS77-3, May 1977.

Ontario Ministry of the Environment/Ontario Ministry of Natural Resources. (1982). Guide to eating Ontario sport fish. Southern Ontario—Great Lakes. Information Services Branch, Ontario Ministry of the Environment, Toronto, 191 pp.

Schwetz, B. A., J. M. Norris, G. L. Sparschu, V. K. Rowe, P. J. Gehring, J. L. Emerson, and C. G. Gerbig. (1973). Toxicology of chlorinated dibenzo-*p*-dioxins. *Environ. Health Persp.* **5:** 87–99.

Shear, H. (1981). Contaminants programs in the North American Great Lakes basin. *Verh. Inernat. Verein. Limnol.* **21:** 1741–1748.

Sonstegard, R. A. (1977). Environmental carcinogenesis studies in fishes of the Great Lakes of North America. *Ann. N.Y. Acad. Sci.* **298:** 261–269.

Stich, H. F., A. B. Acton, and C. R. Forrester. (1976). Fish tumours and sublethal effects of pollutants. *J. Fish. Res. Board Can.* **33:** 1993–2001.

Sundaram, A., D. C. Villeneuve, I. Chu, V. Secours, and G. C. Becking. (1980). Subchronic toxicity of photomirex in the female rat: Results of 28- and 90-day feeding studies. *Drug Chem. Toxicol.*, **3:** 105–134.

Thayer, J. S. (1973). Biological methylation. Its nature and scope. *J. Chem. Educ.* **50:** 390–391.

Van Den Berg, C. M. G., P. T. S. Wong, and Y. K. Chau. (1979). Measurement of complexing materials excreted from algae and their ability to ameliorate copper toxicity. *J. Fish. Res. Board Can.* **36:** 901–905.

Villeneuve, D. C., V. E. Valli, I. Chu, V. Secours, L. Ritter, and G. C. Becking. (1979a). Ninety-day toxicity of photomirex in the male rat. *Toxicology* **12:** 235–250.

Villeneuve, D. C., L. Ritter, G. Felsky, R. J. Norstrom, I. A. Marino, V. E. Valli, I. Chu, and G. C. Becking. (1979b). Short term toxicity of photomirex in the rat. *Toxicol. Appl. Pharmacol.* **47:** 105–114.

Villeneuve, D. C., K. S. Khera, G. Trivett, G. Felsky, R. J. Norstrom, and I. Chu. (1979c). Photomirex: A teratogenicity and tissue distribution study of the rabbit. *J. Environ. Sci. Health* **814:** 171–180.

Weseloh, D. V., P. Mineau, and D. J. Hallett. (1979). Organochlorine contaminants and trends in reproduction in Great Lakes herring gulls, 1974–1978. Transactions of the 44th North American Wildlife and Natural Resources Conference, pp. 543–557.

Wong, P. T. S., Y. K. Chau, and P. L. Luxon. (1975). Methylation of lead in the environment. *Nature* **253**, 263.

Wong, P. T. S., Y. K. Chau, P. L. Luxon, and G. A. Bengert. (1977). Methylation of arsenic in the aquatic environment. In: D. D. Hemphill, Eds., *Trace Substances in Environmental Health—XI*. University of Missouri, St. Louis, 100 pp.

Wong, P. T. S., Y. K. Chau, and P. L. Luxon. (1978a). Toxicity of a mixture of metals on freshwater algae. *J. Fish. Res. Board Can.* **35:** 479–481.

Wong, P. T. S., B. A. Silverberg, Y. K. Chau, and P. V. Hodson. (1978b). Lead and the aquatic biota. In: J. O. Nriagu, Ed., *The Biogeochemistry of Lead in the Environment, Part B.* Elsevier/North Holland Biomedical Press, New York, pp. 279–342.

Wong, P. T. S., Y. K. Chau, O. Kramer, and G. A. Bengert. (1982a). Structure toxicity relationship of tin compounds on algae. *Can. J. Fish. Aquat. Sci.* **39:** 483–488.

Wong, P. T. S., Y. K. Chau, and D. Patel. (1982b). Physiological and biochemical responses of several freshwater algae to a mixture of metals. *Chemosphere* **11**(4): 367–376.

3

ATMOSPHERIC INPUTS OF CHLORINATED HYDROCARBONS TO THE GREAT LAKES

Thomas J. Murphy

Department of Chemistry
DePaul University
Chicago, Illinois 60614

1. INTRODUCTION

The role of the atmosphere as a nutrient source to land areas dates from the early observations of Barrinchins (1674), and a large number of studies have confirmed this role. With the growing awareness of the deleterious effects of many anthropogenic materials on bodies of water in the 1960s, and a growing awareness that large amounts of materials were getting into the atmosphere, a number of investigators began to determine whether the atmosphere could be a significant source of nutrients and other materials to bodies of water.

Perhaps the clearest early appreciation and statement of the possible role of the atmosphere as a source of materials to the Great Lakes, was by Winchester and Nifong (1971). They calculated that a reasonable percentage of the particulate emissions from the Gary–Chicago–Milwaukee metropolitan area could be depositing in Lake Michigan and that this atmospheric contribution could be a significant proportion of the inputs of some elements to the lake. Subsequent measurements by others confirmed the premise that the atmosphere could be an important source of a variety of materials to Lake Ontario (Shiomi and Kuntz, 1973), and of phosphates (Murphy and Doskey, 1976) and lead to Lake Michigan (Edgington and Robbins, 1976).

Interest in chlorinated hydrocarbons (CHCs) in the Great Lakes arose chiefly due to their biological effects. Their chemical properties are usually such that they accumulate in nonpolar environments, such as lipids, in organisms where they can be harmful. Large declines in the herring gull populations on the Great Lakes were shown to be due to accumulations of DDT and PCBs (Gilbertson and Fox 1977); and a precipitous decline in the reproductive success of ranch mink was shown to be due to PCBs in fish from Lake Michigan fed to the mink (Aulerich et al., 1973). While quite small concentrations of CHCs may occur in the water, accumulation of the CHCs by organisms in the water, and magnification through the food chain, can result in quite high concentrations in birds and animals that consume the fish.

The sources of these CHC to the Great Lakes were not known, but it was generally thought to be industrial, municipal, and tributary discharges. It was known, for instance, that runoff from agricultural lands treated with DDT contained high levels of DDT. However, the finding that DDT was present in many areas where it had never been used, and which received no runoff, indicated that it must also be being transported through the atmosphere. Since aerial spraying was a common method of applying DDT, this was thought to be the source of the DDT to the atmosphere (Woodwell, 1967).

Since the mid-70s, there have been a number of studies undertaken to determine the mechanism of transfer and to measure atmospheric inputs of CHC. These studies have contributed some information on atmospheric inputs, chiefly on precipitation inputs. In addition, a lot has been learned about all the mechanisms of CHC exchange between air and water. However, there is still much to be learned about some of the mechanisms of transfer, and many measurements are yet to be made.

This is a report of much of what is presently known about the inputs of CHCs to the Great Lakes from the atmosphere. It is limited to the higher-molecular-weight compounds, those with five or more carbon atoms. Although there is a lot of evidence from a variety of very different studies that the atmosphere is an important source of many CHC compounds to the Great Lakes, there are few studies that clearly show a dominant role of the atmospheric inputs. In part, this is due to incomplete determinations of atmospheric inputs of CHCs, but it is also due to an almost total lack of evidence of CHC inputs from other sources against which to compare the atmospheric inputs.

There are a variety of different mechanisms by which CHCs can exchange between the air and water. What is known about them is discussed here, and their importance to air/water exchange of CHCs is evaluated. Areas where more information and research is needed are pointed out.

2. PHYSICAL PROPERTIES

Although all of the compounds discussed here are different and have a different set of physical properties, their similarities are much greater than their differences. In general, the CHCs have low water solubilities (less than 1 mg/ liter), high solubilities in nonpolar liquids, and low vapor pressures (less than 10^{-5} atm). Information on the physical properties of the CHCs included on the U.S. Environmental Protection Agency's preliminary list of 129 Priority Pollutants are available (U.S. EPA, 1979).

2.1. Fugacities

Rather than discussing and comparing the concentrations and amounts of PCBs in the air, water, and other phases, its conceptually much easier to compare their fugacities. The concept of fugacity as applied to environmental problems recently has been clarified and promulgated by Mackay (1979) and Mackay and Patterson (1981). Fugacity is a measure of the escaping tendency of a material in a particular phase. The attractiveness of fugacity is that for a material distributed between two or more phases, and at equilibrium, the fugacity (f) of the material in each phase is the *same*. In contrast, the concentrations or amounts of the material in each of the phases at equilibrium will be different, with the concentrations (C) present related to the fugacity capacity (Z) for the material in that phase ($C = f \times Z$). For a system not at equilibrium, diffusive transport between phases will be in the direction to equalize the fugacities.

The fugacity of a material in a phase is proportional to the concentration of the material divided by the fugacity capacity for that material ($f = C/Z$). The fugacity of vapor phase materials is easy to determine since Z is equal to 1/RT. At 25°C, 1/RT = 41 mol/m^3 · atm. Thus the fugacity of a gas is equal to its vapor concentration (mol/m^3) divided by 41. For a material in the liquid phase, its fugacity capacity is equal to the reciprocal of the Henry's law constant (HLC; see Section 2.2). Thus, the fugacity of a material in solution is equal to its concentration times its HLC ($f = C \times H$).

Chlorinated Hydrocarbons in water may be dissolved or associated with particulate matter. The fugacity of CHCs in water, however, is determined only by the amount in solution. Because of the low concentration of most of the CHCs in water and the fact that a lot of it may be sorbed by particulates that pass through filters (Means and Wijayaratne, 1982), the fugacity of CHCs in natural waters has been difficult to determine. The concentration of a CHC found in a filtered sample will permit a maximum value for the fugacity of a compound to be determined, but this number will be in error by the amount sorbed on the nonfilterable particulates.

A more involved method for determining the fugacity of materials in water has been described by Murphy et al. (1982a). It involves measuring the vapor concentration, and therefore the fugacity, of a compound in equilibrium with that compound in a water sample. Since the measurement is made at equilibrium, the fugacity of the compound in water is equal to that found for the vapor phase. The actual water concentration can be calculated from the fugacity and the fugacity capacity ($1/H$) This can be compared to the total water concentration to determine the percent of the compound in the water that is in true solution. The results of measurements made on Lake Michigan water, using a single chlorobiphenyl compound (2,2′,5-trichlorobiphenyl) as a tracer, indicated that about 60% of the PCBs present in the water were in solution.

At equilibrium, the fugacity of CHCs on particulates in water is equal to the fugacity of the dissolved CHCs. The concentration in the particulate and dissolved phases are related by a distribution coefficient, K_p, which is equal to the ratio of the fugacity capacities of the particulate and water phases for the CHC in question. Once in a lake, the CHCs on particulates from dry deposition or precipitation scavenging will equilibrate with the CHCs in the water. If the fugacity of the CHCs is higher on the particulates, there should be net transfer of CHCs to the water.

2.2. Distribution between Air and Water: The Henry's Law Constant

A compound will distribute itself between the air and water phases until its fugacity in the two phases is the same. The ratio of the concentrations of the compound in the two phases at equilibrium is the air/water partition coefficient, the Henry's law constant (HLC). The HLC is the ratio of the vapor pressure of the compound (atm) divided by its water solubility (mol/m^3). Given the fugacity of a CHC in the air or water, the use of the HLC permits the equilibrium concentration in the other phase to be determined. If the concentration in the second phase is lower than the equilibrium concentration, the direction of movement will be into that phase, and vice versa.

The HLCs are known for a large number of organic compounds, including many CHCs. Measuring accurate values for these has been difficult chiefly because of the low solubility and vapor pressures of these compounds. The reported values for the water solubilities and vapor pressures often are in wide disagreement. Haque and Schmedding (1975) showed that the apparent solubility of PCBs could be significantly decreased by halting the stirring of a solution, in contact with liquid PCBs, for several weeks before sampling. Presumably, significant amounts of solid or liquid CHC can be kept in suspension in water by normal stirring. The HLC can be calculated if the water solubility and vapor pressure data for the compounds at a particular temperature are available. Henry's law constants for some CHCs are in Table 3.1.

In addition, Mackay et al. (1980) and Chiou and Schmedding (1982) have discussed the problem that many of the measurements of the water solubility and vapor pressures of the PCBs have been made on the individual chlorobiphenyl compounds, which are solids, whereas the PCB mixtures are liquids. Thermodynamic considerations indicate that the vapor pressure and solubility of a chlorobiphenyl compound in a liquid mixture, the PCBs, can be several orders of magnitude higher than the vapor pressure and solubility of the solid. Therefore, HLCs determined by using vapor pressure data from the solid CHCs and solubility data from the liquid CHCs, or vice versa, must

Table 3.1. Henry's Law Constants for Chlorinated Hydrocarbons

Compound	HLC ($atm \cdot m^3$ mol $\times 10^4$)	Reference
2,2′,5,5′-tetrachlorobiphenyl	2.6	Murphy et al., 1982a
2,2′,5,5′-tetrachlorobiphenyl	3.1–5.3	Wescott et al., 1981
2,2′,5,5′-tetrachlorobiphenyl	9.3	Atlas et al 1982
Aroclor 1242	2.2	Murphy et al., 1982a
Aroclor 1242	7.8	Atlas et al., 1982
Aroclor 1242	2	Westcott et al., 1981
Hexachlorobenzene (HCB)	13	Atlas et al., 1982
p,p′-DDE	12	Atlas et al., 1982
trans-Chlordane	*13*	*Atlas et al., 1982*
cis-Chlordane	9	Atlas et al., 1982
γ-hexachlorocyclohexane (Lindane)	0.24	Atlas et al., 1982

be corrected. Also, being very hydrophobic and having a lower surface tension than water, these compounds tend to accumulate at the water surface, which complicates sampling the water.

2.3. Distribution of CHCs in the Atmosphere

Since the mechanisms of exchange between the water and the atmosphere are very different for vapor and particulates, how each CHC in the atmosphere is distributed between the vapor and particulate phases will be the principal determinant of the rate of its exchange with a body of water.

Volatile materials in the atmosphere, such as the CHCs, are distributed between the vapor phase and the aerosol, or particulate, phase. These compounds can be adsorbed on the surface of particulates and absorbed by nonpolar particulates. The amount adsorbed is a function of the surface area of the particulates and the chemical nature of the particulate. The amount absorbed is a function of the fugacity capacity of the particulate for the CHC in question, as well as the amount of the particulate. Junge (1977) has derived an equation which relates ϕ, the fraction of a gaseous compound in the air attached to particulates, to the vapor pressure P_0 of the compound and θ, the amount of surface area on particulates in the air. C is a constant with a value of approximately 0.14.

$$\phi = \frac{C\theta}{P_0 + C\theta}$$

In addition, absorption by (partitioning into) nonpolar particulates could also be significant when there is a large amount of such particulate matter in the atmosphere.

In theory, the distribution of a CHC compound between the vapor and particulate phases for a CHC is easy to determine. Air is drawn through a filter to collect the compound in the particulate phase, and the air is then passed through an absorbant to collect the material in the vapor phase. It is believed, however, that some of the compound adsorbed on the particulates captured on the filter evaporates into the air stream and is collected as vapor (Bidleman and Olney, 1974; Harvey and Steinhauer, 1974). Thus this method may give only a lower estimate of the amount in the particulate phases in the atmosphere. Also, the composition of the CHCs found on the particulates collected may not be the same as their composition while in the air.

In the Great Lakes area such measurements have been reported only for the PCBs, Table 3.2. It can be seen that most of the PCBs seem to be present as vapor. Cautreels and Van Cauwenberghe (1978) have reported similar information for a large number of hydrocarbons. Their information indicates that the polycyclic aromatic hydrocarbon pyrene and the alkane *n*-eicosane are about equally distributed between vapor and the particulate phases in

Table 3.2. Concentrations of PCBs in the Air and on Air Particulates

Location	Total PCB (ng/m^3)	% PCBs on Particles	PCBs on Particles (mg/kg)	% 1242 on Vapor	% 1242 on Part.	Reference
Lake Michigan	1.0 (7)[a]	13	4	75	69	Doskey and Andren, 1981b
	2.8 (5)	<5		70		Rice et al., 1982
Lake Superior	1.2 (29)	<10	<2.8	$[b]		Eisenreich et al., 1982a
Madison, WI	7.7 (2)	3	3	86	25	Doskey and Andren, 1981b
Minneapolis, MN	7.1 (72)					Hollod, 1979
Chicago, IL	7.6 (4)	3	3	85	19	Murphy and Rzeszutko, 1977

[a] Number of samples.

[b] Predominance of 1242 over 1254.

the atmosphere. Compounds more volatile than these should be predominantly in the vapor phase in the atmosphere, while less volatile ones should be predominantly sorbed on particulates. Absorption of CHCs by organic particulates would increase the amount of CHCs on the particles. Thus, vapor exchange with bodies of water will tend to involve the lower-molecular-weight, relatively volatile, compounds while particulate deposition and precipitation scavenging will tend to involve the higher-molecular-weight, less volatile, compounds.

For mixtures of CHCs in the atmosphere, it would be expected that the composition of the adsorbed and absorbed material would be different from the vapor phase material. The vapor/particle compositions in Table 3.2 for the PCBs are in accord with this prediction.

Another way to determine the composition of PCBs on air particulates may be to collect and analyze precipitation. Murphy and Rzeszutko (1977) have pointed out that if the CHCs have HLCs high enough to preclude significant vapor partitioning into precipitation, then essentially all of the CHCs found in precipitation must come from the scavenging of particulate matter. Therefore, the composition of CHCs in precipitation should reflect the CHC composition of the particles scavenged.

The results from precipitation scavenging measurements indicate a very different particulate composition than does air sampling, even for simultaneously collected air and precipitation samples, Murphy and Rzeszutko (1978). This is evidence for the evaporation of compounds collected on filters and collection stages of air particulate samplers. However, it is possible that atmospheric particulates of different sizes and materials have different CHC compositions. To the extent this is true, precipitation scavenging results may not accurately reflect the average composition of CHCs on particles in the atmosphere due to different scavenging efficiencies for different size particles.

Buckley (1982) has recently reported that the PCB concentrations in plant leaves are proportional to ambient PCB vapor concentrations. The proportionality constant is different for different species, but seems to be constant for a given species in different locations. Thus the use of one or more plant species to monitor ambient CHC concentrations over wide areas seems possible.

There are two complications with carrying out such a program, however. The first is that the leaves undoubtedly also collect CHC-containing particulates, whose presence would increase the amount of CHC on the leaf. And, secondly, how is the concentration found in the leaves related to an ambient vapor concentration? With respect to the particulates, Saiki and Maeda (1982) have reported a technique using polar liquids to remove particulates from leaf surfaces. Perhaps this technique or another would permit the particulates to be removed without affecting the CHCs dissolved in the plant lipids.

With respect to comparisons of the CHC concentrations of a particular plant species, leaf concentrations from different locations would give information on the relative ambient CHC concentrations. However, neither the plant species used nor the concentrations found may be important. Presuming that the CHCs in the leaf lipids came to equilibrium with the mean concentration of the CHC in the atmosphere, the fugacity of the CHCs in the leaves should be the same as in the atmosphere. The use of indigenous plants could be a very useful technique for readily mapping the distribution of CHC vapor over large areas, with as much detail as desired, and with minimal sampling effort.

3. ATMOSPHERIC INPUT MECHANISMS

3.1. Dry Deposition

The deposition of particles from the atmosphere on a surface when precipitation is not occurring is called dry deposition. It is an important mechanism for the removal of particulates from the atmosphere. Measurements made for PCBs on atmospheric particles collected by several different methods indicate that atmospheric particles can contain significant concentrations of PCBs, and presumably other CHCs as well (Table 3.2). Thus, dry deposition could be an important source of CHCs to the Great Lakes. However, of the input routes of CHCs to bodies of water, dry deposition is probably the most complex and the least well understood. Slinn et al. (1978) have discussed the details of dry deposition, and Slinn and Slinn (1980) have summarized the relevant details of dry deposition to bodies of water.

The complexity of dry deposition arises chiefly from the fact that atmospheric particulates, aerosols, come in a wide range of sizes. The size of a particulate is the chief determinant of its deposition mechanism and rate. The term deposition velocity is used for all sizes of particles although strictly speaking it may be appropriate only for those large particles which have a net gravitational settling velocity. More often the deposition velocity is derived by measuring the concentration of a material in the atmosphere (gm/m^3) and dividing that into the accumulation rate on a particular surface ($gm/m^2{\cdot}s$). The resulting answer has units of m/s, and is called the deposition velocity, V_d. For a particular material, the V_d is the resultant deposition rate of all of the processes causing deposition of all the forms of that material in the atmosphere.

The smallest particles, with aerodynamic diameters less than 10^{-1} μ, are small enough to be moved about by air molecules and move with the turbulent eddies in the atmosphere. While they are much denser than the air, the forces on these particles due to collisions with air molecules are larger than the gravitational force. Thus they are chiefly transported by Brownian diffusion and the turbulance in the atmosphere. They deposit chiefly by impaction on surfaces. Because there is no gravitational component to their velocity, their velocity is isotropic and the deposition rate should be independent of the orientation of the surface. They can have a high net deposition velocity, generally larger than 0.005 m/s.

The larger particles, larger than 2 μm in aerodynamic diameter, are too massive to be greatly influenced by air molecules, and being more massive, the influence of gravity is more important and imparts a net downward velocity. The net deposition velocity is also generally greater than 0.005 m/s for these particles. The ultragiant particles (diameter greater than 10 μm) have a relatively very large mass, and a higher deposition velocity.

A problem with determining the importance of the ultragiant particulates as a source of materials to bodies of water is that most air particulate sampling methods have a poor collection efficiency for particles larger than 10–15 μm and the number of such particles present is quite small. Thus not a lot is known about the distribution and residence times of the ultragiant particles in the atmosphere. Johnson (1976), however, has reported that the concentration of giant and ultragiant particles upwind of St. Louis was 31 $\mu g/m^3$. This is a significant proportion of the total aerosol mass. Also, dust was collected from horizontal surfaces in several locations around Lake Michigan. This dust should consist chiefly of larger particulates. The average PCB concentration on the dust was 6.7 mg/kg, a concentration very similar to that found on air particulates (Table 3.2; Murphy et al., 1982b).

The intermediate size particles are too big to be much influenced by collisions with air molecules but too small for the force of gravity to overcome the aerodynamic drag forces on them. Thus particles in this range have the lowest average deposition velocity, less than 0.002 m/s.

Slinn and Slinn (1980) have pointed out that particles containing soluble salts, when close to the water surface where the relative humidity is high, may absorb significant amounts of water vapor. In theory, they would absorb water until their ionic strength was equal to that of the body of water. For a body of fresh water with a low ionic strength, like the Great Lakes, and a particle of a soluble salt like ammonium sulfate, the particle may increase in diameter by a factor of 10 (a volume increase of 1000). One result of this water condensation would be to change an intermediate size particle into a large one, thereby greatly increasing its deposition velocity.

However, CHCs in the atmosphere on particulates should be associated chiefly with nonpolar particles, those consisting of or containing organic carbon. These particles should have little tendency to absorb water vapor and grow in size. Andren and Strand (1981) have reported that more than 70% of the total organic carbon (TOC) on air particulates over Lake Michigan is on particles smaller than 1 μm.

This also means, however, that 30% of the TOC is associated with the larger particles, those with the higher deposition velocities and washout coefficients. Some of this TOC could be associated with aged aerosols present in the atmosphere. These aerosols are probably conglomerates of a variety of particles from different sources, and of very different compositions. Thus the CHCs may be associated with particles that will grow with vapor deposition, and this would then be a good mechanism for increasing the dry deposition rate of CHCs associated with particles to the Great Lakes.

Murphy et al. (1982b) reported the results obtained by using horizontally mounted collecting plates coated with a polar (water/diol) mixture for collecting PCBs. Two flat plates were used, held about 0.15 m apart, with the bottom surface of the top plate, and the top surface of the bottom plate coated with the polar liquid mixture. A polar liquid was used because of its low fugacity capacity for the nonpolar CHCs. This should minimize the scavenging of vapor phase CHCs (Murphy et al., 1981). One of the objectives of using the opposed collecting surfaces was to get information on the relative size of the particles depositing on the plates. If the deposition collected was chiefly due to small particles, whose deposition velocity is isotropic, then the two surfaces should have similar amounts. However, to the extent that the deposition was due to large particles that have a net gravitational settling velocity, the plate facing upward should have more material on it.

The results are shown on Table 3.3. The fact that the upward-facing plate collected more PCBs than the lower one near urban areas, was interpreted to mean that much of the dry deposition inputs of PCBs, in near-shore areas of Lakes Huron and Michigan, close to urban centers, is associated with large particles. This sampling method also indicated that the dry deposition inputs to Lakes Huron and Michigan were higher than the wet inputs.

Table 3.3. PCBs on Liquid-Coated Collecting Plates[a]

Location	Amount on Up-Facing Plates / Amount on Down-Facing Plates
Chicago, Il	3.1
Alpena, MI	2
Waukegan, IL	1.8
Sturgeon Bay, WI	1.8
Bridgeman, MI	1
Sebewaing, MI	1
Tawas Point, MI	1

[a] Murphy, et al., 1982b.

Thus, it seems as if there are large particles in the atmosphere with relatively high deposition velocities. These particles contain TOC, PCBs, and presumably, CHCs. Although there are relatively few of them in the atmosphere, and they contain a minor portion of the TOC and PCBs, their mass (the mass of a 20 μm particle is approximately equal to the mass of 64,000 0.5 μm particles) and high deposition velocity could make them the major source of dry deposition inputs of CHC in near shore areas. They could be a significant overall source of CHCs to Lakes Michigan, Erie, and Ontario, the lakes with large urban areas along their shores.

On the other hand, Andren (1982) has used measurements of the lead-210 and PCB inputs to Crystal Lake, an isolated pond in Wisconsin, to calibrate a dry deposition model. That model was then applied to Lake Michigan using meteorological data from Midway Airport in Chicago and the lead inputs to Lake Michigan (Edgington and Robbins, 1976). It was assumed that all of the particulate PCB inputs are associated with 0.5 μm particulates. The model output indicated that dry deposition inputs of PCBs to Lake Michigan are significantly less than the wet inputs.

It is clear that much more information, and probably new techniques and approaches to these measurements, are needed in order to make reasonable estimates of the amounts of CHC being deposited on the Great Lakes by dry deposition.

3.2. Precipitation Inputs

Precipitation scavenging of CHCs in the atmosphere is a complex process involving the scavenging of CHC-containing particulates by cloud droplets, by rain drops in and below the clouds, and by the scavenging of vapor phase CHCs by the rain (Slinn et al., 1978; Scott, 1981). Fortunately, the resultant

Table 3.4. Precipitation Inputs of PCBs to the Great Lakes

Location	Time	PCB Concentration (ng/liter)	Method	Collector Area (m^2)	Reference
Picton; CCIW (L. Ont.)	1976	32 (16)[a]	event	0.36	Strachan and Huneault, 1979
Point Pelee (L. Erie)	1976	9 (6)	event	0.36	Strachan and Huneault, 1979
Goderich; S. Baymouth (L. Huron)	1976	11 (11)	event	0.36	Strachan and Huneault, 1979
Nipigon; Batchawana Bay (L. Superior)	1976	26 (10)	event	0.36	Strachan and Huneault, 1979
Chicago, IL (L. Mich.)	1975–77	104 (39)	event	1.4	Murphy and Rzeszutko, 1977
Chicago, IL (L. Mich.)	1979–80	75 (20)	int. event	0.087	Murphy et al., 1982b
Waukegan, IL (L. Mich.)	1979–80	46 (55)	int. event	0.087	Murphy et al., 1982b
Point Betsie, MI (L. Michigan)	1979–80	12 (63)	int. event	0.087	Murphy et al., 1982b
Whitestone Point, MI (L. Huron)	1977–78	13 (34)	event	1.39	Murphy et al., 1982b
Tawas Point, MI					
Lake Huron vicinity	1976	18	Snow cores	0.025	Strachan and Huneault, 1979
Lake Superior vicinity	1976	38	Snow cores	0.025	Strachan and Huneault, 1979
Lake Ontario vicinity	1976	43	Snow cores	0.025	Strachan and Huneault, 1979
Saginaw Bay (L. Huron)	1978–79	25	Ice cores		Murphy and Schinsky, 1982
Duluth, MN (L. Super.)	1975–76	50 (13)	Snow events		Swain, 1978
Isle Royale (L. Super.)	1975–76	230 (25)	Snow events		Swain, 1978

[a] cm of precipitation.

concentration of CHCs in precipitation is usually relatively easy to determine. At this time, precipitation inputs of some CHCs to the upper Great Lakes are reasonably well known (Tables 3.4, 3.5, and 3.6).

In general, precipitation serves as an excellent mechanism for concentrating and removing the particulate matter found in the atmosphere. The air particulates can serve as condensation nuclei for the formation of cloud droplets and they may also be scavenged by cloud droplets. A rain drop is formed by the coalescence of the cloud droplets in about a liter of air, about 10^6 cloud droplets. This results in the CHCs on particulates being a factor of 10^5 to 10^6 more concentrated in precipitation than in the atmosphere (Scott, 1981). Thus fractions of a ng/m^3 of a CHC on particulates in the atmosphere can result in many ng/liter of that CHC in precipitation. Evidence from experimental programs clearly indicates that much of the CHCs in precipitation are associated with particulates, and that this material is relatively enriched in the early part of the precipitation event (Strachan and Huneault, 1979; Murphy and Rzeszutko, 1977).

In addition, vapor phase compounds can partition into rain drops. This is controlled entirely by the HLC for the material in question. Because of the relatively small amount of water involved, this can be an important source of materials to precipitation only if the material has a low HLC, that is, a high tendency to partition into water. For instance, 1 ng/m^3 of a CHC in the atmosphere with a HLC $= 1 \times 10^{-5}$ atm·m^3/mol and a molecular weight of 250, would have a fugacity of 1×10^{-13} atm, and a concentration of only 2.5 ng/liter in precipitation due to vapor partitioning. For a body of water with annual precipitation inputs of 80 cm/yr, this would be a loading of 2 gm/km^2·yr from vapor partitioning into precipitation. Compounds with higher HLCs would have proportionally smaller inputs.

A number of measurements have been reported for CHCs around the Great Lakes. These measurements have been made using a variety of techniques. The results are summarized in Tables 3.4, 3.5, and 3.6. It can be seen that the concentrations found in precipitation are higher than the open lake concentrations of these compounds, and the concentrations are higher near urban areas.

Andren (1982) has used a washout model for precipitation scavenging to calculate precipitation inputs of PCBs to Lake Michigan. The model is based

Table 3.5. Loading Estimates of CHCs to the Great Lakes from Precipitation

Lake	Average Input (gm/km^2 · yr)	Loading (kg/yr)	Reference
Lake Michigan	17.5	1000	Murphy et al., 1982b
	(11)	650	Andren, 1982
Lake Huron	14	825	Murphy et al., 1982b

Table 3.6. Precipitation Inputs of Other CHCs to the Great Lakes[a]

Compound	Number Samples	Number Positive	Concentration (ng/liter) in Rain to Lake				Mean	Loading[b] ($gm/km^2 \cdot yr$)
			Ontario	Erie	Huron	Superior		
Lindane	50	42	4.7	6.1	6	4.9	5.3 ± 3.9	4.0
γ-BHC	46	27	19.1	10.3	13.3	4.6	12 ± 13	9.6
DDT Residues	50	29	5.6	3.8	2.7	0.8	3.3 ± 4.6	2.4
α-Endosulfan	50	18	3.8	1.6	0.1	0.2	1.3 ± 3.3	1.2
β-Endosulfan	50	26	12.0	2.0	2.1	1.0	4.9 ± 4.4	4.0
Dieldrin	50	24	1.3	2.6	1.0	0.5	1.2 ± 2.0	1.0
Methoxychlor	50	34	8.5	13.1	9.5	1.6	7.7 ± 9.3	6.0
HCB	49	4	nd	nd	nd	2.8	0.8 ± 3.6	0.6

[a] Strachan and Huneault, 1979.
[b] for 80 cm/yr of precipitation.

on the PCB concentration of air particulates from high-volume air samples (Doskey and Andren, 1981b) and assumes that the PCBs are all on 0.5 μm air particulates (Andren and Strand, 1981). The results indicate an input of 650 kg/yr to the Lake (see Table 3.5).

Results obtained from measurements on event precipitation samples indicate that the concentrations found in individual precipitation events can vary by a factor of 10 or more (Murphy and Rzeszutko, 1977) and the standard deviation of the average concentration of PCBs in event samples is ±78%. This variability is due to variations in the amount of particulates and CHCs in the atmosphere, the type of rain event and the rainfall rate, and probably other causes. This variation in the precipitation concentration is the most severe limitation on the precision of precipitation input estimates.

3.3. The Transport Rate of Gases between Water and Air

Vapor is transferred between the air and water phases to equalize any fugacity difference between them. The fugacities can be determined from the measured concentrations and the HLC. What is still needed is the factor that relates the amount of material transferred to the fugacity difference. This transfer is often modeled using the two-film model of Whitman (1923), which assumes that the resistance to transfer of a substance between the air and water occurs chiefly in the thin film of water in contact with the air and in the thin film of air in contact with the water.

As developed by Liss and Slater (1974) and Mackay and Leinonen (1975), the mass transfer between phases will be determined by the fugacity difference between phases and the overall gas (K_{OG}) or liquid (K_{OL}) mass transfer coefficient. If one of these is known, the mass transfer between phases can be calculated by the equations:

$$N = K_{OL}\left(C - \frac{P}{H}\right) \qquad (1)$$

or

$$N = K_{OG}\frac{CH - P}{RT}$$

where N is the flux of the CHC in question; P is its pressure (atm) in the atmosphere; C is the water concentration (mol/m^3); H is the HLC; R is the gas constant (8.2×10^{-5} atm·m^3/mol·°K); and T is the temperature (°K). These can be rewritten in term of fugacities:

$$N = K_{OL}\frac{f_L - f_G}{H} \qquad (2)$$

or

$$N = K_{OG}\frac{f_L - f_G}{RT}$$

where f_G is the fugacity of the CHC in the gas phase and f_L is the fugacity in the water. The overall mass transfer coefficients, K_{OL} and K_{OG}, are related by the dimensionless proportionality factor H/RT, thus $K_{OL} = H/RT\ K_{OG}$. This discussion will consider only K_{OL}.

As mentioned before, the two-phase resistance model looks at the transfer between the two phases as being controlled by the resistance to transfer in each phase. The resistance to transfer is the reciprocal of the individual mass transfer coefficients. Thus the overall resistance to liquid phase mass transfer is $1/K_{OL}$, and it is equal to the sum of the resistances to transfer in each phase:

$$\frac{1}{K_{OL}} = \frac{1}{k_L} + \frac{RT}{Hk_G} \tag{3}$$

where k_G is the individual gas phase mass transfer coefficient, and k_L is the individual liquid mass transfer coefficient.

One can show (Mackay and Yuen, 1981) that for compounds with HLC greater than 10^{-3} atm·m³/mol, the gas phase resistance is small relative to the liquid phase resistance, and the k_G term can be neglected. The transfer in this case is said to be liquid phase controlled. For these cases, Eq. 3 reduces to:

$$\frac{1}{K_{OL}} = \frac{1}{k_L} \tag{4}$$

The resistance to transfer for compounds with HLC less than 10^{-5} atm·m³/mol, can be shown to be principally in the gas phase. The k_L term in Eq. 3 can be neglected for these materials which are said to be gas phase controlled. Thus:

$$\frac{1}{K_{OL}} = \frac{RT}{Hk_G} \tag{5}$$

For compounds with HLC greater than 10^{-5} and less than 10^{-3} atm·m³/mol, both resistances will affect the transfer rate, and Eq. 3 must be used for the calculations. Smith et al. (1980) classify the compounds in each of these categories as high-, low- or intermediate-volatility compounds.

Some recent measurements of HLCs for some CHCs are shown in Table 3.1. A variety of methods were used to determine these HLCs. Although agreement for the HLCs of the PCBs could be better, they are a major improvement over what had been available (Doskey and Andren, 1981a). At present, with HLCs $\sim 5 \times 10^{-4}$ atm·m³/mol (Table 3.1), it looks as if the PCBs are of intermediate volatility at 20°C. Thus both the liquid phase and gas phase rate constants are important in determining their rate of transfer through an air/water interface. Other CHCs, including HCB, *p,p*-DDE and trans-chlordane seem to fall in the high-volatility class where Eq. 4 would be applicable. Lindane, however, with its HLC ~2.4×10^{-5} atm·m^3/mol, would be a low-volatility compound, and Eq. 5 could be used.

Given a difference in fugacity for a particular CHC, the overall mass transfer rate constant, K_{OL} in this case, determines the amount of material transferred with time. These are difficult to measure in the field but some laboratory and field measurements are available (Cohen et al., 1978; Tofflemire et al., 1981; 1982). Usually, measurements are made of the rates of oxygen (liquid phase control) and water (gas phase control) transfer to give k_L and k_G. These constants for oxygen and water are then corrected to give the k_L and k_G for the compound in question (Smith et al., 1980, 1981; Atlas et al., 1982).

In practice, K_G values are much higher than K_L values (Liss and Slater, 1974; Mackay et al., 1982). Thus vapor exchange will occur more rapidly for those compounds with high HLC's, the compounds that are liquid phase controlled (Mackay et al., 1982). For example, if it assumed for Lake Michigan that: the PCBs are liquid phase controlled at 20°C ($K_{OL} = K_L$); K_L = 0.035 m/hr for a wind speed of 3 m/s (Cohen et al., 1978); and the fugacity difference between the air and water is equivalent to 1 ppt of PCBs in the lake (10^{-9} kg/m^3), then an evaporation of 800 ng/day/m^2 of PCBs is calculated. Thus the evaporation of PCBs could be a significant loss mechanism for these compounds from the Great Lakes.

The biggest problem still remaining in this area is determining the mass transfer rate constants under the range of ambient conditions, and properly weighting these to obtain an annual overall mass transfer rate. The problem is that these rate constants are very dependent on the ambient environmental conditions, particularly the wind speed and the temperature difference between the air and water. Most of the measurements that have been made were made during the warm months when the air is warmer than the lakes, the atmosphere is quite stable and the transfer rate constants are low. What is still needed are measurements during the winter months when the water is warmer than the air, the wind is blowing, and the transfer rate constants are high. With the exception of precipitation inputs, most of the air/water exchange of CHCs probably occurs when the water is warmer than the air, atmosphere is unstable, and the wind is blowing, chiefly during the fall and winter.

An interesting cycle is possible involving CHCs on particulates from atmospheric deposition and evaporation of CHCs from a lake surface. Particulates in the atmosphere could be a significant source of CHC compounds to a body of water via precipitation scavenging and dry deposition. With the right combination of HLCs and vapor pressures compared to their fugacity in the air and water, these compounds could partition into the water and then evaporate. Thus the atmosphere could be a net source of CHCs to a lake via particulate inputs, but a sink for CHCs from the lake due to evaporation. There is some evidence that this cycle may be operating for the PCBs in Lake Michigan, (Murphy et al., 1982a).

3.4. The Role of the Surface Microlayer

It is well known that a thin surface film, the surface microlayer, which is nonpolar in nature, is often present on bodies of water. Andren et al. (1976) have reviewed the information on the microlayer on freshwater lakes. The importance of the microlayer in the exchange of CHCs with the atmosphere, however, is not well understood. Much of the attention that has been focused on the surface microlayer is because of the high concentrations of metals, nutrients, and CHCs present, often a factor of 20 or more higher than the bulk water concentrations.

It is not surprising, however, that CHCs are found there in high concentrations, in fact such levels should be expected. The microlayer is a separate phase, containing a high proportion of organic materials and it should be expected to have a fugacity capacity quite different from water, and closer to that of a nonpolar liquid. Even in the absence of a definite surface film, the top layers of water should be enriched with nonpolar materials due to their surface-tension-lowering properties. Measurements in Lake Michigan (Eisenreich et al., 1978; Rice et al., 1982) have found enrichments of PCBs and DDT-related compounds by factors of 4–175 (Table 3.7).

As for the microlayer serving as a sink for atmospheric vapor phase CHCs, or scavenging CHCs from the water phase, the question is whether the fugacity of the CHCs present in the microlayer is higher or lower than the fugacity of those CHCs in the air or water phases. If the material in the surface microlayer comes chiefly from the water column, the fugacities of the compounds in the microlayer will be similar to their fugacities in the water. CHC compounds on air particulates that deposit in the microlayer may end up either in the water or the air. The particles containing the CHCs may become mixed with the water, or the CHCs may partition into the microlayer. Once dissolved in the microlayer, they can partition into the water or evaporate, since their fugacity in the microlayer should now be higher than in the water or atmosphere.

Table 3.7. Surface Layer Concentrations of Chlorinated Hydrocarbons

Compound	Enrichment Factor[a]	Reference
p,p'-DDE	175 (4)[b]	Eisenreich et al., 1978
p,p'-DDT	175 (4)	Eisenreich et al., 1978
Dieldrin	44 (4)	Eisenreich et al., 1978
PCBs (1979)	6 (12)	Rice et al., 1982
(1980)	4 (11)	Rice et al., 1982

[a] Ratio of the concentrations in the surface layer sampled to that in the water.

[b] Number of samples.

Because it is thought to be quite thin, on the order of a micron (10^{-3} mm) or so, the amount of material present in the microlayer may be small relative to the amount exchanged between the water and the atmosphere on a daily basis. The microlayer will then only serve an intermediary role in the air/water exchange of CHCs. Although it cannot change the equilibrium distribution of a compound between air and water or the direction of the net transfer of material, the physical and chemical properties of the microlayer may enhance or reduce the transfer rate.

The surface microlayer can also serve as a source of particulates to the atmosphere. This occurs when bubbles from the water burst at the surface. This bursting ejects droplets of surface microlayer material several centimeters into the atmosphere (MacIntyre, 1968). This bubble ejection phenomena can transport CHCs from the water to the atmosphere and is an additional mechanism for exchange of CHC compounds between the air and water.

The role of transfer by this mechanism is determined by the thickness of the surface layer, the extent of the coverage of the surface of the water with the layer, the number of bubbles breaking at the surface, and the percent of the material ejected that falls back into the Lake. Wu (1981) has shown that droplets from bursting bubbles are the major source of spray from the oceans to the atmosphere under most wind conditions. Thus they are also the major source of microlayer material to the atmosphere. To the extent that CHCs are present in ejected droplets from the Great Lakes, bursting bubbles result in a net transfer of CHCs to the atmosphere.

Mackay (1982) has evaluated the effect of a surface microlayer on air/water exchange rates. He concludes that surface films can have two effects. The first is a damping of the turbulence in the air and water boundary layers which will reduce the mixing of these layers with the bulk phase and lower the transfer rate. The second effect is the additional diffusive resistance of the surface microlayer. He concludes that the additional diffusive resistance in the microlayer will be small for CHCs.

Mackay and Cohen (1976) suggest that a surface microlayer could significantly alter air/water fluxes only during relatively calm periods when the coverage of the surface is essentially complete. At this time then, it seems that the decrease in vapor transport of CHCs due to the presence of a surface microlayer is not large.

3.5. Finding Atmospheric Inputs of CHCs by Mass Balance Techniques

Where the inputs to a lake are known to be from one or only a few sources, and where the amount of inputs from the different sources are not known or cannot be determined, a mass balance approach can often give useful information. This approach involves determining the total amount of the CHC in the lake in question, attempting to determine the annual input rate, and apportioning the inputs between the various known sources.

Eisenreich et al. (1981) have used this approach to determine the atmospheric inputs of PCBs to Lake Superior. Lake Superior is particularly amenable to this approach because, although major discharges or tributary inputs of PCBs to the Lake cannot be totally ruled out, none are known or are thought to exist. The total PCB inputs to Lake Superior are indicated by the mass present in the water and sediments, and evidence from sediment cores indicate the period of time that inputs have occurred (Eisenreich et al., 1980). Thus the relatively large amounts found in the water column and the sediments must be coming principally from the atmosphere.

An advantage of the mass balance method over modeling methods is that the most important information, the amount of CHC in the body of water, is being directly determined. The net inputs are then known, but the apportionment between wet, dry, and vapor sources remains to be determined. Unforeseen or unaccounted for sources or sinks can still affect the results, however. For instance, if PCBs are liquid phase controlled and are evaporating from Lake Superior, then the total wet-plus-dry inputs of PCBs will be higher than the measured accumulation rate as determined from the amount in the water column and sediments.

Another advantage of this method over programs to directly measure the wet, dry, vapor inputs, and modeling methods, is that the short-term variations in input rates are averaged out by the several year residence time of the CHC in the Lake. Thus the higher inputs from urban areas, the variation in the input rate with wind speed and atmospheric stability, and so on, should be reflected in, but not unduly affect, the concentrations found in the water column and sediments.

Recently, Eisenreich and Looney (1982) have reported careful water column profiles taken under low wind and very stable atmospheric conditions. Higher PCB concentrations at the surface indicated a source of PCBs at the surface under those conditions. This source was assumed to be the atmosphere. If the gradient in the water was maintained by vapor transfer, this indicates that the fugacity of the PCBs at that time was higher in the air than in the water. The authors assumed that vapor inputs were maintaining the gradient, and calculated an apparent HLC of 10^{-6} to 10^{-7} atm·m^3/mol. This value for the HLC indicated that the PCBs are low-volatility compounds (see previous discussion). Eisenreich et al. (1982b) have also reported that the surface concentrations of PCBs in Lake Superior are often higher than the concentrations several meters beneath the surface.

The possible role of particulate inputs of PCBs in maintaining the observed gradient was not considered. However, as discussed earlier, CHCs on atmospheric particulates brought into a lake may partition into the water. Eisenreich et al. (1982b) determined the particulate/water partition coefficient, K_p, for Lake Superior: $\log K_p = -1.2 \log(\text{Sus. Solids}) + 5.4$. For the suspended solids concentration (0.1–1 mg/liter) in Lake Superior, and the concentration of PCBs found on air particulates over Lake Michigan (Table 3.2), most of the PCBs on the particulates should partition into the water.

Because the inputs of PCBs and other CHCs to Lake Superior seem to come essentially completely from the atmosphere, the lake is very attractive as a simple system to try to determine real values for dry deposition and vapor inputs. First, however, the wet inputs and PCB composition of the air particulates must be determined. The collection of precipitation from a number of events at several locations over a period of several months or more should yield the desired information. Also, the determination of the concentration and composition of the PCBs and other CHCs on the rain particulates should give needed information on the concentration and of these compounds on air particulates over the lake.

Swain (1978, 1982) also used mass balance techniques to conclude that the atmosphere can be the major source of CHCs to a lake. He made numerous measurements of the CHCs in fish from Siskiwit Lake on Isle Royale (Table 3.8) and on snow collected there (Table 3.4). The concentrations found are significant because Isle Royale is an island in Lake Superior, more than 20 miles from the nearest shore. It has been maintained as a wilderness area since the early 1940s, before there was any significant manufacturing of most of the CHC compounds found there by Swain. The surface of Siskiwit Lake is more than 17 m above the level of Lake Superior. Thus, it seems that the only source of the CHC compounds now found in the fish in Siskiwit Lake, can be the atmosphere.

Table 3.8. Chlorinated Hydrocarbons Found in Siskiwit Lake Trout[a]

	Concentration, Eviscerated Whole Fish (ng/g; ppb)	
Compound	1974–1976	1980
α-Hexachlorocyclohexane (αBHC)	11.0	3.5
γ-Hexachlorocyclohexane (γBHC; Lindane)	—	1.5
Hexachlorobenzene (HCB)	5.0	—
Heptachlor Epoxide	8.0	5.2
Oxychlordane	—	5.3
trans-Chlordane	—	8.8
cis-Chlordane	—	24.4
PCBs	1200	720
p,p'-DDT	455	—
p,p'-DDE	2370	318
p,p'-DDD	36.0	25.0
cis-Nonachlor	—	25.0
Toxaphene-like compounds	—	3200
Dieldrin	11.0	—

In addition, trace amounts of chlorinated dibenzo-*p*-dioxins and 15 parts per trillion (ppt) of chlorinated dibenzofurans (PCDFs) were found. Of the PCDFs found, 10 ppt were tetrachloro compounds.

[a] Swain, 1978; 1982.

There is an additional method which can sometimes be used to obtain an estimate of the *total* net atmospheric inputs to a body of water. It is based on the fact that the wet, dry, and vapor deposition which accumulates on an ice cover during winter, becomes incorporated into the body of water upon ice break-up and melting in the spring. To determine the magnitude of this spring pulse of materials, representative samples of the accumulated deposition on the frozen lake surface need only be taken late in the ice season, and analyzed for the appropriate components. Murphy and Schinsky (1983) have reported such a determination for PCBs in Saginaw Bay of Lake Huron.

4. ESTIMATING ATMOSPHERIC CHC INPUTS

Information on the identity and amounts of CHC compounds in the atmosphere along with measured deposition rates for a number of compounds in a number of locations indicate that the atmosphere is an important source of many CHCs to the Great Lakes. In the absence of any presently unknown, large tributary or effluent sources there is reasonable evidence that the atmosphere is the major source of several CHCs to some of the Great Lakes. This includes PCBs to Lakes Michigan, Superior, and Huron (Tables 3.4 and 3.5); all of the CHCs found in Siskiwit Lake on Isle Royale (Table 3.8); and probably most of the CHCs present in the upper Great Lakes.

In theory it can be quickly ascertained whether atmospheric inputs are a significant source of a particular CHC to one of the Lakes. This can be done as follows:

1. An estimate of the HLC for the compound in question should be obtained.

2. A simple collector or an integrating event collector, should be used to collect precipitation samples from a number of individual precipitation events. The area of the collector used must be large enough to collect sufficient rain for the analysis of the compounds of interest. Care should be taken to clean the collector before the rain event, and to be sure the sample includes the first few millimeters of each precipitation event. The sample would then be cleaned up and analyzed for the CHC of interest. From the precipitation-weighted concentration found (ng/liter) and the annual precipitation (1 mm/yr in precipitation = 1 liter/m^2·yr of water) to the lake, the annual wet loading to the lake for this compound (ng/m^2·yr) can be estimated.

3. If the HLC is less than 10^{-5} atm·m^3/mol, then vapor partitioning into the lake may be important (vapor scavenged by precipitation would have been included in the precipitation measurement). The atmospheric concentration should be measured and the inputs from vapor partitioning estimated using one of the available deposition models (see Section 3.3).

4. At this time, in the absence of a good method for determining or modeling dry deposition, it may be best to assume that the dry deposition inputs are equal to the wet inputs.

5. Sum the wet, vapor, and dry inputs from the atmosphere, compare them to the known inputs for this CHC from discharges, tributaries, and other sources. If information on other sources is not available, determine the concentration of the CHC in the Lake, correct this for the average particulate residence time in the lake, and calculate whether the atmosphere could be supplying a significant proportion of the annual CHC inputs.

There is some information on the atmospheric inputs of several CHCs to the Great Lakes. The limitation in determining the importance of the atmospheric route, however, is that there is little or no information on the inputs of CHCs from tributaries, industrial and municipal discharges, or other sources with which to compare the atmospheric inputs.

A complication in the programs to determine the inputs to the Great Lakes by sampling the various kinds of inputs, and a limitation in modeling efforts, is properly incorporating the higher inputs associated with urban and industrialized areas. The data for air (Table 3.2) and for precipitation (Tables 3.4 and 3.5) clearly show that the atmospheric burdens of PCBs are higher in urban areas. It is also well known that air particulate levels are higher in urban areas. According to Junge (1977), these higher particulate levels should lead to a higher percentage of the CHCs in the atmosphere being on particulates (See Section 2.3). Since CHCs associated with particulates are the principal source of wet and dry atmospheric inputs of CHCs to bodies of water, all of these factors indicate that urban areas probably contribute a disproportionate amount of CHCs to bodies of water. Thus, the Gary–Chicago–Milwaukee area is probably a major source of atmospheric inputs to Lake Michigan; the Cleveland and Detroit areas are major sources to Lake Erie; and Toronto–Hamilton, Buffalo and Rochester areas are major sources to Lake Ontario.

5. RECOMMENDATIONS

Of the information available on inputs of CHCs to the Great Lakes, most of it is related to PCB inputs to the upper Great Lakes. While there has been a lot of progress in determining and understanding atmospheric inputs, there is still a lot to be done. For instance, the present estimates of atmospheric inputs of PCBs are good only to a factor of 2 or 3, at best. The limitations at this time include: the variability in precipitation inputs between individual events and sampling locations; the inability to collect representative samples of dry deposition or to adequately model it; the lack of information on vapor transfer rates; and the lack of knowledge of the role of the microlayer in mediating air/water exchange of CHCs.

Order of magnitude estimates of the atmospheric inputs of other CHC compounds can be made, however, by collecting a number of precipitation samples at a variety of locations. The results from these samples, along with a judicious estimate of vapor and dry deposition inputs, should indicate whether the atmosphere might be an important source of these CHCs to the Great Lakes. In order to significantly improve on these results for each of the Great Lakes, one or more of the following are needed:

1. Good determinations of the HLCs for those compounds for which the HLCs are not known or are poorly known.
2. Year-around information of the meteorology from different locations on the lakes.
3. A means of collecting air particulates and determining the elemental and CHC composition of the different size fractions without the loss of volatile compounds.
4. Information from several different locations around the lake on the air particulates. This would include the size spectrum and the elemental, CHC, and TOC distribution with respect to size.
5. The development of a dry deposition model using the particulate and meteorological information obtained above.
6. A collection technique for measuring dry deposition inputs to lakes to calibrate and verify dry deposition models.
7. Determination of the extent of coverage of the surface of the Great Lakes by a surface microlayer during different meteorological conditions; determination of the fugacity of some CHCs in sets of air, microlayer, and water samples from different locations; and determination of the amount of surface microlayer material ejected into the atmosphere under different sea-state conditions of a lake.

ACKNOWLEDGMENT

The preparation of this paper was supported by a grant from the Faculty Research and Development Fund of the College of Liberal Arts and Sciences, DePaul University.

REFERENCES

Andren, A. W. (1982). Processes determining the flux of PCBs across air/water interfaces. In: D. Mackay, Ed., *Physical behavior of PCBs in the Great Lakes*. Ann Arbor Science, pp. 127–140.

Andren, A. W. and J. W. Strand. (1981). Atmospheric deposition of particulate organic carbon and PAHs to Lake Michigan. In: S. Eisenreich, Ed., *Atmospheric inputs of pollutants to natural waters*. Ann Arbor Press. p. 459–479.

Andren, A. W., A. W. Elzerman, and D. E. Armstrong. (1976). Chemical and physical aspects of surface organic microlayers in freshwater lakes. *J. Great Lakes Res., Suppl.* **1**: 109–111.

Atlas, E., R. Foster, and C. S. Giam. (1982). Air–sea exchange of high molecular weight organic pollutants: Laboratory studies. *Environ. Sci. Technol.* **16:** 283–286.

Aulerich, R. J., R. K. Ringer, and S. Iwamoto. (1973). Reproductive failure and mortality in mink fed on Great Lakes fish. *J. Reprod. Fert. Suppl.* **19:** 365–376.

Barrinchins, O. (1674). Reported in: Hjarne, *Acta et Tentamina Chemica Holmiensia* **2:** 23–25 (1753) (Stockholm).

Bidleman, T. F. and E. J. Christensen. (1979). Atmospheric removal processes for high molecular weight organochlorine compounds. *J. Geophys. Res.* **84:** 7857–7862.

Bidleman, T. F. and C. E. Olney. (1974). Chlorinated hydrocarbons in the Sargasso Sea atmosphere and surface water. *Science* **183:** 517–518.

Buckley, E. H. (1982). Accumulation of airborne PCBs in foliage. *Science* **216:** 520–522.

Cautreels, W. and K. Van Cauwenberghe. (1978). Experiments on the distribution of organic pollutants between airborne particulate matter and the corresponding gas phase. *Atmos. Environ.* **12:** 1133–1141.

Chiou, C. T. and D. W. Schmedding. (1982). Partitioning of organic compounds in octanol-water systems. *Environ. Sci. Technol.* **16:** 4–10.

Cohen, Y., W. Cocchio, and D. Mackay. (1978). Laboratory study of liquid-phase controlled volatilization rates in the presence of waves. *Environ. Sci. Technol.* **12:**553–558.

Doskey, P. V. and A. W. Andren. (1981a). Modeling the flux of atmospheric polychlorinated biphenyls across the air/water interface. *Environ. Sci. Technol.* **15:** 705–711.

Doskey, P. V. and A. W. Andren. (1981b). Concentrations of airborne PCBs over Lake Michigan. *J. Great Lakes Res.* **7:** 15–20.

Edgington, D. N. and J. A. Robbins. (1976). Records of lead deposition in Lake Michigan sediments since 1800. *Environ. Sci. Technol.* **10:** 266–274.

Eisenreich, S. J. and B. B. Looney. (1982). Evidence for the atmospheric flux of PCBs to Lake Superior. In: D. Mackay, Ed., *Physical behavior of PCBs in the Great Lakes.* Ann Arbor Science, pp. 141–156.

Eisenreich, S. J., A. W. Elzerman, and D. E. Armstrong. (1978). Enrichment of micronutrients, heavy metals and chlorinated hydrocarbons in wind-generated lake foam. *Environ. Sci. Technol.* **12:**413–417.

Eisenreich, S. J., G. J. Hollod, T. C. Johnson, and J. Evans. (1980). PCB and other microcontaminant–sediment interactions in Lake Superior. In: R. A. Baker, Ed., *Contaminants and sediments.* Ann Arbor Science, Ch. 4.

Eisenreich, S. J., B. B. Looney, and J. D. Thornton. (1981). Airborne contaminants in the Great Lakes ecosystem. *Environ. Sci. Technol.* **15:** 30–38.

Eisenreich, S. J., B. B. Looney, and G. J. Hollod. (1982a). PCBs in the Lake Superior atmosphere 1978–80. In: D. Mackay, Ed., *Physical behavior of PCBs in the Great Lakes.* Ann Arbor Science, p. 115–126.

Eisenreich, S. J., P. D. Capel, and B. B. Looney. (1982b). PCB dynamics in Lake Superior. In: D. Mackay, Ed., *Physical behavior of PCBs in the Great Lakes.* Ann Arbor Science, p. 181–212.

Gilbertson, M. and G. A. Fox. (1977). Pollution-associated embryonic mortality of Great Lakes herring gulls. *Environ. Pollut.* **12:** 211–216.

Haque, R. and D. Schmedding. (1975). A method of measuring the water solubility of hydrophobic chemicals: Solubility of five PCBs. *Bull. Environ. Contam. Toxicol.* **14:** 13–18.

Harvey, G. R. and W. G. Steinhauer. (1974). Atmospheric transport of polychlorophenyls to the North Atlantic. *Atmos. Environ.* **8:** 777–782.

Hollod, G. J. (1979). Ph. D. dissertation, Univ. of Minnesota.

Johnson, D. B. (1976). Ultragiant urban aerosol particles. *Science* **194:** 941.

Junge, C. E. (1977). Basic considerations about trace constituents in the atmosphere as related to the fate of global pollutants. In: I. H. Suffet, Ed., *Fate of pollutants in the air and water environments*. J. Wiley & Sons, New York, p. 7–25.

Liss, P. S. and P. G. Slater. (1974). Fluxes of gases across the air–sea interface. *Nature* **247:** 181–4.

MacIntyre, F. (1968). Bubbles: A boundary layer "microtome" for micron-thick samples of a liquid surface. *J. Phys. Chem.* **72:** 589–592.

Mackay, D. (1979). Finding fugacity feasible. *Environ. Sci. Tchnol.* **13:** 1218.

Mackay, D. (1982). Effects of surface films on air–water exchange rates. *J. Great Lakes Res.* **8:** 299–306.

Mackay, D. and Y. Cohen. (1976). Chemical and physical aspects of surface organic microlayers in freshwater lakes. *J. Great Lakes Res., Supp.* **1:** 111–113.

Mackay, D. and P. J. Leinonen. (1975). Rate of evaporation of low-solubility contaminants from water bodies to atmosphere. *Environ. Sci. Technol.* **9:** 1178–1180.

Mackay, D. and S. Patterson. (1981). Calculating fugacity. *Environ. Sci. Technol.* **15:** 1006–1014.

Mackay, D., R. Mascarehas, and W. Y. Shiu. (1980). Aqueous solubility of polychlorinated biphenyls. *Chemosphere* **9:** 257–264.

Mackay, D., W. Y. Shiu, J. Billington, and G. L. Huang. (1982). Physical Chemical Properties of PCBs. In: D. Mackay et al., Ed., *Physical Behavior of PCBs in the Great Lakes*. Ann Arbor Press, pp. 50–70.

Means, J. C. and R. Wijayaratne. (1982). Role of natural colloids in the transport of hydrophobic pollutants. *Science* **215:** 968–970.

Murphy, T. J. and P. V. Doskey. (1976). Inputs of phosphorus from precipitation to Lake Michigan. *J. Great Lakes Res.* **2:** 60–70.

Murphy, T. J. and C. P. Rzeszutko. (1977). Precipitation inputs of PCBs to Lake Michigan. *J. of Great Lakes Res.* **3:** 305–312.

Murphy, T. J. and C. P. Rzeszutko. (1978). PCBs in precipitation in the Lake Michigan basin. U.S. EPA report EPA-600/3-78-071, July 1978.

Murphy, T. J. and A. L. Schinsky. (1983). Net atmospheric inputs of PCBs to the ice cover of Lake Huron. *J. Great Lakes Res.,* **9:**92–96.

Murphy, T. J., T. C. Heesen, and D. R. Young. (1981). Evaluation of a technique for measuring dry aerial deposition rates of DDT and PCB residues. *Atmos. Environ.* **15:** 206–207.

Murphy, T. J., J. C. Pokojowczyk, and M. D. Mullin. (1982a). Vapor exchange of PCBs with Lake Michigan: The atmosphere as a sink for PCBs. In: D. Mackay, Ed., *Physical behavior of PCBs in the Great Lakes*. Ann Arbor Science, pp. 49–58.

Murphy, T. J., G. Paolucci, A. W. Schinsky, M. L. Combs, and J. C. Pokojowczyk. (1982b). Inputs of PCBs from the atmosphere to Lakes Huron and Michigan. Report of U.S. EPA Project R-805325. Duluth Environmental Research Laboratory. May.

Rice, C. P., B. J. Eadie, and K. M. Erstfeld. (1982). Enrichment of PCBs in Lake Michigan surface films. *J. Great Lakes Res.,* **8:**265–270.

Saiki, H. and O. Maeda. (1982). Cleaning procedure for removal of external deposition from plant tissues. *Environ. Sci. Technol.* **16:** 536–539.

Shiomi, M. T. and K. W. Kuntz. (1973). Great Lakes precipitation chemistry: Part 1. Lake Ontario basin. Proceedings 16th Conference on Great Lakes Research, pp. 581–602.

Scott, B. C. (1981). Modeling of atmospheric wet deposition. In: S. J. Eisenreich, Ed., *Atmospheric inputs of pollutants to natural waters*. Ann Arbor Science, p. 3–21.

Slinn, S. A. and W. G. N. Slinn. (1980). Prediction for particle deposition on natural waters. *Atmos. Environ.* **14:** 1013–1016.

Slinn, W. G. N., L. Hasse, B. B. Hicks, A. W. Hogan, D. Lal, P. S. Liss, K. O. Munnich, G. A. Sehmel, and O. Vittori. (1978). Some aspects of the transfer of atmospheric trace constituents past the air–sea interface. *Atmos. Environ.* **12:** 2055–2087.

Smith, J. H., D. C. Bomberger, Jr., and D. L. Haynes. (1980). Prediction of the volatilization rates of high-volatility chemicals from natural water bodies. *Environ. Sci. Technol.* **14:** 1332–1337.

Smith, J. H., D. C. Bomberger, Jr., and D. L. Haynes. (1981). Volatilization rates of intermediate and low volatility chemicals from water. *Chemosphere* **10:** 281–289.

Strachan, W. M. and H. Huneault. (1979). Polychlorinated biphenyls and organochlorine pesticides in Great Lakes precipitation. *J. Great Lakes Res.* **5:** 61–68.

Swain, W. R. (1978). Chlorinated organic residues in fish, water and precipitation from the vicinity of Isle Royale, Lake Superior. *J. Great Lakes Res.* **4:** 398–407.

Swain, W. R. (1982). An overview of contaminants in the Lake Superior ecosystem. Presented at the 25th Conference on Great Lakes Res., Sault Ste. Marie, Ontario, May.

Tofflemire, T. J., T. T. Shen, and E. H. Buckley. (1981). Volatilization of PCB from sediment and water: Experimental and field data. Technical Paper #63. New York State Department of Environmental Conservation, Albany, NY, December. 37 pp.

Tofflemire, T. J., T. T. Shen and E. H. Buckley. (1982). Volatilization of PCB from sediment and water: Experimental and field data. In: D. Mackay et al., Ed., *Physical Behavior of PCbs in the Great Lakes,* Ann Arbor Press, pp. 411–422.

U.S. Environmental Protection Agency. (1979). Water-related environmental fate of 129 priority pollutants, 1 and 2. EPA 440/4-79-029 a and b. PB 80204371, PB 80204381.

Westcott, J. W., C. G. Simon, and T. F. Bidleman. (1981). Determination of PCB vapor pressures by a semimicro gas saturation technique. *Environ. Sci. Technol.* **15:** 1375–1378.

Whitman, W. G. (1923). Preliminary experimental confirmation of the two film theory of gas adsorption. *Chem. Metal. Eng.* **29:** 146.

Winchester, J. W. and G. D. Nifong. (1971). Water pollution in Lake Michigan from pollution aerosol fallout. *Water Air Soil Pollut.* **1:** 50–64.

Woodwell, G. M. (1967). Toxic substances and ecological cycles. *Sci. Am.* **216** (3): 24–31.

Wu, J. (1981). Evidence of sea spray produced by bursting bubbles. *Science* **212:** 324–326.

4

DEPOSITION OF AIRBORNE METALS INTO THE GREAT LAKES: AN EVALUATION OF PAST AND PRESENT ESTIMATES

Jill A. Schmidt
Anders W. Andren

Water Chemistry Program
University of Wisconsin
Madison, Wisconsin 53706

1. INTRODUCTION

About 50% of the water entering the Great Lakes falls directly on their surfaces. This is due to low watershed-to-lake area ratios. The chemical composition of this incoming rainwater is therefore not modified by processes in the watershed. This fact provides an immediate appreciation for the potential impact of atmospheric processes on these lakes. Accordingly, it is not surprising that a significant, sometimes predominant, portion of individual chemical constituents in these aqueous systems are atmospherically derived. In the past decade or so, several studies have served to document the importance of this source for nutrients, organic matter, and trace metals. This chapter summarizes results from those efforts designed to evaluate atmospheric loadings of Cd, Cr, Cu, Ni, Pb, and Zn as well as the methods used to derive these estimates for the Great Lakes. In addition, an attempt is made to evaluate the accuracy of these loading estimates and the net impact on water quality by using a mass balance approach. Lake Michigan is used as the example for these calculations.

2. ATMOSPHERIC DEPOSITION PROCESSES AND THEIR EVALUATION

2.1. Wet Deposition

Scavenging of aerosolborne metals by rain and snow encompasses in-cloud nucleation (rainout) and washout, which occurs as the droplet falls. These processes (and resulting rainfall composition) are dependent on the physical and chemical nature of aerosols and type and duration of precipitation events.

Aerosols greater than 1–2 μm diameter are very efficiently scavenged (Radtke et al., 1980), regardless of their chemistry. Those aerosols of a more hygroscopic nature serve as ready nuclei for rain and ice formation. During any particular event rainfall should be selectively enriched in substances associated with large and/or hygroscropic aerosols. Moreover, precipitation falling the first few moments of an event is often more concentrated in atmospheric aerosol constituents than that falling as the event progresses (Gatz and Dingle, 1971). Note that relatively little is known about scavenging by snowfall. However, only about 20% of the total precipitation entering the Great Lakes arrives as snow (Gatz and Chagnon 1976).

The most direct way to measure wet deposition of chemical species over a given time in a particular area is to collect and analyze all rain and snowfall that accumulates in that area in the specified time period. Such measurements can be very reliable on very small temporal and spatial scales. Uncertainty in the estimates increases with increasing area because of sampling density constraints (Granat, 1976). An environment such as a large lake also offers very limited access, further hampering the characterization of its atmospheric deposition processes.

Wet deposition has also been estimated from air concentrations using previously determined relationships between air and rain chemistry. The washout factor W relates concentrations of rain (C_{rain}) to that in air (C_{air}) through the relation;

$$C_{\text{rain}} = C_{\text{air}} W \rho_{\text{air}} \quad (1)$$

where the air density (1200 g/m^3) is included to normalize units. Gatz (1975) and Cawse (1974) have reported washout factors for several trace metals determined from rain concentrations and concurrent air concentrations. Values of Cawse are generally greater, although these data represent samples that were not shielded from dry deposition and are thus likely to be overestimates for the wet deposition process only. Variation in both C_{air} and W for a given species adds significant uncertainty to this approach (Gatz and Chagnon, 1976, suggest factor-of-two accuracy for this estimation). However, it is a particularly useful technique where extensive rainfall collection is not feasible.

2.1 Dry Deposition

Dry deposition as defined here refers to any removal of atmospheric constituents to a surface that is not associated with rain or snowfall. Dry depositional fluxes of aerosols have been shown to be proportional to their concentration in air at some designated reference height z (Chamberlain, 1966) with all of the contributing mechanisms (gravitational setting, diffusion, impaction, electro- and thermophoresis) packed into a proportionality termed the deposition velocity (V_d):

$$F_{\text{dry}} = V_d C_{\text{air},z} \quad (2)$$

V_d is a function of particle size, shape, density, the nature of the receiving surface, and a number of meteorological factors, particularly windspeed. Its theoretical treatment has been the object of much previous study (for reviews see, for example, Sehmel, 1980; Slinn and Slinn, 1980). Problems in describing particle transport theoretically have prompted many investigators to detour the area in favor of a more direct method, that is, analysis of material accumulated on a substrate exposed to the desired depositional environment. Some of these measurements have been made in the controlled conditions of a wind tunnel, where the flux of monodisperse (uniformly sized) artificial aerosols has been determined for various surfaces including smooth metal, filter paper (Clough, 1973), grass (Chamberlain, 1960, 1966) and water (Sehmel and Sutter, 1974). By monitoring the aerosol concentration in air overlying the surface ($C_{\text{air},z}$), and comparing that to the amount of material deposited on the surface (F_{dry}), empirical values of deposition velocity can be found. Similar studies have produced estimates of V_d for ambient aerosol constituents in the field (McMahon and Denison, 1979, and Sehmel, 1980, provide comprehensive reviews of such data). In these studies, dryfall is collected and analyzed in a manner essentially similar to rainfall. However, since dry deposition is not solely governed by gravitation, it is doubtful whether fluxes determined by such measurements adequately represent fluxes to a natural surface, especially an aqueous one. It is useful therefore to consider theoretical as well as empirical information in assessing dry deposition rates for a given system. Schmidt (1982) has compared elemental dry deposition velocities determined from analysis of dryfall collections on land (data from Cawse, 1974) with those determined via a model adapted from Sehmel and Hodgson (1978) and ambient measurements of metal concentrations over Lake Michigan as a function of particle size. The over-lake values are generally about a factor of 2 smaller than corresponding over-land values. Of course, good agreement for these two sets of V_d, determined by different methods for different environments, is not necessarily expected. However, deposition velocities over water are expected to be relatively low from a consideration of surface roughness alone. Profile and eddy flux techniques are presently being explored as a means of measuring dry deposition rates for particulate and gaseous species in the field (Wesely et al., 1977; Sievering, 1982). However, significant difficulties in performing and interpreting this type of measurement over water persist (Wesely and Williams, 1981).

3. APPLICATION OF ATMOSPHERIC WET AND DRY ESTIMATION TECHNIQUES FOR THE GREAT LAKES

3.1. A Historical Summary

Experimental investigations of atmospheric trace metal loadings to the Great Lakes were preceded and probably inspired by some initial estimates produced from consideration of air pollutant dispersion and deposition data collected over land. Perhaps the first to speculate on the transport of metals to the Great Lakes, Winchester and Nifong (1971) estimated the input to Lake Michigan originating from Milwaukee, Chicago, and Gary, Indiana, using an emission inventory of major sources of metal-bearing particulates. Based on an estimate of dispersion and removal rates, input of these emissions to the lake was calculated using an effective net transfer efficiency from source to lake of 10%. This figure was applied for all elements, regardless of their particle size affiliation; moreover, scavenging by precipitation was not rigorously accounted for. Skibin (1973) later reported 25% as a more realistic value for transfer efficiency which includes both wet and dry processes. A study by Sievering (1976), using another model of pollutant dispersion and deposition, yielded values for transfer efficiencies versus particle size and season of the year. Based on the average mean size of particulate metals, the transfer efficiencies were calculated with distinctions made between warm and cold seasons and wet versus dry deposition. According to these figures, transfer occurs with no greater than 12% efficiency for most trace metals and appears dominated by wet precipitation removal.

Gatz (1975) used estimates of air metal concentrations for Chicago and northwest Indiana and treated this region as an area source of metals to the air over Lake Michigan. The dispersal of these emissions over the lake was then modeled, based on average values for windspeed and mixing height reported for the area. In assessing wet and dry removal, Gatz did account for elemental particle size differences. Note that the studies of Winchester and Nifong (1971), Sievering (1976), and Gatz (1975) have quantified input from the southern Lake Michigan basin only.

The Canada Center for Inland Waters (Acres Consulting Services, 1975, 1977) reported some estimates of atmospheric loading to all of the Great Lakes. These were based on a method not unlike that used by Gatz (1975) but expanded to treat the entire shoreline area source. Atmospheric flux to the lakes was determined from air quality data for shoreline locations using a dispersion-deposition model. Dry fallout values were calculated using an overall deposition velocity for all particles without allowance for variation in particle size affiliation for metals. The chosen V_d, 0.1 cm/s, may overestimate this parameter for elements whose mass median diameter (MMD) is less than 0.5 μm (e.g., Pb and SO_4). On the other hand, this value may be too low for elements having MMDs greater than 1 μm (e.g., Al and Ca).

Schmidt (1982) applied wet and dry removal parameters directly to size-fractionated aerosol samples (collected mainly over the southern basin of Lake Michigan) which were analyzed for several metals. Klein (1975) used sediment analyses to identify atmospheric deposition rates using sulfur as a tracer for the atmospheric input. He assumed the lake to be in steady state with respect to the atmospheric load of most elements. The atmospheric load of elements other than sulfur was estimated from their abundance relative to sulfur in aerosols collected in rural northwest Indiana.

Wet and dry deposition collection vessels have probably been most widely used in assessing trace metal atmospheric deposition rates. Several large- and small-scale deposition networks have been designed and implemented for the purpose of evaluating deposition rates on each of the Great Lakes. Perhaps the earliest of these was set up by Shiomi and Kuntz (1973) who used funnel-bottle type collectors around Lake Ontario. Another set of estimates were produced for this lake by Kramer (1975, 1976) from data on rain, high-volume air, as well as bulk precipitation samplers. These data provided estimates of both wet and dry components of atmospheric trace metal loading. Soon after that, Kuntz (1978) and the State of New York (1980) also reported results of bulk precipitation collections made around this lake. Eisenreich et al. (1977) and Eisenreich and Langevin (1978) collected bulk precipitation from sites on and around Lakes Michigan and Superior. The Canadia Centre for Inland Waters sponsored a deposition network study carried out by Acres Consulting Services (1975, 1977) which produced loading values for Lakes Erie, Ontario, Superior, and Huron. More recently, U.S. EPA Region V carried out a bulk precipitation collection study for all lakes except Erie. The results have been reported by Allen and Halley (1980). Most stations used in these studies were land-based. However, several bouy stations have been used, although they were generally less than 10 km from shore and their number is very small compared to the total number of sites.

3.2. Comparison of Trace Metal Loading Results

For ease of comparison, the results of the studies just described are reported in terms of kilograms per hectare per year. If the original publication reported total element mass loading to a lake, or part of a lake, this value was divided by the appropriate area to yield an average aerial loading rate. Many of these loading values were taken from a summary by Allen and Halley (1980). Where a range of loadings was given, the median of the range is presented, as this entity is more meaningful when comparisons between the lakes are made. Figs. 4.1 through 4.7 show results for individual elements for each lake.

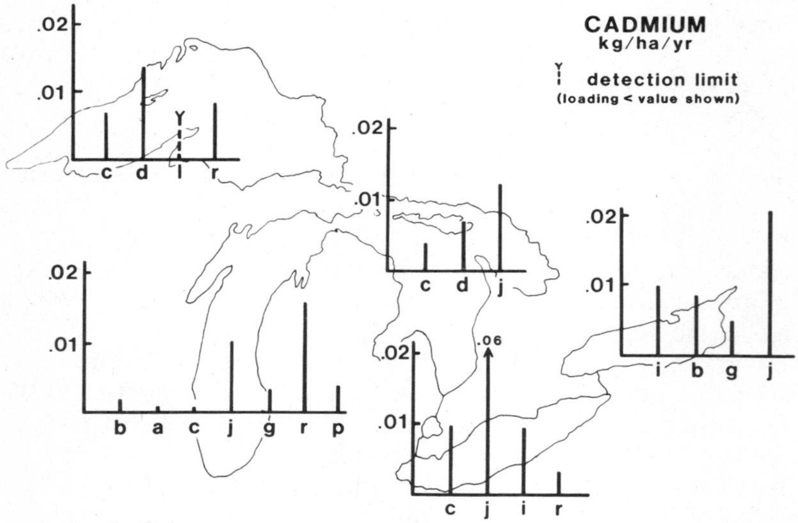

Figure 4.1 Cadmium loading to the Great Lakes. Key to references: a = Winchester and Nifong, 1971; b = Shiomi and Kuntz, 1973; c = Acres Consulting Services, 1975 (model data); d = Acres Consulting Services, 1977 (based on precipitation data); e = Gatz, 1975; f = Klein, 1975; g = Kramer, 1975; h = Kramer, 1976; i = IJC, 1977a; j = IJC, 1977b; k = Eisenreich et al., 1977; l = Eisenreich and Langevin, 1978; m = IJC, 1978; n = Kuntz, 1978; o = Sievering et al., 1979; p = Tisue et al., 1980; q = New York State, 1980; r = U.S. EPA, 1980; s = Schmidt, 1982.

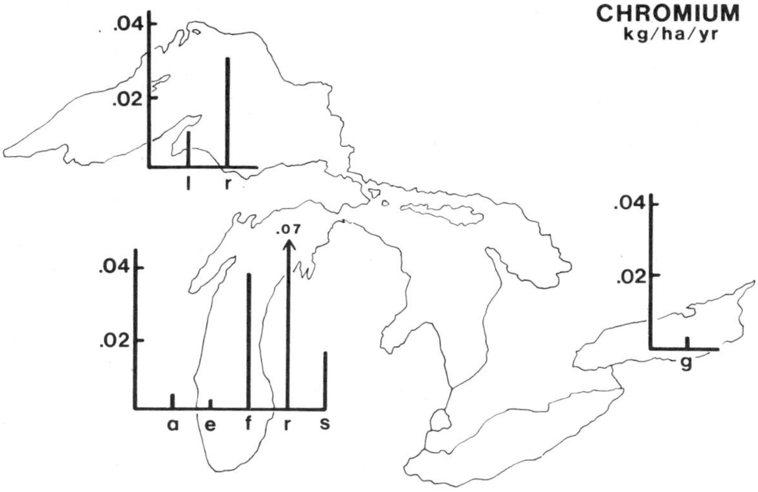

Figure 4.2 Chromium loading to the Great Lakes. For key to references see Fig. 4.1.

Cadmium loadings (Fig. 4.1) appear relatively low for Lake Michigan, with most values <0.005 kg/ha·yr. This is a somewhat unexpected result considering that three of the studies were made for the southern basin, which is bounded by the heavily industrialized Chicago-Gary, Indiana, region. Loading values for the other lakes, with a few exceptions, fall in the range of 0.005–0.01 kg/ha·yr. The relative intensity of loading for Cd, from lowest to highest, seems to be Superior, Huron, Michigan, Ontario, and Erie.

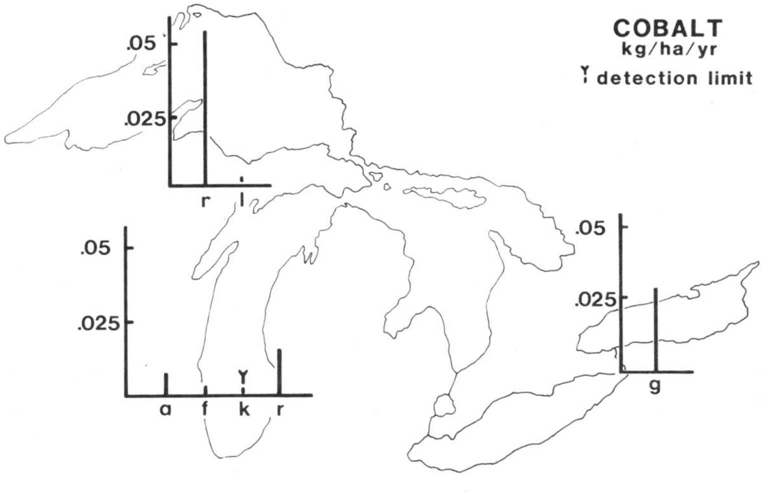

Figure 4.3 Cobalt loading to the Great Lakes. For key to references see Fig. 4.1.

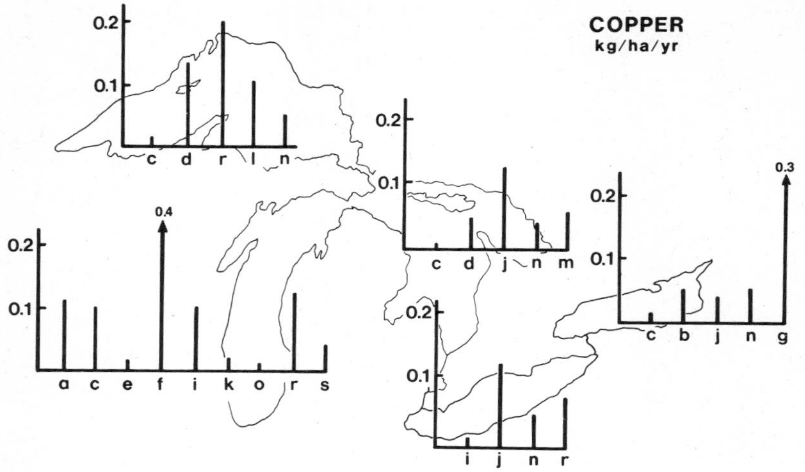

Figure 4.4 Copper loading to the Great Lakes. For key to references see Fig. 4.1.

Relatively few estimates for atmospheric Co and Cr loading in the Great Lakes have been reported. For both elements, estimates range from about 0.002 to 0.7 kg/ha·yr with particularly wide disparity in the Lake Michigan estimates (Figs. 4.2 and 4.3). Estimates for copper are fairly abundant, and cover a general range that is similar for each of the lakes: on the order of 0.05–0.15 kg/ha·yr (Fig. 4.4). Nickel loadings are about 0.01–0.05 kg/ha·yr with the exception of a few values on Lakes Michigan and Ontario that

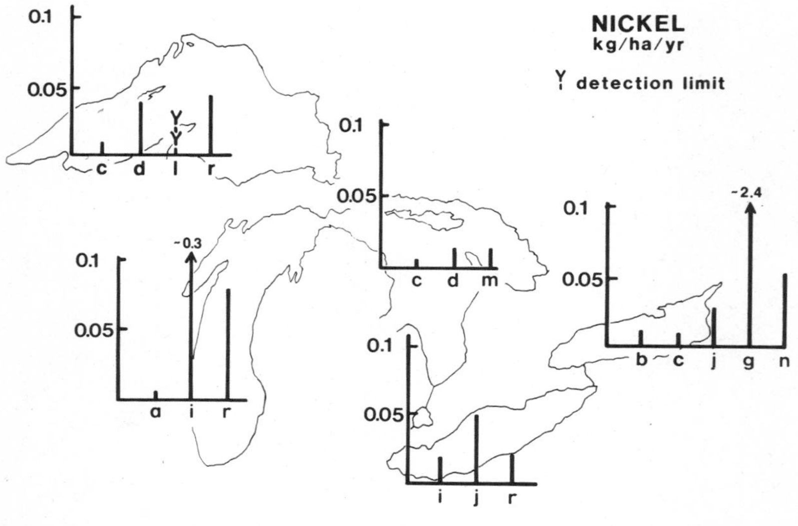

Figure 4.5 Nickel loading to the Great Lakes. For key to references see Fig. 4.1.

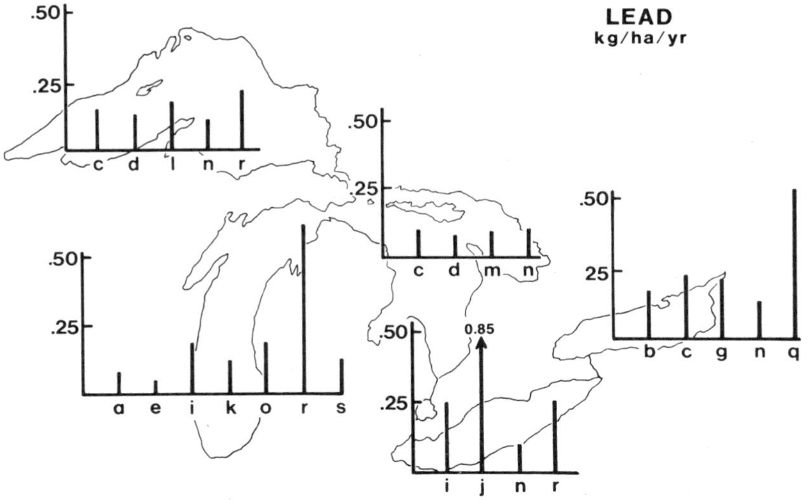

Figure 4.6 Lead loading to the Great Lakes. For key to references see Fig. 4.1.

significantly exceed that range (Fig. 4.5). No relative lake loading assignments between lakes can be made for these elements.

Atmospheric loadings for lead are well represented on each of the lakes, and appear to show the best coincidence (Fig. 4.6). The attention on this contaminant is perhaps a reflection of the fact that vast quantities of lead have been, and still are, emitted to the environment via mining and smelting

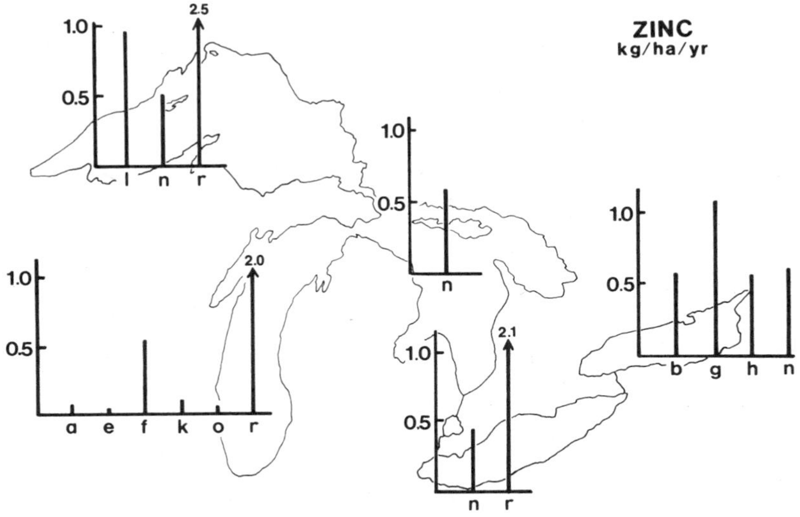

Figure 4.7 Zinc loading to the Great Lakes. For key to references see Fig. 4.1.

operations, industrial processes, and automotive emissions (Nriagu, 1978). Due to these latter emissions, aerosols in urban areas often contain high concentrations (0.5–2%) of this element. Estimates for all lakes are generally in the range of 0.1–0.25 kg/ha·yr, with one value for each of Lakes Michigan, Erie, and Ontario exceeding that range. The order of relative loading intensity for Pb seems to be Huron, Superior, Michigan, Ontario and Erie. The comparative order of this loading does not seem logical if one considers the immediate source strength for each lake. Problems with sampling and analysis (*vide infra*) indicate that it may be premature to accurately rank the lakes in this manner.

Zinc loadings (Fig. 4.7) follow a pattern most similar to Cd, although estimates for Lake Michigan are generally low relative to those for other lakes (about 0.1–0.5 kg/ha·yr with one at 2 kg/ha·yr). Loadings for the remaining lakes range from about 0.5 to 1.0 kg/ha·yr, with two values reported at greater than 2 kg/ha·yr. Sample contamination problems are particularly severe for this element and most likely accounts for much of the observed variation.

Uncertanties involved in making individual loading estimates such as these are substantial, although very difficult to quantify. Some authors who have attempted to evaluate this problem generally concluded that quantifiable uncertainties were on the order of a factor of 2 (Gatz and Chagnon, 1976; Schmidt, 1982; Sievering et al., 1979). However, ranges in estimates summarized here show greater than order-of-magnitude variations in many cases. Further, standard deviations calculated about the mean of all estimates for a particular lake were generally greater than, or about equal to, the mean value. Some loadings appear conspicuously high or low for most of the elements for which estimates are provided, which suggests some type of systematic difference in the method used to produce loading values. One example is Klein's (1975) use of sulfur as a tracer for aerosol input to Lake Michigan: because much sulfur may also derive from SO_2 dry deposition and nonatmospheric sources (Slinn, 1977), the aerosol load for other elements may have been overestimated. Also, certain precipitation networks may have been biased toward urban sampling locations which afford the greatest convenience of operation. Even where there is agreement among estimates derived by similar methods, or with similar assumptions, the question of their accuracy persists. This results from the many uncertainties of the assumptions, such as those surrounding the collection efficiency of dry deposition collection and the utility of wet and dry deposition models.

A critical examination of available lake loading estimates indicate that, even for the same depositional region, wide disparity can exist. Some of this inconsistency is undoubtedly due to real variations in input. However, much is also due to uneven quality and quantity of data. This is of course compounded by the fact that there is no absolute method for measuring the dry component. It is therefore presently not possible to present a rigorous statistical treatment of data in terms of relative lake loading differences between all the Great Lakes. Even if the means existed to produce perfectly accurate loading rates, disagreement among individual estimates would still occur due to annual and long-term variations in such parameters as trace metal discharge rates, dispersion trajectories, and rainfall or snow amounts. For these reasons (and the fact that accuracy in atmospheric load estimates is likely to remain poor for some time), evaluation of long-term average atmospheric loading rates, which is important in assessing water quality effects, will remain difficult because of use of these techniques alone. In aqueous systems such as the Great Lakes, however, evaluation of other input routes may be possible. This information, together with data on sedimentation removal rates, should be used as a crosscheck on the accuracy of present atmospheric loading estimates. Such a mass balance approach can also offer an initial insight into the general aqueous behavior of an element.

4. A MASS BALANCE APPROACH TO EVALUATE ATMOSPHERIC TRACE METAL LOADINGS: LAKE MICHIGAN

A fairly extensive data base on trace metal sources, sediment composition, and hydrology now exists for Lake Michigan. We have therefore chosen this lake to illustrate the mass balance approach. Sources of metals to the lake include direct industrial and municipal discharge, river drainage, shoreline erosion, atmospheric deposition, and release from sediment (post-depositional migration). Removal is attributed to flushing into Lake Huron and to sedimentation. Several previous studies have focused on metals associated with specific modes of transport (for a review see, for example, Sonzogni et al., 1980). However, many questions persist regarding the relative significance of these processes and their net impact on Lake Michigan water quality. Here the mass balance for trace metals in Lake Michigan is constructed and used to: (1) identify the major sources and sinks for these elements; (2) provide a cross check on atmospheric estimates shown earlier; (3) calculate residence times for elements in the water column; and (4) evaluate long-term water quality changes associated with anthropogenic burdens of these elements in the lake. As mentioned earlier, quality and quantity of data, source and sink terms, industrial and municipal discharges, hydrologic regimes, sedimentation/resuspension problems, and geomorphology differ between the various lakes. It may thus be inappropriate to extrapolate the conclusions for Lake Michigan to the other Great Lakes.

4.1. Sources

All estimates of atmospheric trace metal loadings for Lake Michigan were considered. However, owing to problems involved in applying some of these in the mass balance (Schmidt, 1982), the range in values recently reported by Eisenreich et al. (1977) and Schmidt (1982) were used as they were derived for the entire basin (Table 4.1). These seem least affected by major unknown uncertainties.

Stream loadings have been calculated individually for dissolved and particulate trace metal fractions. Dissolved loads were evaluated from data of Robbins et al. (1970) as well as from more recent measurements of Tisue and Fingleton (see Chapter 5) as interpreted by Schmidt (1982). Particulate riverine trace metal concentrations were also measured by Tisue et al. (1980) and used for one estimate of the particulate load. An alternate load was calculated from an independent estimate of the total mass loading to the lake from tributaries as reported by Mildner (1974). The trace element composition of this mass was approximated by data reported for river mouth sediments of Lake Michigan tributaries (Fitchko and Hutchinson, 1975). The range in stream loadings reflect variation in individual estimates of both dissolved and particulate loadings described in Table 4.1.

Monteith and Sonzogni (1976) estimated the particulate metal loading to Lake Michigan by shoreline erosion from analysis of the chemical composition of samples taken from locations along the coast and via a shoreline–lake transport model. Values for elements which were not included in the study (Zn, Cu, Cr) were approximated by assuming eroded material to exhibit average soil composition (Shaklette et al., 1971; Stolzenburg and Andren, 1982). While soil values may not exactly represent eroded matter, a greater uncertainty probably exists in assuming that shoreline constituents are not differentially transported by erosional activity. Results are presented in Table 4.1.

Groundwater intrusion, release from sediment pore waters, and municipal and industrial point sources constitute additional inputs of metals to Lake Michigan. Johnson et al. (1980) have estimated the point source loading of trace metals and found the following annual inputs: Pb, 10 tonnes/yr; Zn, 61 tonnes/yr; and Cu, 15 tonnes/yr. There is little data on which to base an estimate of the role of the first two processes in the lake metal cycles. However, it is likely that their importance as metal sources is small compared to those discussed earlier.

Table 4.1 Loading of Trace Metals to Lake Michigan (tonnes/yr)[a]

Element	Air	Stream	Shoreline (10%)[b]
Cr	52–180	78–250	19–110
Cu	110–950	73–180	11–66
Pb	340–1200	56–130	12–40
Zn	740	250–350	50–440

[a] See references in text for sources.

[b] Figures represent 10% of total estimate.

Table 4.1 shows a comparison for estimated trace metal inputs of major sources to Lake Michigan. No accurate stream and shoreline erosion values could be found for Cd. The large ranges reflect uncertainties in both quantity and chemical composition of mass loadings from each source. In terms of gross loading, shoreline erosion appears to be the largest source of every element except Pb, which originates mainly in atmospheric fallout. Municipal point sources seem relatively minor in terms of their effect on the lake as a whole. Atmospheric Cu, Pb, and Zn loads appear to be greater than the stream loads.

An important distinction between the atmospheric and other sources is that the shoreline and stream inputs are introduced to nearshore waters, whereas the aerosol load is highly dispersed, affecting nearshore and midlake areas nearly alike. Chesters and Delfino (1978) estimated that only about 10% of eroded shoreline soil is likely to be transported past the immediate shoreline region. Further, Fitchko and Hutchinson (1975) concluded from their analysis of sediments in the river mouth–lake boundary that only a small percentage of suspended river sediments are apt to be transported away from shore. These considerations give added importance to the atmospheric load.

Although tributary and eroded particles may not leave the nearshore environment, any substances dissolved from them will have greater potential impact on the entire lake. A comparison of available data on the potential ability of particulate sources to release metals (Armstrong et al., 1979; Monteith and Sonzogni, 1976; Schmidt, 1982) reveals no major differences between availability of metals in aerosols, river particulates, or shoreline material, although the data presented for the latter two were derived from relatively vigorous extraction media which would be expected to yield a higher value of solubilization than was found by Schmidt (1982) for aerosol metals. It is important to note that about half of all Cu and Zn and a significant portion of Cr and Pb stream loads enter the lake in dissolved form.

4.2. Removal Mechanisms

Winchester (1969) calculated that 49 km^3 of water are carried from Lake Michigan to Lake Huron each year. Taking average bulk water metal concentrations from recent measurements as compiled by Torrey (1976) as well as those reported by Erickson et al. (1981), the export of metals via this route has been calculated. As previously noted, reliable metal concentrations for Lake Michigan are sparse, and are restricted mainly to the southern basin which receives more pollutant metal loading. Although an effort was made to extract reliable data from the recent literature (preferring lower values), the transport of trace metals as listed in Table 4.2 may, in some cases, be somewhat overestimated.

The Lake Michigan sedimentary environment is highly variable, ranging

Table 4.2. Removal of Metals to Lake Huron and to Sedimentation (tonnes/yr)[a]

Element	Lake Huron	Sedimentation
Cr	25	330–500
Cu	25–49	160–270
Pb	25–73	460–750
Zn	25–49	860–1700

[a] See text for sources.

from the western scoured region, where glacial till remains largely exposed, to several areas of relatively rapid sediment accumulation. Since sediment composition also varies, it may be inappropriate to assume an average value of sedimentation rate and composition. For example, data on lead reported by Robbins and Edgington (1975), shows that sediment sampled from areas of high sedimentation rates tend to be enriched in trace metals. In estimating the total sedimentary removal rates for metals to Lake Michigan, it is thus important that this heterogeneity be reconciled.

The chemical composition of depositing sediments in Lake Michigan was estimated from the analyses of several cores (Shimp et al., 1971; Robbins and Edgington, 1975) and from the extensive data for metals in the upper 3 cm reported by Cahill (1981). The chemical composition of presently depositing sediments is difficult to ascertain since the top centimeter of a core, usually the smallest depth interval sampled, represents anywhere from 3 to 1000 years of deposition (given the range of sedimentation rates as reported by Robbins and Edgington, 1975). In addition, bioturbation and other post-depositional migration processes make interpretation difficult. Sedimentation trap data should prove useful for future estimates although the resuspension portion must be accounted for.

For this analysis several sediment profiles of Cu, Cr, Pb, and Zn were extrapolated to zero depth to yield an estimate of the most recent sediment composition (Schmidt, 1982). The variation in the zero-depth estimate ranged from 14 to 27% about the average for cores taken from several locations around the southern basin. Since trace metal concentrations exhibit such profound changes with depth in the sediments, it is very likely that much of this variation is more a function of spatially variable sedimentation rates (as well as changing chemical composition over time), than a spatial variation in the depositing sediment composition. Although core data for the northern basin are sparse, the samples of Cahill (1981) reveal that sediment in the middle and northern depositional basins are enriched in Cu, Cr, Pb, and Zn to approximately the same extent as the southern depositional zones. Therefore, the depository sediment composition as derived above, is used for the entire lake.

Sedimentation rates have been determined for several cores using radioisotopes Pb-210 and Cs-137 (Robbins and Edgington, 1975; Christiansen and Chien, 1981). Extensive spatial heterogeneity in sediment accumulation rates is evident from seismic data, however, particularly the Waukegan member thickness, which has been resolved for the southern basin by Wickham et al. (1978) and shown by Edgington and Robbins (1975) to correlate very well with radioisotope-determined sedimentation rates. These authors reported a mean rate of 7 $mg/cm^2 \cdot yr$ for the southern basin. In lieu of sufficient data concerning deposition rates in areas north of the far southern basin, this parameter was estimated by making use of the relationship between sediment lead content (averaged over the top 3 cm) and sedimentation rate. Schmidt (1982) found a good correlation between these two parameters and used the numerous data on lead in sediments given by Cahill (1981) for the entire basin to estimate average sedimentation rates. This method yielded an average for the southern basin of 8.2 $mg/cm^2 \cdot yr$ which compares within 15% of the value reported by Edgington and Robbins (1975). The middle third of the lake actually yielded the highest value, 10 $mg/cm^2 \cdot yr$, and the northern basin averaged 5 $mg/cm^2 \cdot yr$, to yield an overall average for the entire lake of 7.1 $mg/cm^2 \cdot yr$.

Combination of the lakewide average with mean sediment composition yields elemental removal rates which, multiplied by the lake area, yields total annual removal to the sediments (Table 4.2). The ranges presented include uncertainty contributions from sediment composition as well as that for sedimentation rates, as indicated by comparison of the two independent estimates of average sedimentation rates for the southern basin (Schmidt, 1982; Edgington and Robbins, 1975). Results presented in Table 4.2 indicate that it is presently possible to estimate sedimentation removal rates for Cr, Cu, Pb, and Zn to within a factor of two.

4.3. Mass Balance

Table 4.3 compares the total annual input and removal rates for several trace metals in Lake Michigan. The range used in the total source term is summed from values presented in Table 4.1, with the 10% shoreline erosion contribution. The total removal term is the sum of only two components and reflects the entire range of both sedimentation and outflow to Lake Huron. Note that trace metal removal by sedimentation is the sink term of importance in Lake Michigan.

Although the ranges *within* source and sink estimates vary by a factor of about 2, there is fair agreement *between* the two terms for the trace elements considered. Copper exhibits perhaps the most serious imbalance. This may reflect problems in air sampling. Contamination from brushes in the vacuum pump has been shown to occur if appropriate action is not taken. Estimates

Table 4.3. Mass Balance on Trace Metals in Lake Michigan[a]

Element	Input (tonnes/yr)	Output (tonnes/yr)
Cr	160–540	350–525
Cu	160–700	190–320
Pb	420–1300	485–820
Zn	1000–1500	880–1700

[a] See text for details and sources of data.

made using earlier data do indeed result in higher loadings for this element and must thus be considered suspect. Estimates using more recent data indicate the higher range for input should be closer to 350, that is, within the estimated range for the output term.

The concept of elemental residence time, Y_i, has been used extensively in the oceanographic literature. If it is reasonable to assume that mixing times are shorter than Y_i and that steady state conditions exist, an evaluation of Y_i for any metal, i, would indicate the response time likely to be involved when changes in loading occur. Residence time may be expressed as

$$Y_i = \frac{A_i}{dA_i/dt} \tag{3}$$

where A_i = amount of element i in the water column and dA_i/dt = input or output rate of element i.

Concentrations in water (dissolved + particulate) were taken from the compilation of Torrey (1976), and from Erickson et al. (1981) (the quality of data taken from the former source may be questionable, hence lower values were usually preferred). The mass of these elements in the water column was then obtained by multiplying the range in concentration by the volume of the lake (4900 km^3). A range of fluxes was considered for the denominator which covers the variation in both source and sink estimates.

Table 4.4 displays estimated residence times for metals in Lake Michigan. Because it is likely that the lower values of A_i most closely reflect lakewide

Table 4.4. Residence Times

Element	Estimated Content of Lake (tonnes)[a]	Y (yr)
Cr	2500	4.6–18
Cu	2500–4900	3.6–31
Pb	2500–7400	1.9–18
Zn	2500–4900	1.5–5.6

[a] See text for sources of data.

averages, it appears that Y_i for these elements is less than 5–10 years. These figures also incorporate the fact that significant resuspensions of bottom materials occur during late fall turnover (Erickson et al., 1981; Wahlgren et al., 1980). The fact that the residence times of these trace metals may be less than 10 years indicates a strong likelihood that the lake has had sufficient time to respond to the enhanced loadings of these elements which accompanied the urban and industrial expansion of the past 50–100 years. This is in accord with the existence of a reasonable agreement between source and sink estimations (i.e., atmospheric estimates may be good to within a factor of about 2), which would imply that Lake Michigan may be, or very nearly is, in a steady state with respect to these trace elements.

5. ASSESSMENT OF THE ANTHROPOGENIC BURDEN OF METALS IN LAKE MICHIGAN

Trace metal concentrations in Lake Michigan may, according to the foregoing analysis, already have reached a quasi-steady state with respect to loading associated with the surge of urban and industrial expansion 50–100 years ago. The magnitude of this burden, which is of interest because it is responsible for present-day trace metal content of the water column, can be assessed from the mass balance data. Sources and sinks can be evaluated independently to provide separate comparable estimates for major as well as trace metals.

Prior to the onset of significant anthropogenic loadings, it is likely that soil and soil-leached metals comprised the bulk of metallic chemicals transported to the lake via tributaries, the atmosphere, and shoreline erosion. With aluminum as a tracer for soil, the soil-associated component of air and river loading has been estimated by Schmidt (1982) using the soil composition shown in Table 4.5 (Shaklette et al., 1971; Stolzenburg and Andren, 1982).

Table 4.5 Estimate of the Soil-derived Fraction in Tributary and Atmospheric Particulate Matter[a]

Element	Composition of Soil Source[b] (%)	Tributary (%)	Atmosphere (%)
Cr	0.005	27–66	8–9
Cu	0.003	42	2–2.2
Pb	0.0095	10–18	0.2–0.3
Zn	0.0103	40–42	1.1

[a] Assuming that 100% of the Al in tributary and atmospheric particulate matter is soil derived.

[b] From Stolzenburg and Andren (1982) and Shaklette et al. (1971).

Table 4.6. Mass Loading of Nonsoil Components to Lake Michigan (tonnes/yr)

Element	Tributary		Air	
Cr	(34–78)[a]	27–182	(92)	48–167
Cu	(58)	42–104	(98)	120–450
Pb	(82.5–90)	51–75	(100)	340–1200
Zn	(58–60)	145–206	(99)	730

[a] Numbers in parentheses indicate % of total mass loading due to nonsoil sources for each source.

For rivers, the soil-drived portion may be somewhat greater than the values indicated, since most elements considered are more soluble than aluminum. Interestingly, at least half of the particulate portion fo riverine Cd, Ca, and Zn is satisfactorily accounted for by soil. Lead in river particulate matter appears to have a considerable nonsoil source.

The portion of atmospheric loading that can be attributed to soil is small relative to that for tributaries. The shoreline erosion source is assumed to be entirely soil-associated, although its composition differs from soil for some elements (Montieth and Sonzogni, 1976). Table 4.6 compares the *non*-soil portion for the mass loading for atmospheric and stream sources. Because much more of the stream load is attributable to soil, the atmospheric nonsoil load takes on added significance in terms of trace metal loading of suspected anthropogenic origin.

The fact that a singificant portion of an element can be related to soil does not directly translate to its designation as a natural (as opposed to anthropogenic) source. Much of the soil presently being transported to Lake Michigan may have been mobilized through anthropogenic activities such as agriculture and construction. Thus, the nonsoil component of atmospheric and riverine loadings represents a conservative estimate of the load associated with the activities of man.

Table 4.7. A Comparison of Anthropogenic Metal Fluxes to Lake Michigan as Evaluated from the Sedimentary Record and from External Loading Calculations

Element	Loading Calculations (%)	Sediment Record (%)
Cr	55–56	43–57
Cu	62–88	35–83
Pb	92–99	83–91
Zn	62–88	62–87

Perhaps a more accurate record of man's influence on the chemistry of Lake Michigan is contained in the sediments. A comparison of the composition of sediments deposited before about 1880 A.D. (Robbins and Edgington, 1975; Shimp et al., 1971) with that estimated for currently depositing sediment, gives a reasonable estimate of the influence of recent industrial expansion on the trace metal content of the lake. The trace metal content in sediments (and presumably in the water column) have been increased by a factor of from 2 to 8, with Pb showing the most enhancement followed by Zn, Cu, and Cr. Further, it is believed that total mass sedimentation rates have not significantly changed during this time (Robbins and Edgington, 1975). Therefore, changes in sediment composition alone (rather than changes in elemental sedimentation rates) should be a fair indication of changing loading rates. Whether the chemical form of the elements have changed in the water column is presently unknown.

The portion of trace metal fluxes suspected to be associated with anthropogenic activity as evaluated from both the sedimentary record and from external load calculations are shown in Table 4.7. In both cases, the order of the decreasing anthropogenic portion is Pb > Zn > Cu > Cr. Moreover, the ranges of "% anthropogenic" compare quite well, lending credence to the two independently derived estimates. These calculations indicate that roughly half of the Cr and Cu, most of the Zn, and all of the Pb that presently deposits in the bottom sediments of Lake Michigan may be attributable to human activities. Whether the increased metal burden in the water column has any effect on the biota is unknown since no information on the historical change in chemical species exist. However, recent *in situ* bioassay work by Marshall (1980) indicates that very small increases of Cd produce measurable changes in Lake Michigan phytoplankton productivity.

6. CONCLUSIONS

The ratio of watershed to lake surface area is quite low in the Great Lakes. Time constants for elemental removal processes in the water column are also much shorter than hydraulic residence times in these systems. These morphological and hydraulic considerations, together with an intense urbanization of the Great Lakes airshed, cause a significant portion of the material that annually reaches the lakes to come from the atmosphere.

Several attempts to quantify atmospheric loadings to these lakes have appeared in the literature. A critical examination of available estimates indicate that, even for the same depositional region, wide disparity can exist. While some of this inconsistency is due to real spatial and temporal variations, we conclude that much is also due to uneven quality and quantity of data. A rigorous statistical treatment of lake loading rates, based on atmospheric metal data, is therefore not presently possible. However, it is likely that the following ranges should bracket actual metal inputs (in kg/

ha·yr): Cd, 0.005–0.01; Co and Cr, 0.002–0.7; Cu, 0.05–0.15; Ni, 0.01–0.05; Pb, 0.1–0.25; and Zn, 0.5–1.0. The relative order of unit area loading intensity seems to be Erie > Ontario > Michigan > Huron > Superior.

Since difficulties in sampling logistics will persist for some time, we suggest that a variety of approaches must be pursued. The chemical composition of precipitation and aerosols should be measured with as fine a time resolution as possible, at standard heights, with standardized equipment, and with supporting meteorological data (windspeed and atmospheric stability). Mass balance calculations, using elemental data on river and shoreline erosion inputs in combination with sedimentation patterns, should also be vigorously pursued.

Atmospheric loading rates, as calculated via wet and dry deposition models, were compared to those obtained via the mass balance approach. The results indicate that the estimates for Pb, Cu, Zn, and Cr are within a factor of 2 for Lake Michigan (perhaps even better for Pb). We do not know whether the comparisons will be as good for the other lakes.

Calculations based on data from both the sedimentary record and external loadings, indicate that roughly half of the Cr and Cu, most of the Zn (>60%), and almost all of the Pb (>90%) that presently deposits in the bottom sediments of Lake Michigan may be attributable to human activities.

ACKNOWLEDGMENTS

We like to thank Cindy Walder and Roberta Ward for typing assistance and Sue Halverson for drafting. Funding for this research came from the University of Wisconsin NOAA–Sea Grant Program.

REFERENCES

Acres Consulting Services, Ltd. (1975). Atmospheric loading of the upper Great Lakes. Report to Canada Centre for Inland Waters. Vol. 2.

Acres Consulting Services, Ltd. (1977). Atmospheric loading of the lower Great Lakes and the Great Lakes drainage basin. Report to Canada Centre for Inland Waters.

Allen, H. E. and M. A. Halley. (1980). Assessment of airborne inorganic contaminants in the Great Lakes. In: A perspective on the problems of hazardous substances in the Great Lakes basin ecosystem, 1980 annual report, Report to the International Joint Commission, Great Lakes Science Advisory Board. 160 pp.

Armstrong, D. E., M. A. Anderson, J. P. Perry, and D. Flatness. (1979). Availability of pollutants associated with suspended or settled river sediments which gain access to the Great Lakes. Progress Report, EPA No. 68-01-4479. Water Chemistry Laboratory, University of Wisconsin, Madison, Wisconsin. 13 pp.

Cahill, R. A. (1981). *Geochemistry of recent Lake Michigan sediments*. Illinois Institute of Natural Resources, State Geol. Surv. Div., Circular 517, Champaign, Illinois, 35 pp.

Cawse, P. A. (1974). A survey of atmospheric trace elements in the United Kingdom. A.E.R.E. Harwell Report No. R7669. HMSO. London.

Chamberlain, A. C. (1960). Aspects of Deposition of Radioactive Gases and Particles. In: E. D. Richardson, Ed., *Aerodynamic capture of particles*. Pergamon Press, pp. 63–88.

Chamberlain, A. C. (1966). Transport of Lycopodium spores and other small particles to rough surfaces. *Proc. Royal Soc.* **296:** 45–70.

Chesters, G. and J. J. Delfino. (1978). Frequency and extent of wind-induced resuspension of bottom material in the U.S. Great Lakes nearshore waters. U.S. Task D Technical Report to the International Reference Group on Great Lakes Pollution from Land Use Activities. International Joint Commission, Windsor, Ontario, 111 pp.

Christiansen, E. R. and N. K. Chien. (1981). Fluxes of As, Pb, Zn, and Cd to Green Bay and Lake Michigan sediments. *Environ. Sci. Technol.* **15:** 553–558.

Clough, W. S. (1973). Transport of particles to surfaces. *Aerosol Sci.* **4:** 227–234.

Eisenreich, S. J. and S. A. Langevin. (1978). Atmospheric interaction with freshwater surface organic microlayers: Atmospheric deposition of trace metals to Lake Superior and surface accumulations. Final Report, Graduate School of the University of Minnesota. Minneapolis, Minnesota. 180 pp.

Eisenreich, S. J., P. J. Emmling, and A. M. Beeton. (1977). Atmospheric loading of phosphorous and other chemicals to Lake Michigan. *J. Great Lakes Res.* **3:** 291–304.

Erickson, R. E., D. E. Armstrong, and A. W. Andren. (1981). Suspended particulate trace metal distributions in Lake Michigan. Progress Report to Wisconsin Sea Grant Program. 18 pp.

Fitchko, J. and T. C. Hutchinson. (1975). A comparative study of heavy metal concentrations in river mouth sediments around the Great Lakes. *J. Great Lakes Res.* **1:** 46–78.

Gatz, D. F. (1975). Pollutant aerosol deposition into southern Lake Michigan. *Water Air Soil Pollut.* **5:** 239–251.

Gatz, D. F. (1977). Scavenging ratio measurements in METROMEX. Proceedings of a symposium on precipitation scavenging. Technical Information Center, Energy Research and Development Administration. pp. 71–87.

Gatz, D. F. and S. A. Chagnon. (1976). Atmospheric environment of the Lake Michigan drainage basin. Argonne National Laboratory, Rep. ANL/ES-40.

Gatz, D. F. and A. N. Dingle. (1971). Trace substances in rainwater: Concentration variations during convective rains, and their interpretation. *Tellus* **23:** 14–27.

Granat, L. (1976). Principles in network design for precipitation chemistry measurements. *J. Great Lakes Res.* **2:** (Suppl. I):42–55.

International Joint Commission. (1977a). Atmospheric loading of the lower Great Lakes and the Great Lakes drainage basin. Water Quality Board Report to International Joint Commission. Windsor, Ontario, 61 pp.

International Joint Commission. (1977b). Atmospheric loading to the Great Lakes. Technical note to International Reference Group on Pollution of the Great Lakes from Land Use Activities. Windsor, Ontario. 16 pp.

International Joint Commission. (1978). United States Great Lakes tributary loading. Water Quality Board Report to International Joint Commission. Windsor, Ontario, 18pp.

Johnson, M. G., J. C. Comeau, T. M. Heidtke, W. C. Sonzogni, and B. W. Stahlbaum. (1980). Modelling effects of remedial programs to aid Great Lakes environmental management. *J. Great Lakes Res.* **6:** 8–21.

Klein, D. H. (1975). Fluxes, residence times, and sources of some elements to Lake Michigan. *Water Air Soil Pollut.* **4:** 3–8.

Kramer, J. R. (1975). *Fate of atmospheric sulfur dioxide and related substances as indicated by chemistry of precipitation*. Department of Geology, McMaster University, Hamilton, Ontario. 25 pp.

Kramer, J. R. (1976). *Fate of atmospheric sulfur dioxide and related substances as indicated by chemistry of precipitation*. Report from Department of Geology, McMaster University, Hamilton, Ontario. 48 pp.

Kuntz, K. W. (1978). Atmospheric bulk precipitation in the Lake Erie basin. Report No. 56 to Canada Inland Waters Directorate. Environment Canada, Toronto, Canada. 37 pp.

Marshall, J. S. (1980). Population dynamics of *Daphnia galeata mendotae* as modified by chronic cadmium stress. *J. Fish. Res. Bd. Can.* **35:** 461–469.

McMahon, T. A. and P. J. Denison. (1979). Review paper: Empirical atmospheric deposition parameters—A survey. *Atmos. Environ.* **13:** 571–585.

Mildner, W. F. (1974). Assessment of erosion and sedimentation to the U.S. portion of the Great Lakes basin. Task A Report, Vol. 1, Pollution from Land Use Activities Reference Group. International Joint Commission, Windsor, Ontario, 48 pp.

Monteith, T. J. and W. C. Sonzogni. (1976). U.S. Great Lakes shoreline erosion loadings. Pollution from Land Use Activities Reference Group, International Joint Commission. Windsor, Ontario, 211 pp.

New York State. (1980). Unpublished data, New York State Department of Health, Albany, New York.

Nriagu, J. N., Ed. (1978). The biogeochemistry of lead in the environment. Elsevier/North Holland Biomedical Press, New York.

Radtke, L. F., P. V. Hobbs, and M. W. Eltgroth. (1980). Scavenging of aerosol particles by precipitation. *J. Appl. Meteorol.* **19:** 715–722.

Robbins, J. A. and D. N. Edgington. (1975). Determination of recent sedimentation rates in Lake Michigan using Pb-210 and Cs-137. *Geochim. Cosmochim. Acta* **39:** 285–304.

Robbins, J. A., E. Landstrom, and M. Wahlgren. (1970). Tributary inputs of soluble trace metals to Lake Michigan. Proceedings 15th Conference on Great Lakes Research, pp. 270–290.

Schmidt, J. A. (1982). Metal fluxes in Lake Michigan: The atmospheric burden and its effect on water quality. Ph.D. Dissertation. Water Chemistry Program, Univ. of Wisconsin, Madison.

Sehmel, G. A. (1980). Particle and gas dry deposition. A review. *Atmos. Environ.* **14:** 983–1011.

Sehmel, G. A. and W. J. Hodgson. (1978). A model for predicting dry deposition of particles and gases to environmental surfaces. PNL-SA-6721, Battelle, Pacific Northwest Laboratory, Richland, Washington, 242 pp.

Sehmel, G. A. and S. L. Sutter. (1974). Particle deposition rates on a water surface as a function of particle diameter and air velocity. *J. Rech. Atmos.* **8:** 911–920.

Shacklette, H. T., J. C. Hamilton, J. G. Boerngen, and J. M. Bowles. (1971). Elemental composition of surficial materials in the coterminous U.S.A. USGS Professional Paper No. 547D. 70 pp.

Shimp, N. F., J. A. Sleicher, R. R. Ruch, D. B. Heck, and H. V. Leland. (1971). Trace element and organic carbon accumulation in the most recent sediments of southern Lake Michigan. Illinois State Geological Survey Report. Environmental Geology Note No. 41. 24 pp.

Shiomi, M. T. and K. W. Kuntz. (1973). Great Lakes precipitation chemistry: Part 1. Lake Ontario basin. Proceedings 16th Conference of Great Lakes Research, pp. 581–602.

Sievering, H. (1976). Dry deposition on Lake Michigan by airborne particulate matter. *Water Air Soil Pollut.* **5:** 309–318.

Sievering, H. (1982). Profile measurements of particle mass transfer at the air–water interface. In preparation.

Sievering, H., M. Dave, D. Dolske, and P. McCoy. (1978). Cellulose filter high-volume cascade impactor aerosol collection efficiency. A technical note. *Environ. Sci. Technol.* **12:** 1435–1437.

Sievering, H., M. Dave, D. Dolske, R. L. Hughes, and P. McCoy. (1979). An experimental study of lake loading by aerosol transport and dry deposition in the southern Lake Michigan basin. U.S. Environmental Protection Agency final report No. EPA-905/4-79-016. 179 pp.

Skibin, D. (1973). Comment on water Pollution in Lake Michigan from Pollution Aerosol Fallout. *Water Air Soil Pollut.* **2:** 405–407.

Slinn, S. A. and W. G. N. Slinn. (1980). Predictions for particle deposition on natural waters. *Atmos. Environ.* **14:** 1013–1016.

Slinn, W. G. N. (1977). Some approximations for the wet and dry removal of particles and gases from the atmosphere. *Water Air Soil Pollut.* **7:** 513–543.

State of New York. (1980). Unpublished data. Office of Program Development, Planning and Research and Division of Pure Waters, Albany, New York.

Sonzogni, W. C., G. Chester, D. R. Coote, D. N. Jeffs, J. C. Konrad, R. C. Ostry, and J. B. Robinson. (1980). Pollution from land runoff. *Environ. Sci. Technol.* **14:** 148–153.

Stolzenburg, T. and A. W. Andren. (1982). Source reconciliation of atmospheric aerosols. *Water Air Soil Pollut.* **17:** 75–85.

Tisue, T., C. A. Seils, and D. A. Warner. (1980). Concentrations of dissolved and particulate forms of trace metals in Lake Michigan's major tributaries. Radiological and Environmental Res. Div. Annual Report Jan.–Dec. 1980, ANL-80-115, Part III, pp. 58–63.

Torrey, M. S. (1976). Environmental status of the Lake Michigan region, vol. 3, Chemistry of Lake Michigan. Report ANL/ES-40, Argonne National Laboratory, Argonne, Illinois, 418 pp.

United States Environmental Protection Agency. (1980). Interim report on Deposition Monitoring Network. Region V, Chicago. 12 pp.

Wahlgren, M. A., J. A. Robbins, and D. N. Edgington. (1980). Plutonium in the Great Lakes. In: W. C. Hanson, Ed., *Transuranic elements in the environment*. Technical Information Center, U.S. Dept. of Energy, pp. 659–683.

Weseley, M. L., and R. M. Williams. (1981). Field measurements of small ozone fluxes to snow, wet bare soil, and lake water. Environ. Res. Contrib. 80-07. Radiological and Environmental Research Div. Argonne National Laboratory, Argonne, Illinois, 23 pp.

Weseley, M. L., B. B. Hicks, W. P. Dannevik, S. Frisella, and R. B. Husar. (1977). An eddy-correlation measurement of particulate deposition from the atmosphere. *Atmos. Environ.* **14:** 561–563.

Wickham, J. T., D. L. Gross, J. A. Lineback, and R. L. Thomas. (1978). Late Quaternary sediments of Lake Michigan. Illinois State Geological Survey, Environmental Geology Note No. 84. 26 pp.

Winchester, J. W. (1969). Pollution pathways in the Great Lakes. *Limnos* **2:** 20–24.

Winchester, J. W. and G. Nifong. (1971). Water pollution in Lake Michigan by trace elements from pollution aerosol fallout. *Water Air Soil Pollut.* **1:** 50–64.

5

ATMOSPHERIC INPUTS AND THE DYNAMICS OF TRACE ELEMENTS IN LAKE MICHIGAN

Thomas Tisue

Center for Great Lakes Studies
University of Wisconsin—Milwaukee
Milwaukee, Wisconsin 53201

Donald Fingleton

Energy and Environmental Systems Division
Argonne National Laboratory
Argonne, Illinois 60439

Dr. Tisue's present address: Department of Chemistry and Geology, Clemson University, Clemson, South Carolina, 29631.

1. INTRODUCTION

It is commonplace to assert that materials deposited from the atmosphere contribute significantly to the flux of minor and trace elements through various compartments of the biosphere. But although several small watersheds have been studied in detail (Andren and Lindberg, 1977; Fisher et al., 1968), direct experimental determinations of the atmosphere's contribution to large water bodies are rare. The available evidence suggests that atmospheric deposition of trace elements into large lakes is an important, if not the principle, source for some elements.

It can be argued therefore that changing concentrations of these elements in the atmosphere may:

1. Alter significantly their overall mass budgets in large water bodies
2. Affect long-term trends in their concentrations in the water column
3. Influence the magnitude and timing of their seasonal concentration cycles

The purpose of this paper is to explore the extent to which these effects are observable in Lake Michigan. To this end, we have drawn in part on published accounts of the work of others. But we emphasize results obtained through the program of Great Lakes research in the Radiological and Environmental Research Division, Argonne National Laboratory.

2. CONTRIBUTION OF ATMOSPHERIC INPUTS TO TRACE ELEMENT MASS BUDGETS

The pioneering work of Winchester and co-workers (Winchester and Nifong, 1971) alerted Great Lakes scientists to the importance of atmospheric inputs to overall trace element mass budgets. The contribution of atmospheric deposition, both wet and dry, is especially important in the upper lakes, where the water surface comprises a large fraction of the basin area. Lake Michigan's southern basin, which we consider in more detail later, is also in the airshed of one of the nation's largest conurbations and industrial areas.

In order to assess the importance of atmospheric inputs quantitatively, it is necessary to estimate the fluxes of trace elements to the lake or basin from all major sources. In addition to wet and dry deposition from the atmosphere, these inputs include: (1) dissolved and suspended materials in tributaries and direct runoff; (2) point sources, principally industrial and municipal waste discharges; (3) materials eroded from the shoreline; and (4) materials re-entrained in the water by processes affecting the uppermost layers of sediment (Lee et al., 1981; Fisher et al., 1980).

Given the present state of knowledge, none of the fluxes can be estimated with great precision. In Chapter 4, Schmidt and Andren discuss the shortcomings of recent methodologies for establishing mass balance budgets for trace metals in large lakes. Despite these uncertainties, however, the weight of the evidence currently available strongly suggests that the atmosphere is a major source of several trace elements, including several such as Cd, Cu, Pb, and Zn, that are of potential concern as toxicants (Marshall et al., 1981; Parker et al., 1981; Marshall and Mellinger, 1980).

2.1. Mass Balance Budgets for Cd and Zn

In two recent studies, Eisenreich (1980) and Muhlbaier and Tisue (1981) prepared estimates of trace metal inputs into Lake Michigan based on completely independent sets of measurements. Eisenreich used passive bulk collectors located on the shoreline to sample both wet and dry deposition from the atmosphere. From analyses of these samples, he estimated, for example, that 5.4×10^6 g/yr of Cd reach Lake Michigan's southern basin via these routes. Muhlbaier and Tisue collected 29 rainfall events (wet only) at Argonne National Laboratory, and 28 offshore air samples at the 68th St. water intake structure, from August to November, 1978. They combined the measured Cd concentrations in these samples with estimates of total annual precipitation (Jones and Meredith, 1972) and of the mean annual dry deposition velocity (Sievering et al., 1977) to arrive at an annual flux of 6.5×10^6 g/yr of Cd. Thus the two independent estimates do not differ significantly.

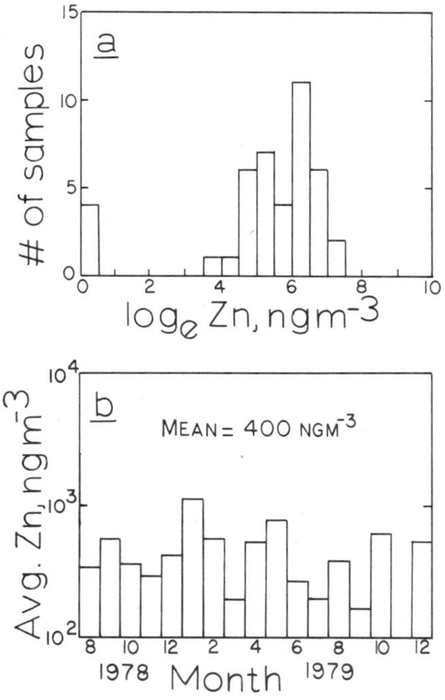

Figure 5.1 Distribution and seasonal variations of zinc concentrations in atmospheric particulate matter collected at Chicago's 68th St. water intake structure in 1978–1979. The top diagram is for air that has just traversed the Gary–Whiting industrial complex, SE to NW.

In more recent work (Tisue et al., 1980a), the Argonne workers collected high-volume air samples at the 68th St. structure for 18 months in 1978–1979. Analyses of these filters by X-ray fluorescence spectrometry produced values for a suite of minor and trace elements, exemplified by the data for Zn shown in Fig. 5.1. Williams used these concentration data, together with meteorological data collected simultaneously, as input to an improved formulation describing particle deposition to a water surface (Williams, 1980). His two-layer model took account explicity of wind speed, relative humidity, surface roughness, particle size, and lateral transfer between smooth and rough patches. This approach led to an estimated annual flux of Zn from the atmosphere to the southern basin of 260×10^6 g/yr due to dry deposition alone.

Using a mean annual Zn concentration of 40 μg/liter (Gatz, 1975; Shiomi and Kuntz, 1973; Parker et al., 1981) in rain and snow, one may estimate that an additional 530×10^6 g/yr reaches the southern basin via wet deposition. The atmospheric loading then totals 790×10^6 g/yr. This figure may be compared with Schmidt's (1982) estimate of about 250×10^6 g/yr, with Winchester and Nifong's (1971) of 1300×10^6 g/yr, with Dolske and Sievering's (1979) of 250×10^6 g/yr, and with Eisenreich's (1980) of 740×10^6 g/yr.

To put these values for Cd and Zn fluxes from the atmosphere into perspective, one must also be able to evaluate fluxes from other sources. We based our budget calculations in part on analyses of samples collected in 1978 and 1979 from Lake Michigan's major tributaries, in the spring and fall, at periods of high and low flow, respectively. The representativeness of such samples is arguable, to be sure, and it is risky to use the mean of a few grab samples to estimate annual transport. However, the major tributaries in Lake Michigan's southern basin are not "event" streams. That is, they do not experience sharp fluctuations in total flow or suspended load as a result of local precipitation events, or the lack thereof. This behavior is due to the influence of the sizeable lakes through which the rivers pass just before debouching into Lake Michigan. These basins act to retain storm water, slowing changes in flow rate, and to increase the retention time of waterborne contaminants, especially those that are bound to particles. The effect of these factors is to reduce the variability of river composition at the mouth, placing fewer demands on sampling frequency and timing.

The Cd and Zn concentrations in river water and suspended solids were determined by atomic absorption spectrophotometry (Muhlbaier and Tisue, 1981) and X-ray fluorescence spectrometry (Tisue et al., 1977; Tisue and Seils, 1982), respectively. By multiplying the mean concentrations times the annual flows for the major tributaries, one arrives at the total annual loadings.

To these sums must be added the contribution of uncharted runoff, that is, water that enters the lake from the land by routes other than the major tributaries. Such routes include direct runoff and minor tributaries. Because

Table 5.1 Mass Balance for Cd in Southern Basin

Sources (10^6 g/yr)		Sinks (10^6 g/yr)	
Atmosphere			
Wet	4.3 (41%)	Outflow	0.4 (10%)
Dry	2.2 (21%)	Sedimentation	3.8 (90%)
Erosion	1.0 (9%)	Total	4.2
Tributaries			
Soluble	1.3 (12%)		
Suspended	1.8 (17%)		
Total	10.6		

these sources were not sampled, their contributions were estimated by assigning them the mean concentration values observed for Cd and Zn in the major tributaries. The volume of uncharted runoff was taken to be the difference between the total runoff to the southern basin, 1.3×10^{13} liters/yr (U.S. Dept. of HEW, 1963), and the sum of the tabulated tributary discharge, 0.94×10^{13} liters/yr (Robbins et al., 1972). This difference amounts to about 30% of the total runoff and thus represents a source of appreciable uncertainty in the budgets.

It is also difficult to gauge the amounts of Cd and Zn that enter the lake's biogeochemical cycle from shoreline erosion. We have attempted to do so by combining values for their average elemental abundances in soils (Wedepohl, 1968; Mason, 1966; Vinogradov, 1959) with Monteith and Sonzogni's (1976) estimates of shoreline erosion rates. However precise these values may be, there is additional uncertainty associated with the behavior of the eroded material. Much of it is probably reworked in the littoral and sublittoral zones without becoming involved in basinwide transport. Also, it is arguable what fraction of the Zn and Cd in the eroding material will be in soluble forms. The contribution of Cd from this source appears relatively small (1

Table 5.2 Mass Balance for Zn in Southern Basin

Sources (10^6 g/yr)		Sinks (10^6 g/yr)	
Atmosphere			
Wet	530 (36%)	Outflow	6 (1%)
Dry	260 (18%)	Sediment	530 (99%)
Erosion	630 (42%)	Total	536
Tributaries			
Soluble	40 (3%)		
Suspended	20 (2%)		
Total	1460		

$\times$ 10^6 g/yr), rendering its precise value immaterial. For Zn, however, because of its much higher concentration in soils (60 ppm vs. 0.1 ppm for Cd), the contribution from erosion is potentially quite large—610 $\times$ 10^6 g/yr by our estimation.

Tables 5.1 and 5.2 show compilations of the rates at which Cd and Zn are estimated to enter the water column of Lake Michigan from the various sources, and the percentage contribution of each. Even allowing for the admitted imprecision of the estimates, these comparisons show that atmospheric inputs of these elements are comparable in magnitude to the sum of all other sources. Indeed, in the case of Zn, if one argues that little of the eroded material reaches the water column in available forms, the atmosphere becomes the *only* significant source.

3. EFFECTS OF VARYING ATMOSPHERIC INPUTS ON LONG-TERM TRENDS IN TRACE ELEMENT CONCENTRATIONS

The other side of the ledger in a mass balance budget consists of those processes that remove material from the water column. For nondegradable, nonvolatile constituents, one must consider at least the following removal mechanisms: (1) outflow, (2) burial within permanent sediments, (3) bubble bursts and spray, and (4) uptake by long-lived organisms. The latter two mechanisms are probably insignificant in Lake Michigan. It has a low surface-to-volume ratio, and the biomass is dominated by short-lived planktonic life forms. To complete the balance sheet for Cd and Zn thus requires evaluation of only two fluxes, outflow and sedimentation.

We used the data on mass sedimentation rates for the southern basin provided by Edgington and Robbins (1976) to calculate Zn and Cd removal rates. For Cd, the calculation was made (Muhlbaier and Tisue, 1981) by multiplying their values for seven ranges of mass sedimentation rate, weighted according to the fraction of the southern basin over which a particular range was observed, times the mean Cd concentration in 40 samples of surficial sediment (Robbins and Edgington, 1976). This method led to a sedimentation rate for Cd of 3.8 $\times$ 10^6 g/yr. This value may be compared to Eisenreich's (1980) independent estimate, 1.9 $\times$ 10^6 g/yr.

We used a different procedure to calculate the annual removal of Zn to the sediments. In this case the analysis rested on our observation that the Zn : Pb ratio is quite constant in both seston (Tisue et al, 1980b) and surficial sediments (Cahill, 1981) from the southern basin, as shown in Fig. 5.2. This result means that Zn and Pb are scavenged and deposited in fairly constant proportions. Therefore, the Zn sedimentation rate may be calculated by multiplying the mean observed Zn : Pb concentration ratio in the solid phases, 2.2, times the Pb mass sedimentation rate, 240 $\times$ 10^6 g/yr, determined previously (Edgington and Robbins, 1976). This procedure leads to a value of 530 $\times$ 10^6 g/yr for Zn removal by sedimentation.

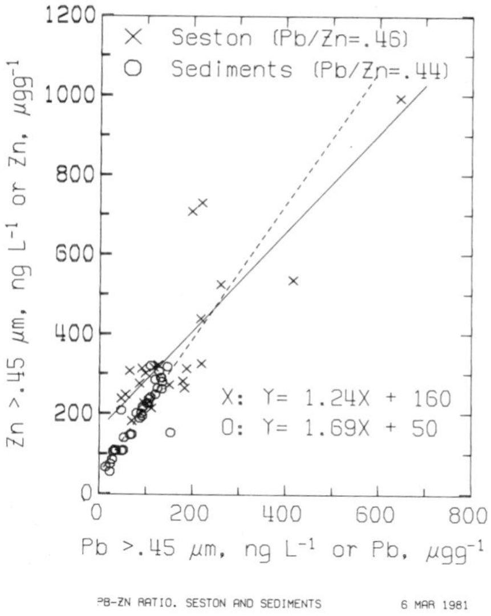

Figure 5.2 Ratio of Pb to Zn concentrations in Lake Michigan seston (X) and surficial sediments (O), obtained by regressing observed concentrations for >0.45 μm Zn against >0.45 μm Pb concentrations (X), or Zn concentrations in surface sediments against sediment Pb concentrations (O).

The rates at which Cd and Zn leave the southern basin in outflowing water may be calculated from the product of the mean elemental concentrations and the net rate of outflow. To make this calculation, we estimated the annual outflow from the basin as the difference between the sum of runoff and precipitation rates, and the evaporation rate (Jones and Meredith, 1972), or 1.07×10^{13} liters/yr.

In our most recent analyses (Muhlbaier et al., 1982) we found a mean total Cd concentration of 36 ± 5 ng/liter, using mass spectrometric isotope dilution analysis. The mean value corresponds to a Cd outflow rate of 0.40×10^6 g/yr.

We analyzed 64 samples of filtered water taken at various depths and times in the southern basin for Zn, using X-ray fluorescence spectrometry following preconcentration with ammonium pyrollidine carbodithioate (Tisue and Seils, 1982). The mean value was 380 ± 340 ng/liter. The material retained by the 0.45 μm filters was analyzed directly using the same detection technique, and exhibited a mean of 205 ± 192 ng/liter. The total mean concentration, 580 ng/liter, corresponds to a Zn outflow rate of 6.0×10^6 g/yr. These values are tabulated in Tables 5.1 and 5.2.

It is clear that, among the processes studied, sedimentation is the only removal mechanism for Zn and Cd of any significance in Lake Michigan's southern basin.

Another process that could lower trace metal concentrations in the southern basin is mixing with other water masses of lower metal content. In Lake Michigan it is easy to imagine that the northern basin, seemingly remote from major sources of pollution, might constitute a vast reservoir of "cleaner" water into which pollutants in the southern basin are constantly being diluted.

Two extreme cases set limits to the importance of this process. At one extreme is the situation where the residence time of a substance in the water column is short with respect to the mixing time between the two basins. In this case, sedimentation will remove the substance before dilution can occur. The second extreme is defined by substances with residence times that are long compared to the mixing time, leading to a relatively uniform distribution throughout the lake. Of course, no dilution is possible when both basins have the same average composition.

For Cd and Zn, the apparent residence times in the southern basin can be calculated from the data at hand, using the relationship

$$t_R = \frac{C_w V_B}{S}$$

where t_R = residence time, yr; C_w = concentration in water; V_B = volume of basin (1.6×10^{15} liters); and S = deposition rate. To make the calculations, we used the mean values for observed Cd and Zn concentrations (36 ng/liter and 580 ng/liter, respectively), together with values for the deposition rates from Figs. 5.2 and 5.3.

The calculated values are 15 yr for Cd, and 2 yr for Zn. The value for Zn is comparable to the corresponding estimates for Cs (Spigarelli and Nelson, 1972) and for Pu (Wahlgren et al., 1978). Cs and Pu are strongly associated with particulate matter and thus exhibit residence times similar to those of the small particles themselves. The significantly longer apparent residence time for Cd reflects its weaker binding to particles (Gardiner, 1974); it behaves as a semiconservative element.

Allender and Saylor (1980) applied a general circulation model (Bennett, 1977) to Lake Michigan, using shore-based meteorological data. The model computed water flows across the midlake sill in good agreement with observations. Simulations over the entire annual cycle indicated a period of about 2 yr is required to reduce an initial interbasin concentration difference to 1% of its original value.

These considerations suggest that zinc's residence time may be comparable to the outmixing time for the southern basin, while cadmium's exceeds it by several-fold. One may conclude that dilution into northern basin waters could be significant for zinc. Cadmium's residence time appears long enough for the interbasin mixing to destroy any north–south concentration gradients that might otherwise exist, precluding any dilution.

3.1. Projections

The data summarized in Tables 5.1 and 5.2 indicate a substantial difference between the annual rates of input for Cd and Zn, and their rates of removal from the water column by sedimentation. The observed discrepancies in the budget might well be artifacts. They do serve, however, to raise the question of the impact on lakewide trace metal concentrations of varying atmospheric inputs. This question is potentially of interest both in public policy arenas and scientific forums.

We used the mathematical formalism developed earlier for Cd (Muhlbaier and Tisue, 1981) to examine the effect on Zn concentrations of increasing atmospheric input rates. Our model uses as variables the estimated total Zn input in any year (I_t); the element's concentration in the water column (C_w); the volume rate of outflow (R_D); the total mass sedimentation rate (R_{ms}); the distribution coefficient of the element between particulate matter and lake water (K_D); and time (t). These factors are related by:

$$C_w = \left(C_{w(0)} \frac{-\beta}{\alpha}\right) e^{-\alpha/t} + \frac{\beta}{\alpha} \qquad (1)$$

where:

$$\alpha = \frac{R_D + (R_{ms}K_D)}{V}$$

$$\beta = \frac{I_t}{V}$$

where $C_{w(0)}$ = Zn concentration at $t = t_0$. This formalism is similar to the one used by Imboden et al. (1980) to reconstruct historical input rates from the observed concentration gradients of chemical species in Swiss lake sediments.

We calculated a value for K_D for zinc in two ways. First we divided its present average concentration in surficial sediments, 195 μg/g (Robbins and Edgington, 1976), by our observed value for its average dissolved concentration in the water column, 380 ng/liter. (The widespread occurrence of resuspension of Lake Michigan's surficial sediments (Lesht et al., 1980) should give ample opportunity for the implied equilibration to take place). Accordingly:

$$K_D = \frac{195 \times 10^3 \text{ ng/g}}{380 \text{ ng/liter}} \times \frac{1 \text{ g}}{10^9 \text{ ng}}$$
$$= 5 \times 10^{-7} \text{ liters/ng}$$

Alternatively, one may compute K_D from our observed values for the average concentrations of its dissolved and particulate forms, 380 and 205 ng/liter, respectively, and the estimated average value for total suspended solids, 1 mg/L. Thus:

$$K_D = \frac{205 \times 10^{-9} \text{ g/liter}}{1 \times 10^{-3} \text{ g/liter}} \times \frac{1}{380 \times 10^{-9} \text{ g/liter}} \times \frac{1 \text{ g}}{10^9 \text{ ng}}$$
$$= 5 \times 10^{-7} \text{ liters/ng}$$

Doubtless the exact agreement of the two calculations is fortuitous.

It is interesting to note that the K_D values the Swiss workers obtained for Cd and Zn in the Greifensee, 0.65×10^{-7} liters/ng and 0.25×10^{-7} liters/ng, lie within an order of magnitude of the corresponding values in Lake Michigan, 1×10^{-7} liters/ng and 5×10^{-7} liters/ng (Imboden et al., 1980). This is especially striking in light of the marked limnological differences in the two lakes, giving some assurance that, in projecting future concentrations, K_D will not vary too sharply with changes in trophic status.

In an earlier application (Muhlbaier and Tisue, 1981), we used Eq. 1 to project future Cd concentrations given various scenarios for increasing input rates. These same projections now also can be made for Zn, using the information summarized in Table 5.2. The results are shown in Fig. 5.3 as plots of Zn concentration vs. time for various values of the annual percent *increase* in the total Zn input rate.

Marshall et al. (1981) demonstrated an unexpected sensitivity of Lake Michigan zooplankton communities to Zn. And current levels of Zn in rain and snow are sufficient to account for most of the effect natural precipitation has on photosynthetic rates of Lake Michigan phytoplankton (Parker et al., 1981). These and other studies (e.g., Gaechter et al., 1979) strongly suggest that current water quality criteria are inadequate for protection of Lake Michigan's aquatic life. If Zn input rates continue to rise, our projections suggest that a century will suffice to bring Zn concentrations to levels where effects on natural populations are already readily observable in short-term experiments.

The sedimentary record in Lake Michigan's southern basin clearly establishes substantial increases in the input rates of anthropogenic substances over the last 150 years (Goldberg et al., 1981). Equation 1 can be used to help quantify the increases in input rates that gave rise to the observed cultural enrichment for Cd and Zn. For a constant input rate, Eq. 1 defines a limiting value for an element's concentration in water,

$$C_{w(\infty)} = \frac{I_t}{R_D + R_{ms}K_D} \tag{2}$$

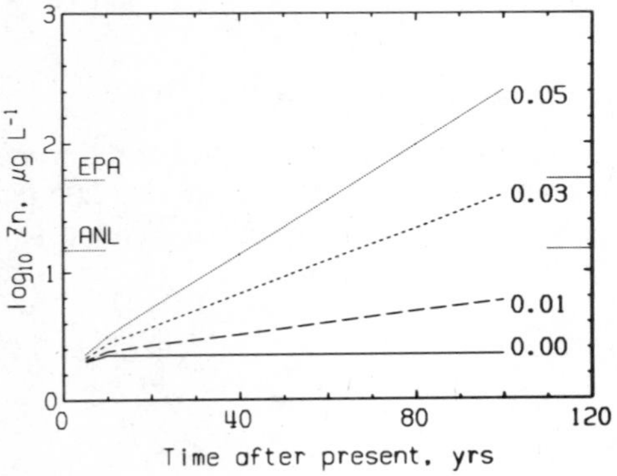

Figure 5.3 Projected future Zn concentrations in Lake Michigan's southern basin for various annual rates of increase in the total input. Erosion inputs are ignored.

so long as the mass sedimentation rate, distribution coefficient, and hydrological factors are invariant. And, because of the relationship between concentration in water and sediment expressed by K_D, the equation also predicts a steady-state value for an element's concentration in the sediments,

$$C_{\text{sed}} = C_w K_D \tag{3}$$

Robbins and Edgington (1976) found average concentrations of Cd and Zn in the present surficial sediments that exceed those in precultural material by factors of 2.3 and 3.3, respectively. Using Eq. 3 and our estimates of K_D for these elements, one calculates corresponding water column concentrations of 12 ng/liter for Cd, and 120 ng/liter for Zn. [Our most recent analyses show that the present values are several-fold higher, 36 ng/liter (300%) for Cd, and 380 ng/liter (320%) for Zn.] Taking the past values as $C_{w(\infty)}$ in Eq. 2, one calculates that the corresponding input rates were 2×10^6 g/yr for Cd, and 8×10^7 g/yr for Zn. These values amount to about 20% and 10% of the current estimated input rates, if erosion inputs are ignored. Put another way, man's activities seem to increase the input rates of Cd and Zn 5- and 10-fold, respectively, over those prevalent in recent precultural times.

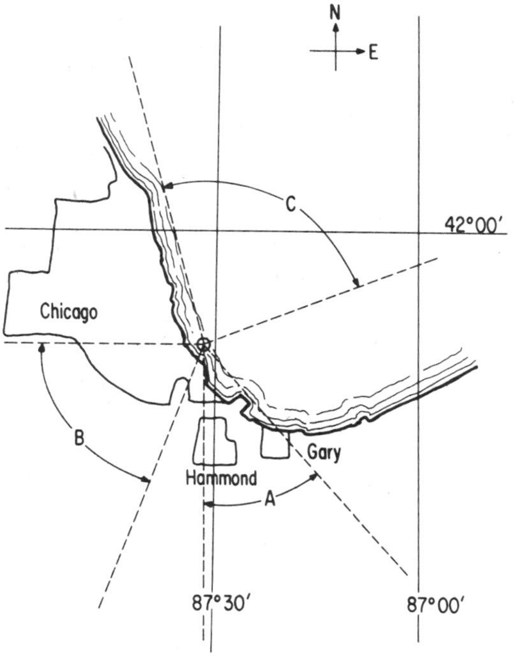

Figure 5.4 Relationship of the three 1978–1979 sampling sectors to major features on the shore. Sector A: Gary–Hammond–Whiting industrial complex; Sector B: south Chicago conurbation; and Sector C: the long fetch of the lake.

If the input rate for Zn were to exhibit a steady increase of 3% per annum, it would be about 80 yr before the rate reached a value 10-fold higher than at present. At that time, our model predicts the water column concentration would be about 12 μg/liter, with epilimnetic values exceeding this by several fold seasonally (see later discussion). These concentrations are ones at which biological effects are discernible even in relatively short-term incubation experiments (Marshall et al., 1981).

A 10-fold increase in the Zn input rate apparently *has* occurred over the last 150 years. How likely is it that this rate of increase will continue? It is not certain that the general level of industrial activity will continue to rise as fast as in the past. And it is reasonable to expect emission controls to become more widespread and effective.

We carried out two experiments, in 1973–1974 (Fingleton, 1976; Fingleton and Robbins, 1980) and in 1978–1979 (Tisue et al., 1980a) whose results bear on the question of air quality trends over southern Lake Michigan. In both experiments, particulate material in the air was collected at regular intervals for 18 months by high-volume samplers located at Chicago's 68th St. water intake structure. In 1978–1979, sampling was switched among three collection devices according to wind direction (Williams and White, 1978). All samples were analyzed by X-ray energy spectrometry (Muhlbaier and Tisue,

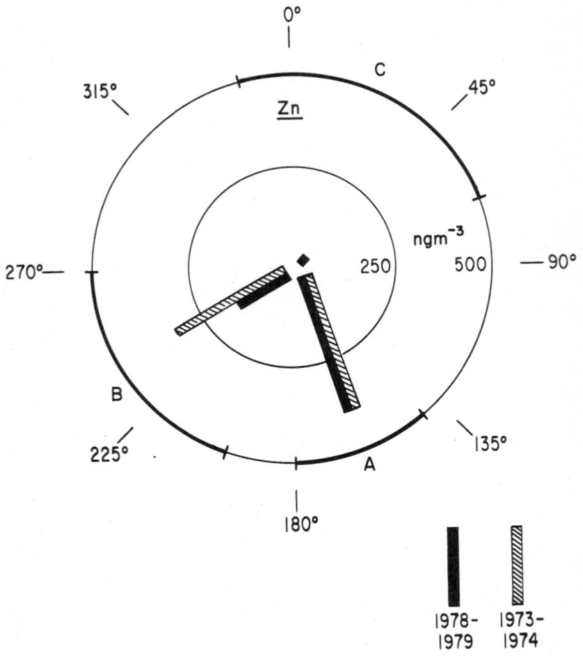

Figure 5.5 Comparison of annual mean Zn concentrations in airborne particulate matter over southern Lake Michigan for 1973–74, and 1978–79. The three sectors are as shown in Fig. 5.4.

1978). Figure 5.4 shows the wind directions included within the three sectors sampled.

In order to compare the two sets of data, the 1973–1974 data were sorted according to the mean wind direction in each sampling interval, then assigned to the corresponding 1978–1979 sector. If the mean wind direction lay outside these sectors, the data were excluded from the averaging.

It is clear from the results shown in Fig. 5.5 that the annual mean Zn concentration in the air over southern Lake Michigan did not increase measurably in the 5 years from 1973–1974 to 1978–1979. In fact, it may have actually decreased. These results point up the difficulties in establishing long-term air quality trends experimentally. Even over an interval of 5 years, the standard deviations of the annual means are large enough to permit changes of 3% per annum to go unrecognized.

4. INFLUENCE OF ATMOSPHERIC INPUTS ON SEASONAL VARIATIONS IN TRACE ELEMENT CONCENTRATIONS

We have seen that atmospheric inputs are a dominant source of Zn and Cd in the offshore waters of Lake Michigan's southern basin. And a simple model of their long-term behavior in the water column predicts that for every change in the rate of atmospheric input there will be a corresponding proportional change in their average lakewide concentrations. In this section we develop the notion that the sequellae of the dominance of atmospheric inputs of Zn include observable effects on its spatial and temporal distribution in the water column.

Figures 5.6 through 5.8 show the distribution of particulate forms of several elements with depth at various times of the year for samples from the southern basin's deepest sounding. In what follows, we argue that these profiles are interpretable in terms of the interplay of these factors: the strong atmospheric source for Pb and Zn, periodic resuspension of surficial sediments, calcium carbonate precipitation, and the annual weather cycle. These data also indicate that the concentration of Zn in the epilimnion exhibits pronounced seasonal cycling.

In the early part of the year, and well into the midyear months in offshore waters, the water column is isothermal, or very nearly so. This circumstance, coupled with the violent weather characteristic of the period October–April leads to homogeneous distribution of both solutes and small particles throughout the water column. Figure 5.6 shows an observation of this situation for particulate forms of Fe, Pb, and Zn at the southern basin's deepest sounding. Although the data show considerable scatter (some of which originates in the analyses, especially at the lowest concentrations), there is no perceivable trend with depth, for either the anthropogenic element Pb, or the mostly naturally occurring Fe.

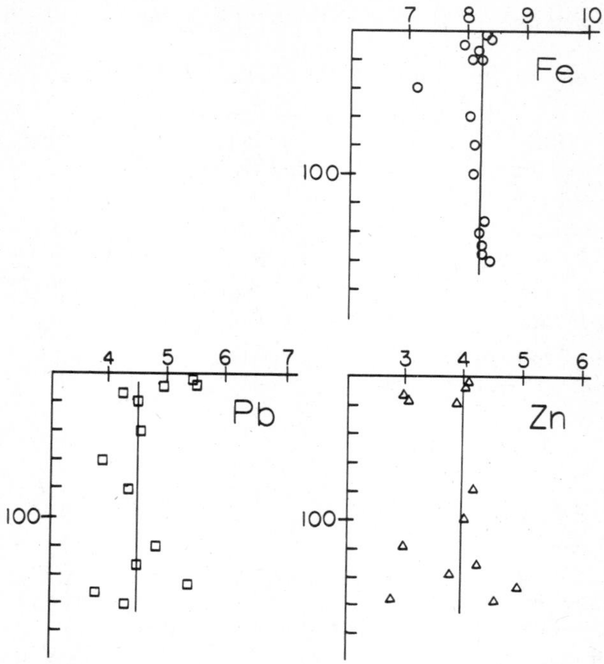

Figure 5.6 Element concentrations for particulate matter retained by 0.45 μm filters. Abscissas: $\log_e$ (concentration, ng/liter); ordinates: sampling depth, meters. *a*, $\log_e$ (mean concentration Fe) = 8.1 ± 0.3; *b*, $\log_e$ (mean concentration Pb) = 4.5 ± 0.8; and *c*, $\log_e$ (mean concentration Zn) = 3.9 ± 0.8. The vertical line on each plot is drawn at the mean value as an aid to visualization. Sampling date, June 1978 (isothermal).

As midyear arrives (Fig. 5.7), conditions become more quiescent; thermal stratification begins. The distributions with depth of particulate forms of Ca, Fe, Pb, and Zn then exhibit the exponential profiles expected as gravitational settling develops in the absence of significant vertical remixing. This situation signifies a significant decrease in eddy diffusivity as energy inputs from weather grow weaker and turbulence in the epilimnion becomes decoupled from the hypolimnion as thermal stratification develops.

The slopes of the semilog plots in Fig. 5.7 may be equated with the product of Stokes settling velocity and the reciprocal of the vertical eddy diffusivity coefficient in the relationship:

$$C_z = C_0 E^{mz} \tag{4}$$

where C_z = concentration of particle-bound species at depth z, C_0 = concentration of $z = 0$ (water surface), E = vertical eddy diffusivity coefficient, $m = W_s E^{-1}$, and W_s = Stokes settling velocity.
Or:

$$\ln C_z = W_s E^{-1} Z + \ln C_0 \tag{5}$$

The equations derived from least-squares fits of the data to this relationship appear in the captions to Figs. 5.7 and 5.8. Our evaluation of W_sE^{-1} for Ca is in good agreement with that reported by Chambers and Eadie (1981), who used sediment traps to study seston accumulation rates as a function of depth.

By September, thermal stratification is fully developed. Figure 5.8 displays the depth profiles observed under these conditions. For Ca, Pb, and Zn, the profiles appear to consist of the sum of two exponentials. Characteristics of the least-squares fits to Eq. 5 for these data appear in the figure caption. In contrast, particulate Fe concentration is constant to about 120m, then rises with a slope similar to the lower segment of the Ca plot.

It is our hypothesis that the segments of these profiles having positive slopes (those below 110 m) are in effect remnants of the profiles shown in Fig. 5.7. That is, they represent particles that have settled from the surface waters into the hypolimnion during the period June–September. The concentrations at the deep minima correspond closely to those observed in the surface waters in June.

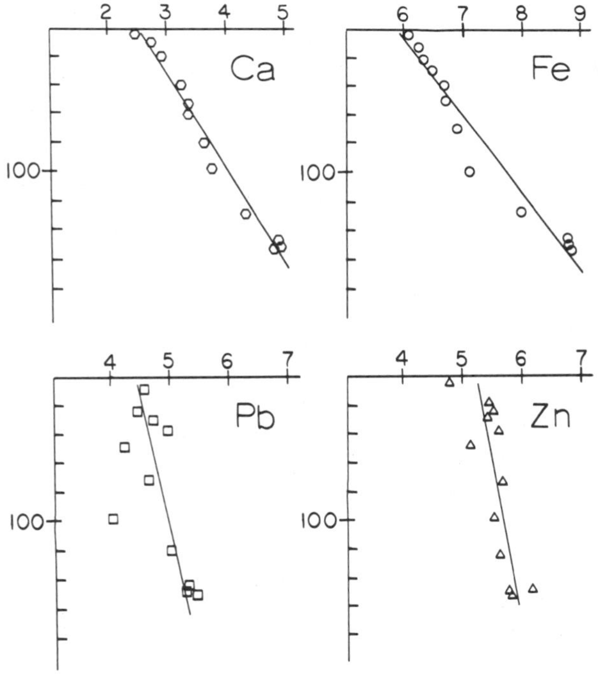

Figure 5.7 Element concentrations for particulate matter retained by 0.45 μm filters. Abscissas: $\log_e$ (concentration, ng/liter); ordinates: sampling depths, meters. *a*, $\log_e$ (Ca, ng/liter) = 0.02 (depth, m) + 2.5; *b*, $\log_e$ (Fe, ng/liter) = 0.02 (depth, m) + 5.8; *c*, $\log_e$ (Pb, ng/liter) = 0.005 (depth, m) + 4.4; and *d*, $\log_e$ (Zn, ng/liter) = 0.004 (depth, m) + 5.2. Sampling date, July 1979 (onset of stratification).

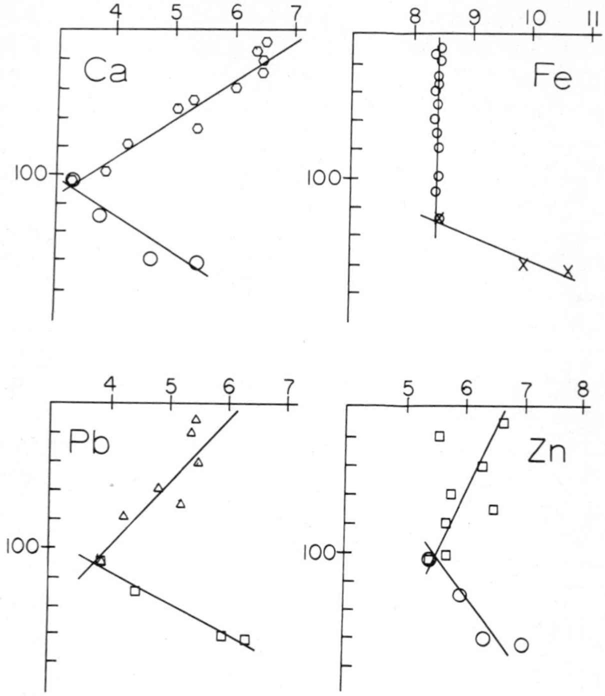

Figure 5.8 Element concentrations for particulate matter retained by 0.45 μm filters. Abscissas: $\log_e$ (concentration, ng/liter); ordinates: sampling depths, meters. *a*, ⬡ : $\log_e$ (Ca, ng/liter) = −0.04 (depth, m) + 7.4; ○: $\log_e$ (Ca, ng/liter) = 0.04 (depth, m) −1.7. *b*, X: $\log_e$ (Fe, ng/liter) = 0.06 (depth, m) − 1.2. *c*, △: $\log_e$ (Pb, ng/liter) = 0.02 (depth, m) + 6.2; □: $\log_e$ (Pb, ng/liter) = 0.05 (depth, m) − 1.7. *d*, □: $\log_e$ (Zn, ng/liter) = −0.01 (depth, m) + 6.6; ○: $\log_e$ (Zn, ng/liter) = 0.02 (depth, m) − 2.8. Sampling date, September 1978 (fully stratified).

At depths shallower than 110 m, the concentrations of particulate forms of Ca, Pb, and Zn increase exponentially, reaching maxima in the surface waters. We theorize that these "reverse" gradients are associated with strong sources of particulate forms of the elements in the epilimnion. In the case of Ca, that source is likely to be the precipitation of calcium carbonate, whose ubiquity has been documented in satellite images (Strong and Eadie, 1978) of the Great Lakes in late summer.

The source of the Pb and Zn is most probably the atmosphere. Figure 5.9 shows the close correspondence between depth profiles of dissolved and particulate Zn observed at this time. This correlation strongly suggests that Zn, and presumably Pb as well, entering the lake from the atmosphere becomes involved in a rapid and reversible equilibrium between dissolved and adsorbed forms. During stratification, these atmospheric inputs cannot be diluted effectively into water masses of lower Pb and Zn concentration by

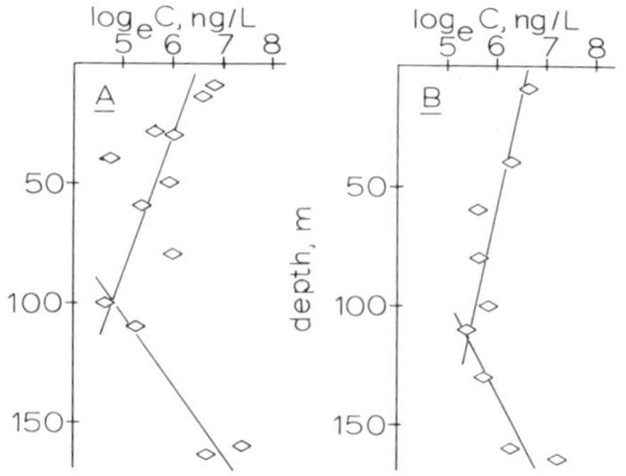

Figure 5.9 *a*, Zn concentration in water passing 0.45 μm filters; *b*, Zn concentration in particulate matter retained by 0.45 μm filters. Sampling date, September 1978 (fully stratified).

downward mixing. Thus, their annual maxima occur in the surface waters during the period of maximum biological productivity.

The data in Figs. 5.6 through 5.8 may be interpreted as fragmentary views of a seasonal cycle of Zn concentration in the epilimnion. In Fig. 5.10, we have plotted the amount of soluble and of total Zn in the uppermost 50 m of a 1 cm^2 cross section of the water column, as a function of the time of year. As the concentration rises, so does the fraction associated with particulate matter. This covariance is attributable to the substantial increase in total suspended solids that occurs in the epilmnion during the warm months as a result of biological production and carbonate precipitation.

The dotted line sketched through the observed data in Fig. 5.10 represents the hypothetical time course of the seasonal concentration cycle for Zn. More detailed observations will be required to confirm its occurrence.

The hypothetical Zn cycle makes an interesting contrast with the seasonal variation of Pu concentration studied in detail by Wahlgren et al. (1980). This contrast is depicted in Fig. 5.11. Although both Pu and Zn concentrations vary seasonally by a factor of 3–4, the timing of the two cycles is out of phase. Zn is at a minimum during isothermal conditions, while Pu is at a

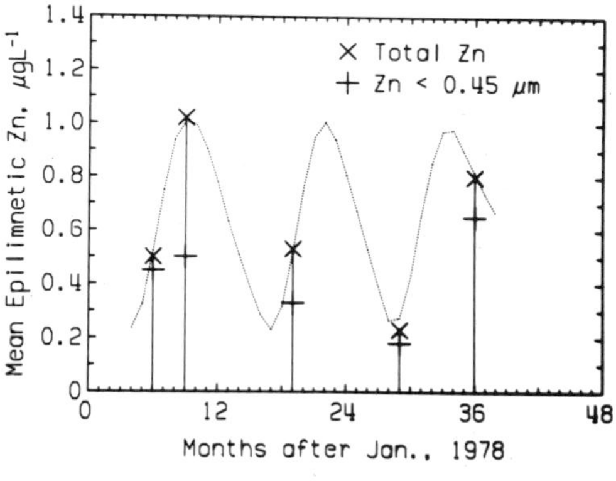

Figure 5.10 Seasonal variations in total epilimnetic zinc concentration. X: sum of soluble and particulate Zn; +: soluble (<0.45 μm) Zn. The smooth curve represents a hypothetical seasonal cycle.

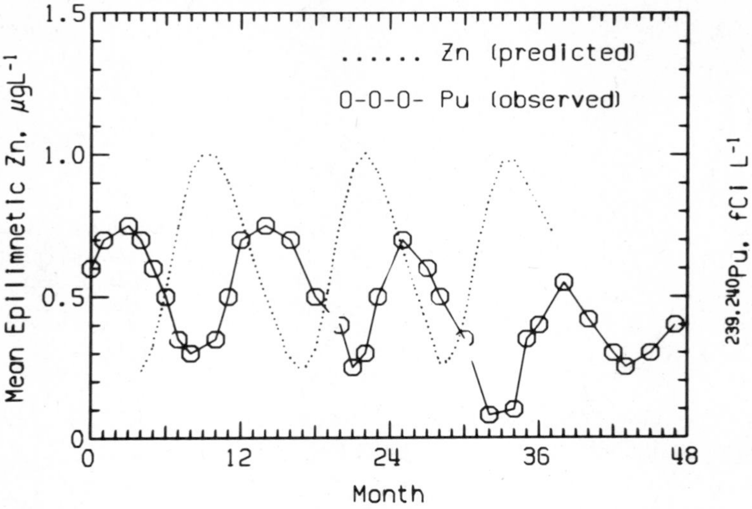

Figure 5.11 Comparison of hypothetical Zn cycling in the epilimnion with the observed seasonal cycle for Pu.

maximum. Towards the end of stratification, Pu is depleted in the surface waters, while Zn reaches its annual maximum.

Both elements are bound rather strongly to particles and are thus subject to the same effects of mixing and settling. The difference lies in the sources. Presently, Pu in the epilimnion originates almost entirely from the resuspension of sediments (Wahlgren et al., 1980), and thus reaches epilimnetic maxima through mixing while the lake is isothermal. Zn in offshore surface waters is derived almost exclusively from the atmosphere, and thus reaches a maximum at times when stratification traps the inputs in the epilimnion. As stratification breaks down, vertical mixing brings Pu-rich waters up from greater depth. Simultaneously, Zn decreases as it is diluted with hypolimnetic waters. The same behavior also should be exhibited by other substances with strong atmospheric sources and a tendency to associate with particulate matter.

ACKNOWLEDGMENTS

We gratefully acknowledge the skilled technical assistance of Charles Seils and Darrick Warner, and the cooperation and hospitality shown us by the captains and crews of the RV *Ekos* and RV *Laurentian*. David N. Edgington encouraged this work at all stages, and supported it in times of scarcity. John A. Robbins stimulated our thinking and encouraged this publication.

This program of Great Lakes research was funded in turn by the U.S. Department of Energy (and its predecessors) and by the U.S. Environmental

Protection Agency. Both authors gratefully acknowledge the support provided by these agencies during our respective periods of affiliation with the Argonne program, as well as that of the Wisconsin Sea Grant Program during Tisue's tenure at the Center for Great Lakes Studies.

REFERENCES

Allender, J. H. and J. H. Saylor. (1980). Model and observed circulation throughout the annual temperature cycle of Lake Michigan. *J. Phys. Oceanogr.* **9**(3): 573–579.

Andren, A. W. and S. E. Lindberg. (1977). Atmospheric input and origin of selected elements in Walker Branch Watershed, Oak Ridge, TN. *Water Air Soil Pollut.* **8:** 199–215.

Bennett, J. (1977). A three-dimensional model of Lake Ontario's summer circulation. I. Comparison with observations. *J. Phys. Oceanogr.* **7:** 591–601.

Cahill, R. A. (1981). Geochemistry of recent Lake Michigan sediments. Illinois Institute of National Resources, State Geological Survey Division, Circular 517, Champaign, Illinois, p. 35.

Chambers, R. L. and B. J. Eadie. (1981). Nepheloid and suspended particulate matter in southeastern Lake Michigan. *Sedimentology* **28:** 439–447.

Dolske, D. D. and H. Sievering. (1979). Trace element loading of southern Lake Michigan by dry deposition of atmospheric aerosol. *Water Air Soil Pollut.* **12:** 485–502.

Edgington, D. N. and J. A. Robbins. (1976). Records of lead deposition in Lake Michigan sediments since 1800. *Environ. Sci. Technol.* **10:** 266–275.

Eisenreich, S. J. (1980). Atmospheric input of trace metals to Lake Michigan. *Water Air Soil Pollut.* **13:** 287–301.

Fingleton, D. J. (1976). Trace element analysis of atmospheric particulates over southern Lake Michigan. Master of Science thesis, Department of Energetics, University of Wisconsin—Milwaukee.

Fingleton, D. J. and J. A. Robbins. (1980). Trace elements in air over Lake Michigan near Chicago during September, 1973. *J. Great Lakes Res.* **6:** 22–37.

Fisher, D. W., A. W. Gambell, G. E. Likens, and F. H. Bormann. (1968). Atmospheric contributions to water quality of streams in the Hubbard Brook Experimental Forest, New Hampshire. *Water Resour. Res.* **4:** 1115–1126.

Fisher, J. B., P. McCall, W. Lick, and J. A. Robbins. (1980). The mixing of lake sediments by the deposit feeder, *Tubifex tubifex*. *J. Geophys. Res.* **85**(C7): 3997–4006.

Gaechter, R. (1979). MELIMEX, an experimental heavy metal pollution study: Goals, experimental design and major findings. *Schweiz. Z. Hydrol.* **41:** 169–176.

Gardiner, J. (1974). The chemistry of cadmium in natural water. II. The adsorption of cadmium on river muds and naturally occurring solids. *Water Res.* **8:** 157–169.

Gatz, D. F. (1975). Pollutant aerosol deposition into southern Lake Michigan. *Water Air Soil Pollut.* **5:** 239–251.

Goldberg, E. D., V. F. Hodge, J. J. Griffin, M. Koide, and D. N. Edgington. (1981). Impact of fossil fuel combustion on the sediments of Lake Michigan. *Environ. Sci. Technol.* **15:** 466–471.

Imboden, D. M., J. Tschopp, and W. Stumm. (1980). Die Rekonstruktion frueherer Stoffrachten in einem See mittels. Sedimentuntersuchungen. *Schweiz. Z. Hydrol.* **42:** 1–14.

Jones, D. M. A. and D. D. Meredith. (1972). Proceedings 15th Conference on Great Lake Research, p. 477.

Lee, D.-Y., W. Lick, and S. W. Kang. (1981). The entrainment and deposition of fine-grained sediments in Lake Erie. *J. Great Lakes Res.* **7:** 224–233.

Lesht, B. M., R. M. Williams, and R. V. White. (1980). Sediment resuspension processes in the Great Lakes. Radiological and Environmental Research Division Annual Report, January–December 1980, ANL-80-115, Part III, pp. 50–52.

Marshall, J. S. and D. L. Mellinger. (1980). Dynamics of cadmium-stressed plankton communities. *Can. J. Fish. Aquat. Sci.* **37:** 403–414.

Marshall, J. S., D. L. Mellinger, and J. I. Parker. (1981). Combined effects of cadmium and zinc on a Lake Michigan zooplankton community. *J. Great Lakes Res.* **7:** 215–223.

Mason, B. J. (1966). *Introduction to geochemistry*. John Wiley and Sons, New York.

Monteith, T. J. and W. C. Sonzogni. (1976). Technical Report of the International Reference Group on Great Lakes Pollution from Land Use, International Joint Commission, Windsor, Ontario.

Muhlbaier, J. L. and G. T. Tisue. (1978). X-ray fluorescence analysis of air filters. Radiological and Environmental Research Division Annual Report, January–December 1978, ANL-78-65, Part III, pp. 100–104.

Muhlbaier, J. and G. T. Tisue. (1981). Cadmium in the southern basin of Lake Michigan. *Water Air Soil Pollut.* **15:** 45–59.

Muhlbaier, J., C. Stevens, D. Graczyk, and T. Tisue. (1982). Determination of cadmium in Lake Michigan by mass spectrometric isotope dilution analysis or atomic absorption spectrometry following electrodeposition. *Anal. Chem.* **54:** 496–499.

Parker, J. I., G. T. Tisue, C. W. Kennedy, and C. A. Seils. (1981). Effects of atmospheric precipitation additions on phytoplankton photosynthesis in Lake Michigan water samples. *J. Great Lakes Res.* **7:** 21–28.

Robbins, J. A. and D. N. Edgington. (1976). The distribution of selected chemical elements in the sediments of southern Lake Michigan. Radiological and Environmental Research Division Annual Report, January–December, 1976, ANL-76-88, Part III, pp. 65–71.

Robbins, J. A., E. Landstrom, and M. A. Wahlgren. (1972). Proceedings 15th Conference on Great Lakes Research, p. 270.

Schmidt, J. (1982). Ph.D. dissertation, University of Wisconsin—Madison, Chapter 3, in preparation.

Shiomi, M. T. and K. W. Kuntz. (1973). Great Lakes precipitation chemistry: Part 1. Lake Ontario Basin. Proceedings 16th Conference on Great Lakes Research, p. 581–602.

Sievering, H., M. Dave, D. Dolske, M. Eason, J. Forst, P. McCoy, N. Sutton, and K. Walther. (1977). Paper presented at 4th Joint Conference Sensing of Environmental Pollutants, New Orleans, Louisiana.

Spigarelli, S. A. and D. M. Nelson. (1972). Radiological and Environmental Research Division Annual Report, January–December 1972, ANL-7960, Part III, p. 25.

Strong, A. and B. J. Eadie. (1978). Satellite observations of calcium carbonate precipitation in the Great Lakes. *Limnol. Oceanogr.* **23:** 877–887.

Tisue, G. T. and C. A. Seils. (1982). Preconcentration of nanomolar amounts of trace elements from natural waters for X-ray energy spectrometric analysis using pyrollidinecarbodithioic acid. In preparation.

Tisue, G. T., C. Seils, and D. Bales. (1977). Determination of ultratrace metals in Lake Michigan and its tributaries by X-ray fluorescence spectrometry. Radiological and Environmental Research Division Annual Report, January–December 1977, ANL-77-65, Part III, pp. 117–120.

Tisue, G. T., D. J. Fingleton, J. A. Robbins, R. Allison, and S. Barr. (1980a). Air quality changes over southern Lake Michigan: 1973–74 compared with 1978–79. Radiological and Environmental Research Division Annual Report, January–December 1980, ANL-80-115, Part III, pp. 39–45.

Tisue, G. T., C. A. Seils, and D. A. Warner. (1980b). Concentrations of dissolved and particulate forms of trace elements in Lake Michigan's major tributaries. Radiological and Environmental Research Division Annual Report, January–December 1980, ANL-80-115, Part III, pp. 58–63.

U.S. Department of Health, Education, and Welfare. (1963). Public Health Service Special Report LM-1.

Vinogradov, A. P. (1959). *The geochemistry of rare and dispersed chemical elements in soil.* Consultants Bureau, New York.

Wahlgren, M. A., D. M. Nelson, K. A. Orlandini, and E. T. Kucera. (1978). Radiological and Environmental Research Division Annual Report, January–December 1978, ANL-78-65, Part III, p. 70.

Wahlgren, M. A., J. A. Robbins, and D. N. Edgington. (1980). Plutonium in the Great Lakes. In: W. C. Hanson, Ed., *Transuranic elements in the environment*, Technical Information Center, U.S. Department of Energy, pp. 659–683.

Wedepohl, K. H. (1968). In: L. A. Ahrens, Ed., *International series of monographs on Earth Science*, Vol. 30, Pergamon Press, Oxford.

Williams, R. M. (1980). A model for the dry deposition of particles on natural water surfaces. Radiological and Environmental Research Division Annual Report, January–December 1980, ANL-80-115, Part III, pp. 37–38.

Williams, R. M. and R. V. White. (1978). Development of a wind-direction controlled aerosol sampling system. Radiological and Environmental Research Division Annual Report, January–December 1978, ANL-78-65, Part III, pp. 92–93.

Winchester, J. W. and G. D. Nifong. (1971). Water pollution in Lake Michigan by trace elements from pollution aerosol fallout. *Water Air Soil Pollut.* **1:** 50–64.

6

THE SURFACE MICROLAYER AND ITS ROLE IN CONTAMINANT DISTRIBUTION IN LAKE MICHIGAN

Robert M. Owen
Philip A. Meyers

Oceanography Program
Dept. of Atmospheric and Oceanic Science
The University of Michigan
Ann Arbor, Michigan 48109

1. INTRODUCTION

A thin (<1 mm) surface film, commonly known as the surface microlayer, is present on all natural water bodies. Based on its gross physical dimensions, that is, its mass and volume, the surface microlayer overlying a lake or ocean obviously is miniscule compared to the bulk water column, and thus can be ignored in many considerations of aquatic processes. On the other hand, the chemistry and biology of the surface microlayer are sufficiently different from the subsurface waters that many now regard it as an important and distinct ecological compartment of the aquatic environment (Rice et al., 1983). For example, Liss (1975) has noted that enrichments of microorganism populations in the surface microlayer on the order of 10^3 are not uncommon. Microorganisms lie at the base of the aquatic food chain and their ability to incorporate and concentrate significant quantities of organic and inorganic substances is well documented. Consequently, the discovery of enriched concentrations of heavy metals and chlorinated hydrocarbons in the surface microlayers of both marine (Seba and Corcoran, 1969; Duce et al., 1972) and freshwater environments (Andren et al., 1976) had led to much concern regarding the relationship between surface microlayer contaminants and ecological cycles.

The first comprehensive literature review of the physical and chemical aspects of freshwater surface microlayers was compiled by Andren et al. (1976). Recently, Meyers et al. (1982) have summarized and updated our knowledge of both natural and pollutant materials in Great Lakes surface microlayers. It is clear from a comparison of the reviews prepared by these authors that studies during the intervening years have added much to our knowledge of the kinds and amounts of microlayer contaminants present in the Great Lakes and smaller lakes in the Midwest. Lake Michigan has received by far the most attention in this regard. Given this data base for Lake Michigan, recent studies have increasingly emphasized a mechanistic approach, that is, they have been aimed at answering critical questions concerning how contaminants become incorporated into the microlayer, how they are distributed, and where they ultimately reside. Our purpose here is to review the progress which has been made toward answering these questions. We also attempt to emphasize some of the major problems that have inhibited research in this area, inasmuch as we believe these problems are the primary obstacles that impede the advancement of surface microlayer research beyond the present, largely descriptive phase.

2. SAMPLING, MEASUREMENT, AND CLASSIFICATION OF SURFACE MICROLAYER CONTAMINANTS

The concept of the surface microlayer is based on the notion that it represents a zone which displays some anomalous property with respect to the bulk water column. The actual thickness of the microlayer is thus defined as the depth to which the chosen property of the surface layer is regarded as being anomalous relative to the subsurface waters. Unfortunately, different theoretical considerations as well as measurements of different physicochemical properties have resulted in thickness estimates that span some six orders of magnitude, ranging from about 10^{-4} μm to more than 10^2 μm (Liss, 1975). Contaminants in the surface microlayer generally are determined by laboratory analysis of samples of surface material rather than by *in situ* measurements. In this case the thickness is operationally defined by the sampling method employed (Duce and Hoffman, 1976). Most studies of this type in the Great Lakes have used either the plate-sampler (Harvey and Burzell, 1972) or the screen-sampler (Garrett, 1965), both of which typically recover the upper $1\text{–}3 \times 10^2$ μm of surface material. Since the true microlayer may extend only to the depth that surface molecules display preferred orientation, roughly 10^{-4} μm to 10^{-1} μm, depending on the strength of surface forces (Horne, 1969), the samples collected by these techniques probably are diluted with large but undeterminable amounts of subsurface waters (Piotrowicz et al., 1972). This unwanted dilution has not been regarded as a major analytical problem because modern analytical techniques are still sensitive enough to detect the various organic and inorganic pollutants which have been examined thus far. On the other hand, the uncertainty about the actual concentrations of contaminants in the microlayer greatly restricts our ability to evaluate the significance of the chemical data. For example, reported enrichments of various pollutants may, in fact, be several orders of magnitude too low.

Contaminants in the surface microlayer are generally classified according to their gross chemistry (organic vs. inorganic) and physical state (dissolved vs. particulate). Some investigators have designed analytical schemes that permit a more detailed classification within these major categories. For example, after filtering surface microlayer samples, Piotiowicz et al. (1972) have extracted the dissolved phase with chloroform in order to isolate a "chloroform extractable" phase, presumably containing metal-organic complexes, from the remaining dissolved "inorganic" phase. Similarly, Meyers and Owen (1980) have measured the organic carbon content of surface microlayer samples before and after filtration in order to estimate the distribution of organic compounds between particulate and dissolved phases. Halogenated hydrocarbons and pesticides have been the most studied groups of organic pollutants in the surface microlayer, and heavy metals have received the most attention among the possible inorganic contaminants. Certain nonpollutant chemical groups, such as lipids and alkali metals, have also been examined because they can provide insight into geochemical processes occurring within the microlayer.

3. DISTRIBUTION OF SURFACE MICROLAYER CONTAMINANTS

Contaminants in Great Lakes surface films are derived from multiple sources of organic and inorganic materials and are incorporated into and removed from the microlayer by a variety of different mechanisms. Contaminants enter the microlayer by spillage, advection, diffusion, adsorption by organisms or onto rising bubbles, rainfall and dry deposition from the atmosphere, and the entrapment of suspended particulate matter. Removal mechanisms include sinking particles which may have adsorbed molecules on their surfaces, advection, diffusion, dissolution of soluble substances by subsurface waters, and transport from the water surface by aerosols, bursting bubbles, and evaporation (Liss, 1975). These transport mechanisms initially were identified in studies of microlayers in marine systems. Subsequent investigations of lacustrine microlayers have noted these same pathways and, further, have shown that their relative importance is related to the specific lake province in which the microlayer occurs. Several investigations in Lake Michigan (Owen et al., 1979; Elzerman and Armstrong, 1979; Meyers and Owen, 1980; Mackin et al., 1980) have shown that the distribution of microlayer and bulk water components can be subdivided into three provinces: (1) the fluvial environment; (2) the nearshore environment within about 8–10 km of the coast; and (3) the open lake environment. A similar zonation probably exists in the other Great Lakes, although this cannot be verified until more detailed studies of these lakes are carried out.

3.1. The Fluvial Environment

Mackin et al. (1980) have examined the composition of microlayer and subsurface (1 m) samples collected from between 0.3 and 1.5 km upstream of the mouths of the 10 largest rivers flowing into Lake Michigan (Fig. 6.1). An R-mode factor analysis of these data was used to determine the relationship between specific variables and to identify key mechanisms that influence the composition of fluvial microlayers. The major finding of this analysis was that the microlayer and subsurface composition of the fluvial sources is governed primarily by the regional geology and the extent and types of weathering processes in individual drainage basins. This point is illustrated in Table 6.1 which compares the compositional data obtained for rivers draining the regions north and south of about latitude 44°30′N. South of this latitude the drainage area is more calcareous and potassic in nature (Robbins et al., 1972) and is more susceptible to physical weathering because of its comparitively low vegetation density. Hence both the particulate and dissolved phase concentrations of Ca and K in the southern river group are significantly higher in the microlayer and subsurface waters. In contrast, the tills and soils in the northern drainage area are stabilized by a dense vegetation cover and are relatively enriched in Fe- and Mn-bearing minerals. These conditions are more conductive to extensive chemical weathering, as is manifested by the higher dissolved phase concentrations of Fe, Mn, and organic carbon. The factor analysis calculated by Mackin et al. (1980) also revealed that 15.8% of the overall variance in the data was due to a strong association between particulate phase Fe and Cu, which was attributed to

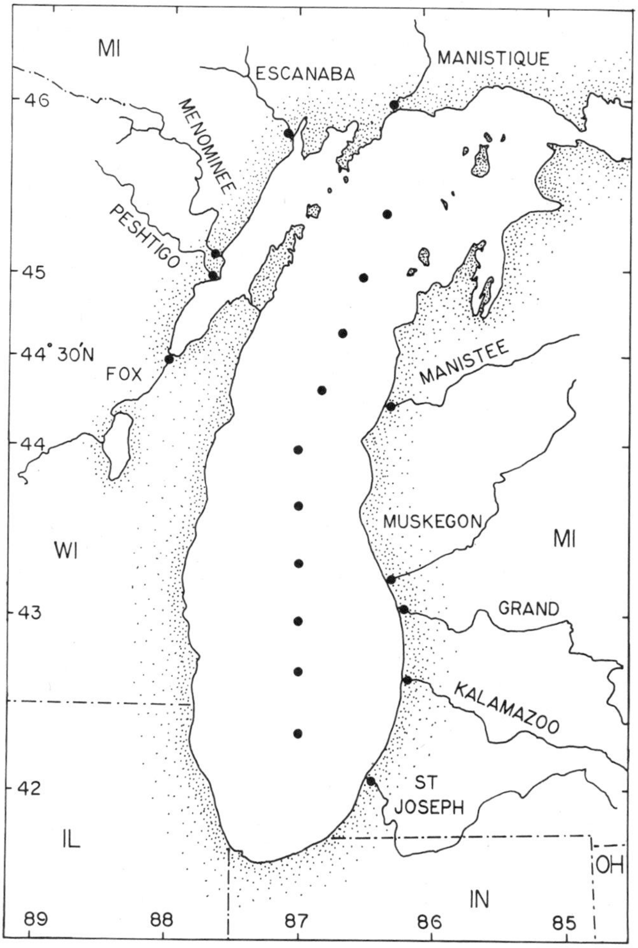

Figure 6.1 Locations of microlayer and subsurface water samples collected near the mouths of the 10 major rivers flowing into Lake Michigan and at 10 open lake stations, August 1978.

scavenging of Cu by hydrous iron oxides commonly formed at weathering sites (see Mackin and Owen, 1979; Gibbs, 1977).

The results of this investigation provide insight into the factors that are likely to influence the initial physical and chemical form of heavy metal contaminants which enter fluvial microlayers. Stability constants for transition metal-organic complexes are relatively high (Stumm and Morgan, 1981) particularily when the organic matter is composed of humic and fluvic acids and proteinaceous material (Andren et al., 1976). These complexes are likely to occur where chemical weathering has produced significant amounts of organic soil acids. The adsorption isotherms of heavy metals indicate they

Table 6.1. Average Composition of Microlayer and Subsurface Fluvial Inputs to Lake Michigan[a]

	Microlayer[b]		Subsurface[b]		Enrichment[c]	
Element	P	D	P	D	FR_P	FR_D
	Northern Rivers[d]					
Ca (mg/liter)	0.28	23.0	1.26	25.9	0.22	0.88
Cu (μg/liter)	1.00	2.85	1.38	5.48	0.72	0.52
Fe (μg/liter)	654	15.1	458	13.4	1.43	1.13
K (μg/liter)	7.98	973	17.9	1148	0.44	0.85
Mn (μg/liter)	12.2	40.7	18.1	35.1	0.67	1.16
Organic C (mg/liter)	0.78	10.5	0.30	9.43	2.60	1.11
	Southern Rivers[e]					
Ca (mg/liter)	21.4	26.9	14.9	30.2	0.80	0.49
Cu (μg/liter)	1.43	2.29	1.45	2.35	0.99	0.97
Fe (μg/liter)	424	5.93	683	2.86	0.62	2.07
K (μg/liter)	174	1253	191	1392	0.91	0.90
Mn (μg/liter)	16.8	25.5	25.4	15.6	0.66	1.63
Organic C (mg/liter)	2.39	6.54	1.76	5.52	1.36	1.18

[a] After Mackin et al., 1980.
[b] P = particulate phase; D = dissolved phase.
[c] FR_P = fractionation ratio, particulate phase; FR_D = fractionation ratio, dissolved phase.
[d] Includes the four rivers North of 44° 31′N (see Fig. 6.1).
[e] Includes the six rivers south of 44° 30′N (see Fig. 6.1).

should readily adsorb onto the surfaces of suspended particulates composed of clay minerals and hydrous iron and manganese oxides and hydroxides. This mechanism probably is important in areas where physical weathering has produced a relatively large detrital load, and where the load itself contains a significant fraction of these mineral groups.

The organic matter content of the fluvial samples summarized in Table 6.1 supports this explanation of trace metal distributions. The four northern rivers have a higher dissolved organic carbon (DOC) load in both the microlayer and in the subsurface region than do the southern rivers. Because dissolved organic carbon in river waters is contained mostly in humic substances (Reuter and Perdue, 1977), the higher concentrations of these organic acids in the northern rivers is consistent with the argument that chemical weathering dominates these watersheds.

In contrast to the dissolved phase, particulate organic carbon (POC) is higher in the six southern rivers than in the northern ones (Table 6.1). DOC/POC ratios are 13.5 in the microlayers and 31.4 in the subsurface waters of the northern rivers but only 2.7 and 3.1 in the corresponding southern river samples. Higher aquatic productivity due to enhanced nutrient loadings in the more populated and agricultural watersheds of the southern rivers may contribute to the higher POC levels, yet the combination of higher POC and lower DOC is best explained by the process of physical weathering, since compositions of POC and DOC in rivers are determined mostly by allochthonous sources (Wetzel, 1975).

Fractionation ratios, defined as the microlayer concentration divided by the subsurface concentration, of DOC and POC are greater than one in both the northern and southern rivers and are greater than all the inorganic materials except for dissolved Fe and Mn (Table 6.1). Organic matter clearly is enriched at the air–water interface and is a major component of Lake Michigan microlayers. However, the data in Table 6.2 fail to reflect the variability inherent in most microlayer data. As shown by Mackin et al. (1980), these fluvial average concentrations are based on widely ranging data, strongly influenced by local physical and biological factors while exhibiting statistically significant geographical differences.

Table 6.2. Average Concentrations (μg/liter) of Heavy Metals in Lake Michigan Surface Waters (film pressure > 1 dyne/cm)[a]

	Element			
Environment[b]	Zn	Cd	Pb	Cu
Rivers and Harbors	28	1.0	15.2	4.4
M	14	0.40	4.9	2.0
S	2.0	2.5	3.1	2.2
FR				
Mixing zones[c]				
M	15.2	0.31	7.4	4.1
S	3.8	0.09	1.2	1.0
FR	4.0	3.4	6.2	4.1
Nearshore[d]				
M	10.9	0.24	8.2	2.0
S	3.2	0.08	1.2	0.9
FR	3.4	3.0	6.8	2.2
Midlake[e]				
M	5.6	0.12	3.8	2.4
S	2.0	0.07	1.3	1.1
FR	2.8	1.7	2.9	2.2

[a] After Elzerman and Armstrong, 1979.

[b] M = microlayer; S = subsurface (~30 cm); FR = fractionation ratio.

[c] River plumes, mixing zones, and near obvious sources of atmospheric particulates.

[d] <8 km from shore and >1 km from river plumes.

[e] >8 km from shore.

3.2. The Nearshore Environment

The distinction between fluvial and nearshore zones is justified by the sharp differences that have been observed in the concentrations of both inorganic and organic materials in the microlayers and subsurface waters of these environments. Table 6.2 is a summary of heavy metal concentrations in surface waters from various locations in Lake Michigan reported by Elzerman and Armstrong (1979). The concentrations of Zn, Cd, Pb, and Cu in both the surface film and subsurface water of river and harbor samples are significantly greater than in any other location. However, the authors note that the microlayer concentrations are lower than what might be expected on the basis of the relatively high subsurface concentrations. In other words, the fractionation ratios of heavy metals in rivers and harbors are lower than those calculated for other environments (Table 6.2). Apparently the microlayer concentrations are not directly related to the subsurface concentrations. Elzerman (1981) has suggested this may be due either to a saturation-level effect, beyond which the microlayer cannot assimilate more trace metals even if the subsurface concentration increases, or else to control by a source other than the bulk water. One possible explanation for this lack of proportionality is that fluvial subsurface waters are strongly influenced by the upward migration of sediment pore fluids and the resuspension of bottom sediments (Mackin et al., 1980). If there is indeed a saturation-level effect, then the higher subsurface concentrations of heavy metals which result from these processes would obviously lead to lower fractionation ratios.

Owen et al (1979) have determined the concentrations of heavy metals in the particulate phase of microlayer and subsurface samples collected from the mouth of the St. Joseph River and from a point about 4 km down plume

Table 6.3. Particulate Phase Concentrations (μg/liter) of Heavy Metals in the St. Joseph River Plume[a]

	Element			
Environment[b]	Cu	Ni	Fe	Mn
River mouth				
M	1.2	1.6	180	43.8
S	0.93	2.2	233	22.1
FR	1.3	0.7	0.8	2.0
Down plume				
M	0.20	0.10	6.1	0.91
S	0.28	0.23	35.4	2.2
FR	0.7	0.4	0.2	0.4

[a] After Owen et al., 1979.

[b] See Table 6.2 for explanation of symbols.

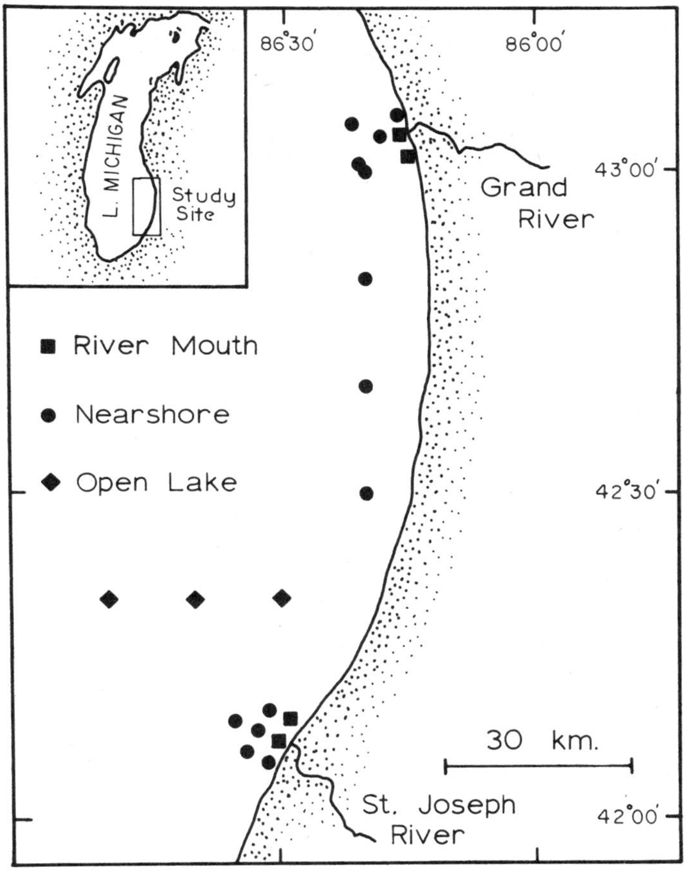

Figure 6.2 Locations of microlayer and subsurface water samples collected in the river plume, nearshore, and open lake provinces of southeastern Lake Michigan, August 1977.

from the river mouth in southern Lake Michigan. As shown in Table 6.3, both the concentrations and fractionation ratios of Cu, Ni, Fe, and Mn decrease significantly between the river mouth and the down plume station. In each case the decrease in fractionation ratio is primarily due to a decrease in the microlayer concentration of the element, although this effect is more pronounced for Fe and Mn than for Cu and Ni. Owen et al. (1979) have suggested this difference is due to differences in the aqueous chemistry and sorption characteristics of the Fe-Mn vs. Cu-Ni elemental pairs. The relatively high ionic potentials of Fe and Mn favor their association with hydrolyzates, whereas the lower ionic potentials of Cu and Ni make them more susceptible to ion-exchange reactions, particularly with clay minerals (Siegel, 1974). These mineralogical groups display different hydraulic characteristics because the smaller grain size and platelike character of clay minerals causes them to settle at a slower rate. Thus the observed segregation of the Fe-Mn and Cu-Ni element pairs is attributed to differential transport of their solid phases.

Meyers and Owen (1980) measured the concentrations of organic carbon, total fatty acids, and total hydrocarbons in the particulate and dissolved phases of microlayer and subsurface samples collected at 20 locations in Lake Michigan. These locations, shown in Fig. 6.2, were selected to represent the three lake provinces described earlier: the fluvial environment, the nearshore environment, and the open lake environment. In the case of Fig. 6.2, the fluvial environment was considered to be locations within the noticeably more turbid river waters at the mouths of the Grand and St. Joseph Rivers and in their plumes in Lake Michigan. Results of the measurements are summarized in Table 6.4.

Organic carbon concentrations in the particulate and dissolved phases are highest in the river plume samples and generally decrease toward open lake locations. Simultaneously, fractionation ratios become greater. These trends mirror those found by Elzerman and Armstrong (1979) for heavy metals, but differs somewhat from the heavy metal patterns found by Owen et al. (1979). Because of the great liklihood of biological involvement with organic materials, it is not surprising that their distributions might differ from those of inorganic materials.

Several patterns in total acid concentrations emerge from the averaged data in Table 6.4. First, particulate acids are consistently at higher levels

Table 6.4. Average Organic Component Compositions of River Plume, Nearshore, and Open Lake Zones in Lake Michigan[a]

	Microlayer[b]		Subsurface		Enrichment[b]	
Compound	P	D	P	D	FR_P	FR_D
	River Plume					
Organic C (mg/liter)	4.73	9.40	4.55	4.78	1.04	1.97
Fatty acids (μg/liter)	74.9	57.5	22.0	11.2	3.40	5.13
Hydrocarbons (μg/liter)	19.0	—	10.1	—	1.88	—
	Nearshore					
Organic C (mg/liter)	2.04	6.87	0.74	3.86	2.76	1.78
Fatty acids (μg/liter)	53.8	40.4	11.2	5.8	4.80	6.97
Hydrocarbons (μg/liter)	245	—	3.2	—	7.66	—
	Open Lake					
Organic C (mg/liter)	1.51	7.30	0.41	4.01	3.68	1.82
Fatty acids (μg/liter)	47.7	40.5	8.4	5.8	5.68	6.98
Hydrocarbons (μg/liter)	14.5	—	3.0	—	4.83	—

[a] After Owen et al., 1979.

[b] From Meyers and Owen, 1980.

than are dissolved acids. This is unlike findings from a variety of marine coastal locations (Daumas et al., 1976; Kattner and Brockman, 1978). Second, concentrations generally decrease with distance from river mouth locations, suggesting riverine input of fatty acids is important to microlayer and subsurface water content. These materials evidently are removed by sinking particles. Third, fractionation ratios increase with distance from shore. Decreased turbulent mixing apparently allows hydrophobic lipid materials to accumulate at the water surface and to become enhanced relative to subsurface concentrations. Combined with this is a general loss of subsurface particulate materials through sinking (Owen et al., 1979) which further enhances fractionation ratios.

Meyers and Owen (1980) note that the qualitative compositions of fatty acids differ in the three lake provinces and also differ between the microlayer and subsurface water. The most likely explanation for these distributions and their differences is that separate combinations of biological sources contribute to and maintain the content of these lake provinces. Evidently, most of the particulate acids transported to Lake Michigan by rivers settles out close to river mouths. This material is partially replenished by organic substances originating from aquatic communities within the lake. Because environmental factors such as temperature, light levels, availability of nutrients, and water turbulence can affect the biological communities and thus their biochemical compositions, the amounts and types of organic materials contributed to microlayers and subsurface waters will vary from one area of the lake to another.

Concentrations of particulate hydrocarbons given in Table 6.4 are in the ranges reported by Marty and Saliot (1976) and Daumas et al. (1976) for marine locations considered to be unpolluted and are about three orders of magnitude less than particulate hydrocarbon levels in a polluted Mediterranean estuary (Daumas et al., 1976). Thus, none of these Lake Michigan stations appears to be grossly contaminated by petroleum hydrocarbons, although evidence for their presence exists.

The most telling evidence of petroleum contamination in these Lake Michigan provinces is given by an unresolved complex mixture (UCM) of hydrocarbons underlying the resolved individual peaks on chromatograms from some samples. This mixture is not found in hydrocarbons from biological sources and is a characteristic of petroleum-derived hydrocarbons. Hydrocarbon patterns having a large UCM similar to those found in the Grand River samples have been reported in marine microlayers (Wade and Quinn, 1975; Daumas et al., 1976) and have been interpreted as being evidence of petroleum pollution.

In these Lake Michigan samples, the UCM dominates several river plume and nearshore locations and differs in the microlayer samples from the fluvial, nearshore, and open lake environments. Furthermore, a concentration gradient exists in which the UCM contribution decreases with distance from the river mouth. These observations indicate that the UCM component of the particulate hydrocarbons originates from land or fluvial sources and is transported to the lake by river flow. Once in the lake, these particulate materials evidently settle our fairly quickly, probably in association with mineral particles. It appears that the UCM component does not have a major aquatic or atmospheric source in Lake Michigan, except from ship activity. Evidence for a shipping source is implied by data from the open lake station farthest offshore from the St. Joseph River (Fig. 6.2). Samples from this station have surprisingly high UCM components whereas the nearshore St. Joseph stations are free of this feature. The most likely source thus is not the river, but rather from ship activities.

3.3. The Open Lake Environment

The inorganic composition of open lake microlayer and subsurface waters in Lake Michigan is summarized in Tables 6.2 and 6.5. Heavy metal concentrations in open lake samples analyzed by Elzerman and Armstrong (1979) were less variable and their fractionation ratios were generally lower than those measured for other lake environments (Table 6.2). The open lake region probably is least influenced by variations in point-source inputs and thus should be relatively more homogenous. The lower surface enrichments observed in the open lake samples were attributed to lower overall trace element inputs. Elzerman and Armstrong (1979) suggest that bubble floation mechanisms (see e.g., Liss, 1975) may play an important role in causing heavy metal accumulations in open lake surface films.

The data summarized in Table 6.5 are the averages of 10 samples collected from the midlike region along the major north–south axis of Lake Michigan (Fig. 6.1). As described previously for the river samples, Mackin et al. (1980)

Table 6.5. Average Composition of Open Lake Surface Waters in Lake Michigan[a]

	Microlayer[b]		Subsurface		Enrichment	
Component	P	D	P	D	FR_P	FR_D
Ca (mg/liter)	5.19	26.8	5.32	25.6	0.98	1.05
Cu (μg/liter)	2.62	7.43	2.27	4.29	1.15	1.73
Fe (μg/liter)	34.3	6.49	10.5	2.42	3.27	2.68
K (μg/liter)	12.6	1200	5.25	1080	2.40	1.11
Mn (μg/liter)	0.84	1.62	0.73	1.51	1.15	1.07
Organic C (mg/liter)	1.01	5.22	0.51	3.32	1.98	1.57
Fatty acids (μg/liter)	53.1	135.4	6.8	12.0	9.4	14.1
Hydrocarbons (μg/liter)	360.1	73.9	13.1	66.1	14.1	8.6

[a] After Mackin et al., 1980, and Meyers and Kawka, 1982.

[b] See Table 6.1 for explanation of symbols.

calculated an R-mode factor analysis for these data to determine the geochemical phases and mechanisms which influence the open lake microlayer. The results of this analysis can be summarized as follows. Geochemical associations within the dissolved phase are controlled primarily by complexation with organic materials (Fe, Cu), carbonate equilibria (Ca, Mg), precipitation of hydrous oxides (Fe, Mn, Cu), and biological uptake (K). Within the particulate phase, the variations of certain metals (Fe, Mn, Cu) are primarily influenced by rapid settling of detrital particulates, and between-element variations reflect the hydraulic characteristics of different mineral groups.

Atmospheric fallout undoubtedly contributes a significant amount of mineralized particles to the open lake (Eisenreich et al., 1977). If this were the dominant enrichment mechanism, then the latitudinal distribution of trace metals should be biased toward the southern portion of the lake. The southern basin of Lake Michigan is subject to the highest loadings of atmospheric particulates (Eisenreich et al., 1977) and, further, these loadings should be augmented by runoff from the large urban–industrial complexes which border southern Lake Michigan. However, Mackin et al. (1980) found that the observed concentrations of particulate phase Mn in the microlayer remained essentially constant at about 0.8 μg/liter in the open lake, whereas particulate phase Fe actually increased from a minimum of 7.2 μg/liter in the southern basin to a maximum of 90.8 μg/liter. The absence of the expected distribution patterns based on a simple atmospheric-fallout model suggests that *in situ* processes such as bubble floatation must also play an important role in controlling the particulate phase enrichments of certain trace elements in the open lake microlayer.

Summaries of the particulate and dissolved organic carbon, total fatty acids, and total hydrocarbons analyses of the samples from the 10 open lake locations are included in Table 6.5. In both the microlayer and the subsurface water, DOC concentrations average about five times greater than the mean POC value. Comparison of mean microlayer and subsurface POC concentrations indicates a slight microlayer enrichment. DOC is also enriched in the microlayer samples and shows much less station-to-station variability than does POC (Meyers and Kawka, 1982).

Like bulk organic carbon, the total fatty acid concentrations of these samples are higher in the dissolved phase than in the particulate phase. In addition, fractionation ratios of the acids are larger than for DOC or POC, and dissolved phase acid fractionation ratio is especially high. Comparison of the lipid fractionation ratios and the DOC and POC values indicates a microlayer enrichment one order of magnitude higher for the hydrophobic hydrocarbons and fatty acids than for the bulk organic matter. Preferential surface enrichments of the lipids may be the result of physical separation due to their low solubility and high hydrophobic character or a result of biological processes dominating the chemical characteristics of the microlayer and subsurface water.

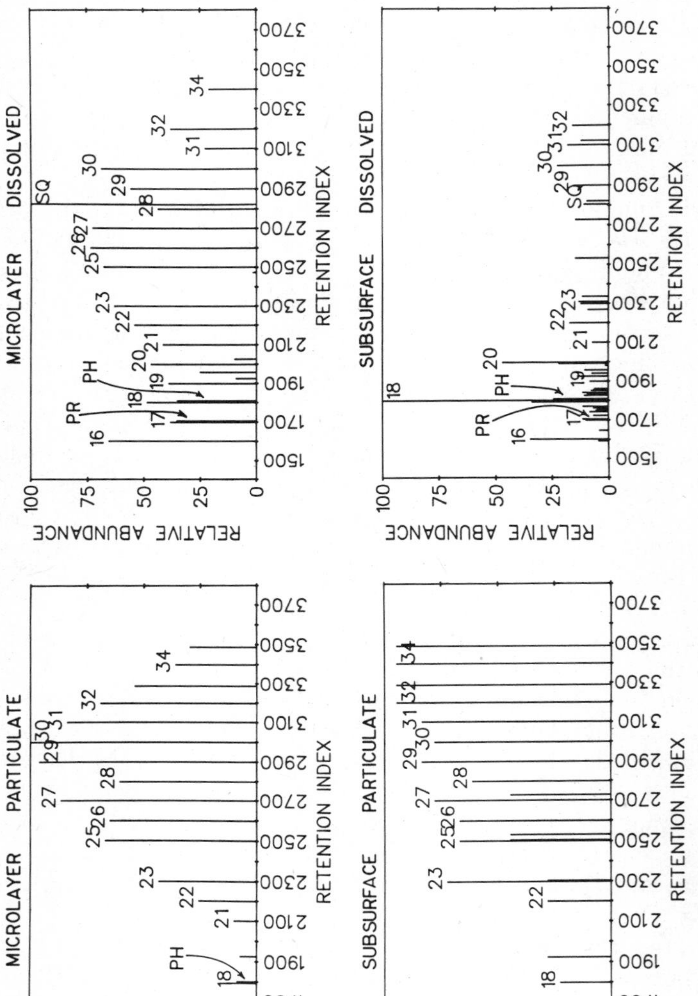

Figure 6.3 Hydrocarbon compositions of the particulate and dissolved phases of surface microlayer and subsurface water samples from Lake Michigan, August 1978. Distributions are presented relative to the major component. *n*-Alkanes are identified by number of carbon atoms, pristane by PR, phytane by PH, squalene by SQ.

The fatty acid compositions of the samples in this study are similar to the microlayer and subsurface compositions reported in Lake Michigan (Meyers and Owen, 1980) and in marine areas (Daumas et al., 1976; Kattner and Brockmann, 1978; Marty and Choiniere, 1979). Chainlength distributions are dominated by the straightchain acids having 14, 16, and 18 carbon atoms typical of planktonic communities. Microlayer and subsurface compositions are not exactly the same, and differ in relative amounts of unsaturated acids indicative of recent biosynthesis (Meyers and Kawka, 1982).

Like the fatty acids, the hydrocarbons present in the microlayer and subsurface samples are somewhat dissimilar. An example of the differences is shown in Fig. 6.3, in which the hydrocarbon distributions in samples from one station are compared. Particulate phase hydrocarbons in both the microlayer and subsurface water are dominated by longchain, terrigenous components. Dissolved phase distributions contain more shortchain, algal components.

A simple comparison of hydrocarbon compositions can be made by using the ratio of *n*-nonacosane to *n*-heptadecane in different samples. This ratio is a measure of land-derived *versus* aquatic hydrocarbons; a high ratio indicates that land-derived hydrocarbons dominate. The particulate hydrocarbon ratios of the samples from the 10 Lake Michigan stations average 22 in the microlayer and 16 in subsurface water and indicate that land-derived components are significant in Lake Michigan hydrocarbon distributions even in areas distant from shore. Dissolved phase hydrocarbons, in contrast, have mean ratios of 1.3 and 1.5 in the microlayer and subsurface samples, respectively. Because fluvial organic matter appears not to be important in open lake areas (Meyers and Owen, 1980), delivery of terrigenous hydrocarbons is probably in association with windborne dust particles similar to those found by Simoneit (1977) over oceanic areas. Rain and snow contain predominantly terrigenous hydrocarbons (Meyers and Hites, 1982), most likely in the form of particles washed from air by the falling precipitation. The strong, land-derived hydrocarbon signature of the particulate phase hydrocarbons in the open Lake Michigan samples is consistent with an eolian input of these organic materials.

4. OTHER FACTORS INFLUENCING CONTAMINANT DISTRIBUTION

Although the lacustrine environment is divided into three provinces on the basis of microlayer composition, a significant degree of compositional variability usually exists within any one of these provinces. A number of factors can contribute to this effect, including the nonuniform distribution of point-source inputs, sporadic zones of high biological productivity, and patches of petroleum hydrocarbons often found in harbors and shipping lanes. One of the most apparent causes of localized compositional variability is the presence of surface slicks. The visual phenomenon commonly known as a "slick" is produced by the change in light reflection properties of surface water resulting from the dampening of capillary waves (wavelength <1.7 cm). Significant dampening of capillary waves occurs when the surface film pressure exceeds about 1 dyn/cm. Laboratory studies indicate surface slicks will form when molecules are compressed by light winds or at frontal convergence zones (Jarvis, 1967), and where the turbulence is low enough so that the slick will be maintained (Liss, 1975). Hence natural slicks are commonly found in coastal areas and protected embayments, and are less likely in open waters where the sea state is usually rougher.

The highest fractionation ratios of trace elements in Lake Michigan have been observed in samples collected from areas with visible slicks or where the film pressure was > 1 dyn/cm (Table 6.2). Elzerman and Armstrong (1979) have noted a moderate positive correlation between trace metal fractionation ratios and surface film pressure. They suggest this relationship may be due either to association of trace metals with the surface accumulated material responsible for creating the film pressure, or to the trapping and retention efficiency of the material in the surface microlayer.

Thus far we have attributed most of the observed distribution of contaminants in the microlayer to the influence of large-scale physical processes in which the microlayer itself plays only an incidental role. It should be emphasized that this bias reflects our present state of knowledge rather than a presumed lack of importance of processes unique to the microlayer. In fact, there is considerable circumstantial evidence that contaminant distribution is affected by micro-scale biochemical or geochemical processes. For example, Elzerman (1981) has cited numerous studies of both freshwater and marine microlayers in which the observed surface enrichment of trace metals appears to be the result of metal-organic complexation. This seems a likely mechanism, although our studies of Lake Michigan microlayers have often failed to observe even a significant statistical correlation between trace metals and different fractions of organic compounds. This difficulty could be due to analytical problems, such as unknown transformations in the sample during solvent extraction, or to the presence of colloidal-sized particulate matter which passes through standard 0.45 μm filters. It is also likely that the organic compounds which have been measured in Lake Michigan thus far (mostly fatty acids and petroleum hydrocarbons) are not the primary compounds responsible for complexation. Simple surfactants such as fatty acid lipids were originally thought to be the major organic legands in the microlayer, but recent studies suggest that components in the uncharacterized portion of the organic matter (polysaccharides, proteins, humic acids, etc.) may be the most important complexing agents (Hunter and Liss, 1981).

Experimental distinction of the dissolved and particulate phases is important to explaining distributions of components of the microlayer and subsurface waters and to learning more about the processes active in microlayer formation. Attempts to use the solubilities of different fatty acids and hydrocarbons (Meyers and Kawka, 1982) and of PCBs (Rice et al., 1983) to explain fractionation of these organic materials between surface films and the underlying water have been unsuccessful. It is probable that the presence of colloidal matter which does not consistently partition into the dissolved or the particulate phase in available separation schemes is responsible for this continuing problem.

5. SUMMARY

The surface microlayer—thin and variable, yet omnipresent—remains a poorly understood feature of Great Lakes waters. Local processes of input and removal appear to control the composition of the microlayer, and physical processes such as turbulence are important in its development, maintenance, and dispersion. Because the microlayer is but a thin film at the water's surface, its small volume prohibits the microlayer from being an important reservoir of contaminants relative to the total lake volume. Nonetheless, enrichments of inorganic and organic materials in the microlayer suggest that this zone does have an important role in the interchange of substances between the atmosphere and the Great Lakes.

Hydrophobic organic material, including high-molecular-weight hydrocarbons such as plant waxes and polychlorinated biphenyls, can accumulate in the microlayer from atmospheric input. Waxes and oils of aquatic organisms are sources of fatty acids and hydrocarbons to this surface film. Heavy metals can be contributed from atmospheric sources, although their major sources appear to be from rivers and from suspended sediments. Once in the microlayer, these materials can be further concentrated into aerosol particles by bursting of bubbles or into sinking particles by compression of the surface film. Furthermore, neustonic communities can incorporate surface microlayer components into their biomass and thereby introduce these materials into aquatic food webs.

ACKNOWLEDGMENTS

Some of the many students who participated in collecting and analyzing some of the samples described in this review include R. Bourbonniere, S. Edwards, K. Erstfeld, H. Iwasiuk, M. Leenheer, J. Mackin, and N. Takeuchi. We appreciate their contributions. We thank the captains and crews of RV *Mysis* and RV *Laurentian* for tolerating the peculiar demands of our microlayer sampling operations. The National Sea Grant Program supported our microlayer research under Grant 04-7-158-44078.

REFERENCES

Andren, A. W., A. W. Elzerman, and D. E. Armstrong. (1976). Chemical and physical aspects of surface organic microlayers in freshwater lakes. *J. Great Lakes Res.* **2** (Suppl. 1): 101–110.

Daumas, R. A., P. L. Laborde, J. C. Marty, and A. Saliot. (1976). Influence of sampling method on the chemical composition of water surface film. *Limnol. Oceanogr.* **21:** 319–326.

Duce, R. A. and E. J. Hoffman. (1976). Chemical fractionation at the air/sea interface. *Annu. Rev. Earth Planet. Sci.* **4:** 187–228.

Duce, R. A., J. G. Quinn, C. E. Olney, S. R. Piotrowicz, B. J. Ray, and T. L. Wade. (1972). Enrichment of heavy metals and organic compounds in the surface microlayer of Narragansett Bay, Rhode Island. *Science* **176:** 161–163.

Eisenreich, S. J., P. J. Emmling, and A. M. Beeton. (1977). Atmospheric loading of phosphorous and other chemicals to Lake Michigan. *J. Great Lakes Res.* **3:** 291–304.

Elzerman, A. W. (1981). Mechanisms of enrichment at the air–water interface. In: S. J. Eisenreich, Ed., *Atmospheric pollutants in natural waters*. Ann Arbor Science, Ann Arbor, pp. 81–97.

Elzerman, A. W. and D. E. Armstrong. (1979). Enrichment of Zn, Cd, Pb, and Cu in the surface microlayer of Lakes Michigan, Ontario, and Mendota. *Limnol. Oceanogr.* **24:** 133–144.

Garrett, W.D. (1965). Collection of slick-forming materials from the sea surface. *Limnol. Oceanogr.* **10:** 602–605.

Gibbs, R. J. (1977). Transport phases of transition metals in the Amazon and Yukon rivers. *Geol. Soc. Am. Bull.* **88:** 829–843.

Harvey, G. W. and L. A. Burzell. (1972). A simple microlayer method for small samples. *Limnol. Oceanogr.* **17:** 150–157.

Horne, R. A. (1969). *Marine chemistry*. A. D. Little, New York, p. 340.

Hunter, K. A. and P. S. Liss. (1981). Principles and problems of modeling cation enrichment at natural air–water interfaces. In: S. J. Eisenreich, Ed., *Atmospheric pollutants in natural waters*. Ann Arbor Science, Ann Arbor, pp. 99–127.

Jarvis, N. L. (1967). Adsorption of surface-active material at the sea–air interface. *Limnol. Oceanogr.* **12:** 213–221.

Kattner, G. G. and U. H. Brockman. (1978). Fatty-acid composition of dissolved and particulate matter in surface films. *Mar. Chem.* **6:** 233–241.

Liss, P. S. (1975). Chemistry of the sea surface microlayer. In: J. P. Riley and G. Skirrow, Eds., *Chemical oceanography*, Vol. 2, 2nd ed. Academic Press, New York, pp. 193–243.

Mackin, J. E. and R. M. Owen. (1979). The geochemistry of sediments from Little Traverse Bay, Lake Michigan. *Can. J. Earth Sci.* **16:** 532–539.

Mackin, J. E., R. M. Owen, and P. A. Meyers. (1980). A factor analysis of elemental associations in the surface microlayer of Lake Michigan and its fluvial inputs. *J. Geophys. Res.* **85:** 1563–1569.

Marty, J. C. and A. Choiniere. (1979). Acides gras et hydrocarbures de l'ecume marine et de la microcouche de surface. *Naturaliste Can.* **106:** 141–147.

Marty, J. C. and A. Saliot. (1976). Hydrocarbons (normal alkanes) in the surface microlayer of seawater. *Deep-Sea Res.* **23:** 863–873.

Meyers, P. A. and R. A. Hites. (1982). Extractable organic compounds in Midwest rain and snow. *Atmos. Environ.* **16:**2169–2175.

Meyers, P. A. and O. E. Kawka. (1982). Fractionation of hydrophobic organic materials in surface layers. *J. Great Lakes Res.* **8:** 288–298.

Meyers, P. A. and R. M. Owen. (1980). Sources of fatty acids in Lake Michigan surface microlayers and subsurface waters. *Geophys. Res. Lett.* **7:** 885–888.

Meyers, P. A., C. P. Rice, and R. M. Owen. (1983). Input and removal of natural and pollutant materials in the surface microlayer on Lake Michigan. In: R. O. Hallberg, Ed., Proceedings Fifth International Symposium on Environmental Biogeochemistry, Ecological Bulletin (Stockholm) Volume 35, pp. 519–532.

Owen, R. M., P. A. Meyers, and J. E. Mackin. (1979). Influence of physical process on the concentration of heavy metals and organic carbon in the surface microlayer. *Geophys. Res. Lett.* **6:** 147–150.

Piotrowicz, S. R., B. J. Ray, G. L. Hoffman, and R. A. Duce. (1972). Trace metal enrichment in the sea-surface microlayer. *J. Geophys. Res.* **77:** 5243–5254.

Reuter, J. H., and E. M. Perdue. (1977). Importance of heavy metal–organic matter interactions in natural waters. *Geochim. Cosmochim. Acta* **41:** 325–334.

Rice, C. P., P. A. Meyers, and G. S. Brown. (1983). Role of surface microlayers in the air–water exchange of PCBs. In: D. Mackey, Ed., *Physical aspects of PCB cycling in the Great Lakes*. Ann Arbor Science Publishers, Ann Arbor, Michigan, pp. 157–179.

Robbins, J. A., E. Landstrom, and M. Wahlgren. (1972). Tributary inputs of soluble trace metals to Lake Michigan. Proceedings 15th Conference on Great Lakes Research, pp. 270–290.

Seba, D. B. and E. F. Corcoran. (1969). Surface slicks as concentrators of pesticides in the marine environment. *Pest. Monit. J.* **3:** 190–193.

Siegel, F. R. (1974). *Applied geochemistry*, J. Wiley & Sons, New York, pp. 29–58.

Simoneit, B. R. T. (1977). Organic matter in eolian dusts over the Atlantic Ocean. *Mar. Chem.* **5:** 443–464.

Stumm, W. and J. J. Morgan. (1981). *Aquatic chemistry*, 2nd ed. Wiley-Interscience, New York, pp. 374–386.

Wade, T. L. and J. G. Quinn. (1975). Hydrocarbons in the Sargasso Sea surface microlayer. *Mar. Pollut. Bull.* **6:** 54–57.

Wetzel, R. G. (1975). Organic carbon cycle and detritus. In: *Limnology*, Chapter 17, pp. 538–621. Saunders, Philadelphia.

7

PETROLEUM CONTAMINANTS IN THE GREAT LAKES

Philip A. Meyers

Oceanography Program
Department of Atmospheric and Oceanic Science
The University of Michigan
Ann Arbor, Michigan 48109

1. INTRODUCTION

Contamination of the Great Lakes by petroleum and petroleum products is a common and virtually unavoidable result of modern society. Petroleum is introduced to the waters of the lakes from numerous leaks and spills from pleasure craft, commercial ships, fuel tankers, and their harborside facilities. It also enters the lakes from sources more distant in the forms of land runoff, river inputs, and airborne materials. Such non-point-source contributions, while individually small, are continually being added and are collectively a major concern. An additional potential source is from offshore wells, presently not allowed in waters of the United States but actively producing in Canadian parts of Lake Erie.

The history of commercial use of substantial volumes of petroleum in the Great Lakes region goes back over a century. The first oil wells in this region were dug in 1858 near Oil Springs, Ontario, and drilled wells soon followed in 1860 (Malcolm, 1915). The earliest reported oil spill of major proportions occurred at Oil Springs in 1862 when overpressured strata were drilled into and the resulting uncontrollable flow of oil lasted for a week; some of this petroleum reached at least as far as Lake St. Clair (Caley, 1945). Since these first wells, others have been drilled into the Ordovician, Silurian, and Devonian reservoirs underlying much of the central Great Lakes area. These continue to produce petroleum and contribute to some of the large demands of the industrial, agricultural, and societal sectors for fuel, lubricants, and chemical feedstocks. Prior to the first wells, oil seeps were known to early European settlers and, even earlier, to pre-European Americans.

Petroleum is composed primarily of hydrocarbons and is considered a contaminant because of its significant differences from hydrocarbon mixtures made by organisms, even though petroleum is a natural substance. The most notable difference between petroleum hydrocarbons and those of plants and animals is the great diversity of molecular structures in petroleum. Included in this vast mixture are a number of homologous series of straight-chain, branched, cyclic, and aromatic hydrocarbons (Farrington and Meyers, 1975) which are the products of long-term geochemical transformations of the original, biological precursors of petroleum components. In contrast, biological hydrocarbons are rather limited in their variety and, in particular, do not commonly contain cyclic or aromatic structures (Douglas and Eglinton, 1966). Because of the differences between petroleum and biological hydrocarbons, organisms cannot tolerate certain petroleum components, even at trace levels. Such intolerance is especially true of aromatic hydrocarbons, which have lethal and sublethal effects on a wide range of aquatic organisms (Blumer, 1969).

This discussion of petroleum contaminants in the Great Lakes is divided into four parts—the problem of spilled petroleum, the extent of chronic petroleum contamination, the record of natural petroleum seeps, and the fate of petroleum contaminants—even though many of the processes involved cross these arbitrary divisions. Where appropriate, examples from marine studies are used to augument the limited information about petroleum in these freshwater bodies.

2. PETROLEUM SPILLS

In his review of petroleum transport on the Great Lakes by tankers and barges, Scher (1979) makes three important observations. First, the total volume of transported petroleum products has held more-or-less steady over the past two decades. Second, the spillage record of Great Lakes tankers and tankbarges has been a smaller percentage than of marine carriers. Third, the combination of the smaller volumes and sizes of the lakes as compared to the oceans and the absence of tidal flushing increases the potential impact of petroleum spills upon the Great Lakes environment.

In view of the continuing transport of petroleum on the Great Lakes and the ensuing potential for accidental spillage, it is important to discuss the fate of spilled oil. Predicting the fate of oil spilled into the Great Lakes environment is made difficult by several factors. One of these is that crude oil and its distillation fractions are mixtures of literally thousands of hydrocarbon and hydrocarbonlike compounds. These mixtures vary from oil to oil and with time within a single spilled oil, causing the physicochemical characteristics to change. A second problem is that the environmental conditions can be quite different from one part of the lakes to another and from one season to another. A third, perhaps most complicated problem is that a variety of disparate processes act upon spilled oil to alter its character and to disperse it. Some of the major processes active in marine spills have been summarized by Farrington (1980) and Jordan and Payne (1980) and include evaporation, dissolution, degradation, sedimentation, and incorporation into biota. The relative contribution of each of these processes to the eventual fate of an oil spill is not constant during the dispersion of a spill. In the case of light fuel oils and gasoline, evaporation and dissolution may be the major processes which disperse the spill. However, sedimentation to the bottom is the only one of these processes which has the potential of prolonging the impact of spilled petroleum while having the seemingly positive effect of removing this contaminant from the overlying water. For this reason, the sedimentation process will be examined in greater detail.

Incorporation of spilled oil into bottom sediments is the result of the complex interaction of sorption/desorption of oil components onto sinking particles, solubility of these oil components, downward transport of oiled particles, and possible dissolution/release from the bottom. None of the mechanisms has been adequately studied, yet important information exists about each.

Field data from sediment traps in the Bedford Basin, Nova Scotia, have shown the presence of petroleum hydrocarbons in silty sediments but the absence of discrete oil particles (Hargrave and Phillips, 1975), suggesting that petroleum residues can be concentrated on fine-sized sediment particles. Laboratory studies agree with this field observation. For example, Meyers and Quinn (1973) and Zürcher and Thuer (1978) demonstrate that clay minerals sorb up to one half the fuel oil from oil–water mixtures. Preferential uptake of less-soluble alkanes and higher-molecular-weight oil components was observed in both studies. Experiments show that removal of longchain alkanes from seawater is greater than that of shortchain alkanes or aromatic hydrocarbons (Meyers and Quinn, 1973; Meyers and Oas, 1978), indicating the significant role that hydrocarbon solubility or accomodation in water has in the sedimentation process.

The true solubilities of petroleum-type hydrocarbons in seawater are on the order of 1 ppb for alkanes (Button, 1976; Sutton and Calder, 1974) and of 10 ppb for aromatics (Boylan and Tripp, 1971). These decrease as molecular weights become larger within a homologous series of compounds. Complications exist, however, concerning the behavior of petroleum components in natural seawater. Dissolved organic matter appears to enhance the solubilities of alkanes in seawater but to have little effect on aromatic hydrocarbons (Boehm and Quinn, 1973). Furthermore, petroleum can exist in water as truly dissolved components, and as accomodated droplets of intact oil (Boehm and Quinn, 1974). These complications present real difficulties in predicting the response of a spilled oil to sedimentation in both oceanic and Great Lakes situations.

Malinky and Shaw (1979) find that sorption of *n*-decane and biphenyl onto glacial flour to be less than onto pure clay minerals. This finding agrees with the experiments done by Meyers and Quinn (1973) using Narragansett Bay sediment and implies that the mineral character of settling particles is another important factor in the sedimentation process which has been little studied. In addition, the presence of indigenous organic matter on natural sediments appears to decrease their sorbtive capacity towards petroleum mixtures (Meyers and Quinn, 1973), although it may actually enhance the sorption of polycyclic aromatic hydrocarbon components (Meyers and Quinn, 1974).

Transport of spilled oil to the bottom can take two major routes: settling of intact oil globules or sinking of sediment particles with associated oil. Either mechanism involves weathering of the oil, so that the sedimented oil can be quite different from initial spill material. As noted by Jordan and Payne (1980), the nonvolatile fraction of Kuwait crude has a specific gravity 20% higher than the total oil and virtually the same as that of water. Loss of volatile components, coupled with incorporation of a small amount of mineral matter, can increase the density of oil globules and tar balls to the point that they sink. Indeed, intact tar balls have been recovered from Gulf of Mexico sediments in over 2000 m of water (L. Jeffries, personal communication). Petroleum hydrocarbons have been found in suspended particulate matter in a number of coastal marine locations (Schultz and Quinn, 1977; Crisp *et al.*, 1979; Prahl et al., 1980) and in Lake Michigan (Meyers et al., 1981). It is clear that these petroleum components can be carried from the surface to the underlying sediments, yet the proportion of total spilled petroleum that can become incorporated into the bottom is not adequately known.

Estimates of the fraction of marine spills that become buried in the sea bottom vary widely. Jordan and Payne (1980) summarize some of these attempts at a mass balance. Estimates of the fraction of spilled oil that is sedimented range from a low of 1% in the case of an oil well blowout to 10% for the Amoco Cadiz spill, but up to half the oil cannot be accounted for. Gearing et al. (1979) estimate that between 7 and 16% of the fuel oil added to a controlled ecosystem study entered the bottom sediments, primarily as the lower-solubility components of the original material. In contrast, Knap and Williams (1982) report that 70% of the refinery effluent hydrocarbons added to laboratory enclosures rapidly became incorporated in sediments and that this was the major removal process from estuarine waters. The disagreement in these various mass balance estimates underscores the prediction problems that result from differences in types of spilled oil, in local environmental conditions, and in the approaches used by individual investigators.

That portion of an oil spill that becomes incorporated into the bottom can be buried for long periods of time. For instance, fuel oil spilled in Buzzard's Bay, Massachusetts, was present in sediment samples collected 5 years later (Teal and Farrington, 1977). Although *n*-alkanes had experienced degradation, branched and cyclic compounds evidently are resistant to breakdown after burial in sediments. This was earlier noted by Blumer and Sass (1972), who also observed that petroleum degradation under natural conditions proceeds slower than laboratory simulations might indicate. Indeed, Blumer et al. (1973) find that spilled petroleum is remarkably persistent in the coastal marine environment. However, resuspension of bottom sediments by bioturbation, storm stirring, or dredging has the potential of returning petroleum components to overlying waters as a form of chronic contamination, thereby extending the period of deleterious impact of an oil spill.

3. CHRONIC PETROLEUM CONTAMINATION

Although direct input of petroleum to the Great Lakes from spills is of great concern, it is probable that most petroleum contaminants enter the lakes from a multitude of small, indirect sources. Stormwater runoff, municipal sewage effluents, industrial wastewaters, and atmospheric inputs commonly contain low concentrations of petroleum hydrocarbons. These sources, essentially all from land, help maintain a low but chronic level of petroleum contamination in the Great Lakes. Evidence of these contaminants can be seen at the air–water surface, in the water column, and in sediments of the bottoms of the lakes.

In their study of the surface microlayer on Lake Michigan, Meyers and Owen (1980) analyzed the hydrocarbon contents of particulate matter of samples from the top 200 μm of the lake and from water 1 m below the surface. A significant feature in the hydrocarbon content of some of the samples is an unresolved complex mixture (UCM) of hydrocarbons underlying the resolved individual peaks on the chromatograms. This mixture is not found in hydrocarbons from biological sources and is a characteristic of petroleum-derived hydrocarbons (Farrington and Meyers, 1975). Hydrocarbon patterns having a large UCM similar to those found in samples from

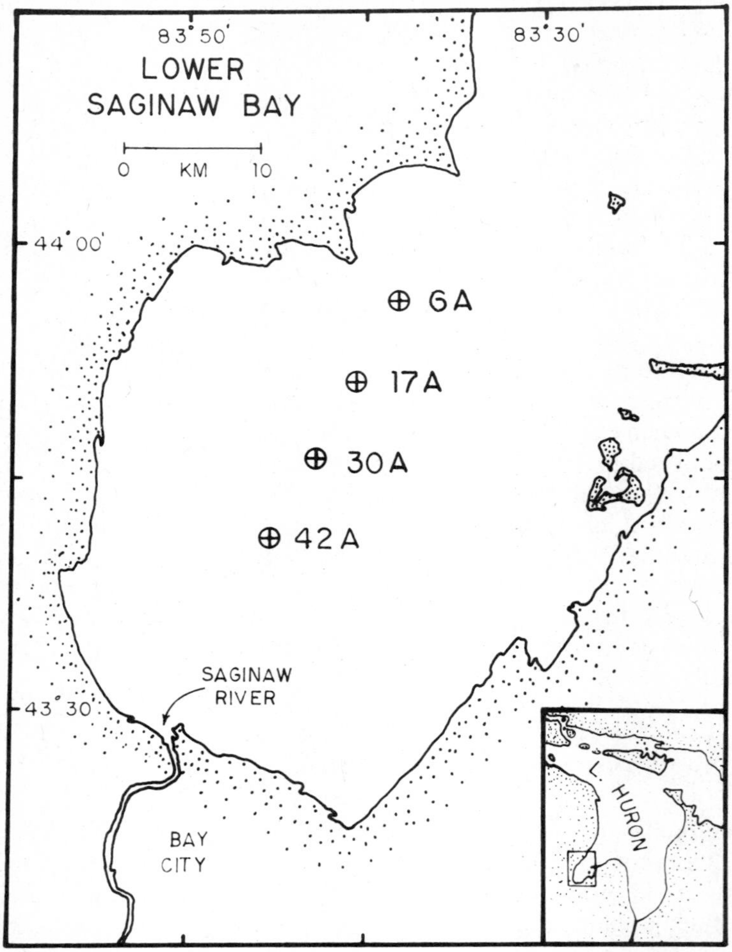

Figure 7.1 Locations of sediment samples obtained from Saginaw Bay in 1975 and studied by Meyers et al. (1980a) and Meyers and Takeuchi (1981).

the mouth of the Grand River have been reported in marine microlayers (Wade and Quinn, 1975; Daumas et al., 1976) and have been interpreted as being evidence of petroleum pollution.

The UCM in the Lake Michigan samples dominates several river plume and nearshore locations and differs in the microlayer samples from riverine, nearshore, and open lake environments. Furthermore, a concentration gradient exists in which the UCM contribution decreases with distance from the river mouth. These observations suggest that the UCM component of the particulate hydrocarbons originates from land or riverine sources and is transported to the lake by river flow. Once in the lake, these particulate materials evidently sediment out fairly quickly, probably in association with mineral particles.

Meyers and Owen (1980) note an interesting contrast in the hydrocarbon contents of samples collected from near Grand Haven and from St. Joseph several days later. The Grand Haven samples were collected after a period of heavy rain and contain a large UCM contribution. In contrast, the St. Joseph samples contain no UCM contribution and were sampled after a rainless period of over 24 hours. Wakeham (1977b) has concluded that most of the hydrocarbon input to Lake Washington is from urban stormwater runoff from roads and other paved surfaces. The contrast described here indicates that the major source of petroleum hydrocarbons to Lake Michigan is similarly riverborne land runoff.

Evidence of petroleum contaminants have been reported in sinking particulate material collected in sediment traps in Lake Michigan (Meyers et al., 1981). As further noted by Meyers et al. (in press), hydrocarbon distributions in sediment traps above the thermocline contain weathered petroleum, but near-bottom traps lack these hydrocarbon components. Because these hydrocarbons are relatively resistant to further degradation, their absence indicates that the particles with which they are associated do not sink into the deeper waters of Lake Michigan. In fact, sediments from the lake bottom directly under these sediment traps do not contain detectable amounts of petroleum contamination.

Sediments from most areas of the Great Lakes are generally not noticeably contaminated by petroleum. Hydrocarbon distributions of sediments from offshore areas of Lakes Huron and Michigan, for example, are made up of biological hydrocarbons (Meyers and Takeuchi, 1979; Meyers et al., 1980b; Leenheer, 1981). However, areas close to land sources of petroleum often show contamination. An example of such an area is Saginaw Bay, as shown by Meyers and Takeuchi (1979) and Meyers et al. (1980a).

In the sediments of Saginaw Bay, high concentrations of hydrocarbons resolvable by gas–liquid chromatography and of UCMs are evidence of substantial contributions of petroleum contaminants (Meyers and Takeuchi, 1979). Comparison of hydrocarbon components present in surficial sediments from four different locations in Saginaw Bay (Fig. 7.1) is given in Table 7.1. With increasing distance from the mouth of the Saginaw River, concentrations of total biogenic *n*-alkanes change little from a mean value of 24 ppm. However, UCM concentrations drop dramatically over the 25 km distance covered by these sampling locations. This decrease is best shown by the ratio of UCM to total *n*-alkanes, since this compensates for any variability in sediment character and organic matter content. The ratio drops over a factor of 8 between Station 42A, 16 km from the river mouth, and Station 6A, 41 km distant, yet remains higher than the value of 8 found in similarly textured sediments from the Goderich Basin of Lake Huron (Meyers and Takeuchi, 1979).

Table 7.1 Hydrocarbon Content of Saginaw Bay Sediments. Station Locations Shown in Fig. 7.1

Station	Distance from Saginaw River (km)	Total *n*-Alkanes[a] (ppm)	UCM[a] (ppm)	UCM / *n*-Alkanes
42A	16	26	2669	103
30A	25	23	608	26
17A	32	20	469	23
6A	41	28	349	12

[a] From Meyers and Takeuchi, 1979.

The relationship between distance from the mouth of the Saginaw River and the amount of petroleum contamination in sediments of Saginaw Bay is similar to findings from other lakes and from coastal marine locations. For example, in Lake Zug, Switzerland, the concentration and complexity of nearshore sediment hydrocarbons are highest close to cities and lowest near less populated shores (Giger et al., 1974). In some coastal marine areas, concentrations of petroleum hydrocarbons have been found to be higher close to rivers and appear to be related to the relative contribution of riverborne particles derived from various types of land runoff (Farrington and Quinn, 1973a,b; Gearing et al., 1976; Van Vleet and Quinn, 1978; Hurtt and Quinn, 1979).

A variety of sources may contribute to the input of petroleum hydrocarbons to Saginaw Bay. Direct inputs of fuel oil, lubricants, outboard motor fuel, and partially combusted fuels are likely from commercial and recreational vessels on the bay and the rivers flowing into it. Wastewater effluents have been shown to contain a significant concentration of petroleum-derived hydrocarbons (Farrington and Quinn, 1973b), mostly in association with suspended solids which can settle out and become incorporated into local sediments (Van Vleet and Quinn, 1977). In the case of Lake Washington, which presently receives no sewage effluent, Wakeham (1977a,b) has determined that urban stormwater runoff and bridge runoff account for 85% of the total hydrocarbon input. Rainfall and dustfall may add to this predominantly petroleum-type total in modern sediments of Lake Washington but probably not in an important degree to Saginaw Bay, inasmuch as Midwest rain generally contains few detectable petroleum hydrocarbons (Meyers and Hites, 1982). In view of these numerous potential sources of petroleum hydrocarbons, the materials in Saginaw Bay most likely are of multiple, anthropogenic origins, and reach the Bay largely from fluvial transport.

Major depth-related changes have been reported by Meyers et al. (1980a) and Meyers and Takeuchi (1981) in the character and concentration of hydrocarbons present in sediments collected at Station 30A in Saginaw Bay (Fig. 7.1). Some of these changes are summarized in Fig. 7.2, which shows the percent contributions of the UCM to total aliphatic hydrocarbon concentrations and of normal alkanes to total resolvable alkane concentrations, respectively. Except for the topmost sections, the UCM comprises more than 90% of the total above 35 cm, a depth corresponding to a depositional date of 1876. This contribution then decreases rapidly with depth but still amounts to about 40% at a depth of 60 cm in the sediments. A similar pattern has been found from a study of the hydrocarbon composition of sediments from Lake Washington (Wakeham, 1976; Wakeham and Carpenter, 1976). In addition, concentrations decrease with depth in the sediments of Narragansett Bay, Rhode Island, and give a historical record of the use of petroleum and its products in the surrounding urban and suburban regions (Van Vleet and Quinn, 1977; Hurtt and Quinn, 1979). Such distributional patterns record the advent and continued use of petroleum hydrocarbons beginning about a century ago, a conclusion strengthened by anomalously old radiocarbon ages of surfical sediment organic matter and by comparison of hydrocarbon distributions in Lake Washington to those in a pristine lake (Wakeham, 1976).

At the 35 cm depth in Saginaw Bay sediments (Fig. 7.2), the percentage of *n*-alkanes in the resolved hydrocarbon fraction changes drastically. Above

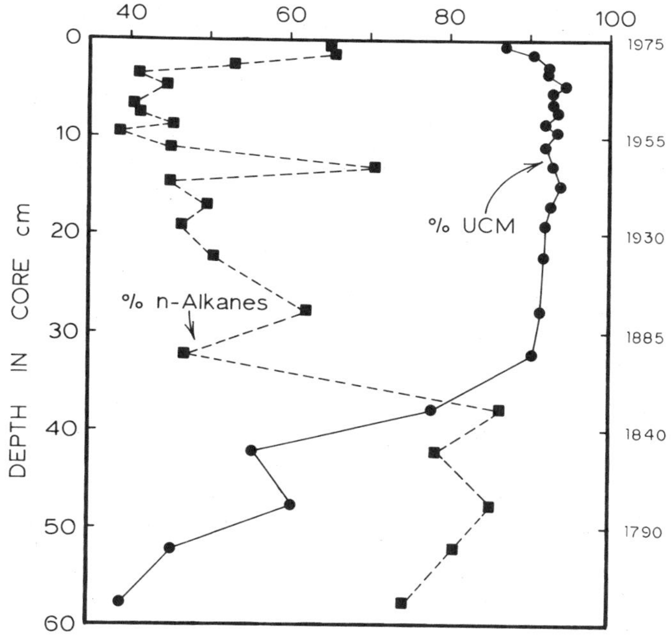

Figure 7.2 Evidence of petroleum hydrocarbons in Saginaw Bay sediments. Percentage of normal alkanes is of resolved hydrocarbon fraction and of unresolved complex mixture (UCM) is of total aliphatic hydrocarbon fraction of sediment extract. Deposition ages prior to 1876 (32 cm) extrapolated from ^{210}Pb measurements. After Meyers and Takeuchi (1981).

this depth nearly half the resolved alkanes are branched and cyclic compounds that are common in petroleum but minor components of biological hydrocarbons (Farrington and Meyers, 1975). Deeper in this core, straight-chain compounds more typical of biosynthesis (Douglas and Eglinton, 1966) are dominant. These changes in amount and type of hydrocarbons in Saginaw Bay sediments record the chronic input of petroleum contaminants beginning about 100 years ago and continuing to today.

4. NATURAL PETROLEUM SEEPS

Seeps occur where petroleum escapes from underground reservoirs through cracks and fissures which reach to the surface. In the Great Lakes area, petroleum seeps existed around the present towns of Oil Springs and Petrolia, Ontario (Fig. 7.3), prior to release of subsurface pressures by drilling. Documentation of underlake seeps is nonexistent, yet circumstantial evidence indicates that their presence is likely although their magnitude is prob-

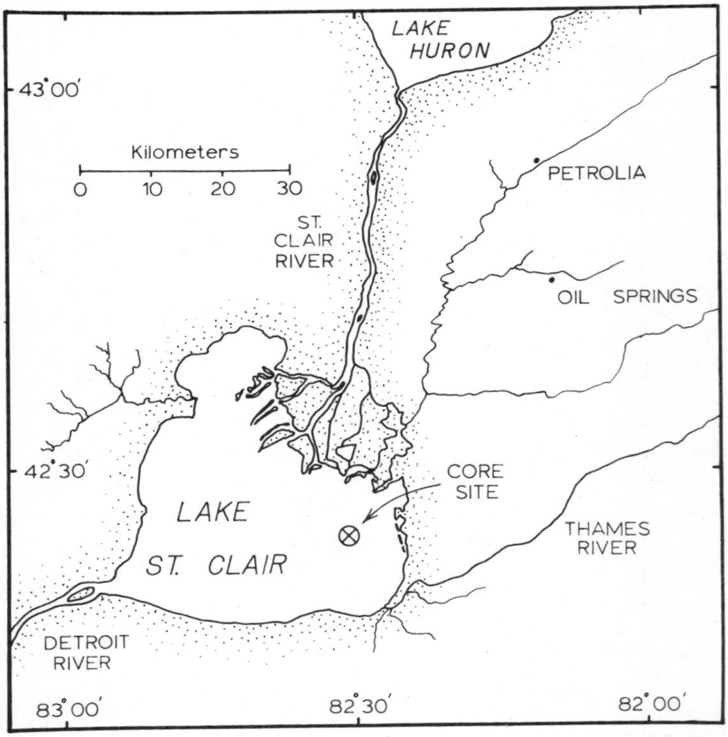

Figure 7.3 Locations of Oil Springs and Petrolia, Ontario, and of sediment core location sampled in 1978 in Lake St. Clair, by Moore and Meyers (1980).

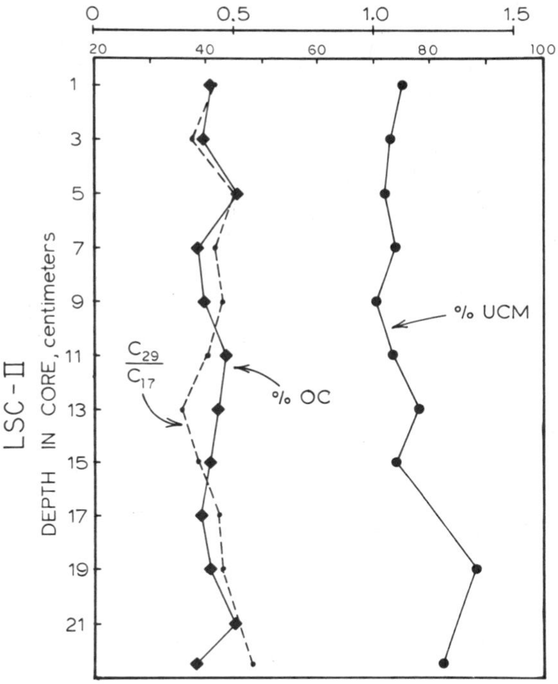

Figure 7.4 Selected organic matter constituents of sediment core from Lake St. Clair (Fig. 7.3). Percent organic carbon (% OC) and ratio of ^{29}C and ^{17}C *n*-alkanes given by upper scale. Percent unresolved complex mixture of hydrocarbons (% UCM) given by lower scale.

ably much less than the quite obvious seepage at such undersea locations as off Coal Oil Point near Santa Barbara, California (see Allen et al., 1970).

Sediment samples from a core obtained from Lake St. Clair (Fig. 7.3) contain petroleum hydrocarbons as indicated by distributions of *n*-alkanes and by the presence of an important UCM contribution to total hydrocarbons (Meyers and Moore, 1980). Some of the downcore patterns in organic matter characteristics of samples from a short core of sediment from Lake St. Clair are shown in Fig. 7.4. Throughout this core, the percent of total organic carbon is about 0.5, a surprisingly low value in view of the clay-size texture of these sediments, inasmuch as most fine-fixed sediments in the Great Lakes have organic carbon values in the range of 2–5% (Kemp, 1971; Meyers and Takeuchi, 1979; Rea et al., 1980; Johnson et al., 1982). The ratio of concentrations of land-derived ^{29}C *n*-alkane to aquatic ^{17}C *n*-alkane is also lower than is commonly found in postglacial Great Lakes sediments (Meyers et al., 1980b; Meyers et al., 1981; Leenheer and Meyers, 1983). However, the contribution of UCM hydrocarbons to the total hydrocarbon content remains between 70 and 90% throughout the core and is comparable to values present in petroleum-contaminated modern sediments from Saginaw Bay (Fig. 7.2).

The data in Fig. 7.4 present an interesting interpretive challenge which is best resolved by the hypothesis that the sediments of Lake St. Clair are glaciolacustrine deposits laid down under former Lakes Maumee and Whittlesey during glacial retreat between 12,500 and 14,000 years ago and that during deposition they incorporated petroleum hydrocarbons from seeps in the present Oil Springs area. Several lines of information point to this conclusion. First, the shallowness of Lake St. Clair (maximum depth 8 m) combined with strong wave turbulence and a high rate of discharge through the Detroit River would work against permanent deposition of modern sediments. Second, the sediment core obtained by Moore and Meyers (1980) closely resembles glaciolacustrine clay in its texture, in its organic carbon content, and in its $^{29}C/^{17}C$ *n*-alkane ratios (Meyers et al., 1980b). Third, radiocarbon dating of the top half of a sister core of Lake St. Clair sediment from the location shown in Fig. 7.3 gives an organic matter age of 24,170 ± 1540 years (Meyers and Leenheer, unpublished), which is several millenia older than is usually found in the top meter of Great Lakes sediments (Rea et al., 1980; Leenheer, 1981). Although none of these pieces of information conclusively proves that the sediments of Lake St. Clair are relict, their consistent agreement is persuasive.

The overall picture that emerges from the Lake St. Clair study is that the sediments contain evidence of a former underlake petroleum seep. The location of the seep was probably in the region of Oil Springs, Ontario, when this area was covered by one of the lakes that predate the present Great Lakes and were formed during times of glacial retreat. It is possible that the relatively fast removal of glacial overpressure on oil-bearing strata may have resulted in greater rates of seepage than in more recent times.

There is virtually no evidence of modern petroleum seeps under the Great Lakes. One sediment sample collected from the Goderich Basin flank of the Ipperwash Scarp in southern Lake Huron had an obvious petroliferous odor and may have contained petroleum seep hydrocarbons (Bourbonniere, personal communication). Because the Ipperwash Scarp is an underlake extension of the geologic structures underlying Oil Springs and Petrolia, Ontario, the presence of petroleum is likely and the possibility of underlake seepage exists.

5. FATE OF PETROLEUM IN THE GREAT LAKES

Many processes act to disperse, conceal, or destroy petroleum that enters aquatic environments, yet the very existence of this material is a reminder that it represents the relatively refractory residuum of organic matter produced by life in distant times. Because of its resistant character, petroleum does not degrade as readily as does the biosynthesized material from which it is originally derived. Evaporation, dissolution, dispersion, and sedimentation merely relocate petroleum within environments; petroleum components continue to be present. Nonetheless, degradation does evidently proceed via photochemical, oxidative, and enzymatic routes, although few of these are adequately understood.

As shown by the presence of petroleum-derived hydrocarbons in the sediments of Saginaw Bay (Fig. 7.2) and of Lake St. Clair (Fig. 7.4), petroleum can remain in the Great Lakes environment for hundreds and even thousands of years after being introduced. During these long periods of time, the opportunities for petroleum components and their degradation products to become incorporated into food webs are many. Hyland and Schneider (1975) review the lethal and sublethal effects of petroleum components on a wide variety of aquatic lifeforms. In view of the many deleterious reactions of organisms presented in their review, it seems that the probable persistent nature of even a small fraction of the petroleum spilled in the Great Lakes should be of greater concern than are attempts to estimate the contributions of various processes on the fate of petroleum.

The uptake of petroleum components by marine organisms is well documented, and similar uptake in the Great Lakes is reasonable to assume. In addition to lethal and sublethal effects, petroleum commonly imparts an oily off-taste to edible fish. Hence, petroleum contaminants can have a harmful, direct impact upon sports and commercial fishing in the Great Lakes.

ACKNOWLEDGMENTS

Professor Wayne E. Moore of Central Michigan University graciously provided the unpublished information about Lake St. Clair. I thank him for bringing this interesting problem to my attention. Some of the studies described in this chapter were performed with the valued participation of R. Bourbonniere, S. Edwards, O. Kawka, M. Leenheer, J. Mackin, and N. Takeuchi. I am grateful to them and to the captains and crews of RV *Laurentian*, RV *Mysis*, and RV *Shenehon* for their help. Support for these studies originated from various grants from the U.S. Environmental Protection Agency, the National Oceanographic and Atmospheric Administration, the National Science Foundation, the National Sea Grant Program, and the Petroleum Research Fund, administered by the American Chemical Society.

REFERENCES

Allen, A. A., R. S. Schlueter, and P. G. Mikolaj. (1970). Natural oil seepage at Coal Oil Point, Santa Barbara, California. *Science* **170:** 974–977.

Blumer, M. (1969). Oil pollution of the ocean. In: D. P. Hoult, Ed., *Oil on the sea*. Plenum Press, pp. 5–13.

Blumer, M. and J. Sass. (1972). Oil pollution: Persistence and degradation of spilled fuel oil. *Science* **176:** 1120–1122.

Blumer, M., M. Ehrhardt, and J. H. Jones. (1973). The environmental fate of stranded crude oil. *Deep-Sea Res.* **20:** 239–259.

Boehm, P. D. and J. G. Quinn. (1973). Solubilization of hydrocarbons by the dissolved organic matter in sea water. *Geochim. Cosmochim. Acta* **37:** 2459–2477.

Boehm, P. D. and J. G. Quinn. (1974). The solubility behaviour of No. 2 fuel oil in sea water. *Mar. Pollut. Bull.* **5:** 101–104.

Boylan, D. B. and B. W. Tripp. (1971). Determination of hydrocarbons in seawater extracts of crude oil and crude oil fractions. *Nature* **230:** 44–47.

Button, D. K. (1976). The influence of clay and bacteria on the concentration of dissolved hydrocarbon in saline solution. *Geochim. Cosmochim. Acta* **40:** 435–440.

Crisp, P. T., S. Brenner, M. I. Venkatesan, E. Ruch, and I. R. Kaplan. (1979). Organic chemical characterization of sediment-trap particulates from San Nicolas, Santa Barbara, Santa Monica and San Pedro Basins, California. *Geochim. Cosmochim. Acta* **43:** 1791–1801.

Daumas, R. A., P. L. Laborde, J. C. Marty, and A. Saliot. (1976). Influence of sampling method on the chemical composition of water surface film. *Limnol. Oceanogr.* **21:** 319–326.

Douglas, A. G. and G. Eglinton. (1966). The distribution of alkanes. In: T. Swain, Ed., *Comparative phytochemistry*. Academic Press, pp. 57–77.

Farrington, J. W. (1980). An overview of the biogeochemistry of fossil fuel hydrocarbons in the marine environment. In: L. Petralis and F. T. Weiss, Eds., *Petroleum in the marine environment*. Advances in Chemistry Series, No. 185, American Chemical Society, pp. 1–23.

Farrington, J. W. and P. A. Meyers, (1975). Hydrocarbons in the marine environment. In: G. Eglinton, Ed., *Environmental chemistry*, Vol. I. The Chemical Society, London, pp. 109–136.

Farrington, J. W. and J. G. Quinn. (1973a). Petroleum hydrocarbons in Narragansett Bay. I. Survey of hydrocarbons in sediments and clams (*Mercenaria mercenaria*). *Est. Coastal Mar. Sci.* **1:** 71–79.

Farrington, J. W. and J. G. Quinn. (1973b). Petroleum hydrocarbons and fatty acids in wastewater effluents. *J. Water Pollut. Control Fed.* **45:** 704–712.

Gearing, J. N., P. J. Gearing, T. Wade, J. G. Quinn, H. B. McCarty, J. W. Farrington, and R. F. Lee. (1979). The rates of transport and fates of petroleum hydrocarbons in a controlled marine ecosystem and a note on analytical variability. Proceedings of the 1979 Oil Spill Conference, Los Angeles, pp. 555–564.

Gearing, P., J. N. Gearing, T. F. Lytle, and J. S. Lytle. (1976). Hydrocarbons in 60 northeast Gulf of Mexico shelf sediments: a preliminary survey. *Geochim. Cosmochim. Acta* 40, 1005–1017.

Giger, W., M. Reinhard, C. Schaffner, and W. Stumm. (1974). Petroleum-derived and indigenous hydrocarbons in recent sediments of Lake Zug, Switzerland. *Environ. Sci. Technol.* **8:** 454–455.

Hargrave, B. T. and G. A. Phillips. (1975). Estimates of oil in aquatic sediments by fluorescence spectroscopy. *Environ. Pollut.* **8:** 193–215.

Hurtt, A. C. and J. G. Quinn. (1979). Distribution of hydrocarbons in Narragansett Bay sediment cores. *Environ. Sci. Technol.* **13:** 829–836.

Hyland, J. L. and E. D. Schneider. (1975). Petroleum hydrocarbons and their effects on marine organisms, populations, communities, and ecosystems. In: *Sources, effects and sinks of hydrocarbons in the aquatic environment*. American Institute of Biological Sciences, Washington, pp. 463–506.

Johnson, T. C., J. E. Evans, and S. J. Eisenreich. (1982). Total organic carbon in Lake Superior sediments: Comparisons with hemipelagic and pelagic marine environments. *Limnol. Oceanogr.* **27:** 481–491.

Jordan, R. E. and J. R. Payne. (1980). *Fate and weathering of petroleum spills in the marine environment*. Ann Arbor Science, Ann Arbor, 174 pp.

Kemp, A. L. W. (1971). Organic carbon and nitrogen in the surface sediments of Lakes Ontario, Erie, and Huron. *J. Sed. Petrol.* **41:** 537–548.

Knap, A. H. and P. J. LeB. Williams. (1982). Experimental studies to determine the fate of petroleum hydrocarbons from refinery effluent on an estuarine system. *Environ. Sci. Technol.*, **16:** 1–10.

Leenheer, M. J. (1981). Use of lipids as indicators of diagenetic and source-related changes in Holocene sediments. Ph.D. dissertation, The University of Michigan, 246 pp.

Leenheer, M. J. and P. A. Meyers. (1983). Comparison of lipid compositions in marine and lacustrine sediments. In: M. Bjorgy, Ed., *Advances in organic goechemistry 1981*. John Wiley, pp. 309–316.

Malinky, G. and D. G. Shaw. (1979). Modeling the association of petroleum hydrocarbons and sub-arctic sediments. Proceedings of the 1979 Oil Spill Conference, Los Angeles, pp. 621–624.

Meyers, P. A. and R. A. Hites. (1982). Extractable organic compounds in Midwest rain and snow. *Atmos. Environ.* **16:** 2169–2175.

Meyers, P. A. and W. E. Moore. (1980). Petroleum seep hydrocarbons in Lake St. Clair sediments. Geol. Soc. Am. North-Central Mtg, Abs. with Progr., p. 251.

Meyers, P. A. and T. G. Oas. (1978). Comparison of associations of different hydrocarbons with clay particles in simulated seawater. *Environ. Sci. Technol.* **12:** 934–937.

Meyers, P. A. and R. M. Owen (1980). Sources of fatty acids in Lake Michigan surface microlayers and subsurface waters. *Geophys. Res. Lett.* **7:** 885–888.

Meyers, P. A. and J. G. Quinn. (1973). Association of hydrocarbons and mineral particles in saline solution. *Nature* **244:** 23–24.

Meyers, P. A. and J. G. Quinn. (1974). Organic matter on clay minerals and marine sediments—effect on adsorption of dissolved copper, phosphate, and lipids from saline solutions. *Chem. Geol.* **13:** 63–68.

Meyers, P. A. and N. Takeuchi. (1979). Fatty acids and hydrocarbons in surficial sediments of Lake Huron. *Organic Geochem.* **1:** 127–138.

Meyers, P. A. and N. Takeuchi. (1981). Environmental changes in Saginaw Bay, Lake Huron, recorded by geolipids in sediments deposited since 1800. *Environ. Geol.* **3:** 257–266.

Meyers, P. A., N. Takeuchi, and J. A. Robbins. (1980a). Petroleum hydrocarbons in sediments of Saginaw Bay, Lake Huron. *J. Great Lakes Res.* **6:** 315–320.

Meyers, P. A., R. A. Bourbonniere, and N. Takeuchi. (1980b). Hydrocarbons and fatty acids in two cores of Lake Huron sediments. *Geochim. Cosmochim. Acta* **44:** 1215–1221.

Meyers, P. A., S. J. Edwards, and B. J. Eadie. (1981). Fatty acid and hydrocarbon content of settling sediments in Lake Michigan. *J. Great Lakes Res.* **6:** 331–332.

Meyers, P. A., M. J. Leenheer, B. J. Eadie, and S. J. Maule. Organic geochemistry of suspended and settling particulate matter in Lake Michigan. *Geochim. Cosmochim. Acta* In press.

Moore, W. E. and P. A. Meyers. (1980). Fingerprinting an 1862 oil spill into Lake St. Clair. 23rd Conference on Great Lakes Research, Program with Abstracts, p. 57.

Prahl, F. G., J. T. Bennett, and R. Carpenter. (1980). The early diagenesis of aliphatic hydrocarbons and organic matter in sedimentary particulates from Dabob Bay, Washington. *Geochim. Cosmochim. Acta* **44:** 1967–1976.

Rea, D. K., R. A. Bourbonniere, and P. A. Meyers. (1980). Southern Lake Michigan sediments: Changes in accumulation rate, mineralogy and organic content. *J. Great Lakes Res.* **6:** 321–330.

Schultz, D. M. and J. G. Quinn. (1977). Suspended material in Narragansett Bay: Fatty acid and hydrocarbon composition. *Organic Geochim.* **1:** 27–36.

Scher, R. M. (1979). Petroleum transport on the Great Lakes. Technical Report, MICHU-SG-79-217, Michigan Sea Grant Program, 80 pp.

Teal, J. M. and J. W. Farrington. (1977). A comparison of hydrocarbons in animals and their benthic habitats. *Rapp. P.-v. Reun. Cons. Int. Explor. Mer.* **171:** 79–83.

Van Vleet, E. S. and J. G. Quinn. (1977). Input and fate of petroleum hydrocarbons entering the Providence River and upper Narragansett Bay from wastewater effluents. *Environ. Sci. Technol.* **11:** 1086–1092.

Van Vleet, E. S. and J. G. Quinn. (1978). Contribution of chronic petroleum inputs to Narragansett Bay and Rhode Island Sound sediments. *J. Fish. Res. Board Can.* **35:** 536–543.

Wakeham, S. G. (1976). A comparative survey of petroleum hydrocarbons in lake sediments. *Mar. Pollut. Bull.* **7:** 206–211.

Wakeham, S. G. (1977a). A characterization of the sources of petroleum hydrocarbons in Lake Washington. *J. Water Pollut. Control Fed.* **49:** 1680–1687.

Wakeham, S. G. (1977b). Hydrocarbon budgets for Lake Washington. *Limnol. Oceanogr.* **22:** 952–957.

Wakeham, S. G. and R. Carpenter. (1976). Aliphatic hydrocarbons in sediments of Lake Washington. *Limnol. Oceanogr.* **21:** 711–723.

Zürcher, F. and M. Thuer. (1978). Rapid weathering processes of fuel oil in natural waters. Analysis and interpretations. *Environ. Sci. Technol.* **12:** 838–843.

8

TOXAPHENE IN THE GREAT LAKES

Clifford P. Rice
Marlene S. Evans

Great Lakes Research Division
The University of Michigan
Ann Arbor, Michigan 48109

1. INTRODUCTION

Toxaphene is a technical mixture of more than 177 polychlorinated diterpenes (Durkin et al., 1980). As a broad-spectrum insecticide, toxaphene is used on more than 168 agricultural commodities. However, its primary use is on cotton with extensive use also on soybeans, peanuts, and cattle (Korte et al., 1979). Toxaphene is used most extensively in the south and southeast United States (Carey et al., 1978, 1979) where most cotton farming is located. Toxaphenelike compounds are used worldwide as pesticides in such countries as Mexico, Romania, Hungary, Federal Republic of Germany, Poland, and the USSR (Ribick et al., 1982). Such compounds also are called camphechlor, polychlorocamphenes, and Strobane.

Although the current major application of toxaphene is as an insecticide, in the 1960s it had a second application. Fisheries managers in Canada and the United States used toxaphene as a fish poison to rid lakes of undesirable fish (Lee et al., 1977; Webb, 1980). However, this application was discontinued when researchers discovered that toxaphene was a persistent compound that sometimes prevented for years the successful restocking of treated lakes (Johnson et al., 1966).

Toxaphene has been produced as a pesticide since 1947 with more than 180 companies holding Federal Registers on this technical mixture: there are more than 817 registered toxaphene products (Toxaphene Working Group, 1977). When DDT was banned in the early 1970s, toxaphene replaced DDT as a major agricultural insecticide (Schmitt et al., 1981).

A large number of studies have investigated the environmental characteristics of toxaphene. Many investigations were conducted in the 1960s to determine the tolerance of organisms to this pesticide. The focus of these studies was the quantification of short-term tolerances to high toxaphene concentrations. Such information was required to investigate the implications of using toxaphene as a piscicide. Information also was needed on the environmental effects of toxaphene in water bodies surrounding areas of high toxaphene application, particularly cotton fields. These studies determined that toxaphene was a relatively toxic and persistent compound. More recent studies have determined that, in addition to being persistent, toxaphene is highly mobile. Zell and Ballschmiter (1980b) discovered that toxaphene is bioaccumulated by biota inhabiting regions hundreds and thousands of kilometers away from toxaphene usage. These observations and recent discoveries of toxaphene's carcinogenic and mutagenic properties (Reuber, 1975; Hooper et al., 1979) have further increased the concern for the worldwide environmental effects of toxaphene usage.

The Great Lakes have had a relatively long history of environmental degradation. In recent times, sport and commercial fisheries have been threatened because of unacceptably high concentrations of mirex, mercury, PCBs, and DDT in fish inhabiting certain regions of these lakes (IJC, 1978; Delfino, 1979). Given an extensive history of toxic organic contamination and the recent discoveries that toxaphene is a major atmospheric contaminant (Bidleman and Olney, 1975), Great Lakes researchers are becoming concerned that the lakes also may be contaminated by these polychlorinated diterpenes. Consequently researchers are initiating programs to quantify the extent of toxaphene contamination.

This chapter presents the most recent evidence for toxaphene occurrence in the Great Lakes and discusses probable routes by which toxaphene enters the lakes, its environmental pathways, and possible implications to human health. These discussions are based on an extensive review of the toxaphene literature. We also include a discussion of the physicochemical characteristics of this technical mixture and explain some of the difficulties in analyzing toxaphene in environmental samples. Researchers investigating the environmental characteristics of toxaphene have experienced difficulties in quantifying this complex mixture of more than 177 isomers and its various breakdown products. This has contributed to our relatively poor understanding of the environmental nature of toxaphene despite its broad and heavy usage and its persistence.

2. PHYSICAL AND CHEMICAL PROPERTIES OF TOXAPHENE

The movement of a chemically persistent compound such as toxaphene through various environmental compartments is strongly affected by three of its physicochemical properties: its vapor pressure, water solubility, and octanol–water partition coefficient. Bidleman and Christianson (1979) reported that toxaphene has a vapor pressure of 3×10^{-7} mm Hg at 20°C, which is similar to the vapor pressure of 2.5×10^{-7} mm Hg at 25°C for DDT (Atkins and Eggleton, 1971). Others report the vapor pressure of toxaphene to be 0.17–0.4 mm Hg at 25°C (Brooks, G. T. 1974). Korte et al. (1979) reported a vapor pressure of 10^{-6} mm Hg. Unlike DDT, which consists of one compound, toxaphene consists of more than 177 isomers, each of which have their own vapor pressures. In addition, toxaphene isomer composition varies with the manufacturer and probably between batches. This may account for the differing estimates of its vapor pressure.

Water solubility of toxaphene has been estimated to be 500 μg/liter (Guyer et al., 1971; Toxaphene Working Group, 1977) and Sanborn et al. (1976) reported a solubility of 400 μg/liter. However, Korte et al. (1979) reported a value of 1 mg/liter at room temperature. Differences in reported solubility may be due to differences in the toxaphene analyzed or the temperature at which the tests were conducted. Toxaphene is relatively soluble in water. By comparison, DDT has an estimated solubility at 25°C of only 0.2 μg/liter (O'Brien, 1967) whereas chlordane has a solubility of 56 μg/liter (Sanborn et al., 1976).

Toxaphene has an estimated log (octanol–water coefficient) of 6.44 (Magnuson et al., 1979). This is only slightly lower than the value of 6.93 for 1,1′, 4,4′-tetrachlorobiphenyl (a homoloque of PCB) and considerably higher than a value of 6.11 (3.76–6.11) for *p,p′*-DDT, 6.19 for *p,p′*-DDD, and 6.19 for *p,p′*-DDE. Sanborn et al., (1976) reported that toxaphene has a higher octanol–water partition coefficient than chlordane, another widely used insecticide.

Thus, toxaphene is a relatively volatile and water-soluble compound. Furthermore, because it has a relatively high octanol–water partition coefficient, it is very soluble in lipids and tends to accumulate in the various fats of organisms.

3. DIFFICULTIES IN ANALYZING FOR TOXAPHENE

Toxaphene is made up of a mixture of at least 177 compounds which vary in chlorine content (Holmstead et al., 1974). Discussions of the specific chemistry of this complex can be found in Nelson and Matsumura (1975a, b), Turner et al (1977), and Seiber et al. (1975).

Persistant chlorinated organics most commonly have been analyzed using packed-column chromatography to separate the various chemical fractions,

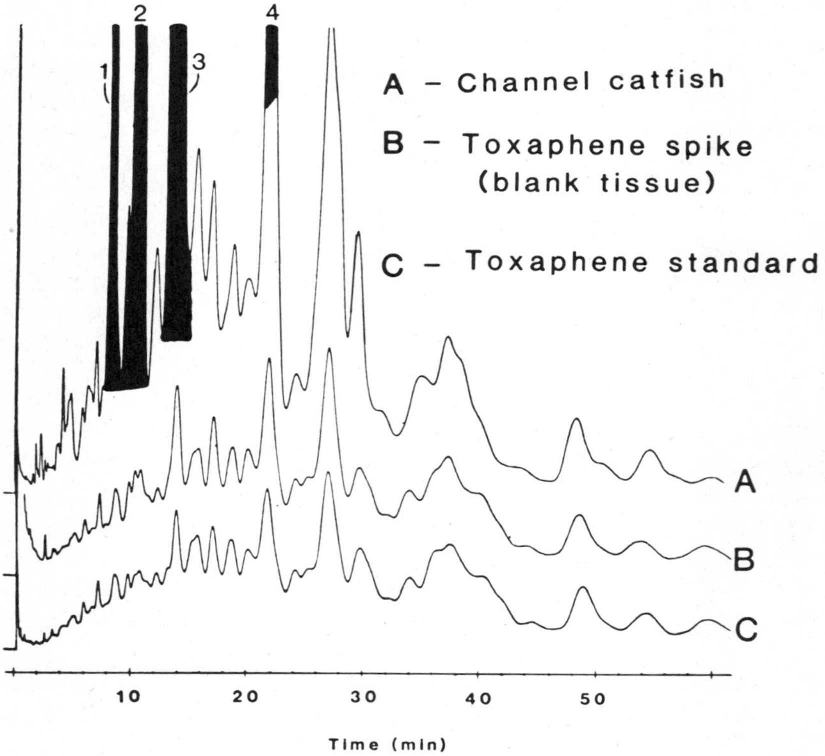

Figure 8.1 Packed-column chromatograms of technical toxaphene (chromatogram B and C) compared to natural residues of toxaphene and other pesticide impurities (peaks 1–4) in a channel catfish (*Icatalurus punctatus*) from the Arroyo Colorado floodway (chromatogram A). (From Ribick et al., 1982).

and electron capture detection which is highly sensitive and selective for chlorinated organic molecules. It is very difficult to analyze for toxaphene using this basic method. Electron capture sensitivity to toxaphene is less than for other chlorinated compounds such as, *p,p'*-DDT (McMahon, 1977). Furthermore, the more than 177 compounds comprising the toxaphene mixture are difficult to identify and quantify using a packed column which does not provide high resolution separation between the various compounds. In addition, many of the toxaphene components have long retention times and form late-eluting gas chromatographic (g/c) peaks (Fig. 8.1). These peaks are broad and are more difficult to distinguish above natural baseline drift. As a consequence of these factors, the detection limits for toxaphene are lower than for other compounds. Pollack and Kilgore (1978) reported a detection limit of 10 ppb for most of the organochlorine residues in estuarine molluscs, but a detection limit of 250 ppb for toxaphene. Klein and Link (1970) also reported difficulties in the simultaneous analysis for toxaphene and DDT. When PCBs and toxaphene co-occur in environmental samples, toxaphene peaks may be masked by PCB peaks creating a chemical interference (Swain et al., 1982).

Environmental samples may contain toxaphene that differs substantially in composition from the original parent mixture. Such differences are due to photolysis, chemical degradation, and the metabolic activities of living organisms (Korte et al., 1979). As a result of such processes, some toxaphene peaks are lost or reduced while others increase in height relative to that of the parent compound. In addition, certain of these breakdown products do not occur in the original formulation. Thus, it is difficult to identify peaks in environmental samples as corresponding to particular peaks when the parent mixture is run as a standard. As a consequence of these methodological difficulties, toxaphene has been extremely difficult to detect in the environment. It is only recently, with the advent of improved technology, that it has become possible to identify toxaphene in environmental samples at low concentrations and in the presence of other interfering contaminants.

Capillary rather than packed columns are now being used to separate the numerous toxaphene components (Zell and Ballschmitter, 1980a; Ribick et al., 1982). This method separates many more toxaphene peaks than packed-column chromatography and allows for a better matching of standard and environmental sample peaks (Fig. 8.2). Capillary-equipped negative chemical ion mass spectrometry is especially sensitive for toxaphene analysis, particularly when interference from chlordane and DDT occurs (Ribick et al., 1982). This method also allows partial structural elucidation. Ziranski et al. (1982) and Stalling et al. (1982) are independently investigating methods whereby complex capillary chromatograms produced from the analysis of environmental samples are matched against combinations of pure standards and partially weathered standards. The relative similarities and nonsimilarities in the resulting retention times, coupled with their matching peak intensities, are subjected to various statistical matching routines such as

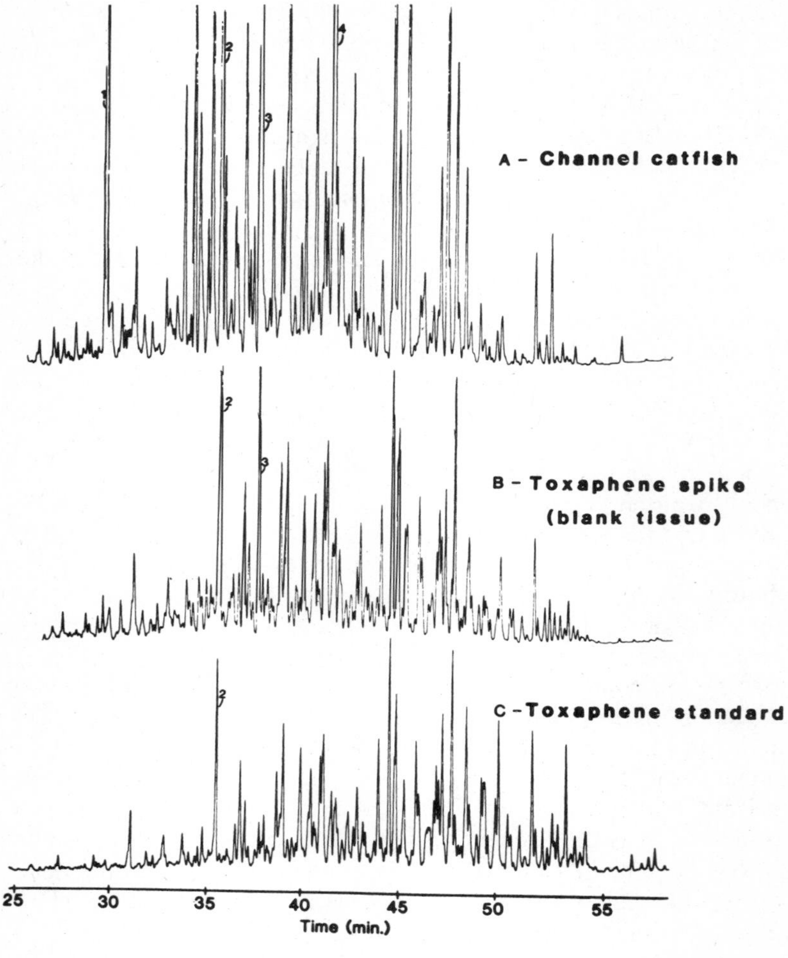

Figure 8.2 Capillary-column chromatograms of technical toxaphene (chromatograms B and C) compared to natural residues of toxaphene and other pesticide impurities (peaks 1–4) in a channel catfish (*Icatalurus punctatus*) from the Arroyo Colorado floodway (chromatogram A). (From Ribick et al., 1982).

SIMCA (Wold and Sjostrom, 1977) or an *n*-dimensional Euclidean distance routine (Ziranski et al., 1982). These results can be plotted to show consistent similarities between more- and less-weathered toxaphene residues.

4. TOXAPHENE IN THE GREAT LAKES

The earliest indication that toxaphene may represent a problem in the Great Lakes occurred when toxaphene was detected in fish. Rappe et al. (1979)

were the first to present this information to the scientific community. Subsequently the International Joint Commission (1981) added toxaphene to its list of chemicals found in the Great Lakes which may potentially adversely affect human health.

This section focuses on the historical toxaphene data base for the Great Lakes and highlights some of the difficulties encountered in its detection. Published data on toxaphene residues in the Great Lakes are very sparse, partially due to the recent discovery of the problems with toxaphene and because of historical difficulties in analyzing for the mixture. Much of the information is new and has not yet been published.

The Environmental Protection Agency is attempting to ban the use of toxaphene and has had it on its Rebuttable Presumption Against Registration (RPAR) list since May 1977. Toxaphene is now one of 34 current chemicals on an original list of 190 chemicals which is still pending determination (U.S. EPA, 1982). While a final decision has yet to be made, efforts continue to place restrictions on toxaphene use.* The discovery of toxaphene in pristine regions of the world (Zell and Ballshmiter, 1980b) and the recent confirmation of its presence in the Great Lakes suggest that research on toxaphene will continue to be an emerging and active need for the Great Lakes.

4.1. Air, Rain, Water, and Sediments

Very few data are available for toxaphene values in any of these Great Lakes environmental compartments. The International Joint Commission (1981) reported that toxaphene levels in Lake Superior and Lake Huron water range from 0.1 to 1.0 ng/liter. These values are slightly lower than the Swain et al. (1982) estimate of 1.6 ng/liter for the open waters of Lake Huron in 1980 and 1981. This level was 3.3 times higher than the 0.5 ng/liter level of PCBs which was measured in these samples.

We have measured toxaphene in air and rain samples along the eastern shore of Lake Michigan. Air concentrations ranged from non-detectable to 5.5 ng/m^3. One composite rain sample collected on the eastern shore of Lake Michigan (Bridgman, Michigan) contained 9.2 ng/liter. Swain et al. (1982) reported toxaphene concentrations of 7.3–108.3 ng/liter for samples collected in 1980 and 1981 along the shore of Lake Huron.

No published results for toxaphene in sediment are available. However, toxaphene has been detected in samples of Lake Superior sediment (S. Eisenreich, 1982, personal communication).

4.2. Plankton

A plankton sample collected at a station 16 km offshore of Grand Haven, Michigan, in May 1982 contained 85.0 ng/g dry weight (8.5 ng/g fresh weight

* In October of 1982, the U.S.E.P.A. banned the sale of toxaphene in the United States. Existing supplies can be used until 1986.

approximately). Some more recent data from this location gave values ranging from 510–560 ng/g dry weight for plankton and 190–432 ng/g dry weight for mysids (*mysis relicta*).

4.3. Fish

The presence of suggested toxaphene peaks was noted in 1974 by an analyst working for the United States Fish and Wildlife Service in Ann Arbor, Michigan. However, this was only confirmed 5 years later (Rappe et al., 1979).

Veith et al. (1977) reported the presence of "toxaphene-like" GC/MC (gas chromatograph–mass spectrometric) fragments in Lake Superior lake trout (*Salvelinus namaycush*) collected in 1973. They also reported "toxaphene-like components" in whole fish extracts of lake trout from Coppermine and Apostle Islands, Lake Superior, and in burbot from the Straits of Mackinac (joining Lakes Michigan and Huron) in 1975; values ranged from 0.1 to 1.0 μg/g (ppm). No "toxaphene-like" material was found in 1974 fish collections from Goderich, Lake Huron (IJC, 1977).

The National Pesticide Monitoring Program has been conducting toxaphene analyses of fish from nine sites in the Great Lakes since 1970. Table 8.1 shows the mean concentration of toxaphene by lake and by sampling interval for the 1970–1979 period. Mean concentrations were less than 0.25 μg/g for the 1970–1974 period and have increased in subsequent years. Such an increase in mean concentration of toxaphene in the early 1970s was also noted in other regions of the United States (Schmitt et al., 1981). Concentrations tended to be highest in Lake Michigan. Table 8.2 shows the residue concentrations by site, location, and time. The highest residue level (9.0 μg/g) was detected in lake trout (*Salvelinus namaycush*) collected in 1977 near Alpena, Lake Huron. The second highest value (7.80 μg/g) also was a lake trout sample; it was collected from Lake Michigan (near Sheboygan) in 1977. Highest toxaphene concentrations tended to occur in lake trout.

Table 8.1. Geometric Mean Concentration of Toxaphene in Fish Collected at Nine National Pesticide Monitoring Program Sites on the Great Lakes (Schmitt et al., 1983)

	Average Toxaphene Concentration by Sampling Period (μg/g)		
Location	1970–1974	1977	1979
Lake Michigan	0.22 (37)[a]	3.66 (12)	3.93 (9)
Lake Superior	0.05 (28)	2.35 (10)	1.92 (9)
Lake Huron	0.02 (24)	0.51 (9)	0.70 (6)

[a] Numbers in parentheses are the number of individual analytical results that were averaged.

Table 8.2. Toxaphene Residues in Fish from the U.S. Fish and Wildlife National Pesticides Monitoring Program Stations on the Great Lakes (1970–1979) (Schmitt et al., 1983)

Year	Species	Average Length (kg/cm)	Average Weight (kg/cm)	Percent Lipid	Toxaphene Concentration (μg/g wet wt.)
	Lake Michigan at Sheboygan, Wisconsin				
1970	One composite fish sample of two that were collected had measurable toxaphene. Both of these were reanalyses.				
	Bloater reanalyzed	28.4	0.3	29.9	2.30
1971	One composite fish sample of five that were collected had measurable toxaphene. Two of these were reanalyses.				
	Bloater reanalyzed	23.6	0.2	23.9	4.53
1972	Two composite fish samples of six that were collected had measurable toxaphene. Two of these were reanalyses.				
	Bloater	27.2	0.2	20.7	3.00
	Lake Trout	48.8	1.1	12.3	4.00
1973	No toxaphene was measured in six analyses of six composite fish samples; one was a re-analysis.				
1974	One composite fish sample of five which were collected had measurable toxaphene. Two of these were reanalyses.				
	Bloater reanalyzed	24.6	0.2	22.4	2.73
1977	Bloater	25.7	0.2	24.6	3.50
	Bloater	25.4	0.3	25.8	3.30
	reanalyzed	25.4	0.3	27.7	0.00
	Lake trout	69.8	3.0	15.3	7.80
	reanalyzed	69.8	3.0	12.0	3.50
	Lake trout	60.5	2.0	20.6	6.40
1979	Bloater	28.4	0.2	24.7	2.70
	Bloater	26.9	0.3	26.1	2.30
	Lake trout	61.2	2.2	17.8	7.10
	Lake Michigan at Saugatuck, Michigan				
1974	One composite fish sample of four that were collected had measurable toxaphene. Two of these were reanalyses.				
	Yellow perch reanalysis	27.7	0.3	10.0	0.02
1977	Bloater	29.0	0.3	23.9	3.70
	Bloater	28.7	0.3	7.1	0.59
	Lake trout	62.7	2.1	119.0	7.00
	reanalyzed	62.7	2.1	18.1	3.60

(Continued)

Table 8.2. (*Continued*)

Year	Species	Average Length (kg/cm)	Average Weight (kg/cm)	Percent Lipid	Toxaphene Concentration (μg/g wet wt.)
1979	Bloater	28.7	0.2	23.1	3.40
	Bloater	29.7	0.3	27.4	3.40
	Lake trout	66.3	2.7	18.6	6.80
	Lake Michigan at Beaver Island				
1977	Bloater	28.7	0.3	22.5	6.4
	Lake trout	65.3	2.7	23.5	8.10
1979	Bloater	30.7	0.3	28.9	3.30
	Bloater	28.2	0.0	28.2	3.10
	Lake trout	61.0	2.0	18.6	5.50
	Lake Huron (Saginaw Bay) at Bay Port, Michigan				
1970–1973	No toxaphene measured in 16 analyses of 14 composite fish samples. Two were reanalyses.				
1974	One composite fish sample of four that were collected had detectable toxaphene. One fish was a reanalysis.				
	Carp reanalysis	46.0	1.4	21.2	0.58
1977	Two composite fish of five analyzed had detectable toxaphene in them. One was a reanalysis.				
	Yellow perch	19.3	0.1	4.0	0.4
	Yellow perch	19.3	0.1	3.3	0.3
1979	Carp	47.	1.5	12.5	3.70
	Carp	48.4	1.1	8.8	0.20
	Lake Huron at Alpena, Michigan				
1974	No toxaphene measured in five analyses of four composite fish samples. One was a reanalysis.				
1977	Lake trout	59.4	2.0	25.9	9.0
	White sucker	48.3	1.3	8.4	0.5
	White sucker	41.4	0.9	6.6	0.5
	White sucker	32.3	0.4	4.6	0.00
1979	White sucker	32.5	0.5	5.5	0.00
	White sucker	34.5	0.6	5.4	0.30
	Yellow perch	23.6	0.2	6.2	1.50
	Lake Superior at Bayfield, Wisconsin				
1971–1974	No toxaphene measured in 23 analyses of 20 composite fish samples. Three were reanalyses.				

Table 8.2. (*Continued*)

Year	Species	Average Length (kg/cm)	Average Weight (kg/cm)	Percent Lipid	Toxaphene Concentration (μg/g wet wt.)
1977	Lake trout[a]	62.7	2.3	20.4	5.20
	reanalyzed	62.7	2.3	13.0	0.00
	reanalyzed	62.7	2.3	20.0	3.80
	Lake whitefish	51.6	1.3	10.6	2.3
1979	Lake trout	63.0	2.1	13.7	2.20
	reanalyzed	63.0	2.1	15.7	—
	Lake whitefish	33.5	0.3	4.7	0.90
	reanalyzed	33.5	0.3	4.7	—
	Lake whitefish	32.8	0.3	6.1	0.60
	Lake Superior at Keweenaw Point, Michigan				
1974	One composite fish sample of four collected had measurable toxaphene. One was a reanalysis.				
	Lake trout				
	reanalyzed	61.5	2.5	32.0	3.33
1977	Bloater	26.4	0.3	15.1	2.80
	Bloater	26.2	0.3	13.3	2.90
	Lake trout	55.9	2.0	16.2	3.00
1979	Bloater	27.7	0.2	19.3	3.60
	Bloater	29.5	0.3	14.7	2.80
	Lake trout	60.5	2.4	31.1	7.30
	Lake Superior at Whitefish Point, Michigan				
1977	Lake trout	60.7	2.3	22.4	3.10
	Lake whitefish	48.8	1.4	14.4	1.40
	Lake whitefish	50.3	1.3	13.5	2.10
1979	Lake trout	56.1	1.9	21.3	1.90
	Lake whitefish	56.6	1.6	9.7	1.20
	Lake whitefish	49.5	1.2	7.9	0.70

Concentrations also were high in bloaters (*Coregonus hoyi*). Lowest toxaphene concentrations were detected in yellow perch (*Perca flavescens*) and white suckers (*Catostomus commersonni*). Carp (*Cyprinus carpio*) also tended to contain relatively low toxaphene concentrations. The relatively low toxaphene residue levels in yellow perch, carp, and white suckers suggest that these fish accumulate less of this organochloride than lake trout and bloater. The fact that many of these less efficient toxaphene accumulators were included in with the Lake Huron average (Table 8.1) may explain the low relative level of toxaphene in the Lake Huron fish data.

Table 8.2 also contains a number of reanalysis values conducted as part of the Columbia National Fisheries Laboratory, United States Fish and Wildlife Service quality assurance program for the National Pesticide Monitoring Program. Differences in values reflect the inherent difficulties in quantifying toxaphene levels. Results generally are most reproducible when analyses are conducted in the same laboratory (e.g., comparable reanalyses results in Table 8.3). With further research into methodology for analyzing toxaphene, these difficulties should be minimized.

In Lake Michigan, lake trout toxaphene concentrations typically range from 5 to 10 μg/g (Schmitt et al., 1983). This is further shown in Table 8.3, which presents the results of toxaphene contamination studies conducted by the United States Fish and Wildlife Service, Ann Arbor. Values ranged from 6.1 to 10.9 μg/g. There was no apparent trend for toxaphene values to increase with time or to be consistently higher in one region of the lake than another.

Table 8.3. Concentration of Toxaphene-like Compounds in Composited Samples of Lake Michigan Lake Trout. Fish Collected and Analyzed by Great Lakes Fishery Laboratory, U.S. Fish and Wildlife Service (Data from R. Hesselberg, U.S. Fish and Wildlife Service, Ann Arbor, MI; personal communication)

Year	Species	Mean Length (mm)	Toxaphene Concentration (μg/g whole fish)
		Charlevoix, Michigan	
1977	Lake trout	697	7.2
	Lake trout reanalysis	697	6.1
1978	Lake trout	671	8.4
	Lake trout reanalysis	671	8.1
		Sturgeon Bay, Wisconsin	
1978	Lake trout	717	10.5
	Lake trout reanalysis	717	10.9
1979	Lake trout	688	6.5
	Lake trout reanalysis	688	6.9
		Saugatuck, Michigan	
1977	Lake trout[a]	653	6.6
	Lake trout[a] reanalysis	653	7.3
1978	Lake trout	701	7.7
	Lake trout reanalysis	701	7.4
1979	Lake trout	694	7.3
	Lake trout reanalysis	694	6.5

[a] Eighteen fish were composited, otherwise twenty fish were used.

Table 8.4. "Apparent" Toxaphene Concentration in Fall run Coho Salmon Collected in 1981 by Great Lakes National Program Office, U.S. EPA (DeVault et al., 1982).

	"Apparent" Toxaphene Concentration (μg/g)			
	Individual Composite[a] Result			
Sample Location	Sample 1	Sample 2	Sample 3	Average
Lake Michigan				
Kellogg Creek, IL	0.8	1.2	0.8	0.9
Trail Creek, IN	1.4	1.4	1.7	1.5
St. Joseph River, MI	1.0	1.4	1.6	1.3
Platt River MI	1.4	1.0	1.6	1.3
Grand Mean				1.3
Lake Superior				
Pine Creek, WI	0.6	T(0.13)[b]	0.4	0.4
Sheboygan River, WI	1.2	1.7	1.4	1.4
Grand Mean				0.9
Lake Huron				
Tawas River, MI	1.4	1.6	1.4	1.5
Grand Mean				1.5
Lake Ontario				
Spring Brook, NY	0.5	0.8	1.0	0.8
Grand Mean				0.8
Lake Erie				
Detroit River, MI	0.4	0.4	0.3	0.4
Huron River, OH	T(0.13)	T(0.13)	T(0.13)	0.13
Chagrin River, OH	T(0.13)	T(0.13)	T(0.13)	0.13
Trout Run Trib., PA[c]	ND(0)[d]	ND(0)	ND(0)	0.0
Grand Mean				0.2

[a] Composites of 15 fish fillets for each sample number (except Platt River where only 12 were collected).

[b] Detection limits for toxaphene were 0.25 μg/g. Therefore T (trace) was estimated to be ½ of this value.

[c] Trout Run Tributary fish were supplemented with 2-year-old fish, not 3-year-old, as were collected at all the other sites.

[d] ND = none detected. This was presumed to be equal to a zero concentration of "apparent" toxaphene.

The Great Lakes National Program Office, U.S. EPA, Chicago, has reported what it has identified as "apparent toxaphene" in 1980 fall run coho salmon (*Oncorhynchus kisutch*) from each of the Great Lakes (Devault et al., 1982). These data (Table 8.4) are based on fillets whereas the previous U.S. Fish and Wildlife Service data are on whole fish. Lakewide mean concentrations ranged from 0.21 μg/g for Lake Erie fillets to 1.5 μg/g for Lake

Huron fillets. Comparisons of these data with data presented in Tables 8.1 and 8.2 are meaningless because different locations were sampled, different species were analyzed, the fish were prepared differently, and different laboratories conducted the analyses. Thus, the toxaphene data base is not sufficiently strong to determine whether or not some species of fish are more likely to contain high concentrations of toxaphene than others or whether toxaphene contamination is a more serious problem in any particular lake.

Recent findings by the Environmental Protection Agency's Large Lakes Laboratory at Grosse Ile, Michigan, further illustrate the difficulties in quantifying PCBs and toxaphene when these pesticides co-occur in environmental samples. In addition, these studies suggest that some of the recent PCB data for environmental samples from the Great Lakes may be erroneously high as a result of reporting laboratories not separating interfering toxaphene components prior to their quantitation (Swain et al., 1982). Swain et al. (1982) cited an example where a Siskiwit Lake (a landlocked lake on Isle Royale, Lake Superior) lake trout (*Salvelinus namaycush*) collected in 1980 was analyzed by three laboratories for PCB, first by conventional packed-column separation which gave a mean total PCB concentration of 4.3 ppm, and secondly by using high-resolution capillary separation techniques which yielded a mean PCB concentration of 0.7 ppm, or approximately 16% of the original estimate. Furthermore, the peaks incorrectly identified as PCB were found to match many of the retention times of peaks in technical toxaphene standards. This discovery also appeared to solve an additional problem that the EPA analysts were experiencing with their time series data for PCB levels in Lake Superior fish. In most of the Great Lakes, PCB levels in fish were declining with time as a result of the curtailed use and discharge of PCBs by American and Canadian industries (Delfino, 1979). However, in Siskiwit Lake, PCB levels in fish apparently were increasing with time. In 1974, PCB levels in fish averaged 1.2 ppm, and by 1980 the mean had increased to 4.3 ppm. The puzzle apparently was solved when it was determined that approximately 84% of the apparent PCB in the 1980 samples resulted from the occurrence of toxaphenelike interfering contaminants. The most recent data (Swain, 1982) indicate that lake trout caught from Siskiwit Lake have toxaphene-derived residues which range from 1.7 to 4.5 ppm. In addition, there was a rather consistent ratio of toxaphene:PCB of 4:1.

4.4. Other Vertebrates

Toxaphene was reported to be present in eggs of waterfowl nesting in islands in northwestern Lake Michigan near Green Bay (Haseltine et al., 1981). Detectable values (Table 8.5) ranged from 0.05 μg/g for mallards (*Anas platyrhynchos*) in 1977 to 0.27 μg/g for red-breasted mergansers (*Mergus serrator*) in 1978. PCBs, DDE, and dieldrin were more frequently present in these eggs and at higher concentrations.

Table 8.5. Toxaphene Measured in Eggs of Waterfowl Nesting on Islands of Lake Michigan, U.S. Fish and Wildlife Service (Haseltine et al., 1981)

Species	Year	Number of Eggs Analyzed	Number >0.0	Toxaphene Concentration (μg/g) Range	Geometric Mean	PCB	DDE
Red-breasted merganser	1977	114	57	ND–0.89	0.14	20.0	7.4
Red-breasted merganser	1978	92	88	ND–0.64	0.27	19	7.6
Common merganser	1978	2	2	0.2–0.28	0.24	40	19
Mallard	1977	22	2	ND–0.12	0.05	2.0	0.89
Mallard	1978	5	ND	ND	ND	1.1	1.0
Gadwall	1977	4	ND	—	—	1.1	0.35
Gadwall	1978	5	ND	—	—	1.3	0.80
Black Duck	1977	3	ND	—	—	2.2	0.77

Heinz et al. (1980) also examined snakes on these islands. Toxaphene was not detected. Nor was it detected in nestling birds and earthworms, the major food items for the tested snakes.

5. SOURCES OF TOXAPHENE TO THE GREAT LAKES

In the United States, toxaphene is used most extensively in the southern and southeastern states. Alabama, Arkansas, Georgia, Louisiana, Mississippi, North Carolina, and South Carolina are the states reporting the most frequent and heaviest application (1.12–26.23 kg/ha) of toxaphene to agricultural areas. California and Illinois, while utilizing toxaphene, have lower percentages of sites (0.7–1.9% versus 1.12–26.23%) reporting toxaphene application (0.45–3.36 kg/ha). Toxaphene is applied most frequently to cotton (38.9% of sites), followed by soybeans (1.9%) (Carey et al., 1978). A representation of the 1972 usage patterns of toxaphene in the United States is displayed in Figure 8.3 (Von Rumker et al., 1974).

Toxaphene is used in a variety of formulations including wettable powders, emulsifiable concentrates, dust, granules, and baits (Korte et al., 1979). LaFleur et al. (1973) estimated a soil (Dunbar soil, South Carolina) half-life of 1–2 years although values as high as 10–12 years have been reported. There are several loss routes for toxaphene after it has been applied to crops. Some toxaphene leaches down to greater depths in the soil. LaFleur et al. (1973) reported that some toxaphene leached to 91 cm below the soil surface and was found in ground water a year after application. However, this represented only 5% of the toxaphene originally applied to the soil.

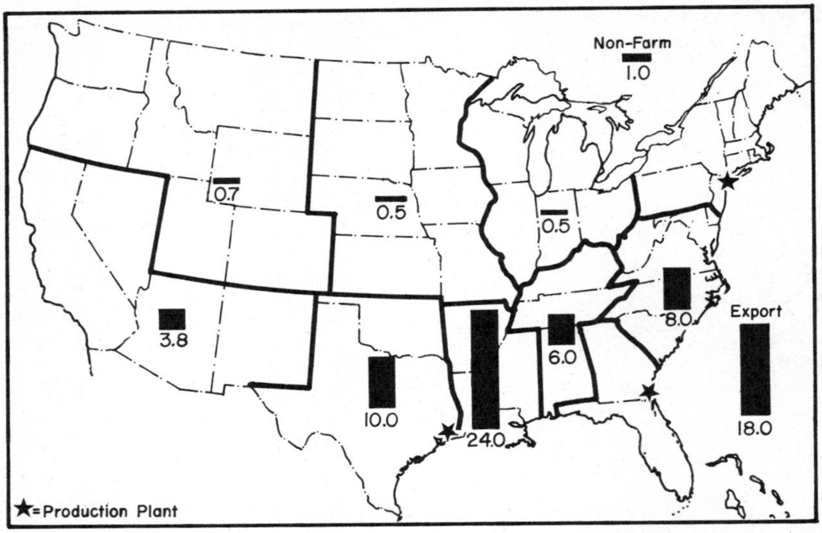

Figure 8.3 Geographical distribution of the usage of toxaphene (in millions of pounds) in 1972. (Adapted from Von Rumker et al., 1974).

Some of the toxaphene applied to soils undergoes degradation. Smith and Willis (1978) determined that soil degradation occurred rapidly under anaerobic conditions and that microbes either directly or indirectly stimulated this degradation. Much of the degradation involved the sequential disappearance of the slower-eluting, more highly chlorinated toxaphene compounds.

Some toxaphene is carried with runoff to drainage ditches and eventually to water bodies. In the case of the southern and southeastern states, such waters eventually drain into the Gulf of Mexico or Atlantic Ocean. McDowell et al. (1981) found that there was a linear relationship between sediment yield and toxaphene yield in a Mississippi delta site. Furthermore, 93% of the toxaphene was associated with the sediment fraction with the strongest association occurring with organic matter. These authors estimated a 16,000–670,000 toxaphene concentration factor between particulate organic matter and water. Bradley et al. (1972) also found that toxaphene was associated with the sediment fraction of runoff, although with a somewhat lower percentage (75%) than observed by McDowell et al. (1981); such differences may be related to differences in toxaphene partitioning between water and various soils (LaFleur, 1974). Bradley et al. (1972) also estimated that only 0.36% of the toxaphene applied to the study cotton plot was lost through runoff. While LaFleur et al. (1973) did not estimate loss due to runoff, they observed that after one year 17% of the applied toxaphene remained in the soil, more than 43% was apparently degraded, and approximately 40% was transported elsewhere. Volatilization of gaseous toxaphene and erosion of dust particles with sorbed toxaphene probably are the major transport routes for this insecticide (Nash et al., 1977; Seiber et al., 1979). Swoboda et al. (1971) found that less than 22% of the toxaphene applied to a Houston black clay over a 10-year period remained in the top 5 feet of soil and suggested that volatilization was the major loss route: they also found that toxaphene loss was greater than DDT (84% loss). However, Bradley et al. (1972) found that toxaphene was less persistant than DDT; these differences in the persistance of toxaphene and DDT were related to the greater water solubility and volatility of toxaphene.

For the Great Lakes, atmospheric deposition probably is the major source of toxaphene. Relatively little toxaphene is used in the drainage basin, and river runoff probably contributes negligible amounts of this insecticide; future research should address this question. Most of the toxaphene entering the Great Lakes probably originates from the south and southeast, where usage is heaviest, and is transported to the lakes by north and northwesterly winds. The presence of toxaphene in lake trout (*Salvelinus namaycush*) from Siskiwit Lake (Swain, 1982) provides further evidence that the atmosphere is the major source of toxaphene to the Great Lakes. This lake is far removed from the population centers of Lake Superior. Thus, the only apparent link to cultural and/or industrial activities attributable to man is through the atmosphere.

Atmospheric toxaphene deposition cannot be well quantified for the Great Lakes because of the small data set of air and rain toxaphene concentrations. Undoubtedly, toxaphene input varies seasonally. Arthur et al. (1976) investigated atmospheric levels of toxaphene in the Mississippi delta and found considerable seasonal variation. Levels increased in February and March as farmers began to till the cotton fields, and then peaked in September and October when toxaphene was applied to the crops. In 1974, concentrations ranged from 9.7 ng/m^3 (February) to 903.6 ng/m^3 (September). For the Great Lakes region, air concentrations ranged from less than detectable to a maximum of 5.5 ng/m^3. These values are considerably lower than in the atmosphere over the region of toxaphene application. These values are similar to Bidleman and Olney's (1974) estimates for atmospheric toxaphene concentrations over the North Atlantic in 1973 and 1974; these authors estimated concentrations ranging from 0.04 ng/m^3 to a maximum of 5.2 ng/m^3.

Toxaphene may be present in the atmosphere both in the particulate phase and in vapor form. Because toxaphene is relatively soluble in water, rain may be important in washing toxaphene from the atmosphere (Bidleman and Christensen, 1979). The relative role of dry and wet deposition needs to be evaluated. However, researchers have determined higher toxaphene concentrations in rain than in the air overlying the Great Lakes (see previous section).

Toxaphene undergoes considerable chemical, metabolic, and photolytic degradation which alters the composition of the parent compound and creates further difficulties in detecting the weathered product (Durkin et al.,

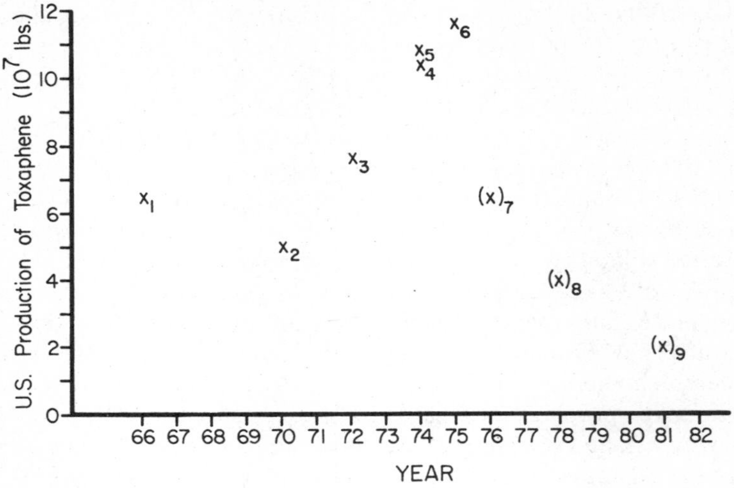

Figure 8.4 United States production of toxaphene from 1965 to 1981. Values (X) were estimated from data on usage. The references for these data are as follows: 1, 2, 3 Von Rumker et al., 1974; 4, 6 Toxaphene Working Group, 1977; 5 Durkin et al., 1980; 7 Ribick et al., 1982; 8, 9 Calvin Menzie, personal communication.

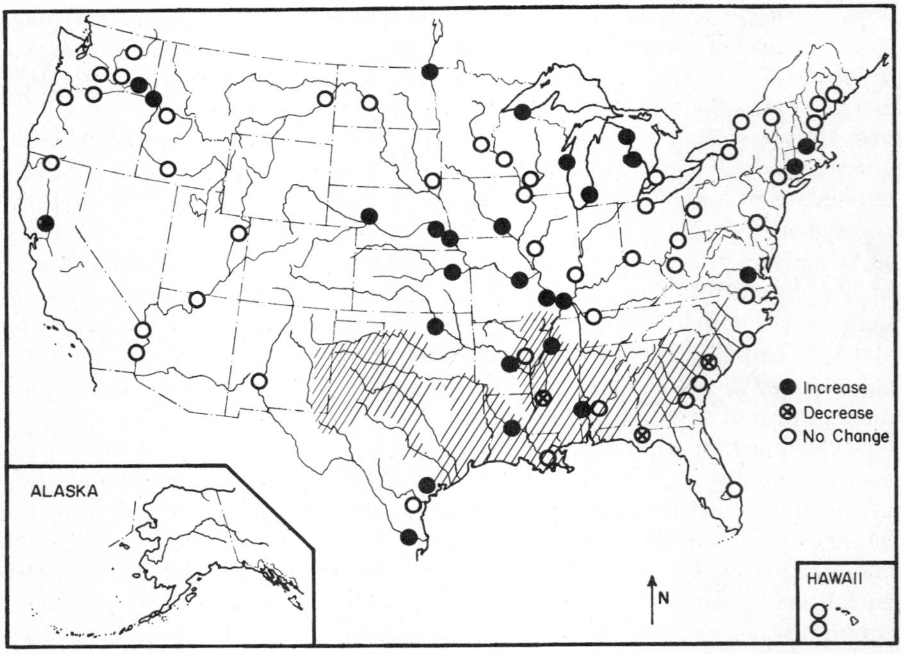

Figure 8.5 Locations of monitoring stations for National Pesticide Monitoring Program where changes in lipid-weight residues of toxaphene in fish are compared over time, 1974–1979. (Schmitt et al., 1983).

1980). This weathering has not been quantified for the Great Lakes. Its major characteristic probably is the loss of the more highly chlorinated and slower gas chromatograph (g/c) eluting compounds and an increase in the less highly chlorinated and faster g/c eluting compounds.

Annual atmospheric deposition of toxaphene to the Great Lakes undoubtedly has varied over the past several decades as toxaphene usage patterns have changed. More than 5.5×10^8 pounds of toxaphene were produced in the United States between 1947 and 1977. This is only slightly lower than the 1940–1974 DDT production figure of 3.3×10^9 (Woodwell et al., 1971). After DDT was banned, toxaphene usage increased and higher concentrations of this compound were detected in atmospheric samples (Arthur et al., 1976) and in fish collections (Schmitt et al., 1981). In recent years, toxaphene usage apparently has begun to decline. Figure 8.4 shows estimated United States toxaphene usage between 1965 and 1981. The values for 1976, 1978, and 1981 are production estimates based on agricultural toxaphene use figures that were available. Toxaphene production peaked in 1974 and has been declining since. This decline may be related to reduced effectiveness of the insecticide, fewer major boll weevil outbreaks, the introduction of new synthetic pyrethrins which are more effective in dealing with cotton pests (C. Menzie, 1982, personal communication), and the recent efforts by the Environmental Protection Agency to ban toxaphene.

Despite the fact that toxaphene use apparently is declining, it has become more widespread and concentrations in the biota have increased in many areas. Figure 8.5 shows the regions of most extensive toxaphene use and relative changes in toxaphene concentrations in fish over the last several years (Schmitt, et al., 1983). While fish toxaphene concentrations have decreased at a few sites between 1974 and 1979, a large number of sites have shown an increase. This includes the Great Lakes region, an area far removed from any major toxaphene source. Atmospheric transport undoubtedly is the major source of this contaminant.

6. TOXAPHENE TOXICITY TO AQUATIC ORGANISMS

Given the highly limited data on toxaphene in the Great Lakes, pathways and effects have not been adequately investigated. Nevertheless, the toxaphene literature is rich, allowing us to describe what we believe are the major toxaphene pathways and their potential effects on the biota.

Once toxaphene enters the Great Lakes, it has several fates. Some remains dissolved in the water. Measured concentrations range from 0.1 to 1.6 ng/liter and are considerably less than the estimated solubility of 400 μg/liter. Toxaphene then can be incorporated into the biota and move through the food web. For the biota, there are two basic questions. Are toxaphene concentrations in the water sufficiently high to exert lethal effects on the biota over short intervals (hours) of time? What are the long-term effects of sublethal toxaphene concentrations?

Toxaphene consists of a large number of compounds which differ in their relative toxicity (Landrum et al., 1976; Saleh et al., 1977) and sorption characteristics. A large number of studies have determined short-term toxicity values for a variety of plankton, invertebrates, and fish. The results of all these studies indicate that toxaphene concentrations (a few parts per trillion) in the dissolved phase of the water column does not have short-term lethal effects on the biota. Stadnyk et al. (1971) determined that concentrations of 0.1–1.0 mg/liter inhibited cell numbers of the algae *Scenedesmus guadricaudata*. Sanders and Cope (1966) determined a 48-hr LC_{50} of 15 μg/liter for the cladocerans *Daphnia pulex* and 19 μg/liter for *Simocephalus serrulatus*; toxaphene was more toxic to these organisms than chlordane and dieldrin. Naqvi and Ferguson (1968) tested a mixed assemblage of cyclopoid copepods and found that 58% of a population collected from a pesticide-free area were killed by a 48-hr exposure to 15 μg/liter toxaphene, while a similar assemblage collected from an area of heavy toxaphene use was more tolerant; after 48-hr only 28.3% of the animals exposed to 45 μg/liter were dead. Clams (*Eupera singleyi*) and snails (*Physa gyrina*) were approximately 10 times more resistant to toxaphene, and again, species inhabiting areas where toxaphene was applied heavily had higher resistances than similar species collected from nonagricultural areas. Naqvi and Ferguson (1970) determined that the 24-hr LC_{50} for the freshwater shrimp *Palaemontes kadiokensis* ranged from 20.9 to 229.0 μg/liter, with higher resistance associated with animals collected from cotton field areas. Schimmel et al. (1977) determined that several estuarine organisms exhibited a 96-hr LC_{50} ranging from 0.5 μg/liter (pin fish, *Lagodon rhombiodes*) to 4.4 μg/liter (grass shrimp, *P. pugio*). Given plankton and invertebrate toxaphene lethal tolerances of a few parts per million, it is highly unlikely that toxaphene water concentrations of a few parts per trillion are lethal to Great Lakes organisms.

Fish have toxaphene lethal tolerances similar to the previously described invertebrates and plankton. Henderson et al. (1959) determined a 48-hr LC_{50} of 7.5 μg/liter for fathead minnows (*Pimephales promelas*), 3.8 μg/liter for bluegills (*Lepomis macrochirus*), 6.8 μg/liter for goldfish (*Carassius auratus*), and 24 μg/liter for guppies (*Lebistes reticulatus*). Katz (1961) determined a similar 48-hr LC_{50} of 3.3 μg/liter for chinook salmon (*Oncorhynchus tshawytscha*), 10.5 μg/liter for coho salmon (*O. kisutch*), and 8.4 μg/liter for rainbow trout (*Salmo gairdneri*). Macek and McAllister (1970) showed a 96-hr LC_{50} ranging from 2 μg/liter for bass to 18 μg/liter for bluegill (*Lepomis macrochirus*). Tolerance to toxaphene decreases with increasing temperature (Macek et al., 1969). Macek (1975) also showed that toxaphene toxicity was enhanced in the presence of other pesticides (DDT, malathion, and parathion) when bluegill were used as the test organism. Fish inhabiting regions of heavy toxaphene use may have greater tolerances to toxaphene than fish collected in nonagricultural areas (Ferguson et al., 1964; Burke and Ferguson, 1969). The results of these fish studies suggest that fish inhabiting the Great Lakes are unlikely to experience significant immediate (hours) mortality from dissolved toxaphene concentrations.

Amphibians tend to have higher resistance to toxaphene than invertebrates and fish. Ferguson and Gilbert (1967) determined a 36-hr LC_{50} of 0.5–5.4 mg/ml for the northern cricket frog (*Acris crepitans*) and 0.6–50.0 mg/ml for the Fowler toad (*Bufo woodhousei fowleri*). In these tests, amphibians were placed on filters soaked with toxaphene. Amphibians with highest toxaphene tolerances were collected from drainage ditches surrounding cotton fields. Hall and Swineford (1980) conducted tests with larval and subadult southern leopard frogs (*Rana spehocephala*); frogs were maintained in toxaphene-containing water. Toxaphene was less toxic to subadults (0.378–0.790 mg/liter LC_{50}) than larvae (0.032–0.054 mg/liter). As amphibians have relatively high tolerances to toxaphene, this suggests that amphibians inhabiting the wetland regions of the Great Lakes are unlikely to experience short-term (hours) stress resulting from dissolved toxaphene concentrations.

Thus, given the low concentration of toxaphene dissolved in Great Lakes waters (a few parts per trillion) and chronic toxicity values ranging from a few parts per billion to a few parts per million, toxaphene is unlikely to have any immediate lethal effect on the Great Lakes aquatic community. Of more concern is the movement of toxaphene through the food web and its subsequent biological amplification.

A number of studies have documented toxaphene bioaccumulation. Leard et al. (1980) detected toxaphene concentrations of 0.22–7.68 μg/liter (ppb) in water samples and bivalve concentrations of 0.44–2.11 μg/g (ppm). Hughes et al. (1970) observed toxaphene values of 2.2 μg/liter in Ottman Lake water, a lake that had been treated with toxaphene to remove undesirable fish species. Plankton contained 317 μg/g toxaphene and fish contained 4.4 μg/g toxaphene, indicating significant bioaccumulation from the water. Johnson et al. (1966) also observed relatively high concentrations of toxaphene in aquatic plants, invertebrates, and fish collected from Wisconsin lakes that had been treated with toxaphene. For example, in Miller Lake, rainbow trout (*Salmo gairdneri*) bioaccumulated 10–12 μg/g toxaphene although the uptake route (diet, water) was not identified. In the laboratory, Sanborn et al. (1976) observed bioaccumulation factors of 890 for mosquito larvae (*Culex* sp.), 4247 for fish (*Gambusia* sp.), 6902 for algae (*Oedogonium*), and 9600 for snails (*Physa*) in their model ecosystem study. Schimmel et al. (1977) observed concentrations factors of 3,100–20,600 for fish (*Cyprindon variegatus, Lagodon rhomboides*) and oysters (*Crassostrea virginica*) and 400–1200 for shrimp (*Penaeus duorarum, Palaemonetes pugio*). Paris et al. (1977) observed that the freshwater fungus *Aspergillus* sp. bioaccumulated toxaphene by a factor of 1700, and two species of bacteria (*Flavobacterium harrisonii, Bacillus subtilis*) had concentration factors of 3400–5200: the algae *Chlorella pyrenoidosa* had a concentration factor of 1700. They also found that the microorganisms tested in their bioconcentration study tended to absorb relatively high amounts of the slower g/c eluting, less water soluble, more highly chlorinated toxaphene compounds.

Toxaphene thus has been shown to be bioaccumulated by a wide variety of organisms including marine and freshwater plants and animals. Thus it is not surprising that the limited data set for the Great Lakes presented earlier in this chapter confirms this trend. Although toxaphene concentrations for lake waters are in the parts per trillion range, the plankton sample contained toxaphene in the parts per billion range (8.5 ng/g wet weight) and fish generally contained toxaphene in the parts per million range. These data suggest that toxaphene undergoes significant bioaccumulation and trophic transfer within the Great Lakes ecosystem.

Long-term sublethal effects of toxaphene have been investigated for only a few aquatic organisms. Sanders (1980) determined that *Daphnia magna* fecundity was reduced by 21 day exposures to toxaphene levels of 0.12 μg/liter and higher. Midge (*Chironomus plumosus*) emergence was inhibited by toxaphene concentrations of 3.2 μg/liter and higher and scud (*Gammarus pseudolimnaeus*) growth was reduced by exposures to 0.25 μg/liter and higher. Conversely, Gardner and Miller (1981) determined that toxaphene concentrations of 0.5 mg/liter did not affect *Daphnia magna* excretion rates of amino acids and ammonia over a 4- to 6-hr period. A longer-term study may have shown more subtle effects.

Most of the research documenting long-term toxaphene effects on aquatic organism has been conducted on fish. Mehrle and Mayer (1975a) determined that toxaphene concentrations ranging from 39 to 502 ng/liter did not affect the hatchability of brook trout (*Salvelinus fontinalis*) eggs. However, all fry exposed to concentrations of 288 ng/liter and 502 ng/liter died within 60 days. Growth was significantly reduced at 139 and 288 ng/liter by 30 days after hatching and at all concentrations by 90 days after hatching. Whole-body and backbone collagen decreased and calcium and phosphorus backbone concentrations increased 30, 60, and 90 days after hatching for all exposure levels. Fifteen days after hatching, larvae had concentrated toxaphene from the water by a factor of 67,000–76,000. There was a slight decline after 15 days, possibly suggesting some metabolic breakdown of accumulated toxaphene, and then levels increased again.

Mehrle and Mayer (1975b) conducted a similar study using fathead minnows (*Pimephalis promelas*). Fish were exposed to 55–1230 ng/liter toxaphene for 150 days. Growth was reduced at all exposure levels by the end of the study. Backbone collagen content decreased, amino acid composition changed, and calcium concentrations increased at all exposure levels. This change in biochemical composition was evidenced by exposed fish exhibiting the "broken-back" syndrome. Such fish had relatively fragile backbones which were broken easily under stress (electric shock). After 150 days exposure, toxaphene residues ranged from 5.9 μg/g in fish exposed to the 55 ng/liter dose to 52 μg/g for the fish exposed to 621 ng/liter toxaphene.

In a later paper, Mayer et al. (1978) related changes in backbone composition of channel catfish (*Ictalurus punctatus*) to a toxaphene-induced vitamin C deficiency in bone. Vitamin C was hypothesized to have a major role in detoxifying toxaphene in the liver. When fish were exposed to high toxaphene doses, body vitamin C was used for detoxification and an insufficient amount remained for normal bone development.

Desaiah and Koch (1975) determined that toxaphene concentrations as low as 29 μM inhibited ATPase activity in channel catfish (*Ictalurus punctatus*) brain. Toxaphene had a lesser effect on kidney, liver, and gill ATPases.

These studies indicate that relatively low concentrations of toxaphene (a few ng/liter) can exert subtle, long-term effects on invertebrates and fish. They suggest that there is the potential for toxaphene to adversely affect Great Lakes biota either through the direct uptake of dissolved toxaphene or through dietary intake of organic matter containing relatively high toxaphene concentrations.

7. TOXAPHENE DEGRADATION

Aquatic organisms have the capability to metabolize sorbed toxaphene. In a model ecosystem study, Isensee et al. (1979) found after 32-day exposure to sublethal levels of toxaphene, that snails (*Helisoma* sp.) had metabolized a considerable amount of this mixture of compounds while bluegills (*Lepomis macrochirus*) apparently had metabolized little of the accumulated toxaphene. Schaper and Crowder (1976) found little evidence of toxaphene metabolism in mosquito fish (*Gambusia affinis*) after an 8-hr exposure to lethal concentrations of toxaphene. Mayer et al. (1978) provided evidence which suggests that channel catfish (*Ictalurus punctatus*) are able to metabolize toxaphene through vitamin-C-mediated processes, although the specific pathways were not elucidated. Hall and Swineford (1980) concluded from their southern leopard frog (*Rana sphenocephala*) study that amphibians have an ability to metabolically degrade toxaphene which is intermediate to that of fish (on the low end) and that of higher vertebrates which tend to show rapid breakdown of toxaphene.

In soil, toxaphene is metabolized under anaerobic and aerobic conditions. Smith and Willis (1978) found that toxaphene was degraded in Mississippi soil under anaerobic conditions: addition of alfalfa meal increased the rate of degradation suggesting that microbial activity was involved. The authors noted the sequential disappearance of the more highly chlorinated toxaphene compounds and hypothesized that an Fe^{2+} poryphyrin system was instrumental in this degradation which was most efficient at low electron potentials. Parr and Smith (1976) conducted a similar study, testing Crowley silt loam and obtained comparable results.

Williams and Bidleman (1978) found substantial toxaphene degradation in anoxic estuarine sediments. Again, there was a selective loss of the more highly chlorinated compounds and an increase in peak height of the faster g/c eluting, less highly chlorinated compounds. Williams and Bidleman (1978) concluded that a reductive chlorination mechanism was involved which required a Fe^{2+}/Fe^{3+} system and could occur in sterile and nonsterile sediments. They also reported that a rain sample collected near the estuary contained relatively high concentrations of toxaphene. Oysters collected in the estuary contained toxaphene that was similar to the parent compound; however, an altered toxaphene mixture was found in the sediments, suggesting that oysters were accumulating toxaphene from atmospheric deposition and not from the sediment and also that toxaphene was being degraded in the sediments. The authors also suggested that these degradation products had been overlooked when the sample was analyzed. Munson (1976) also noted differences in toxaphene fingerprint characteristics, in this case between a water sample (13 ng/liter) and a zooplankton sample (1.7 μg/g) and suggested that degradation had occurred.

Microbial degradation has been examined by a number of researchers. Weber and Rosenberg (1980) tested the bacteria *Vibrio* sp. and concluded that it was unable to metabolize toxaphene. Clark and Matsumura (1979) demonstrated that bacterial sediment toxaphene metabolism can occur under aerobic conditions. They tested the camphor-degrading pseudomonad *Pseudomonas putida* and found that anaerobic degradation favored dechlorination. No degradation occurred when aquatic sediments were autoclaved to kill the microbial community. The authors suggested that if toxaphene degradation occurred primarily through reductive processes, one would expect to see large numbers of recalcitrant residues. As such residues are not widely observed, they proposed that other degradation products must be present in the environment. They suggested that the first step in toxaphene degradation is initial dechlorination under anaerobic conditions, followed by oxidative action on the less chlorinated products under aerobic conditions.

Toxaphene can persist for long periods of times in lakes. Persistence appears to be longer in deep, oligotrophic lakes than in shallow, eutrophic lakes (Johnson et al., 1966) and longer in stratified than in well-mixed lakes (Veith and Lee, 1971). Hughes and Lee (1973) suggested that decreases in toxaphene concentration in the water column is strongly related to sorption on suspended matter that eventually reaches the sediments (Veith and Lee, 1971). Once on the sediments, toxaphene can be mixed to greater depths by the biological activity of benthic organisms, where it may undergo both anaerobic and aerobic degradation. Toxaphene concentrations within the sediment vary not only with depth but as a function of the concentration of organic matter (Gallagher et al., 1979). Lee et al. (1977) related detoxification of toxaphene applied to lakes as a piscicide to the partial degradation of sedimentary toxaphene.

Thus, for the Great Lakes, toxaphene is bioaccumulated by the biota, although this requires further quantification. Uptake of toxaphene will occur through the water and through dietary ingestion, although the relative roles of these two sources are unknown and undoubtedly vary with the taxonomic group. There is insufficient information to predict whether or not current toxaphene levels exert stress on the biota. Once the biota accumulate toxaphene, they will metabolize it to varying extents with the major processes involving the selective loss of the more highly chlorinated compounds. Toxaphene will be transported to the sediments with sinking particulates where it will undergo further degradation. As the less highly chlorinated compounds tend to be more water soluble, they may tend to move from the sediments back into the water column. However, Veith and Lee (1971) were unable to leach toxaphene from sediments into the water column under laboratory conditions. Nevertheless, movement of toxaphene from the sediments back into the water colum is likely either through food web processes (ingestion of benthic organisms by fish) or resuspension of sediments. Toxaphene may persist for longer periods of time in a deep, stratified, oligotrophic lake such as Superior than in a shallow, eutrophic lake such as Erie. However, there are not sufficient data to test these hypothesis.

8. OTHER CONSIDERATIONS

Toxaphene, in addition to potentially stressing the aquatic community, may have adverse impacts on waterfowl. The evidence for this is less clearcut. Klaas et al. (1980) did not detect toxaphene residues in the eggs of clapper rail (*Rallus longirostris*), common gallinule (*Gallinulla chloropas*), purple gallinule (*Porphyrula martinica*), and limpkin (*Aramus guarauna*) collected from eastern and southern United States from 1972 to 1974. However, Blus et al. (1977, 1979) detected toxaphene from the tissues and eggs of brown pelicans (*Pelecanus occidentalis*) from South Carolina (1969–1973) and Louisiana (1971–1976). Geometric mean concentrations in eggs ranged from 0.11 μg/g (1976) to 0.49 μg/g (1973). These values are only slightly lower than the range of toxaphene values (0.05–0.27 μg/g) reported by Haseltine et al. (1981) for waterfowl eggs from northwestern Lake Michigan. Toxaphene also has been detected in bald eagles (*Haliaeetus leucocephalus*) collected from various parts of the United States (Prouty et al., 1977).

Toxicity studies conducted on 1-day-old White Leghorn chicks for 32 weeks and fed food containing 5, 50, and 100 μg/g toxaphene exhibited sternal deformation; birds fed the 100 ppm dose had renal lesions (Bush et al., 1977). Poultry apparently concentrate chlorinated insecticides in their tissues to an extent several times greater than that for sheep, hogs, and cattle. The authors determined that 32-week-old chicks accumulated toxaphene in their tissues to a level comparable to that in their diet. Toxaphene concentrations decreased in older birds, suggesting that these chickens had a better developed microsomal enzyme system for metabolizing toxaphene. In addition, toxaphene apparently was dissipated faster from eggs and adipose tissue than was DDT or dieldrin.

Mehrle et al. (1979) fed black ducks (*Anas rubripes*) 0, 10, 20 μg/g toxaphene prior to egg laying and through the reproductive period. Reproduction and survival were not affected but ducklings exhibited reduced growth and impaired backbones; collagen concentration decreased and calcium concentration increased. Duckling carcasses contained only slightly lower toxaphene concentrations than in their diet, suggesting toxaphene metabolism.

Haseltine et al. (1980) conducted a 19-month study of black ducks (*Anas rubripes*), exposing them to 0, 10, 20, or 50 μg/g toxaphene in their diet. Carcasses contained toxaphene at 50–100% of the dietary level, suggesting that both young and adults were able to metabolize toxaphene. Toxaphene was found in the liver of all birds ingesting toxaphene. The authors suggested that mammals and birds apparently are able to metabolize toxaphene and tolerate ingestion for long periods of time without serious effect. Thus, for the Great Lakes, high concentration of toxaphene is more likely to be of environmental concern in fish than in birds. It is interesting to note, with the limited data set for toxaphene concentrations in the Great Lakes biota, that higher values were recorded in fish than in bird eggs.

The results of laboratory experiments using mammals as test animals suggest that humans risk affecting their health by ingesting toxaphene-contaminated fish. Chernoff and Carver (1976) studied the effects of toxaphene on fetal toxicity of rats and mice. Test animals were incubated daily with 15, 25, or 35 mg toxaphene/kg body weight·day. Rats exhibited dose-related reduction in maternal weight gain and fetal weight gain. In addition, dose-related reductions occurred in the average number of fetal sternal and caudal ossification centers. Mice also exhibited a dose-related reduction in maternal weight gain and an increase in the liver:body weight ratio. However, no effects were observed in fetal weight gain or number of caudal and sternal ossification centers.

Reuber (1975) reported the results of a study in which dogs were fed a daily dose of 5–50 mg toxaphene per kg body weight; death resulted from doses as low as 18 mg/kg. In subacute studies involving 4 mg/kg daily doses, histopathological changes were noted in the liver and kidney. The author concluded that the single daily dose of toxaphene had a cumulative effect and that the chemical was excreted or metabolized with difficulty.

Toxaphene effects on humans are poorly understood. Korte et al. (1979) reported an estimated acute oral toxicity dose of 60 mg toxaphene/kg body weight. Furthermore, they reported that toxaphene has seldom been detected in human tissues. This may be related to the rare occurrence of toxaphene in human food (see Carey et al., 1978) and/or to quick elimination of toxaphene. However, there is still an unknown health hazard from the ingestion of large amounts of fish contaminated with high (ppm) concentrations of toxaphene. The Toxaphene Working Group (1977) reported that chromosomal changes occurred in the leukocytes of women exposed to polychlorocamphene (toxaphene). In addition, they reported deviations from the norm in the menstrual cycle and estrogen levels of women exposed to toxaphene. Hooper et al. (1979) recently demonstrated that toxaphene has mutagenic properties, and Reuber (1975) summarized the results of several studies that demonstrated that toxaphene was carcinogenic to rats and mice. Durkin et al. (1980) discussed the health effects of toxaphene related to humans and laboratory mammals and concluded that the potential for carcinogenic effects was of major concern.

9. CONCLUDING REMARKS

This chapter has presented evidence to show that toxaphene is present in the Great Lakes ecosystem reaching concentrations as high as a 10 ppm in lake trout. The environmental behavior of toxaphene in the Great Lakes appears to be similar to that observed in other areas, that is, it is accumulated by the biota and is distributed throughout all environmental compartments. The atmosphere is the major transport route of toxaphene to the Great Lakes. The environmental effects of toxaphene on the Great Lakes ecosystem are not well understood nor are the probable implications to human health. However, because of toxaphene's known carcinogenic potential, it is vital that research continues on toxaphene in the Great Lakes. Such research should quantify toxaphene levels, transport routes, and environmental pathways. More research is required to improve our understanding of toxaphene inputs and losses (degradation, sedimentary burial, volatilization, photolysis). In addition, because of its chemical complexity and the historical problems in identifying and quantifying this mixture, more research must be conducted to characterize the chemical behavior and composition of toxaphene in the Great Lakes.

ACKNOWLEDGMENTS

This is contribution number 362 of the Great Lakes Research Division, University of Michigan. Support for preparation of this chapter was provided by Michigan Sea Grant.

REFERENCES

Arthur, R. D., J. D. Cain, and B. F. Barrentine. (1976). Atmospheric levels of pesticides in the Mississippi Delta. *Bull. Environ. Contam. Toxicol.* **15:** 129–134.

Atkins, D. H. F. and A. E. J. Eggleton. (1971). Studies of atmospheric washout and deposition of BHC, dieldrin, and *p,p'*-DDT using radiolabeled pesticides. Rep. SM/142a/32, p. 521, Int. Atomic Energy Agency, Vienna, Austria.

Bidleman, T. F. and E. J. Christensen. (1979). Atmospheric removal processes for high molecular weight organochlorines. *J. Geophys. Res.* **84:** 7857–7862.

Bidleman, T. F. and C. E. Olney. (1974). Chlorinated hydrocarbons in the Sargasso Sea atmosphere and surface water. *Science* **188:** 516–518.

Bidleman, T. F. and C. E. Olney. (1975). Long range transport of toxaphene insecticide in the atmosphere of the Western North Atlantic. *Nature* **257:** 475–477.

Blus, L. J., S. N. Burkett, Jr., T. G. Lamont,and B. Mulhern. (1977). Residues of organochlorines and heavy metals in tissues and eggs of brown pelicans, 1969–73. *Pest. Monit. J.* **11**(1): 40–53.

Blus, L., E. Cromartie, L. McNease and T. Joanen. (1979). Brown pelican: Population status, reproductive success, and organochlorine residues in Louisiana, 1971–1976. *Bull. Environ. Contam. Toxicol.* **22:** 128–135.

Bradley, J. R., Jr., T. J. Sheets, and M. D. Jackson. (1972). DDT and toxaphene movement in surface water from cotton plots. *J. Environ. Quality* **1:** 102–105.

Brooks, G. T. (1974). Chlorinated insecticides, Vol I: Technology and applications. Cleveland Ohio: CRC Press. 232 pp.

Burke, D. W. and D. E. Ferguson. (1969). Toxicities of four insecticides to resistant and susceptible mosquitofish in static and flowing solutions. *Mosquito News* **29:** 96–101.

Bush, P. B., J. T. Kiker, R. K. Page, N. H. Booth, and O. J. Fletcher. (1977). Effects of graded levels of toxaphene on poultry residue accumulation, egg production, shell quality, and hatchability in White Leghorns. *J. Agric. Food Chem.* **24**(4): 928–932.

Carey, A. E., J. A. Gowen, H. Tai, W. G. Mitchell, and G. B. Wiersma. (1978). Pesticide residue levels in soils and crops, 1971—National Soils Monitoring Program(III). *Pest. Monit. J.* **12**(3): 117–136.

Carey, A. E., P. Douglas, H. Tai, W. G. Mitchell, and G. B. Wiersma. (1979). Pesticide residue concentrations in soils of five United States cities, 1971—Urban Soils Monitoring Program. *Pest. Monit. J.* **13**(1): 17–22.

Chernoff, N. and B. D. Carver. (1976). Fetal toxicity of toxaphene in rats and mice. *Bull. Environ. Contam. Toxicol.* **15**(6): 660–664.

Clark, J. M. and F. Matsumura. (1979). Metabolism of toxaphene by aquatic sediment and a camphor degrading pseudomonad. *Arch. Environ. Contam. Toxicol.* **8:** 285–298.

Delfino, J. J. (1979). Toxic substances in the Great Lakes. *Environ. Sci. Technol.* **13**(12): 1462–1468.

Desaiah, D. and R. B. Koch. (1975). Toxaphene inhibition of ATPase activity in catfish, *Ictalurus punctatus*, tissues. *Bull. Environ. Contam. Toxicol.* **13**(12): 238–244.

DeVault, D., R. J. Bowden, J. C. Clark, and J. Weishaar. (1982). Results of contaminant analysis of fall run coho salmon, 1980. Presented at 25th Conference on Great Lakes Research, International Association for Great Lakes Research, Sault Ste. Marie, Ontario.

Durkin, P. R., P. H. Howard, J. Saxena, S. S. Lande, J. Santodonato, J. R. Strange, and D. H. Christopher. (1980). Reviews of the environmental effects of pollutants. X, Toxaphene. Oak Ridge National Lab. Health Effects Research Lab. Cincinnati, OH. Dept. of Energy Washington, D.C. Report no. EPA-600/1-79-044, 500 pp.

Ferguson, D. E. and C. C. Gilbert. (1967). Tolerances of three species of Anuran amphibians to five chlorinated hydrocarbon insecticides. *Miss. Acad. Sci.* **13:** 135–138.

Ferguson, D. E., D. D. Culley, W. D. Cotton, and R. P. Dodds. (1964). Resistance to chlorinated hydrocarbon insecticides in three species of freshwater fish. *Bioscience* **14:** 43–44.

Gallagher, J. L., S. E. Robinson, W. J. Pfeiffer, and D. M. Seliskar. (1979). Distribution and movement of toxaphene in anaerobic saline marsh soils. *Hydrobiology* **63**(1): 3–9.

Gardner, W. S. and W. H. Miller, III. (1981). Intracellular composition and net release rates of free amino acids in *Daphnia magna*. *Can. J. Fish. Aquat. Sci.* **38:** 157–162.

Guyer, G., P. Adkisson, K. DuBois, C. Menzie, and H. P. Nicholson. (1971). Toxaphene status report; Special report. U.S. EPA, Washington, D.C., Hazardous materials advisory committee. PB-251 576/5ST, 171 pp.

Hall, R. J. and D. Swineford. (1980). Toxic effects of endrin and toxaphene on the southern leopard frog *Rana sphenocephala*. *Environ. Pollut.* Ser. A. **23:** 53–65.

Haseltine, S. D., M. T. Finley, and E. Cromartie. (1980). Reproduction and residue accumulation in black ducks fed toxaphene. *Arch. Environ. Contam. Toxicol.* **9:** 461–471.

Haseltine, S. D., G. H. Heinz, W. L. Reichel, and J. F. Moore. (1981). Organochlorine and metal residues in eggs of waterfowl nesting on islands in Lake Michigan off Door County, Wisconsin, 1977–78. *Pest. Monit. J.* **15**(2): 90–97.

Heinz, G. H., S. D. Haseltine, R. J. Hall, and A. J. Krynitsky. (1980). Organochlorine and mercury residues in snakes from Pilot and Spider Islands, Lake Michigan—1978. *Bull. Environ. Contam. Toxicol.* **25:** 738–743.

Henderson, C., Q. H. Pickering, and C. M. Tarzwell. (1959). Relative toxicity of ten chlorinated hydrocarbon insecticides to four species of fish. *Trans. Am. Fish. Soc.* **88**(1): 23–32.

Holmstead, R. L., S. Khalifa, and J. E. Casida. (1974). Toxaphene composition analyzed by combined gas chromatography-chemical ionization mass spectrometry. *Agric. Food Chem.* **22**(6): 939–944.

Hooper, N. K., B. N. Ames, M. A. Saleh, and J. E. Casida. (1979). Toxaphene, a complex mixture of polychloroterpenes and a major insecticide, is mutagenic. *Science* **205**: 591–593.

Hughes, R. A. and G. F. Lee. (1973). Toxaphene accumulation in fish in lakes treated for rough fish control. *Environ. Sci. Technol.* **7**(10): 934–939.

Hughes, R. A., G. D. Veith, and G. F. Lee. (1970). Gas chromatographic analysis of toxaphene in natural waters, fish and lake sediments. *Water Res.* **4**: 547–558.

International Joint Commission. (1977). The waters of Lake Huron and Lake Superior. Vol. II (part B). International Joint Commission, 235 pp.

International Joint Commission. (1978). International reference group on Great Lakes pollution from land use activities. International Joint Commission, 115 pp.

International Joint Commission. (1981). Committee on the assessment of human health effects of Great Lakes water quality. International Joint Commission, 142 pp.

Isensee, A. R., G. E. Jones, J. A. McCann, and F. G. Pitcher. (1979). Toxicity and fate of nine toxaphene fractions in an aquatic model ecosystem. *J. Agric. Food Chem.* **27**(5): 1041–1046.

Johnson, W. D., G. F. Lee, and D. Spyridakis. (1966). Persistence of toxaphene in treated lakes. *Air Water Pollut. Int. J.* **10**: 555–560.

Katz, M. (1961). Acute toxicity of some organic insecticides to three species of salmonids and to the threespine stickleback. *Trans. Am. Fish. Soc.* **90**(1): 264–268.

Klaas, E. E., H. M. Ohlendorf, and E. Cromartie. (1980). Organochlorine residues and shell thicknesses in eggs of the clapper rail, common gallinule, purple gallinule, and limpkin (Class Aves), Eastern and Southern United States, 1972–74. *Pest. Mont. J.* **14**(3): 90–94.

Klein, A. K. and J. D. Link. (1970). Elimination of interferences in the determination of toxaphene residues. *J. Assoc. Anal. Chem.* **53**(3): 524–529.

Korte, F., I. Scheonert, and H. Parlar. (1979). Toxaphene (Camphlechlor), a special report. *Internat. Union Pure Appl. Chem.* **51**: 1583–1601.

LaFleur, K. S. (1974). Toxaphene–soil–solvent interactions. *Soil Sci.* **117**(4): 205–210.

LaFleur, K. S., G. A. Wojeck, and W. R. McCaskill. (1973). Movement of toxaphene and fluometuron through Dunbar soil to underlying ground water. *J. Environ. Quality* **2**(4): 515–518.

Landrum, P. F., G. A. Pollock, and J. N. Seiber. (1976). Toxaphene insecticide: Indentification and toxicity of a dihydrocamphene component. *Chemosphere* No. 2, pp. 63–69.

Leard, R. L., B. J. Grantham, and G. F. Pessoney. (1980). Use of selected freshwater bivalves for monitoring organochlorine pesticide residues in major Mississippi stream systems, 1972–73. *Pest. Monit. J.* **14**(2): 47–52.

Lee, G. F., R. A. Hughes, and G. D. Veith. (1977). Evidence for partial degradation of toxaphene in the aquatic environment. *Water Air Soil Pollut.* **8**: 479–484.

Macek, K. J. (1975). Acute toxicity of pesticide mixtures to bluegills. *Bull. Environ. Contam. Toxicol.* **14**(6): 648–652.

Macek, K. J. and W. A. McAllister. (1970). Insecticide susceptibility of some common fish family representatives. *Trans. Am. Fish. Soc.* No. 1, pp. 20–27.

Macek, K. J., C. Hutchinson, and O. B. Cope. (1969). The effects of temperature on the susceptibility of bluegills and rainbow trout to selected pesticides. *Bull. Environ. Contam. Toxicol.* **4**(3): 174–183.

Magnuson, V., D. Harriss, W. Maanum, and M. Fulton. (1979). ISHOW user's manual. Information system for hazardous organics in water. Department of Chemistry, Univ. of Minnesota, Duluth.

Mayer, F. L., P. M. Mehrle, and P. L. Crutcher. (1978). Interactions of toxaphene and vitamin C in channel catfish. *Trans. Am. Fish. Soc.* **107**(2): 326–333.

McDowell, L. L., G. H. Willis, C. E. Murphree, L. M. Southwick, and S. Smith. (1981). Toxaphene and sediment yields in runoff from a Mississippi delta watershed. *J. Environ. Quality* **10**(1): 120–125.

McMahon, B. (1977). Gas chromatograph response factors and relative retention times. Special addendum to: Pesticide analytical manual, Vol. I. Food and Drug Administration, Dept. of Health and Human Services, Wash., D.C., 1968.

Mehrle, P. M. and F. L. Mayer, Jr. (1975a). Toxaphene effects on growth and development of brook trout (*Salvelinus fontinalis*). *J. Fish. Res. Board Can.* **32:** 609–613.

Mehrle, P. M. and F. L. Mayer, Jr. (1975b). Toxaphene effects on growth and bone composition of fathead minnows, *Pimephales promelas*. *J. Fish. Res. Board Can.* **32:** 593–598.

Mehrle, P. M., M. T. Finley, J. L. Ludke, F. L. Mayer, and T. E. Kaiser. (1979). Bone development in black ducks as affected by dietary toxaphene. *Pest. Biochem. Physiol.* **10:** 168–173.

Munson, T. O. (1976). A note on toxaphene in environmental samples from the Chesapeake Bay region. *Bull. Environ. Contam. Toxicol.* **16**(4): 491–494.

Naqvi, S. M. and D. E. Ferguson. (1968). Pesticide tolerances of selected freshwater invertebrates. *Miss. Acad. Sci.* **14:** 121–127.

Naqvi, S. M. and D. E. Ferguson. (1970). Levels of insecticide resistance in fresh-water shrimp. *Palaemonetes kadiakensis*. *Trans. Am. Fish. Soc.* **4:** 696–699.

Nash, R. G., M. L. Beall, Jr., and W. G. Harris. (1977). Toxaphene and 1,1,1-trichloro-2,2-*bis*(*p*-chlorophenyl)ethane (DDT) losses from cotton in an agroecosystem chamber. *J. Agric. Food Chem.* **25**(2): 336–341.

Nelson, J. O. and F. Matsumura. (1975a). Separation and comparative toxicity of toxaphene components. *J. Agric. Food Chem.* **23**(5): 984–990.

Nelson, J. O. and F. Matsumura. (1975b). A simplified approach to studies of toxic toxaphene components. *Bull. Environ. Contam. Toxicol.* **13**(4): 464–470.

O'Brien, R. D. (1967). Insecticides action and metabolism. Academic Press, New York and London, 332 pp.

Paris, D. F., D. L. Lewis, and J. T. Barnett. (1977). Bioconcentration of toxaphene by microorganisms. *Bull. Environ. Contam. Toxicol.* **17**(5): 564–571.

Parr, J. F. and S. Smith. (1976). Degradation of toxaphene in selected anaerobic soil environments. *Soil Sci.* **121**(1): 52–57.

Pollock, G. A. and W. W. Kilgore. (1978). Toxaphene. *Residue Rev.* **50:** 87–140.

Prouty, R. M., W. L. Reichel, L. N. Locke, A. A. Belisle, E. Cromartie, T. E. Kaiser, T. G. Lamont, B. M. Mulhern, and D. M. Swineford. (1977). Residues of organochlorine pesticides and polychlorinated biphenyls and autopsy data for bald eagles, 1973–74. *Pest. Monit. J.* **11**(3): 134–137.

Rappe, C., D. L. Stalling, M. Ribick, and G. Dubay. (1979). Identification of chlorinated 'toxaphene-like' compounds in Baltic seal fat and Lake Michigan fish extracts by CI-GC/MS. 177th National Meeting of the American Chemical Society.

Reuber, M. D. (1975). Carcinogenicity of toxaphene: A review. *J. Toxicol. Environ. Health* **5:** 729–748.

Ribick, M. A., G. R. Dubay, J. D. Petty, D. L. Stalling, and C. J. Schmitt. (1982). Toxaphene residues in fish: Identification, quantification, and confirmation at part per billion levels. *Environ. Sci. Technol.* **16**(6): 310–318.

Saleh, M. A., W. V. Turner, and J. E. Casida. (1977). Polychlorobornane components of toxaphene: Structure–toxicity relations and metabolic reductive dechlorination. *Science* **198:** 1256–1258.

Sanborn, J. R., R. L. Metcalf, W. N. Bruce, and P. Lu. (1976). The fate of chlordane and toxaphene in a terrestrial–aquatic model ecosystem. *Environ. Entomol.* **5:** 533–538.

Sanders, H. O. (1980). Sublethal effects of toxaphene on daphnids, scuds, and midges. USEPA-600/3-80-006.

Sanders, H. O. and O. B. Cope. (1966). Toxicities of several pesticides to two species of cladocerans. *Trans. Am. Fish. Soc.* **95**(2): 165–169.

Schaper, R. A. and L. A. Crowder. (1976). Uptake of ^{36}Cl-toxaphene in mosquito fish, *Gambusia affinis. Bull. Environ. Contam. Toxicol.* **15**(5): 581–587.

Schimmel, S. C., J. M. Patrick, Jr., and J. Forester. (1977). Uptake and toxicity of toxaphene in several estuarine organisms. *Arch. Environ. Contam.* **5:** 353–367.

Schmitt, C. J., J. L. Ludke, and D. F. Walsh. (1981). Organochlorine residues in fish: National Pesticide Monitoring Program. *Pest. Monit. J.* **14:** 136–206.

Schmitt, C. J., M. A. Ribick, J. L. Ludke, and T. W. May. (1983). Organochlorine residues in freshwater fish, 1976–1979: National Pesticide Monitoring Program. U. S. Fish and Wildlife Service, Resource Publication No. 152.

Seiber, J. N., P. F. Landrun, S. C. Madden, K. D. Nugent, and W. L. Winterlin. (1975). Isolation and gas chromatographic characterization of some toxaphene components. *J. Chromatogr.* **114:** 361–368.

Seiber, J. M., S. C. Madden, M. M. McChesney, and W. L. Winterlin. (1979). Toxaphene dissipation from treated cotton field environments: Component residual behavior on leaves and in air, soil, and sediments determined by capillary gas chromatography. *J. Agric. Food Chem.* **27**(2): 284–291.

Smith, S. and G. H. Willis. (1978). Disappearance of residual toxaphene in a Mississippi delta soil. *Soil Sci.* **126**(2): 87–93.

Stadnyk, L., R. S. Campbell, and B. T. Johnson. (1971). Pesticide effect on growth and ^{14}C assimilation in a freshwater alga. *Bull. Environ. Contam. Toxicol.* **6**(1): 1–8.

Stalling, D. L., M. Ribick, S. Wold, and E. Jonansson. (1982). Characterization of toxaphene residues in fish samples analyzed by capillary gas chromatography using SIMCA, multivariate analysis. Abstracts American Chemical Society, Kansas City, Mo.

Swain, W. R. (1982). An overview of contaminants in the Lake Superior ecosystem. Presented at the 25th Conference on Great Lakes Research, International Association for Great Lakes Research, Sault Ste. Marie, Ontario.

Swain, W. R., M. D. Mullin, and J. C. Filkins. (1982). Refined analysis of residue forming organic substances in lake trout from the vicinity of Isle Royale, Lake Superior. 25th Conference on Great Lakes Research, International Association for Great Lakes Research, Sault Ste. Marie, Ontario.

Swoboda, A. R., G. W. Thomas, F. B. Cady, F. W. Braid, and W. G. Knisel. (1971). Distribution of DDT and toxaphene in Houston Black clay on three watersheds. *Environ. Sci. Technol.* **5**(2): 141–145.

Toxaphene Working Group, U.S. Environmental Protection Agency. (1977). Toxaphene: Position Document I. Report No. EPA-SPRD-80/55 Arlington, Va., Special Pesticide Review Div.

Turner, W. V., J. L. Engel, and J. E. Casida. (1977). Toxaphene components and related compounds: Preparation and toxicity of some hepta-, octa-, and nonachlorobornanes, hexa- and heptachlorobornenes, and hexachlorobornadiene. *J. Agric. Food Chem.* **25**(6): 1394–1401.

U.S. Environmental Protection Agency. (1982). Status report on rebuttable presumption against registration (RPAR) or special reviews. Office of Pesticide Programs Washington, D.C.

Veith, G. D. and G. F. Lee. (1971). Water chemistry of toxaphene—role of lake sediments. *Environ. Sci. Technol.* **5**(3): 230–234.

Veith, G. D., D. W. Kuehl, F. A. Puglisi, G. E. Glass, and J. G. Eaton. (1977). Residues of PCB's and DDT in the western Lake Superior ecosystem. *Arch. Environ. Contam. Toxicol.* **5**: 487–499.

Von Rumker, R., A. W. Lawless, A. F. Meiners, K. A. Lawrence, G. C. Kelsco, and F. Horay. (1974). Production, distribution, use, and environmental impact potential of selected pesticides. U.S. Nat. Tech. Inform. Serv. PB 238 795, Springfield, Va.

Webb, D. W. (1980). The effects of toxaphene piscicide on benthic macroinvertebrates. *J. Kansas Entomol. Soc.* **53**(4): 731–744.

Weber, F. H. and F. A. Rosenberg. (1980). Biological stability of toxaphene in estuarine sediment. *Bull: Environ. Contam. Toxicol.* **25**: 85–89.

Williams, R. R. and T. F. Bidleman. (1978). Toxaphene degradation in estuarine sediments. *J. Agric. Food Chem.* **26**: 280–282.

Wold, S. and M. Sjostrom. (1977). SIMCA: A method for analyzing chemical data in terms of similarity and analogy. In: B. Kowalski, Ed., *Chemometrics: Theory and application.* ACS Symposium Series, No. 52, Am. Chem. Soc., Washington, D.C.

Woodwell, G. M., P. P. Craig, and H. A. Johnson. (1971). DDT in the biosphere: Where does it go? *Science* **174**: 1101–1107.

Zell, M. and K. Ballschmiter. (1980a). Baseline studies of the global pollution. II. Global occurrence of hexachlorobenzene (HCB) and polychlorocamphenes (toxaphene) (PCC) in biological samples. *Frensenius Z. Anal. Chem.* **300**: 387–402.

Zell, M. and K. Ballschmiter. (1980b). Baseline studies of the global pollution. III. Trace analysis of polychlorinated biphenyls (PCB) by ECD glass capillary gas chromatography in environmental samples of different trophic levels. *Frensenius Z. Anal. Chem.* **304**: 337–349.

Ziranski, M. T., V. J. Homer, and T. F. Bidleman. (1982). Comparison of toxaphene residues in environmental samples with laboratory weathered standards using capillary gas chromatography and computer assisted pattern recognition. 1982 Pittsburg Conference, Atlantic City, March 3–13.

9

DISTRIBUTION OF POLYCYCLIC AROMATIC HYDROCARBONS IN THE GREAT LAKES

Brian J. Eadie

National Oceanic and Atmospheric Administration
Great Lakes Environmental Research Laboratory
2300 Washtenaw Avenue
Ann Arbor, Michigan 48104

1. INTRODUCTION

The polycyclic aromatic hydrocarbons (PAH) are a class of compounds with a basic structure consisting of carbon and hydrogen atoms arranged in two or more fused aromatic (benzene) rings. The term also covers fused aromatic systems containing a cyclopentene ring or hetero atoms of sulfur or nitrogen. This chapter concerns itself with only the hydrocarbons that range from the two-ring compound napthalene ($C_{10}H_8$) to the seven-ring compound coronene ($C_{24}H_{12}$). Permutations in the spatial orientation of the rings and multiple types of substitution lead to a large number of PAH isomers. Of this

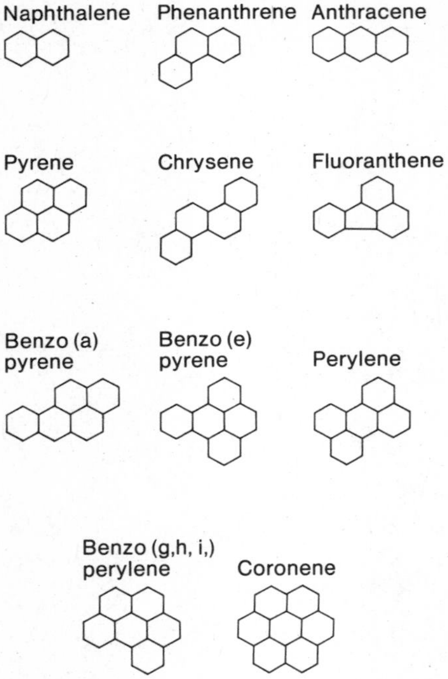

Figure 9.1 Ring structures of the most common PAH in environmental samples. By convention, symbols representing the aromatic character of the rings are omitted.

large group, those most often reported in environmental samples are illustrated in Fig. 9.1; the mental gymnastics associated with nomenclature are defined in the Handbook of Chemistry and Physics (CRC, 1974) and discussed by Neff (1979).

PAHs are primarily products of the incomplete combustion of organic materials such as forest fires or the burning of fossil fuels. As is to be expected from such commonly available source material, PAHs are ubiquitous in the environment (Hites et al., 1980), with elevated concentrations reported near urban areas (Laflamme and Hites, 1978; Wakeham et al., 1980a). The predominant compounds found in atmospheric samples (Gordon, 1976; Strand and Andren, 1980) and sediments (Laflamme and Hites, 1978) are the unsubstituted parent compounds shown in Fig. 9.1. These are primarily generated at temperatures in excess of 400–500°C. At lower combustion temperatures, alkyl-substituted PAHs begin to predominate. This trend is carried to the extreme in the low temperature (100–150°C) maturation of oil which contains complex mixtures of substituted PAH (Youngblood and Blumer, 1975).

In the atmosphere, PAHs generated in the combustion process are primarily associated with fine particles (Neff, 1979). Recent work indicates that most of the mass of PAH is attached to submicron particles (Miguel and Rubenich, 1979). This size class of particles could be expected to have an atmospheric residence time of weeks to months, although this is considerably reduced by washout during rain events. This still leaves time (days to weeks) for long-range atmospheric transport of PAH, accounting for their wide distribution.

There are other sources for some of the PAH. In sediments, perylene concentration increases with depth within the core and often becomes the most abundant PAH (Wakeham et al., 1980b). Laflamme and Hites (1978) describe some possible quinone pigment precursors that might form perylene in reducing sedimentary environments. Alkylated phenanthrenes occasionally appear in large quantities, and are postulated to have two sources in addition to combustion; terpenes associated with pine forests (Laflamme and Hites, 1978) and dehydrogenation of steroids (Wakeham et al., 1980b) within the sediment.

2. BEHAVIOR OF PAH IN THE AQUATIC ENVIRONMENT

Solubility dominates and behavior and fate or persistent organic contaminants in the Great Lakes (and other aquatic systems). The distribution of the contaminent within the system is defined as its equilibrium partition coefficient K_p.

$$K_p = \frac{\text{concentration in particulate phase (ppm)}}{\text{concentration in dissolved phase (ppm)}}$$

Several investigators have recently published on this topic (Chiou et al, 1977; Herbes, 1977; Means et al., 1979; and Karickhoff et al., 1979), the most comprehensive being a review by Kenaga and Goring (1979) who derived the following relationship:

$$\log K_{oc} = 3.64 - 0.55 \log WS$$

$$n = 106 \text{ compounds} \qquad r = -0.84$$

where WS = solubility in water (ppm), and r is the correlation coefficient
$K_{oc} = (100 \times K_p)/\%$ organic carbon of the substrate

Reported solubilities of PAH in distilled water are tabulated in Neff (1979) and range from approximately 1 ppm for phenanthrene to less than 1 ppb for benzo(a)pyrene (BaP). Values for six and seven-ring PAH were not found but are presumably lower than 1 ppb.

The equilibrium distribution of PAH in the water column of the Great Lakes, calculated from this information, is illustrated in Fig. 9.2. The fraction of contaminant associated with the dissolved phase is equal to:

$$f_d = \frac{1}{1 + K_p\text{*TSM}}$$

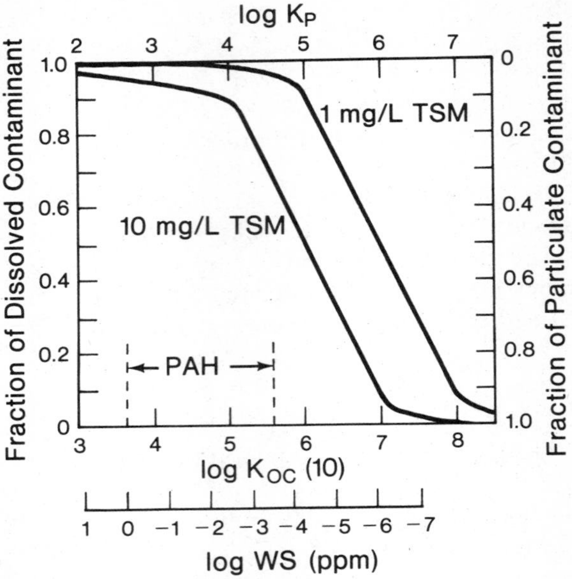

Figure 9.2 The equilibrium distribution of PAH (and other organic contaminants) within the water column of the Great Lakes. The fraction of dissolved contaminant is defined as $f_d = 1/(1 + K_p\text{*TSM})$. K_{oc} (10) represents the partition coefficient assuming 10% substrate organic carbon, equivalent to 10 $*K_p$. The water solubility is calculated from the expression of Kenaga and Goering (1979), see text. The region of Pah solubility from Neff (1979) is indicated. At equilibrium, the PAH are predominately in the dissolved form.

where TSM represents the concentration of total suspended matter (mg/liter). The Great Lakes vary in their open lake TSM concentrations as shown in the following (Bell, 1982):

Lake	Approximate TSM (mg/liter)
Superior	0.5–1
Michigan	1–2
Huron	1–2
Erie	4–8
Ontario	2–4

In shallow nearshore regions, TSM concentrations are generally greater by a factor of 2 or more. The range of TSM illustrated in Fig. 9.2 (1–10 ppm) covers most Great Lakes situations. In the upper three lakes, approximately 90% of the water column inventory of PAH is in the dissolved phase. Only in nearshore regions or western Lake Erie, is a substantial fraction of the low-solubility PAH associated with particles.

Decomposition of PAH in the water column appears to be slow and occurs primarily through photooxidation (NAS, 1972). BaP, for example, has a half-life of less than 10 hr when exposed to sunlight; however, the near UV light which supplies the energy for these reactions is rapidly attenuated within the first few meters of the water column. Fecal pellets have been identified as the primary transport vehicle for PAH in a marine bay (Prahl and Carpenter 1979). Preliminary evidence suggests that fecal pellets play a similar role in Lake Michigan (M. Evans, University of Michigan, personal communication). These are relatively large and have settling speeds of tens of meters per day, rapidly removing PAH from the euphotic zone, thereby reducing the importance of photodecomposition. Biological decomposition is at a maximum near the sediment–water interface (Gardner et al., 1979; Lee et al., 1981) and competes with burial as the predominant removal mechanism in the Great Lakes.

Laflamme and Hites (1979) and Hites et al. (1980) describe the distribution of PAH in various sedimentary environments around the world. Although the distribution of compounds is complex, fluoranthene and pyrene are usually about equal in concentration and are the most abundant PAH, which is consistant with the hypothesis of a combustion source. Concentrations of these compounds range from a few parts per billion in regions remote from urban environments up to approximately 1 part per million in sediments from the New York Bight. Similarly high concentrations were reported for sediments from Lake Washington and three Swiss lakes (all close to urban sources) by Wakeham et al. (1980a). These high sediment concentrations indicate that organisms which live or feed in surficial sediments will be exposed to very high concentrations of PAH.

The primary concern regarding PAHs is their known ability to cause cancer (Jones and Leber, 1979; Gelboin and Ts'o, 1978). Acute toxicity from these compounds is of significance in the immediate area of an oil spill. In addition, Leversee et al. (1983) recently described a high mortality rate for sunfish exposed to ppb concentrations of anthracene in the presence of sunlight. The limited amount of work on chronic response of aquatic organisms has been summarized by Neff (1979) and Malins and Hodgins (1981). Both reviews agree that measurable effects of chronic exposure are probably limited to polluted coastal environments or locations of oil spills and that these locations are usually contaminated with a wide variety of pollutants making cause and effect discriminations very difficult. Hose et al. (1982) determined that benzo(a)pyrene adversely affected the hatching and early development of flatfish. In order to better estimate the levels of PAH stress in fish, Payne and Fancey (1982) advocate the monitoring of mixed-function oxidase which they find elevated in fish exposed to moderate levels of petroleum hydrocarbons. In general, fish and some invertebrates have inducible mixed-function oxidase systems capable of oxidizing PAH, resulting in low bioconcentration factors. Some invertebrates may lack the appropriate enzymes for biotransformation, resulting in large bioconcentration factors. Neff (1979) summarizes a great deal of information on biological uptake, accumulation, and degradation in the aquatic environment.

Are PAHs causing cancer in fish? There appears to be some conflicting evidence. There are documented studies in which increased levels of neoplasia or tumorous lesions are found in regions contaminated with PAH (Malins and Hodgins, 1981; Sonstegard, 1977; Black et al., 1980a,b) but as mentioned previously, these regions are usually contaminated with multiple pollutants. Lesions have been induced by PAH in laboratory studies with fish (Jones and Hoffman, 1957) and fish enzyme systems have been shown to produce carcinogenic metabolites. Thus, circumstantial evidence exists for both chronic effects and carcinogenic/mutagenic responses in fish populations exposed to PAH, and these are most severe near urban areas and for fish that live or feed on the bottom.

3. PAH IN THE GREAT LAKES

The Great Lakes are the focus of a heavily populated and industrialized region and, as previously described, such areas are expected to receive large loads of PAH. Estimates of the atmospheric load of PAH to the Great Lakes are shown in Table 9.1. These loads are similar to maximum estimates of PCB and DDT loads to the Great Lakes. Load estimates are based on a very sparse data base, and thus are subject to improvement as more information becomes available. In addition to atmospheric input, the input from tributaries can be estimated using suspended solids loads data of Sonzogni et al. (1979) and Sullivan et al. (1980) and an estimate of 50–100 ppb fluoranthene or pyrene from soil from Hites et al. (1980). It turns out that approximately 10–25% as much PAH enters via tributaries as compared to atmospheric input. Shoreline erosion and diffuse sources would also contribute small amounts but direct atmospheric input appears to be the major source of PAH to the Great Lakes.

Table 9.1. Atmospheric Flux of PAH to the Great Lakes (metric tonnes/yr)

	Lake					
Compound	Superior[a]	Michigan[a]	Michigan[b]	Huron[a]	Erie[a]	Ontario[a]
Phenanthrene	4.8	3.4	2.1	3.5	1.5	1.1
Anthracene	4.8	3.4	2.1	3.5	1.5	1.1
Fluoranthene	—	—	3.6	—	—	—
Pyrene	8.3	5.9	4.0	6.1	2.6	1.9
Benzo(a)anthracene	4.1	2.9	3.3	3.0	1.5	1.1
Benzo(a)pyrene	7.9	5.6	4.0	5.8	2.5	1.8
Perylene	4.8	3.3	2.1	3.4	1.5	1.1

[a] Eisenreich et al., 1981.
[b] Andren and Strand, 1981.

Table 9.2. Concentration of PAH in Great Lakes Surficial Sediments (ng/g dry)

	Lake				
Compound	Superior[a]	Michigan[b]	Huron	Erie[c]	Ontario[d]
Phenanthrene	34	533 ± 382	272	346 ± 92	58.5
Fluoranthene	88	754 ± 444	487	569 ± 442	615 ± 394
Pyrene	53	607 ± 399	356	391 ± 91	647 ± 594
BaP	28	480 ± 246	294	255 ± 152	—
n	1	9	1	3	2–4

[a] Gschwend and Hites, 1981.
[b] Eadie, 1983; depositional basins.
[c] Eadie et al., 1982.
[d] IJC, 1977.

Because of their low solubility and concentration, there are very limited data on PAH in the water column. Neff (1979) summarizes the worldwide information with concentrations of BaP ranging from approximately 0.1 to 100 ng/liter. In Lake Erie, near Buffalo, Basu and Saxena (1979) found 0.3 ng/liter BaP and 4.7 ng/liter total PAH. Williams et al. (1982) extracted large volumes of municipally treated drinking water taken from 12 plants using Great Lakes water and found relatively high concentrations of pyrene (11.2 ± 20.0 and 3.9 ± 10.2 ng/liter) and fluoranthene (9.2 ± 12.0 and 10.6 ± 25.0 ng/liter). The numbers in parentheses represent the mean ±1 standard deviation for winter and summer samples, respectively. Eadie (1983) found 15 ± 9 ng/liter of fluoranthene and 14 ± 6 ng/liter of pyrene and BaP in filtered offshore waters of southern Lake Michigan. The concentration of these compounds on suspended particles was 2–4 μg/g. At a concentration of 1 mg/liter of TSM, greater than 75% of these PAH were in the dissolved phase.

PAH concentrate in sediments and there have been several analyses of Great Lakes sediments for these compounds. These results have been compiled in Table 9.2. A fine-grained Lake Ontario sediment sample analyzed in our laboratory had PAH concentrations as follows: Ph (205.0), Fl (474.), Py (434.0) and BaP (306 ng/g) which are generally within the range of the Ontario data in Table 9.2.

The Lake Superior sediment is an order of magnitude lower in PAH than the lower lakes, and Lake Michigan sediments are highest in PAH concentration. This observation supports the hypothesis of localized urban sources, because the Lake Superior region is heavily forested and undeveloped whereas the region around southern Lake Michigan (where eight of the sediments were collected) is heavily industrialized. The isomeric distribution of measured PAH in Great Lakes sediments is similar in a gross sense to sediments reported by Hites et al. (1980), Laflamme and Hites (1978), Wak-

Table 9.3. Lake Michigan Sediment PAH (ng/g dry)

Station	Phen	Fl	Py	C + T	BaP
T1	28	34	30	36	33
T2	809	906	733	—	450
T3[a]	1268	1664	1430	1128	944
T4	308	445	363	319	251
T5	537	794	665	522	248
T6	19	62	38	41	26
T7	298	620	421	554	572
T8	245	419	319	348	571
T11	263	413	318	543	324

[a] Coefficients of variation (%) for replicate analyses of replicate extracts from station T3 are: phen., 27.6; Fl., 13.0; Py., 11.6; C + T, 19.4; BaP, 18.5.

eham et al. (1980a), Tan and Heit (1981), and Eadie et al. (1982a,b) for environments that receive PAH from anthropogenic sources. Characteristically, these are highest in unsubstituted fluoranthene and pyrene, with large concentrations of phenanthrene, chrysene (+ triphenylene), and benzo(a)pyrene.

More detailed information on the distribution of PAH in the sediments of Lake Michigan is presented in Table 9.3. Stations T1 and T6 are located in regions where sediments do not accumulate (Cahill, 1981); surficial sediments consist of glacial till and/or bedrock. The other seven stations are located in regions of recent deposition. Sediment fluoranthene concentrations for these nine stations are plotted in Fig. 9.3. The distribution is in general agreement with the map of chlorinated organic contaminents (whose source is also primarily atmospheric) published by Frank et al. (1981). It appears that the final sedimentary distribution of PAH and other hydrophobic organics will be controlled by the processes that affect the movement of the fine-grained, organic-rich sediments to which they are attached.

Using the information from the previous discussion, a simple steady-state mass balance for PAH in the Great Lakes can be estimated. Figure 9.4 illustrates such a calculation for BaP in Lake Michigan. The loss of 2.8 tonnes (MT) of BaP per year (calculated by difference) indicated that approximately 50% of the load is being decomposed. If the concentration in the water is divided by the load, one calculates an apparent residence time of approximately 10 yr for BaP in Lake Michigan water. The large reservoir of BaP in the surficial sediments may act as an important source to the water column through resuspension and diffusion. The load of 5 tonnes/yr is equivalent to 3.5 $ng/cm^2 \cdot yr$ which is higher (by a factor of 10) than that calculated by Gschwend and Hites (1981) from a core in Lake Superior, but is much lower than their reported fluxes for urban sites.

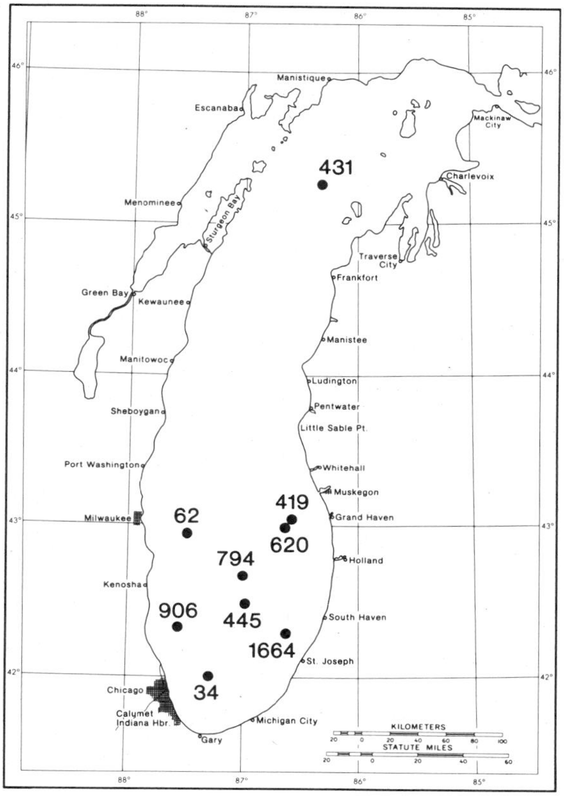

Figure 9.3 The distribution of fluorathene (ng/g) in the surfacial sediment of Lake Michigan.

The complex mixtures and relatively high concentration of PAH in lake sediment has raised questions regarding the exposure of benthic organisms to these compounds, resultant bioconcentration factors, and the transfer of these compounds up the food web to fish. Because fish have the ability to enzymatically oxidize PAH, monitoring the ambient levels of PAH in fish is not a measure of their exposure. Recent analyses of oligochaete worms and chironomids from Lake Erie (Eadie et al., 1982b) have shown that the PAH concentration in these organisms is similar to that of the fine-grained fraction of their sedimentary environment. Bioconcentration factors in *Pontoporeia hoyi,* the most abundant benthic organism (by mass) in Lake Michigan, ranged from 10^4 to 10^5 with respect to the overlying water concentrations of seven measured PAH. Recent laboratory experiments with this organism confirm these bioconcentration factors for anthracene and BaP (Landrum, 1982; Landrum, personal communication). In *Pontoporeia* from recent (fine-grained) sediments, concentrations of phenanthrene, pyrene, flouranthene, chrysene, and BaP exceeded 1 ppm (wet wt.).

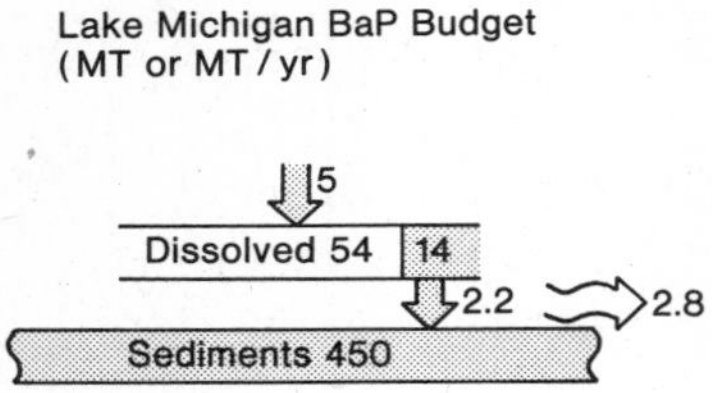

Figure 9.4 Lake Michigan benzo(a)pyrene budget. Shaded areas represent particle-bound BaP. The input (5 tonnes/yr) is from Table 9.1. Concentration in the water column (14 ng/liter, 80% dissolved) is from Eadie (1983) times the volume of the lake. Sedimentation is estimated from recent accumulation data (Robbins and Edgington, 1975) and trap fluxes (Chambers and Eadie, 1981), both of which are approximately 7 mg/cm^2·yr. This was multiplied by the lake area and a BaP concentration of 500 ng/g (Table 9.2). Sediment inventory was estimated for the upper 3 cm (well mixed, 2.6 g/cm^3, 80% porosity). The losses of approximately 2.8 tonnes/yr were calculated by difference.

Do these concentrations present a serious threat to Great Lakes fish? Sonstegard (1977) found very high incidences of tumorous tissue in Great Lakes fish that feed primarily on bottom organisms. In an analysis of a museum collection of similar Great Lakes fish collected in 1952 he didn't detect any such malformities. The implication is that some carcinogenic or mutagenic agent(s) have been introduced into the Great Lakes since the early 1950s. There are many compounds or possibly combinations of compounds which could be responsible for this, PAH are certainly among them. Black et al. (1980a,b) found high incidences of tumorous tissue in the carp and goldfish hybrids of Lake Erie tributaries to be correlated with concentrations of PAH in sediments.

4. THE FUTURE OF PAH IN THE GREAT LAKES

The atmospheric concentrations of BaP in the Great Lakes region over the period 1966 through 1980 are presented in Fig. 9.5. Lines for the U.S. cities are annual averages from the National Air Surveillance Network (NASN) as reported in Faoro and Manning (1981, Appendix). Data beyond 1977 were kindly supplied by Mr. Jerry Ackland of the U.S. EPA. Quarterly averages of this data show a great deal of seasonality with maxima in the cold winter months often an order of magnitude higher than the summer values. The decreasing trend appears to be real and is attributable to the reduced use of residential coal and wood for heating. Concentration estimates from the early seventies to the present all hover around 1 ± 0.5 ng BaP/m^3, which is approximately the concentration used in estimating the loads reported in Table 9.1. It appears that recent concentrations for Youngstown and Milwaukee are again on the increase. This is to be expected from the recent increase in residential wood burning, which is related to increased oil prices (Peters et al., 1981). Two Department of Energy reports summarized by Norman (1982) state that residential usage of wood has doubled in the last decade. This, coupled with increasing population and consequent increased energy consumption will result in increasing loads to the Great Lakes. Regional

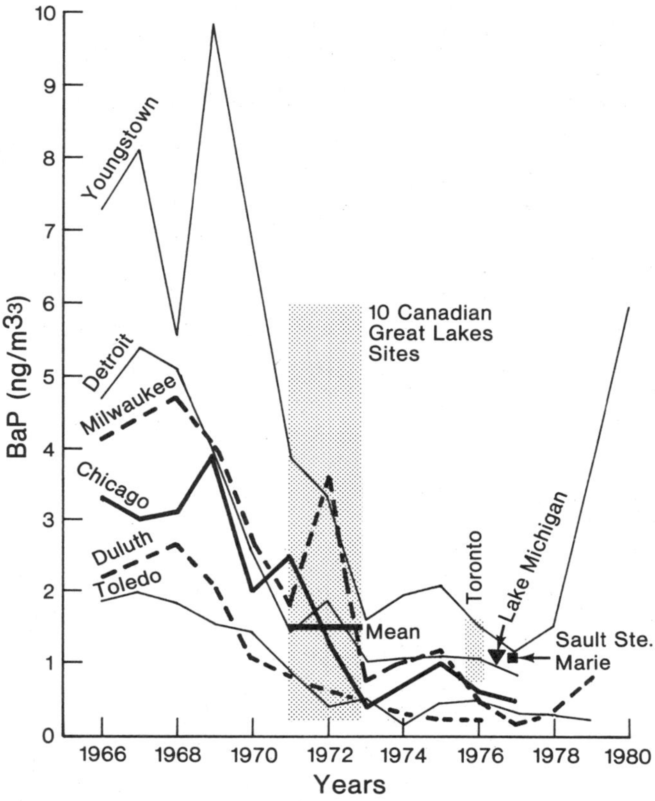

Figure 9.5 Atmospheric concentration of benzo(a)pyrene in the Great Lakes region. Data for U.S. cities are from the U.S. EPA/NASN (Faoro and Manning, 1981; Ackland, 1982). Data for 10 Canadian cities on the Great Lakes are from MOE (1979); all of the 2-yr averages fall within the gray zone, the grand mean for all the data is also shown. The Toronto data for 1976 is from Katz et al. (1978). The Lake Michigan data is from Strand and Andren (1980). The Sault St. Marie point is from MOE (1979). Youngstown and Milwaukee are showing recent increases in BaP concentrations.

population data are given in Table 9.4. Population estimates for the year 2000 are approximately 10% larger than the current population, a somewhat slower rate of growth than the country as a whole. Similar growth on the Canadian side can be expected.

A study by the National Academy of Sciences (NAS, 1979, p. 204) has concluded that due to its availability, coal will be a key element in the U.S. energy policy well beyond the end of the century. They estimate that three times today's production or about 45 quads (1 quad = 10^{15} BTU) annually will provide one third to one half of the nation's energy compared to 18% (14 of 78 quads) in 1978. A regional study (GLBC, 1981) estimates that by the year 2000, the use of coal will increase with respect to oil and gas, and

Table 9.4. Population Data for the Great Lakes States (Millions)

Location	1970[a]	1980[b]	1990[c]	2000[c]	2050[d]
U.S. Total	203,000	222,000	243,000	260,000	316,000
Illinois	11,113	11,418	12,015	12,491	—
Indiana	5,196	5,490	5,804	6,069	—
Ohio	10,657	10,797	11,570	11,999	—
Michigan	8,882	9,258	10,302	10,970	—
New York	18,241	17,557	18,528	18,816	—
Pennsylvania	11,801	11,913	12,272	12,465	—
Minnesota	3,806	4,077	4,382	4,637	—
Wisconsin	4,418	4,705	5,156	5,476	—
Great Lakes states	73,394	75,215	80,029	82,923	—

[a] U.S. Bureau of Census, 1979.
[b] U.S. Bureau of Census, 1981.
[c] U.S. Bureau of Census, 1979, Projection IIA.
[d] Schurr et al., 1979, p. 110.

that energy production within the basin will increase from 3.1 quads (1975) to 3.9 quads by 1990.

Information summarized for the United States in the late 1960s (NAS, 1972) indicates that more than 1200 tonnes/yr of BaP were emitted. This amount was apparently reduced to approximately 300 tonnes/yr by the mid-1970s as shown in Table 9.5. This 75% reduction in emissions supports the

Table 9.5. Estimated PAH Emissions (tonnes/yr)

	Location and Time of Emissions			
Source	BaP U.S.[a] late 1960s	BaP U.S.[b] 1975	BaP Ontario[c] 1976	Total PAH U.S.[d] mid-1970s
Heat and power				
Coal	431	27	0.013	184.0
Oil	2	2.7	0.052	9.0
Gas	2	1.0	0.048	12.2
Wood	40	73	—	7674
Open burning	450	>52	—	2547
Coke production	20	110	9.33	632
Forest fires	140	11	8.16	1478
Mobile sources	22	13	0.323	2266
Grams emitted per capita	6.6	1.4	2.2	70.5

[a] NAS, 1972.
[b] Faoro and Manning, 1981.
[c] MOE, 1979.
[d] Peters et al., 1981.

NASN estimate of approximately a 75% reduction in the atmospheric concentration of BaP at 26 urban sites throughout the United States from 1966 to 1977 (Faoro and Manning, 1981). Considering changes in sample collection and analysis techniques, it is not clear that the magnitude of the reported decline in PAH is real or an artifact of the methods employed. Potvin et al. (1981) found no significant difference in atmospheric BaP measured 1971–1975 and those from 1975–1979 at a site in northern Ontario. What is obvious from the four emission estimates in Table 9.5 is that there is considerable disagreement about the relative importance of even major sources. In the late 1960s, the major sources were perceived to be open burning of refuse followed closely by the combustion of coal for heat and power. By the mid-1970s, this was altered to coke production in both the United States and Ontario. In the case of total PAH estimates, the major source for this period was the combustion of wood for heat and power. These estimates are rather weak because sampling procedures and analytical intercomparisons are not yet standardized nor uniformly efficient for various environments. Bennett et al. (1979) found order-of-magnitude ranges in the PAH concentrations of oil-fired power plant emissions (23–2550 ng PAH/m^3) and coal-fired power plant emissions (24–378 ng PAH/m^3), thus the use of an average emission factor for each source type is only a gross approximation. A literature review by Junk and Ford (1980) found 109 organics in coal combustion products and a total of 331 in coal and refuse combustion, but state that there are probably many more as yet uncollected or unidentified compounds. Many of these could have caused interferences in earlier, less specific, analytical procedures.

At this time it is unclear to what extent PAH emissions will increase with projected increased fossil fuel combustion. The fact that PAH are associated with submicron particles makes them very difficult to remove from stack gases. Increased relative use of "dirty" fuels such as coal and wood will further increase the potential for PAH production. If regional energy production increases by 25% (3.1 to 3.9 quads) as estimated by the GLBC (1981) report, and this increase is primarily in the form of coal and wood combustion, then based on the figures in Table 9.5 we can conservatively estimate an increase of 25% in the loads of PAH to the lake. If the rates of removal from the aquatic ecosystem remain the same, or possibly decrease due to the reduction in primary productivity caused by the Great Lakes phosphorus abatement program, then concentrations within the Great Lakes ecosystem will increase.

ACKNOWLEDGMENTS

I would like to thank the crew of the R/V *Shenehon* for their assistance in sample collection; Warren Faust, Nancy Moorehead, and Shiela Nihart for their analytical assistance; and Drs. Peter Landrum, John Robbins, and

Wayne Gardner for helpful comments. This work was jointly funded by the Office of Marine Pollution Assessment, NOAA and the Great Lakes Environmental Research Laboratory, NOAA. GLERL Contribution No. 338.

REFERENCES

Ackland, J. (1982). Personal communication: computer printout of the NASN BaP data 1975–81. U.S. EPA, Res Tri Park North Carolina.

Andren, A. W. and J. W. Strand. (1981). Atmospheric deposition of particulate organic carbon and polyaromatic hydrocarbons to Lake Michigan. S. J. Eisenreich, Eds., *Atmospheric pollutants in natural waters*. Ann Arbor Sci., Ann Arbor, Michigan, pp. 459–479.

Basu, D. K. and J. Saxena. (1979). Polynuclear aromatic hydrocarbons in selected U.S. drinking waters and their raw water sources. *Environ. Sci. Technol.* **12:** 795–798.

Bell, G. H. (1982). Personal communication from his partially completed task of combining and synthesizing all available TSM and organic carbon data.

Bennett, R. L., K. T. Knapp, P. W. Jones, J. E. Wilkerson, and P. E. Strup. (1979). Measurement of polynuclear aromatic hydrocarbons and other hazardous organic compounds in stack gases. In: P. W. Jones and P. Leber, Eds., *Polynuclear aromatic hydrocarbons*. Ann Arbor Sci., Ann Arbor, Michigan, pp. 419–428.

Black, J. J., P. P. Dymerski, and W. F. Zapisek. (1980a). Environmental carcinogenesis studies in the western New York Great Lakes aquatic environment. pp. 215–225 in D. R. Branson and K. L. Dickson, Eds., *Aquatic Toxicology and Hazard Assessment,* Proceedings of the Fourth Annual Symposium on Aquatic Toxicology, ASTM Special Technical Publication 737, American Society for Testing and Materials, 1916 Race St., Philadelphia, Pa. 19103.

Black, J. J., M. Holmes, P. P. Oymerski, and W. F. Zapisek. (1980b). Fish tumor pathology and aromatic hydrocarbon pollution in a Great Lakes estuary. In: B. K. Afghan and D. Mackay, Eds., *Hydrocarbons and halogenated hydrocarbons in the aquatic environment*. Plenum Press, New York, pp. 559–565.

Cahill, R. A. (1981). Geochemistry of recent Lake Michigan sediments. Ill. State Geol. Survey, Circular 517.

Chambers, R. L. and B. J. Eadie. (1981). Nepheloid and suspended particulate matter in southeastern Lake Michigan. *Sedimentology* **28:** 439–447.

Chiou, C. T., V. F. Freed, D. W. Schmedding, and R. L. Kohnert. (1977). Partition coefficient and bioaccumulation of selected organic chemicals. *Environ. Sci. Technol.* **11:** 475–478.

CRC. (1974). Handbook of chemistry and physics, 54th ed. Section C. CRC Press, Cleveland, Ohio.

C&EN. (1980). Report touts expanded use of coal. p. 6 in *Chem. Eng. News,* May 19, 1980, Am. Chem. Soc.

Eadie, B. J. (1983). Partitioning of polycyclic aromatic hydrocarbons in sediments and pore waters of Lake Michigan. *Geochim. Cosmochim. Acta*, in press.

Eadie, B. J., W. Faust, W. S. Gadner, and T. Nalepa. (1982a). Polycyclic aromatic hydrocarbons in sediments and associated benthos in Lake Erie. *Chemosphere* **11:** 185–191.

Eadie, B. J., P. F. Landrum, and W. Faust. (1982b). Polycyclic aromatic hydrocarbons in sediments, pore water and the amphipod *Pontoporeia hoyi* from Lake Michigan. *Chemosphere* **11:** 847–858.

Eisenreich, S. J., B. B. Looney, and J. D. Thronton. (1981). Ariborne organic contaminants in the Great Lakes ecosystem. *Environ. Sci. Technol.* **15:** 30–38.

Faoro, R. B. and J. A. Manning. (1981). Trends in benzo(a)pyrene, 1966–77. *J. Air Pollut. Control Assoc.* **31:** 62–64.

Frank, R., R. L. Thomas, H. E. Braun, D. L. Gross, and T. T. Davies. (1981). Organochlorine insecticides and PCB in surficial sediments of Lake Michigan (1975). *J. Great Lakes Res.* **7**(1): 42–50.

Gardner, W. S., R. F. Lee, K. R. Tenore, and L. W. Smith. (1979). Degradation of selected polycyclic aromatic hydrocarbons in coastal sediments: Importance of microbes and polychaete worms. *Water Air Soil Pollut.* **11:** 339–347.

Gelboin, H. V. and P. O. P. Ts'o. (1978). *Polycyclic hydrocarbons and cancer.* Academic Press, San Francisco.

GLBC. (1981). Great Lakes Regional Water and Energy Study. Great Lakes Basin Commission; c/o Great Lakes Commission, Ann Arbor, Michigan, 245 pp.

Gordon, R. J. (1976). Distribution of airborne polycyclic aromatic hydrocarbons throughout Los Angeles. *Environ. Sci. Technol.* **10:** 370–373.

Gschwend, P. M. and R. A. Hites. (1981). Fluxes of PAHs to marine and lacustrine sediments in the northeastern United States. *Geochim. Cosmochim. Acta.* **45:**2359–2368.

Herbes, S. E. (1977). Partitioning of polycyclic aromatic hydrocarbons between dissolved and particulate phases in natural waters. *Water Res.* **2:** 493–496.

Hites, R. A., R. E. Laflamme, and J. G. Sindsor, Jr. (1980). Polycyclic aromatic hydrocarbons in marine/aquatic sediments: Their ubiquity. In: L. Petrakis and F. T. Weiss, Eds., *Petroleum in the marine environment.* Advances in Chemistry Series 185, Am. Chem. Soc., Washington, D.C., pp. 289–311.

Hose, J. E., J. B. Hannah, D. DiJulio, M. L. Laudolt, B. S. Niller, W. T. Iwaoka, and S. P. Felton (1982). Effects of benzo(a)pyrene on early development of flatfish. *Arch. Env. Contam. Tox.* **11:** 167–171.

International Joint Commission. (1977). Great Lakes water quality—Appendix E: Status report on the persistent toxic pollutants in the Lake Ontario basin. International Joint Commission, Windsor, Ontario.

Jones, R. W. and M. N. Hoffman. (1957). Fish embryos as bio-assay material in testing chemicals for effects on cell division and differentation. *Trans. Am. Mic. Soc.* **76:** 177–183.

Jones, P. W. and P. Leber. (1979). Polynuclear aromatic hydrocarbons. Third International Symposium on Chemistry and Biology—Carcinogenesis and Mutagenesis. Ann Arbor Sci., Ann Arbor, Michigan.

Junk, G. A. and C. S. Ford. (1980). A review of organic emissions from selected combustion processes. *Chemosphere* **9:** 187–230.

Karickhoff, S. W., D. S. Brown, and T. A. Scott. (1979). Sorption of hydrophobic pollutants on natural sediments. *Water Res.* **13:** 241–248.

Katz, M., T. Sakuma, and A. Ho. (1978). Chromatographic and spectral analysis of PAH—quantitative distribution in air of Ontario cities. *Environ. Sci. Technol.* **12:** 909–915.

Kenaga, E. E. and C. A. I. Goring. (1979). Relationship between water solubility, soil sorption, octanol–water partitioning and concentration of chemicals in biota. Pp. 78–115 in *Aquatic toxicology.* Proceedings 3rd Symposium on Aquatic Toxicology, ASTM Spec. Tech. Publ. 707, Philadelphia, Pennsylvania.

Konasewich, D., W. Traversy, and H. Zar. (1978). Great Lakes water quality: Status report on organic and heavy metal contaminants in Lakes Erie, Michigan, Huron and Superior basins. Great Lakes Water Quality Board, Windsor, Ontario.

Laflamme, R. E. and R. A Hites. (1978). The global distribution of polycyclic aromatic hydrocarbons in recent sediments. *Geochim. Cosmochim. Acta* **42:** 289–303.

Landrum, P. F. (1982). Uptake, depuration, and biotransformation of anthracene by the scud, *Pontoporeia hoyi. Chemosphere,* (in review).

Lee, R. F., K. Hinga, and G. Almquist. (1981). Fate of radiolabelled PAH and pentachlorophenol in enclosed marine ecosystems. In: G. Grice and M. R. Reeve, Eds., *Marine mesocosms*. Springer-Verlag, New York, pp. 123–136.

Leversee, G. J., J. W. Bowling, P. F. Landrum, and J. P. Giesy. (1983). Acute mortality of anthracene contaminated fish exposed to sunlight. *Aquatic Toxicol.* **3:** 79–90.

Malins, D. C. and H. O. Hodgins. (1981). Petroleum and marine fishes: A review of uptake, deposition and effects. *Environ. Sci. Technol.* **11:** 1273–1280.

Means, J. C., J. J. Hassett, S. G. Wood, and W. L. Banwart. (1979). Sorption properties of energy related pollutants and sediments. In: P. W. Jones and P. Leber, Eds., *Polynuclear aromatic hydrocarbons*. Ann Arbor Sci., Ann Arbor, Michigan.

Miguel, A. H. and L. M. S. Rubenich. (1979). Submicron size distributions of particulate polycyclic aromatic hydrocarbons in combustion source emissions. In: A. Bjorseth and A. J. Dennis, Eds., *Polynuclear aromatic hydrocarbons: Chemistry and biological effects*. Battelle Press, Columbus, Ohio.

Ministry of the Environment. (1979). Polynuclear aromatic hydrocarbons—a background report including available Ontario data. ARB-TDA-Report 58–79. Ministry of the Environment, Ontario, Canada.

National Academy of Sciences. (1972). *Particulate polycyclic organic matter*. National Academy of Sciences, Washington, D.C., 361 pp.

National Academy of Sciences. (1979). *Energy in transition 1985–2010*. Final Report of the Committee on Nuclear and Alternate Energy Systems. National Research Council. National Academy of Sciences, Washington, D.C., 677 pp.

Neff, J. M. (1979). *Polycyclic aromatic hydrocarbons in the aquatic environment*. Applied Sci. Publ., London, England.

Norman, C. (1982). OPEC gives a boost to U.S. firewood use. *Science* **217:** 1017.

Payne, J. F. and L. L. Fancy. (1982). Effects of long term exposure to petroleum on mixed function oxygenases in fish: Further support for the use of the enzyme system in biological monitoring. *Chemosphere* **11:** 207–213.

Peters, J. A., D. G. DeAngelis, and T. W. Hughes. (1981). An environmental assessment of POM emissions from residential wood fired stoves and fireplaces. In: M. Cooke and A. J. Dennis, Eds., *Polynuclear aromatic hydrocarbons: Chemical analysis and biological fate*. Battelle Press, Columbus, Ohio, pp. 571–581.

Potvin, R. R., E. G. Adamek and D. Balsillie. (1981). Ambient PAH levels near a steel mill in northern Ontario. In: M Cooke and A. J. Dennis, Eds., *Polynuclear aromatic hydrocarbons: Chemical analysis and biological fate*. Battelle press, Columbus, Ohio.

Prahl, F. G. and R. Carpenter. (1979). The role of zooplankton fecal pellets in the sedimentation of polycyclic aromatic hydrocarbons in Dabob Bay, Washington. *Geochim. Cosmochim. Acta*. **43:** 1959–1972.

Robbins, J. A. and D. N. Edgington. (1975). Determination of recent sedimentation rates in Lake Michigan using Pb-210 and Cs-137. *Geochim. Cosmochim. Acta*. **39:** 285–304.

Schurr, S. H., J. Darmstadter, H. Perry, W. Ramsay, and M. Russell. (1979). *Energy in America's future, the choices before us*. Resources for the Future. Johns Hopkins Univ. Press, Baltimore, Maryland, 555 pp.

Sonstegard, R. A. (1977). Environmental carcinogenesis studies in fishes of the Great Lakes of North America. PP. 261–268. Annals of the NY Acad. of Sci. Vol. 298. Aquatic Pollutants and Biologic Effects with Emphasis on Neoplasia.

Sonzogni, W. C., T. J. Monteith, W. E. Skimin, and S. C. Chapra. (1979). Critical assessment of U.S. land derived pollutant loadings to the Great Lakes. Great Lakes Basin Commission, Ann Arbor, Michigan.

Strand, J. W. and A. W. Andren. (1980). Polyaromatic hydrocarbons in aerosols over Lake Michigan, fluxes to the lake. In: A. Bjorseth and A. J. Dennis, Eds., *Polynuclear aromatic hydrocarbons: Chemistry and biological effects*. Battelle Press, Columbus, Ohio, pp. 127–137.

Sullivan, R. A. C., T. J. Monteith, and W. C. Sonzogni. (1980). Post-pluarg evaluation of Great Lakes water quality management studies and programs. Vol. II. EPA-905/9-80-C06-B, September 1980.

Tan, Y. L. and M. Heit. (1981). Biogenic and abiogenic polynuclear aromatic hydrocarbons in sediments from two remote Adirondack lakes. *Geochim. Cosmochim. Acta* **45:**2267–2279.

U.S. Bureau of the Census. (1979). Illustrative Projections of State Populations by Age, Race, and Sex: 1975 to 2000. U.S. Bureau of the Census, Current Population Reports, Series P-25, No. 7, U.S. Government Printing Office, Washington, D.C., 1979.

U.S. Bureau of the Census. (1981). Advance reports. U.S. Dept. of Commerce.

U.S. Environmental Protection Agency. (1980). Ambient water quality criteria for polynuclear aromatic hydrocarbons. U.S. EPA EPA-440/5-80-069. Washington, D.C., 202 pp.

Wakeham, S. G., C. Schaffner, and W. Giger. (1980a). Polycyclic aromatic hydrocarbons in recent lake sediments. I. Compounds having anthropogenic origins. *Geochim. Cosmochim Acta* **44:** 403–413.

Wakeham, S. G., C. Schaffner, and W. Giger. (1980b). Polycyclic aromatic hydrocarbons in recent lake sediments. II. Compounds derived from biogenic precursors during early diagenesis. *Geochim. Cosmochim. Acta* **44:** 415–429.

Williams, D. T., E. R. Nestmann, G. L. LeBel, F. M. Benoit, and R. Otson. (1982). Determination of mutagenic potential and organic contaminants of Great Lakes drinking water. *Chemosphere* **11:** 263–276.

Youngblood, W. W. and M. Blumer. (1975). Polycyclic aromatic hydrocarbons in the environment: Homologous series in soils and recent marine sediments. *Geochim. Cosmochim. Acta* **39:** 1303–1314.

10

CYCLING OF POLYNUCLEAR AROMATIC HYDROCARBONS IN THE GREAT LAKES ECOSYSTEM

Douglas J. Hallett

Scientific Advisor
Department of the Environment
Ontario Region
55 St. Clair Avenue East
Toronto, Ontario

Ronald W. Brecher

Chemist
Canadian Wildlife Service
Ontario Region
National Wildlife Research Centre
100 Gamelin Boulevard
Hull, Quebec

1. SOURCES

Polynuclear Aromatic Hydrocarbons (PAH) are organic compounds composed of two or more fused rings. The most commonly studied PAH found in the environment include benzo(a)pyrene (BaP), anthracene, benzo(a)anthracene, fluoranthene, indeno(1,2,3-cd) pyrene, phenanthrene, perylene, and pyrene. Many other PAH have been recognized. PAH are of concern in aquatic environments not only because they are ubiquitous contaminants but also because many, particularly BaP, are well-documented carcinogens. PAH may be formed during any incomplete combustion of organic matter. Important industrial sources of PAH include coal gasification and liquification processes, combustion of fossil fuels (gasoline, kerosene, coal, natural gas, diesel fuel), waste incineration, as well as production of coke, carbon black, coaltar pitch, asphalt and petroleum cracking (Hangebrauck et al., 1967). Waste effluents of certain industrial processes, for example electrolytic reduction of magnesium and ferro alloy smelting, provide a direct route of entry for PAH to aquatic ecosystems (Lunde and Bjorseth, 1977; Bjorseth et al., 1979).

Other sources of PAH which may be important in terms of human exposure are foods that have absorbed PAH, agricultural burning, and the use of items such as gasoline-powered lawn mowers. Cigarette smoking is the most significant source of direct personal exposure in humans. A one-pack-a-day smoker who inhales 1 μg per day of BaP receives 50 times as much as a non-smoker counterpart in an average North American "coke oven city," where the concentration of BaP is 1.2 ng/m^3 (MOE, 1979).

Fossil fuels such as crude or bunker oil contain PAH which are spilled accidently into aquatic environments including the Great Lakes. High-molecular-weight PAH fractions of several crude oils have been shown to resist weathering (Law, 1980; Hallett et al., 1983) and PAH in bunker oil have been found to concentrate after weathering in the St. Lawrence River (Palm and Alexander, 1978). PAH are also produced naturally, for example, during forest fires and volcanic eruptions. One species of bacteria (*Platymonas* sp.) synthesized 1.15 μg/liter of BaP during 4 months of culturing in normal medium (Erhardt, 1972). This result may be controversial in that few environments are free of PAH, and it has been suggested that BaP may have been accumulated rather than synthesized.

Estimates of BaP emission in Ontario indicate that of the 17,930.28 kg of BaP, more than 97% was from coke production and forest fires, with coke production alone accounting for more than 52% of the total (Statistics Canada, 1978; OMNR, 1977).

2. CHEMISTRY

2.1. Formation

The chemistry of PAH formation is complex, being a function of many parameters, but the general mechanism is quite well understood. At usual combustion temperatures (500–800°C) aliphatic carbon–carbon and carbon–hydrogen bonds break easily. The free radicals formed during pyrolysis recombine to produce PAH which accumulate as pyrosynthesis proceeds. Aromatic ring systems are the most stable at these temperatures and in the chemically reducing atmospheres found at the center of flames (Badger et al., 1960, 1964a,b).

2.2. Physical State of Atmospheric PAH

Based on the relatively high boiling and melting points of PAH and their generally low vapor pressures at ambient temperature, atmospheric PAH are thought to exist almost exclusively adsorbed on airborne particulates (for example fly ash) (Commins, 1962; Thomos et al., 1968). The size of the PAH-bearing particles during human exposure may be the key factor in determining health risk. Small particles are able to penetrate deep into the alveoli of the lung, whereas larger aerosols gain only limited entry to the respiratory tract. A Canadian study showed that 70–90% of PAH collected in Toronto was adsorbed to particles with aerodynamic diameter less than 5 μm (Pierce and Katz, 1975). Other BaP-containing aerosols had average mass median diameters of 0.65 μm (Albaglia et al., 1974). More recently, a new type of particulate has been identified, making up about 1% of fly ash. This carbonaceous material is like a honeycomb in form and is relatively large compared to other particulates. This type of particle accounts for more than 25% of the total surface area of fly ash and has twice the affinity for

PAH as fly ash. It is estimated that about 70% of total PAH is adsorbed on this material (Natusch, 1983). Adsorption on this type of carbonaceous material renders PAH less susceptible to photodegradation, thus permitting relatively long atmospheric residence times. On the other hand, the large size of these particles restricts their access to the lungs.

2.3. Physical State of Aquatic PAH

PAH gain entry to aquatic systems through several routes, including precipitation and fallout from the atmosphere, wastewater discharges (residential and industrial), runoff from asphalt-covered surfaces (e.g., roads), and oil spills.

PAH have a very low solubility in water due to their hydrophobic nature and are thus thought to exist primarily adsorbed to organic particles. The solubility of benzo(a)pyrene in water is 0.004 mg/liter at 27°C. Certain synthetic detergents and such natural products as humic acid may aid in solubilization of PAH.

3. TRANSPORT AND TRANSFORMATION OF PAH

3.1. Transport in Air

To a first approximation, particle size is the key factor affecting the transport behavior of PAH-containing particulates. Particle sizes generally range in diameter from 0.01 to 10 μm with the bulk of PAH associated with particles having a diameter less than 5 μm. Although the size concentration distribution varies in space and time, particles larger than about 0.1 μm in diameter tend to follow the relation

$$\frac{dN}{dD_p} = k\phi D_p^{-4} \tag{1}$$

where N is the number of particles, D_p is the particle diameter, and ϕ is the volume fraction of particles. The constant k has the value of about 0.40 (Winchester, 1980).

Particle shape and density are also important to the transport behavior of particulate-associated PAH. Larger aerosols are selectively removed from air by gravitational settling or impaction; turbulent air causes larger, coarser particles to be impacted on rough surfaces (Winchester, 1980).

Stoke's Law is useful to help visualize the gravitational settling of particulates. The velocity of a particle settling in still air, V_d, is given by

$$V_d = \frac{2}{9}\frac{(\rho - \rho_0)g}{\eta}\left(\frac{d}{2}\right)^2 \tag{2}$$

where g is the gravitational acceleration (approximately 9.8 m/s^2), $\rho - \rho_0$ is the difference between the particle and air densities, and η is the viscosity of air. In this equation, d represents the equivalent aerodynamic diameter (the diameter of a unit density sphere of equivalent aerodynamic characteristics) (Winchester, 1980). At STP (0°C, 1 atm, $\rho_0 = 1.3 \times 10^{-3}$ g/cm^3, $\eta = 171$ μpoisse), particles as small as 2 μm in diameter would settle less slowly than air would rise, and would thus remain suspended.

Dispersion of PAH in the atmosphere occurs by diffusion, convection, and advection, all of which depend on meteorological considerations. Such dispersion has the effect of diluting emission close to the source, but provides a means for the appearance of PAH at locations distant from the source. At least 20 PAH have been shown to be transported over long distances (Winchester, 1980). Additionally, atmospheric residence times of various PAH were estimated, taking into account photocatalytic transformation. These varied from a few minutes to 500 hr. On this basis it was predicted that PAH may travel from a few to several thousand kilometers (Blau and Gusten, 1983).

3.2. Transport in Water

As in air, transport of PAH occurs by both diffusion and convection. Sedimentation is probably the most important route for clearance of PAH from lakes and, as in the atmosphere, depends largely on particle size. In most rivers, settling or sedimentation is counterbalanced by resuspension of particles and net sedimentation is zero.

3.3. Chemical Transformation in the Environment

3.3.1. Atmospheric

In the environment, photochemical reactions are among the most important, because the fate of PAH is influenced largely by its photochemical behavior. PAH containing three or more rings are able to absorb solar radiation at wavelengths above 300 nm. The excited PAH reacts with O_2 to produce highly reactive singlet oxygen molecules which react with ground state PAH (NAS, 1972; Radding et al., 1976; Geacintov, 1973; Katz et al., 1979; Tosine, 1978). The process can be schematically represented as

$$M \xrightarrow{h\eta} {}^1M^* \xrightarrow{{}^3O_2} {}^3M^* \xrightarrow{{}^3O_2\ {}^1O^{*}{}_2} M$$

$$M + {}^1O_2^* \rightarrow MO_2$$

where M, ${}^1M^*$, and ${}^3M^*$ are the ground, excited singlet, and excited triplet states of the PAH, respectively. The symbols ${}^3O_2$ and ${}^1O_2^*$ represent triplet (ground state) and singlet (excited state) oxygen, respectively.

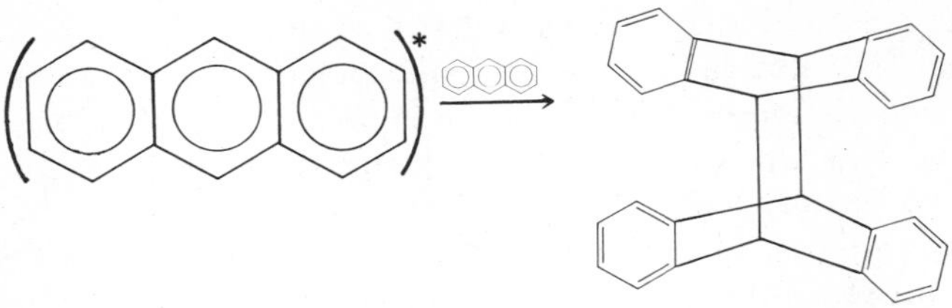

Figure 10.1. Anthracene photodimer formation.

PAH adsorbed to surfaces of aerosols may be photolyzed rapidly, but depending on the affinity of the photooxidation products for the particulate, underlying layers of PAH may be protected from further exposure to solar radiation (NAS, 1972). PAH found in weathered oils may persist for considerable periods, likely due to the physical matrix present (Hallett et al., 1983).

Photoexcited singlet PAH may react with another PAH in its ground state to form a photodimer. For anthracene the reaction is illustrated in Fig. 10.1.

3.3.2. Aquatic

Differences in photolysis rates are due primarily to differences in the absorption spectra of PAH. Volatilization also plays a key role in environmental persistence. Volatilization rates decrease by factors of 3–10 with each additional benzene ring. With the exception of naphthalene, the rates of photolysis of dissolved PAH are higher than the rates of volatilization (Table 10.1) although the physical state of bound or adsorbed PAH will render these processes negligible.

Table 10.1. Volatilization and Photolytic Half-Lives of PAH

Compound	Number of Rings	Half-life (hr)	
		Volatilization[a]	Photolysis[b]
Naphthalene	2	16	71
Anthracene	3	62	0.75
Phenanthrene	3	—	8.4
Benz(a)anthracene	4	500	0.59
Chrysene	4	—	4.4
Benz(a)pyrene	5	1500	0.54

[a] Calculated for river water 1 m deep, water velocity 0.5 m/s, wind velocity 1 m/s from gas–water partition coefficients (Henry's law coefficients) and river reaeration rate expressions (Southworth, 1979).

[b] Calculated for surface waters in midsummer at 40°N latitude (Zepp and Schlotzhauer, 1969).

Table 10.2. Rates of Microbial Degradation and Predicted Rates of Direct Adsorption to Sediments[a]

Compound	Number of Rings	Microbial Degradation Rate Constant (hr^{-1})	Sediment Sorption Coefficient, *k*
Naphthalene	2	0.23	40
Anthracene	3	0.035	80
Benzo(a)anthracene	4	0.005	5000
Benzo(a)pyrene	5	<0.005	3000

[a] Herbes et al., 1980.

Because PAH solubilities are low (on the order of μg/liter), their tendency to adsorb to particulate matter is high and increases with molecular weight. Their persistence in this adsorbed state is, however, affected by microbial degradation. Soil microorganisms can degrade PAH (Shabad, 1968) and some degradation by aquatic bacteria also occurs (Poglazova et al., 1972; Zobell, 1971; Gibson, 1975). Microbial degradation is 8 to 20 times higher in sediment than in natural waters studied and the rate varies with molecular size (decreasing with increasing molecular weight), and is inversely related to the concentration of PAH present. Although degradation rates in sediment decrease as PAH increase in molecular weight, the increasing tendency of high-molecular-weight PAH to adsorb tends to counterbalance this trend. Therefore, half-lives calculated for river water considering adsorption to bottom sediments and microbial degradation as the only removal processes are similar for two to four-ringed PAH. Note in Table 10.2 that the adsorption rate for the five-ringed BaP is extremely high, with a negligible microbial degradation rate.

In summary, PAH appear to be persistent in aquatic systems. Microbial degradation appears to be relatively unimportant. Sorption and sedimentation prevail in aquatic systems with rapid settling rates, as in lakes (but not rivers). Volatilization and photolysis appear to be the most important processes for dissolved PAH. These latter two processes are severely hindered if PAH are complexed in aquatic systems by adsorption to particulates, either before or after entering the water. One must bear in mind that most sources relate to some form of combustion or coking. The highly adsorptive properties of carbonaceous materials prevalent in these two sources would bind PAH unless these particles remained suspended for long periods, where desorption would become prevalent, as in river systems.

4. BIOAVAILABILITY

The effect of PAH adsorption on particulate organic matter is important in determining PAH bioavailability. Using ^{14}C-BaP in particle-free water and in water containing autoclaved yeast cells (used to adsorb the ^{14}C-BaP) this effect was investigated in *Daphnia magna*. Bioconcentration (i.e., bioavailability) decreased with increasing fraction of PAH adsorbed (McCarthy, 1983). One must therefore bear in mind that, in river systems, more PAH will be present in solution and therefore be bioavailable. In lakes and bays adsorption and sedimentation will prevail (Herbes et al., 1980) and although PAH may be obviously present in sediment, it is not necessarily bioavailable throughout the water column.

5. BIOCONCENTRATION AND BIOACCUMULATION

Bioconcentration, or direct uptake from air and water into terrestrial and aquatic organisms, has been demonstrated for many species. These include soil and aquatic bacteria, marine and freshwater algae, many invertebrates, fish, and some higher plants (Shabad et al., 1971; Hallett et al., 1977; Lee et al., 1972; Lu et al., 1977). The relationship between the log K_{ow} (where K_{ow} represents the octanol–water partition coefficient) of PAH and their environmental magnification factors predicted in model aquatic ecosystems indicate that PAH such as BaP should bioaccumulate similar to chlorinated aromatic hydrocarbons such as hexachlorobenzene and DDT. Lu et al. (1977) state that ''Benzo(a)pyrene is highly lipophilic and can be bioaccumulated to high and potentially hazardous levels. This problem is intensified in organisms, such as snails, deficient in microsomal oxidase or in the presence of mixed function oxidase inhibitors.'' This theory pertains correctly to bioconcentration, and perhaps to bioaccumulation at the far lower ends of food chains. However, we will propose another hypothesis: that rapid degradation of PAH in higher species possessing mixed-function oxidase systems capable of metabolizing PAH limits bioaccumulation at the higher trophic levels.

The metabolism of naphthalene and anthracene throughout various tissues of coho salmon has been demonstrated (Roubal et al., 1977). Up to 46% of anthracene in the fish flesh was found in a metabolized form 144 hr after injection. The data of Lu et al. (1977) show that the bioaccumulation factor for BaP in mosquitofish after three days exposure was less than 1, whereas for snails the corresponding value was 2177. The monooxygenase activity, as measured by BaP metabolism, is undetectable in marine bivalves (Vandermeulen and Penrose, 1978), whereas trout readily metabolize BaP. Relative monooxygenase activities for fish are roughly one order of magnitude below those of amphibians, birds, and mammals, including primates (Walker, 1978).

This data supports the theory that PAH will only bioaccumulate up the lower levels of aquatic food chains into molluscs but not into higher fish, birds, and mammals. We therefore do not expect to find high levels of PAH in fish and birds monitored in the Great Lakes. The limited data available support this.

Table 10.3. Abundances of Polycyclic Aromatics in Lake Ontario Sediment (LAT. 43°39′, LONG. 78°12′) μ/g Dry Sediment [a]

Aromatic	0–5 cm	10–15 cm	20–25 cm	30–35 cm	55–60 cm	70–75 cm
Biphenyl	0.014	0.007	0.009	0.004	0.004	0.004
Tetrahydropyrene	0.056	0.029	—	—	—	—
Fluoranthene	0.281	0.058	—	—	—	—
Pyrene	0.056	0.029	—	—	—	—
1,2-Benzanthracene						
Chrysene	0.225	0.088	0.052	—	—	—
Triphenylene						
Dimethyl chrysene	0.112	—	—	—	—	0.018
2,3-Benzofluoranthene	0.450	0.029	0.017	0.017	0.020	0.009
Methyl benzofluoranthene	0.056	—	—	—	—	—
Benzpyrenes	0.337	—	0.017	0.034	0.010	0.009
Perylene	0.056	0.029	0.017	0.034	0.30	0.046
Methyl benzpyrene	0.056	—	—	—	—	—
Methyl perylene	0.112	—	—	—	0.010	0.027
20-Methyl cholanthrene	0.337	—	—	—	—	0.018
Benzperylene	0.225	—	—	—	—	—
Coronene	0.562	—	—	—	—	—
Total aromatics	2.935	0.269	0.112	0.089	0.084	0.131

[a] IJC, 1976.

6. PAH IN THE GREAT LAKES ECOSYSTEM

Some 27 PAH were identified in sediments taken from Hamilton Harbour throughout an east–west transit of Lake Ontario (Strosher and Hodgson, 1976; IJC, 1976). Total concentrations of PAH reached a maximum of 54 μg/g of dry sediment. Perylene, benzopyrenes, and benzofluoranthenes were commonly present. A typical list of PAH found off Toronto is shown in Table 10.3. Few PAH were detectable in Lake Ontario water. Methyl naph-

Table 10.4. Benzo(a)pyrene Concentrations Commonly Found in Water[a]

Sample Type	Benzo(a)pyrene (ng/liter)
Drinking water	0.1–23.4
Surface water	0.6–114.0
Sewage water	34,500
Industrial effluents	1.0–1840

[a] IARC, 1973.

Table 10.5. Polynuclear Aromatic Hydrocarbons (ng/kg) Fresh Weight Fillet[a]

Fish	Perylene	Benzo(k)fluoranthene	Benzo(a)pyrene	Coronene
		Hamilton Harbour		
Carp				
1	46	8	108	300
2	140	40	160	360
3	26	12	96	210
4	160	40	200	400
5	40	16	108	220
6	74	12	160	60
7	40	8	144	300
8	142	68	268	320
9	nd[b]	nd	nd	20
Pike				
1	90	48	154	240
2	64	32	128	200
3	40	12	70	200
4	32	10	54	220
5	58	20	74	200
6	40	12	64	170
7	34	8	60	152
8	40	12	54	140
9	nd	nd	nd	nd
10	20	12	34	100
		Detroit River		
Carp				
1	16	10	40	80
2	nd	nd	nd	60
3	40	14	40	nd
4	26	10	40	40
5	nd	nd	nd	nd
6	nd	nd	nd	nd
7	nd	nd	nd	120
8	nd	nd	nd	80
9	nd	nd	nd	nd
10	nd	nd	nd	nd
Pike				
1	34	26	40	20
2	20	14	14	40
3	18	8	34	40
4	20	8	20	44
5	68	26	128	290
6	18	10	24	40
7	20	6	30	30
8	nd	nd	nd	nd
9	46	24	70	120
10	52	26	100	120

[a] Hallett et al., 1978.

[b] nd = nondetectable.

Table 10.6. Polynuclear Aromatic Hydrocarbons Identified by Mass Spectrometry in Great Lakes Fish[a,b]

PAH	Hamilton Harbour		Detroit River	
	Carp	Pike	Carp	Pike
Naphthalene	X	X		X
2-methyl Naphthalene	X	X		X
1-methyl Naphthalene	X	X		X
Biphenyl	X	X		X
Acenaphthene		X		X
Dimethyl naphthalene		X		X
Fluorene		X		X
Anthracene	X	X		X
Phenanthrene	X	X		X
1-phenyl Naphthalene	X	X		X
1-methyl Phenanthrene	X	X		X
1-methyl Anthracene	X	X		X
2-methyl Anthracene	X	X		X
2-methyl Phenanthrene	X	X		X
9-methyl Anthracene				X
Fluoranthrene	X	X		X
Pyrene	X	X		X
1,2-benzofluorene		X		X
2,3-benzofluorene		X		X
Chrysene	X	X		X
Benzo(a)pyrene		X		X
Perylene		X		X
Dibenz(a,h)anthracene	X	X		X
Coronene	X	X		X

[a] Hallett et al., 1978.
[b] X = detected.

thalenes, biphenyl, and methyl antracenes were the only PAH detectable at concentrations from 3 to 600 ng/liter (Strosher and Hodgson, 1976). Organochlorines such as DDT, DDE, and PCBs were one to two orders of magnitude higher in concentration than PAH in Lake Ontario sediments (IJC, 1976). Similar types and concentrations of PAH were found in western Lake Erie sediments, near the mouth of the Raisin River, associated with a large coal-fired power plant (Eadie et al., 1982). PAH were also detected in worms and midges. PAH are commonly found in drinking water and surface water throughout the world (IARC, 1973). Sewage waters and industrial effluents can be heavily contaminated. General ranges of levels found are shown in Table 10.4.

Plant effluents discharging into Hamilton Harbour on Lake Ontario have shown a range of PAH concentrations from 0.02 to 66 μg/liter with

Table 10.7. Identification of Polynuclear Aromatic Hydrocarbons in Great Lakes Herring Gull Lipid[a,b]

	Concentration (μg/kg)		
Compounds	Pigeon Island	Kingston	Mass Spectral Confirmation
Naphthalene	0.050	0.054	+
2-methyl Naphthalene	0.036	0.005	+
1-methyl Naphthalene	0.043	0.009	+
Biphenyl	0.151	0.017	+
Acenaphthene	0.038	0.007	+
4-methyl Biphenyl	0.061	0.010	+
Fluorene	0.044	0.003	+
Anthracene	0.152	0.024	+
Phenanthrene	nd	0.002	+
1-phenyl Naphthalene	0.008	0.008	
2-methyl Phenanthrene	0.021	0.007	+
1-methyl Phenanthrene	0.010	0.015	+
9-methyl Anthracene	0.011	0.025	+
3,6-dimethyl Phenanthrene	nd	0.012	+
Fluoranthrene	0.082	0.017	+
Pyrene	0.076	0.015	+
1-aza Pyrene	a	a	
9-acetylanthracene	a	a	
DDE	—	—	+
1,2-benzofluorene	a	a	+
2,3-benzofluorene	a	a	+
1-methyl Pyrene	a	a	+
2-acetyl Phenanthrene	a	a	
1,1-binaphthyl	a	a	
Chrysene	0.053	a	+
Benz(e)pyrene	0.026	0.021	+
Benz(a)pyrene	0.038	0.030	+
Perylene	0.053	0.026	
9-dichloromethylene Fluorene	b	b	+b
Dimethyl biphenyl	b	b	+b

[a] Hallett et al., 1977.

[b] a = PCB interference; b = Standards of compounds unavailable, compounds identified by mass spectra; nd = not detected.

benzo(a)pyrene concentrations ranging from 0.03 to 30 μg/liter (Smillie et al., 1978).

Fish collected from Hamilton Harbour and the Detroit river, including carp (a bottom feeder), were found to contain low levels of PAH in ng/kg (ppt) range as shown in Table 10.5 (Hallett et al., 1978). Although the types of PAH found in Lake Ontario were similar to those found in water and sediment (Table 10.6), the levels reflect the lack of bioaccumulation of the PAH from water to fish.

Similarly, herring gulls from Lake Ontario at or near Kingston, Ontario (Table 10.7) contained levels of PAH slightly lower than those found in the fish (Hallett et al., 1977). The herring gull, being primarily a piscivorous species, demonstrates the lack of bioaccumulation at the highest trophic level of the aquatic ecosystem.

It seems obvious that the Great Lakes are heavily exposed to PAH, as is much of the fresh water of the world. It is interesting to note that the PAH most commonly reported in sediment and water in the so-called "ambient" environment are not the methylated series of PAH associated with petroleum hydrocarbons. Rather, the nonmethylated forms indicative of combustion sources predominate. It is also obvious that PAH are accumulated by lower organisms associated with contaminated lake sediments that form the basis of the food web for fish and herring gulls. However, PAH levels are extremely low in those fish and herring gulls. It is also worth noting that the same fish and herring gulls contained stable organochlorine compounds such as DDE and PCBs at levels 10^5 to 10^6 time higher than PAH levels although sediment levels for PAH and organochlorines were similar. It appears then, in the case of fish, that the monooxygenase system responsible for PAH metabolism is not the same enzyme system responsible for metabolism of DDT, PCBs, and other organochlorine contaminants prevalent in Lake Ontario fish.

7. METABOLISM AND METABOLIC ACTIVATION

PAH are lipid-soluble, hydrophobic compounds. Since such molecules are difficult to excrete; they are first metabolized to more polar compounds which can then be excreted (Lehninger, 1974). However it would appear that nature has made a serious error in metabolizing chemical carcinogens such as benzo(a)pyrene.

The enzymes responsible for these conversions are collectively known as aryl hydrocarbon hydroxylase (AHH) or mixed-function oxidase (MFO). This system allows elimination of aromatic hydrocarbons by catalyzing their ring hydroxylation to polar metabolites (Heidelberger, 1975). AHH has been found in all species examined to date, including bacteria, algae, higher plants, invertebrates, crustaceans, amphibians, reptiles, birds, and mammals (Marquardt, 1977). Ironically, this enzyme system also results in conversion of many precarcinogenic PAH to their ultimate carginogenic forms.

The AHH system converts PAH to polar compounds through epoxide intermediates. These are then further metabolized in one of several ways. The epoxide may isomerize to yield a phenol or it may combine with glutathione.

Alternately, the enzyme epoxide hydrase may convert the epoxide to a dihydrodiol. Dihydrodiols and phenols then form glucuronides and sulfates. These compounds and mercapturates (produced by conjugation of the epoxide with glutathione) are the usual nontoxic excretion products of PAH (Heidelberger, 1975).

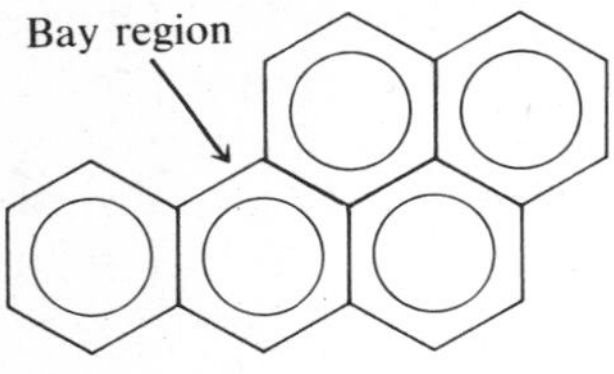

Figure 10.2. Bay region structure in BaP.

It is widely held that the carcinogenic and mutagenic activity of PAH is due to the interaction between the electron-deficient reactive epoxide metabolites and critical nucleophilic biomacromolecules, such as DNA (Jerina et al., 1977). Epoxide hydrase activity may be crucial in determining the sensitivity of a particular species to PAH-induced carcinogenesis or mutagenesis. Species that are able to metabolize the epoxides rapidly are relatively insensitive, because there is less chance of interaction between the epoxide and key biomolecules (Oesch, 1977). Alternately, sensitive cells may produce reactive metabolites faster or have reduced capacity to conjugate carcinogenic intermediates.

PAH that include a "bay region" in their structure and an unsubstituted carbon in the *peri* position are usually carcinogenic (Jerina et al., 1977; Yang, 1983). This structure is shown for BaP, in Fig. 10.2.

Since all studied PAH are metabolized in a similar manner, BaP will serve to illustrate the production of an ultimate carcinogen from a bay region PAH. BaP is first converted via AHH and epoxide hydrase to benzo(a)pyrene-7,8-dihydrodiol. Further epoxidation at the 9,10 position produces the ultimate carcinogen, benzo(a)pyrene-7,8-dihydrodiol-9,10-epoxide. That the *peri* position must be unsubstituted becomes apparent when one considers that such a substitution would prevent the formation of the 7,8-dihydrodiol, the procarcinogen of BaP. On the other hand, substitution at other positions may increase production of the 7,8-dihydrodiol-9,10-epoxide, since fewer metabolic sites are available, making Bay region metabolism more likely. The ultimate carcinogen is detoxified by conversion to a tetrahydrotetrol, which can be eliminated via the metabolic pathways discussed previously. Alkylated PAH are usually metabolized by hydroxylation of the alkyl side chain, however, bay region metabolites of alkyl-PAH have been identified as well (Marquardt et al., 1977).

AHH or cytochrome P_{450} is an inducible enzyme system; that is, it can be induced or repressed by other chemicals, for example 2,3,7,8-tetrachlorodibenzo-*p*-dioxin. Since other enzymes involved in PAH metabolism may not be induced, production of active metabolites (e.g., bay region dihydrodiol epoxides) may exceed the organisms ability to eliminate these compounds, allowing a buildup of carcinogens. Conversely, substances like butylated hydroxyanisole (BHA) are anticarcinogenic because they inhibit or alter the regioselectivity of AHH. BHA is also known to induce a wide variety of detoxifying enzymes.

8. INDUCTION OF MFO SYSTEMS

8.1. General

The cytochrome P_{450} system has been examined in detail by Nebert et al. (1981). Many substrates, highly variable in size and shape induce monoxygenase activity. Either monooxygenase induction is a relatively nonspecific phenomenon involving a single form of P_{450} and any of a number of inducers, or a more specific event where each enzyme in a family of monooxygenases has only one or a few specific inducers. P_{450} inducers have long been classified as being of two major types: 3-methylcholanthrene-like (cytochrome P_{448}), and phenobarbital-like (cytochrome P_{450}) (Conney, 1967). However, about 60 P_{450} inducers have now been found which are not precisely 3-methylcholanthrene-like or phenobarbital-like (Nebert et al., 1981). Compounds within this class are shown in Table 10.8.

The induction by polycyclic hydrocarbons (3-methylcholanthrene, benzo(a)pyrene, and TCDD) or cytochrome P_{448}-like inducers of numerous

Table 10.8. Toxic Chemicals Present in the Great Lakes Which Stimulate Their Own Metabolisms and/or Induce Unique Forms of P_{450}

Toxic Chemical	Reference
PAH	
Benzo(a)pyrene	Conney et al., 1957
3-Methyl cholanthrene	Nebert and Gelboin, 1968
Benzo(a)anthracene	Nebert and Gelboin, 1968
7,12-Dimethyl benzo(a)anthracene	Snyder et al., 1966
Methylated benzenes and naphthalenes	Fabacher and Hodgson, 1977
Chlorinated hydrocarbons	
pp'-DDT	Morello, 1965
Benzene	Snyder et al., 1966; Fabacher and Hodgson, 1977; Morello, 1965; Tunek and Oesch, 1979
Mirex	Fabacher and Hodgson, 1976; Kaminsky et al., 1978
Polybrominated biphenyls	Moore et al., 1978; Robertson et al., 1980
TCDD	Mikol and Decloitre, 1979; Guenthner et al., 1979; Parkinson et al., 1980a,b; Ahotupa and Aitio, 1980
Lindane	Mikol and Decloitre, 1979; Parkinson et al., 1980a,b; Ahotupa and Aitio, 1980
Polychlorinated biphenyls	Parkinson et al., 1980a,b
Halogenated naphthalenes and terphenyls	Ahotupa and Aitio, 1980

drug-metabolizing enzymes has been shown to be regulated by the murine Ah locus. The Ah complex is considered to be a combination of regulatory, structural, and probably temporal genes which may or may not be linked (Nebert et al., 1981).

Induction of the Ah locus commences with the passive transfer of the inducer (e.g., BaP) across the cell membrane and the highly specific binding to a cytosolic Ah receptor, regarded as the major regulatory gene product. Following apparent translocation of the inducer–receptor complex to the nucleus, the response includes the activation of numerous structural genes. The induction of metabolic enzymes from these structural genes then leads to the formation of reactive intermediates and/or detoxified products for elimination (Nebert et al., 1981). Thus, induction of the Ah locus allows for many variations among the regulatory and structural genes of various species. These variations will ultimately affect the propensity of substrates to bind to the regulatory gene product and to the induction of the Ah locus (Nebert et al., 1981).

8.2. Selectivity of MFO Induction in Fish

It was noted earlier that trout and salmon efficiently metabolize PAH, particularly BaP (Roubal et al., 1977). Trout and salmon possess an inducible form of cytochrome P_{448} that converts BaP to benzo(a)pyrene-7,8-oxide (Ahokas et al., 1979; Elcombe and Lech, 1979; Lech and Bend, 1980). Rainbow trout were induced by TCDD, which is considered to be a cytochrome P_{448}-like inducer, but were not induced by mirex or kepone, both considered to be P_{450} inducers (Vodicnik et al., 1981). Neither *p,p'*-DDT nor *p,p'*-DDE caused MFO induction in trout. These compounds are both P_{450} inducers. The P_{448}-type inducers, 3-methylcholanthrene, and PCB (Arochlor 1254) strongly induced some components of the MFO system of trout, but gave overall response (Willis et al., 1978).

Thus, fish appear to be the first species in the aquatic food web that are capable of metabolizing significant quantities of PAH under natural conditions, allowing for their elimination and lack of bioaccumulation. On the other hand, fish are not induced by P_{450}-like organochlorines such as the DDT types, mirex, or kepone, and are poorly induced by PCB isomers, likely allowing for their bioaccumulation to higher levels in the food chain.

9. CONSEQUENCE OF PAH METABOLISM BY FISH

The metabolism of PAH by fish is not without consequence. Brown bullheads treated with repeated applications of a PAH-containing organic solvent extract obtained from a heavily polluted river sediment developed pro-

Table 10.9. Distribution of Lesions among Specimens Received

Phylogenetic Group	Neoplastic	Nonneoplastic	Not Diagnosed	Total
Reptiles	54	93	2	149(10%)
Amphibians	91	51	3	145(10%)
Bony fish	386	238	37	661(45%)
Sharks	5	8	0	13(1%)
Lampreys and hagfish	37	16	0	53(4%)
Tunicates	0	11	0	11(1%)
Molluscs	191	97	3	291(20%)
Arthropods	7	78	0	85(6%)
Annelids	0	24	0	24(2%)
Other (echinoderms, nematodes, protozoa, fungi, acanthocephia, platyhelminths, coelenterates, porifera)	0	29	7	36(2%)
Totals	771(53%)	645(44%)	52(4%)	1468(100%)
Percentage of those diagnosed	(54)	(46)		

gressive skin alterations including epidermal hyperplasia and papillomas (Black, 1983). Brown bullheads from eastern Lake Erie and the upper Niagara River where sediments and water contain elevated levels of PAH were observed to possess epidermal neoplasms including one observation of a highly invasive carcinoma. The anatomic location of the lesions near the mouth supported the hypothesis that contaminants encountered in the act of feeding may have been a causal factor (Black, 1982).

Benzo(a)pyrene has recently been shown to be extensively metabolized in bile, liver, muscle, and gonads of adult and juvenile English sole. Benzo(a)pyrene-7,8-dihydrodiol, benzo(a)pyrene-9,10-dihydrodiol, and 1- and 3-hydroxy benzo(a)pyrene were found, indicating that toxic metabolites were formed.

Many lesions have been found in marine and freshwater species and a tabulation from the registry of tumors in lower animals (Harshbarger, 1977) is shown in Table 10.9.

Sonstegard surveyed some 50,000 Great Lakes fish between 1973 and 1976 (Sonstegard, 1977). Gonadal tumors in carp, goldfish, and hybrids of these species were observed as well as papillomas in white suckers with a significant tumor frequency clustering in the Oakville–Burlington region of Lake Ontario. One hundred percent of fish collected in heavily polluted waters possessed neoplasms having tumors exclusively associated with the lips; the upper lip, which has direct contact with bottom sediments, was most affected.

PAH are of concern toxicologically due to their ability to induce carcinogenic responses in mammals. In addition mutagenicity and teratogenecity have been reported. A summary of toxic effects in mammals for PAH commonly found in water is shown in Table 10.10 (CDHW, 1979).

Table 10.10. PAH Commonly Found in Water and Reported Toxic Effect in Mammals

Compound	Relative[a] Carcino- genicity	Toxic Effects (Dose, Duration, Admin. by, Species, Effect)	Reference[b]
Anthracene	(?)	1. 18g/kg, 78 weeks, oral, rat, carcinogenic	R
		2. 3.3g/kg, 33 weeks, subcutaneous, rat, neoplastic effect	R
		3. 2.5 μmoles-TPA[c], —, skin, mouse, papillomas	Scribner (1973)
Benzo(a)-anthracene	(+)	1. 240 mg/kg, 5 weeks, skin, mouse, carcinogenic	R
		2. 2 mg/kg, —, subcutaneous, mouse, carcinogenic	R
		3. 8 mg/kg, —, *parenteral*, mouse, carcinogenic	R
		4. 80 mg/kg, —, implant, mouse, carcinogenic	
		5. *LDLO*[d]: 10 mg/kg, —, intravenous, mouse, —	R
		6. 2.2 μmole-TPA[c], —, skin, mouse, papillomas	Scribner (1973)
Benzo(b)-fluoranthene	(+ +)	1. 40 mg/kg, —, skin, mouse, carcinogenic	R
		2. 72 mg/kg, 9 weeks, subcutaneous, mouse, carcinogenic	R
Benzo(j)-fluoranthene	(+ +)	1. 288 mg/kg, 24 weeks, skin, mouse, carcinogenic	R
Benzo(k)-fluoranthene	(−)	1. 2640 mg/kg, 44 weeks, skin, mouse, nepolastic effect	R
		2. 72 mg/kg, 9 weeks, subcutaneous, mouse, carcinogenic	R
Benzo(a)-pyrene	(+ + +)	1. 4563 mg/kg, 52 weeks continuous, oral, rat, neoplastic effect	R

Table 10.10. (*Continued*)

Compound	Relative[a] Carcino-genicity	Toxic Effects (Dose, Duration, Admin. by, Species, Effect)	Reference[b]
		2. 55 mg/kg, 22 weeks, skin, rat, carcinogenic	R
		3. 16 mg/kg, —, intraperitoneal, rat, carcinogenic	R
		4. 250 μg/kg, —, subcutaneous, rat, neoplastic effect	R
		5. 39 mg/kg, 6 days, intravenous, rat, neoplastic effect	R
		6. 5 mg/kg, —, intramuscular, rat, carcinogenic	R
		7. 22 mg/kg, —, intracerebral, rat, neoplastic effect	R
		8. 7 g/kg, 20 weeks, oral, mouse, carcinogenic	R
		9. 2 μg/kg, week, skin, mouse, carcinogenic	R
		10. 1 mg/kg, —, subcutaneous, mouse, carcinogenic	R
		11. 10 mg/kg, —, intraperitoneal, mouse, carcinogenic	R
		12. 10 mg/kg, —, intravenous, mouse, carcinogenic	R
		13. 2 mg/kg, —, subcutaneous, monkey, carcinogenic	R
		14. 17 mg/kg, 57 weeks, skin, rabbit, carcinogenic	R
		15. 216 mg/kg, 26 weeks, oral, hamster, carcinogenic	R
		16. 21 mg/kg, —, subcutaneous, hamster, carcinogenic	R
		17. 432 mg/kg, 36 weeks, intratracheal, hamster, carcinogenic	R
		18. *TDLO*[f]: 750 mg/kg, —, intraperitoneal, mouse, *mutagenic* effects	R
		19. *LDLO*[d]: 500 mg/kg, —, intraperitoneal, mouse, —	R
		20. *LD50*[e]: 50 mg/kg, —, subcutaneous, rat, —	R
		21. *TDLO*[f]: 240 mg/kg, 11–15 days, intraperitoneal, *pregnant mouse*, *teratogenic* effect	R

Table 10.10. (*Continued*)

Compound	Relative[a] Carcino-genicity		Toxic Effects (Dose, Duration, Admin. by, Species, Effect)	Reference[b]
		22.	*TDLO*[f]: 240 mg/kg, 11–15 days, subcutaneous, mouse, *teratogenic* effect	R
		23.	*TDLO*[f]: —, —, *parenterally*, mouse, *mutagenic* effect	Epstein (1968)
Benzo(e)-pyrene	(+)	1.	516 mg/kg, 45 weeks, skin, mouse, carcinogenic	R
		2.	160 mg/kg, —, subcutaneous, mouse, neoplastic effect	R
		3.	140 mg/kg, 32 weeks, subcutaneous, guinea pig, neoplastic effect	R
		4.	10 μmole-TPA[e], —, skin, mouse, papillomas	Scribner (1973)
Benzo(ghi)-perylene	(−)			
Chrysene	(+)	1.	4.4 μmoles-TPA[e], —, skin, mouse, papillomas	Scribner (1973)
		2.	200 mg/kg, —, subcutaneous, mouse, carcinogenic	R
		3.	99 mg/kg, 31 weeks, skin, mouse, neoplastic effect	R
Dibenzo(a,h)-anthracene	+ + +[g]	1.	500 μg/kg, —, subcutaneous, rat, carcinogenic	R
		2.	360 mg/kg, 22 weeks, oral, mouse, carcinogenic	R
		3.	6 μg/kg, —, skin, mouse, neoplastic effect	R
		4.	76 μg/kg, —, subcutaneous, mouse, carcinogenic	R
		5.	*LDLO*[d]: 10 mg/kg, —, intravenous, mouse, —	R
		6.	10 mg/kg, —, intravenous, mouse, carcinogenic	R
		7.	80 mg/kg, —, implant, mouse, neoplastic effect	R
		8.	30 mg/kg, —, intravenous, guinea pig, neoplastic effect	R
Fluoranthene	(−)	1.	*LD50*[e]: 2000 mg/kg, —, oral, rabbit, —	R
		2.	*LD50*[e]: 3100 mg/kg, —, skin, rabbit, —	R
Indeno(1,2,3-	(+)	1.	72 mg/kg, 9 weeks,	R

Table 10.10. (*Continued*)

Compound	Relative[a] Carcino-genicity	Toxic Effects (Dose, Duration, Admin. by, Species, Effect)	Reference[b]
cd)pyrene		subcutaneous, mouse, carcinogenic	
Phenanthrene	(?)	1. *LD50*[d]: 700 mg/kg, —, oral, mouse, —	R
		2. 71 mg/kg, —, skin, mouse, neoplastic effect	R
		3. 10 μmole-TPA[e], —, skin, mouse, papillomas	Scribner (1973)
Perylene	(−)		
Pyrene	(−)	1. 10 g/kg, 3 weeks, skin, mouse, neoplastic effect	R
		2. 10 μmole-TPA, —, skin, mouse, papillomas	Scribner (1973)
Fluorene	(?)	No data	

[a] Relative activity on mouse epidermis: + + + active; + + moderate; + weak; (?) unknown; (−) inactive. Wynder and Hoffman (1962).

[b] R indicates data taken from Register of Toxic Effects of Chemical Substances, 1976 Edition, Christensen, H.E., Ed., U.S. Department of Health, Education and Welfare, Rockville, Md.

[c] TPA = 12-o-tetradecanoylphorbol-13-acetate used as a promoter.

[d] LDLO = lowest published lethal dose.

[e] LD50 = lethal dose 50% kill.

[f] TDLO = lowest published toxic dose.

[g] Carcinogenic rating from NAS, 1972.

11. CURRENT REGULATORY STANDARDS

11.1. Air Quality

Occupational standards are usually expressed as Threshold Limit Values (TLV) and refer to concentrations of substances to which repeated exposure can occur without adverse effects.

A TLV of 0.2 ng/m^3 for coaltar pitch volatiles (benzene-soluble fraction) aimed to minimize exposure to anthracene, BaP, phenanthrene, acridine, chrysene, and pyrene was adopted by the U.S. Federal Standard under the Occupational Safety and Health Act of 1970 and by the Ontario Ministry of Labour.

A coke oven emmission standard (benzene-soluble fraction) with a TLV of 0.15 ng/m^3 over 8 hr was also promulgated by the U.S. Occupational Safety and Health Administration. It has recently been recommended that this be lowered to 0.1 ng/m^3 as a cyclohexane-soluble fraction.

The Soviet Union has set an occupational exposure limit for BaP of 0.15 μg/m^3 maximum allowable concentration. It is the only country to have set a maximum allowable concentration as high as 1 ng/m^3 for BaP in ambient air. No standards for ambient air levels of PAH as such have been established in any other country.

The bulk of atmospheric PAH are associated with airborne particulate matter and PAH regulatory standards are considered to parallel the latter. Current ambient air quality criteria for suspended particulates in Ontario are 60 μg/m^3 as an annual geometric mean, or 120 μg/m^3 averaged over 24 hr. The corresponding U.S. figures established as National Primary Ambient Air Quality Standards are 75 and 260 μg/m^3, respectively, while the desirable secondary standards are 60 and 150 μg/m^3 for annual and daily measurements, respectively. Ontario regulations limit particulate emmissions at 100 μg/m^3 measured as a half-hour average concentration at the point of impingement.

11.2. Drinking Water

The World Health Organization (WHO) has recommended a maximum of 200 μg/liter of PAH in drinking water, expressed as a sum of six easily measurable PAH: fluoranthene, benzo(b)fluoranthene, benzo(k)fluoranthene, BaP, benzo(g,h,i)perylene and indeno(1,2,3-cd)pyrene. Recently, the WHO has proposed a more stringent limit of 0.01 μg/liter for BaP in drinking water.

Although the U.S. Environmental Protection Agency has set no specific limit for PAH in drinking water, the agency opted to require the best available technology, namely treatment with granular activated carbon, to minimize exposure to potentially harmful chemicals including PAH. Such a procedure was deemed necessary "as a reasonable insurance policy to protect against what the agency perceives to be a carcinogenic hazard of low levels of organic chemicals in drinking waters" (Costle, 1978). The proposed cost of such treatment was estimated to be U.S.$4.00 to U.S.$6.00 per annum per family (Epstein and Shafner, 1968). Although, two thirds of PAH concentration can be eliminated in modern treatment plants by sedimentation, floculation, and filtration (Borneff, 1974), mutagenic activity of drinking water, as demonstrated by *Salmonella typhimurium* strains TA98 and TA100, can be eliminated after virgin granular activated carbon treatment, and by greater than 87% after exhausted granular activated carbon treatment (Meier et al., 1982).

REFERENCES

Ahokas, J. T., H. Saarni, D. W. Nebert and O. Pekkonen. (1979). *Chem. Biol. Interact.* **26:** 103.

Ahotupa, M and A. Aitio. (1980). *Biochem. Biophys. Res. Commun.* **93:** 250.

Albaglia, A., H. Oja, and L. Dubois. (1974). *Environ. Lett.* **6:** 241.

Badger, G. M., R. W. L. Kimber, and T. M. Spotswood. (1960). *Nature* **187:** 663.

Badger, G. M., R. W. L. Kimber, and T. M. Spotswood. (1964a). *Aust. J. Chem.* **17:** 771.

Badger, G. M., R. W. L. Kimber, and T. M. Spotswood. (1964b). *Aust. J. Chem.* **17:** 778.

Bjorseth, A., J. Knutzen, and J. Skei. (1979). *Sci. Total Environ.* **13:** 71.

Black, J. J. (1982). Twenty-fifth Anniversary Conference on Great Lakes Research on the International Association for Great Lakes Research, Sault Ste. Marie, Quebec.

Black, J. J. (1983). In: M. Cooke and A. Dennis, Eds., Seventh International Symposium on Chemical Analysis and Biological Fate of Polynuclear Aromatic Hydrocarbons. Battelle Press, Columbus, Ohio.

Blau, L. and H. Gusten. (1983). In: M. Cooke and A. Dennis, Ed., Proceedings of the Sixth International Symposium on Polynuclear Aromatic Hydrocarbons. Battelle Press, Columbus, Ohio. In press.

Borneff, J. (1974). Polycyclic aromatics in surface and groudwater (Polyzyklische Aromatie in Oberflaechen und Grundwasser). U.S. Environ. Protection Agency, Report No. EPA-TR-498-74, Research Triangle Park, N. Carolina, 22 pp.

Canada Department of Health and Welfare. (1979). Polycyclic aromatic hydrocarbons. Environmental Health Criteria Document 80-EHD-50. Environmental Health Directorate, Health Protection, Branch, DHW, Canada.

Commins, B. T. (1962). *Natl. Cancer Inst. Monogr.* **9:** 225.

Conney, A. H. (1967). *Pharmacol. Rev.* **19:** 317.

Conney, A. H., E. C. Miller, and J. A. Miller. (1957). *J. Biol. Chem.* **228:** 753.

Costle, D. M. (1978). *Chem. Engin. News* **7.** January 30.

Eadie, B. J., W. Faust, W. S. Gardner, and T. Nalepa. (1982). *Chemosphere* **11:** 185.

Elcombe, C. R. and J. J. Lech. (1979). *Toxicol. Appl. Pharmacol.* **49:** 437.

Epstein, S. S. and H. Shafner. (1968). *Nature* **219:** 285.

Erhardt, J. P. (1972). *L'Evolut. Med.* **16:** 269.

Fabacher, D. L. and E. Hodgson. (1976). *Appl. Pharmacol.* **38:** 71.

Fabacher, D. L. and E. Hodgson. (1977). *J. Toxicol. Environ. Health* **2:** 1141.

Geacintov, N. E. (1973). U.S. EPA Report No. EPA-65011-74-010 (PB-238294). U.S. EPA. Washington, D.C.

Gibson, D. T. (1975). Effects and sinks of hydrocarbons in the aquatic environment. American Institute of Biological Sciences, Arlington, Virginia, p. 225.

Guenthner, T. M., J. M. Fysh, and D. W. Nebert. (1979). *Pharmacology* **19:** 12.

Hallett, D. J., R. J. Norstrom, F. I. Onuska, and M. E. Comba. (1977). Proceedings of the Second International Symposium on Glass Capillary Chromatography, R. E. Kaiser, Ed. Bad Durkheim, Germany.

Hallett, D. J., R. D. Smillie, D. T. Wang, F. I. Onuska, M. E. Comba, and R. Sonstegard. (1978). International Symposium on the Analysis of Hydrocarbons and Halogenated Hydrocarbons. Canada Centre for Inland Waters, Burlington, Ontario.

Hallett, D. J., F. I. Onuska, and M. E. Comba. (1983). *Mar. Environ. Res.* **8:** 73–85.

Hangebrauck, R. P., D. J. Von Lehmden, and J. E. Meeker. (1967). Sources of polynculear hydrocarbons in the atmosphere. U.S. Dept. of Health Education and Welfare, 999-AP-33.

Harshbarger, J. C. (1977). *Ann. N.Y. Acad. Sci.* **298:** 289.

Heidelberger, C. (1975). *Ann. Rev. Biochem.* **44:** 79.

Herbes, S. E., G. R. Southworth, D. L. Shaeffer, W. H. Griest, and M. P. Maskarinec. (1980). In: H. Witschi, Ed., *The scientific basis of toxicity assessment*. Elsevier Biomedical Press, p. 113.

IARC. (1973). Monographs on the Evaluation of Carcinogenic Risk of Chemicals to Man, 3: Certain polycyclic aromatic hydrocarbons and heterocyclic compounds. IARC, Lyons.

International Joint Commission. (1976). Great Lakes water quality, Appendix E: Status report on the persistent toxic pollutants in the Lake Ontario basin. International Joint Commission, Windsor, Ontario.

Jerina, D. M. et al. (1977). In: H. H. Hiatt, J. D. Watson, and J. S. Winston, Eds., Origins of human cancer. Book B: Mechanisms of carcinogenesis. Cold Spring Harbour Laboratory, p. 639.

Kaminsky, L. S., L. J. Piper, D. N. McMartin, and M. J. Fasco. (1978). *Toxicol. Appl. Pharmacol.* **43:** 327.

Katz, M., A. Ho, C. Chan, and H. M. Tosine. (1979). Atmospheric reactions of PAH. Report submitted to ARB for FY 1978.

Law, R. J. (1980). *Sci. Total Environ.* **15:** 37.

Lech, J. J. and J. R. Bend. (1980). *Environ. Health Persp.* **34:** 115.

Lee, R. F., Sauerheber and G. H. Dobbs. (1972). *Mar. Biol.* **17:** 201.

Lehninger, A. L. (1974). *Biochemistry*, 2nd ed. Worth Publishers.

Lu, P. Y., R. L. Metcalf, N. Plummer, and D. Mandel. (1977). *Arch. Environ. Contam. Toxicol.* **6:** 129.

Lunde, G. and A. Bjorseth. (1977). *Nature* **268:** 518.

Marquardt, H. (1977). Air pollution and cancer in man. IARC Scientific Publication No. 16, Lyon, p. 309.

Marquardt, H., S. Baker, B. Tierney, P. L. Grover, and P. Sims. (1977). *Intl. J. Cancer* **19:** 828.

McCarthy, J. F. (1983). Proceedings of the Sixth International Symposium on Polynuclear Aromatic Hydrocarbons. In press.

Meier, J. R., S. Monarca, R. J. Bull, and F. C. Kopfler. (1982). Thirteenth Annual Meeting of the Environmental Mutagen Society, Boston, Massachusetts, March 1982.

Mikol, Y. B. and F. Decloitre. (1979). *Toxicol. Appl. Pharmacol.* **47:** 461.

Ministry of Environment. (1979). Polynuclear aromatic hydrocarbons: A background report including available Ontario data. MOE, ARB-TDA Report 58-79.

Moore, R. W., S. D. Sleight, and S. D. Aust. (1978). *Toxicol. Appl. Pharmacol.* **44:** 309.

Morello, A. (1965). *Can. J. Biochem.* **43:** 1289.

National Academy of Science. (1972). Particulate polycyclic organic matter. NAS, Washington, D.C.

Natusch, D. F. S. (1983). In: M. Cooke and A. Dennis, Eds. Proceedings of the Sixth International Symposium on Polynuclear Aromatic Hydrocarbons. Battelle Press, Columbus, Ohio.

Nebert, D. W. and H. V. Gelboin. (1968). *J. Chem. Biol.* **243:** 6242.

Nebert, D. W., H. J. Ecsein, M. Negishi, M. A. Lang, L. M. Hjelmeland, and A. B. Okey. (1981). *Annu. Rev. Pharmacol. Toxicol.* **21:** 431.

Oesch, F., D. Raphael, H. Schwind, and H. R. Glatt. (1977). *Arch. Toxicol.* **39:** 97.

Ontario Ministry of Natural Resources. (1977). Annual Report, p. 24.

Palm, D. J. and M. Alexander. (1978). Damage assessment studies following the Nepco 140 oil spill of June 27, 1976, on the St. Lawrence River, New York. U.S. EPA, New York, St. Lawrence—Eastern Ontario Commission, Watertown, New York.

Parkinson, A., R. Cockertine, and S. Safe. (1980a). *Biochem. Pharmacol.* **29:** 259.

Parkinson, A., R. Cockertine, and S. Safe. (1980b). *Chem. Biol. Interact.* **29:** 277.

Pierce, R. C. and M. Katz. (1975). *Environ. Sci. Technol.* **9:** 347.

Poglazova, M. N., A. Y. Khesina, G. E. Fedoseava, M. N. Meisel, and L. M. Shabad. (1972). *Dokl. Akad. Nauk SSSR* **204** (1):222.

Radding, S. B. et al. (1976). The environmental fate of selected polynuclear aromatic hydrocarbons. U.S. EPA, EPA-560/2-75-009.

Robertson, L. W., A. Parkinson, and S. Safe. (1980). *Biochem. Biophys. Res. Commun.* **92:** 175.

Roubal, W. T., K. C. Tracy, and D. C. Malias. (1977). *Arch. Environ. Contam. Toxicol.* **5:** 513.

Shabad. L. M. (1968). *Z. Krebsforsch* **70:** 204.

Shabad, L. M. et al. (1971). *J. Natl. Cancer Inst.* **47:** 1179.

Smillie, R. D., D. Robinson, and D. T. Wang. (1978). OTC Report No. 7809, Ontario Ministry of the Environment.

Snyder, R., E. Bromfeld, F. Uzuki, S. Kang, and A. Wells. (1966). *Pharmacologist* **8:** 218.

Sonstegard, R. A. (1977). *Ann. N.Y. Acad. Sci.* **298:** 261.

Southworth, G. R. (1979). Proceedings of the Second Annual Symposium of Aquatic Toxicology. ASTM, Philadelphia, Pennsylvania.

Statistics Canada. (1978). Coke and Coal Statistics, December 1977. Cat. No. 45-002, June 1978, Table 9.

Scribner, J. (1973). *J. Nat. Cancer Inst.* **50:** 1717–1719.

Strosher, M. T. and C. W. Hodgson. (1976). Special Technical Publication 573, American Society for Testing and Materials, p. 259.

Thomos, J. F., M. Mukai, and B. B. Tebbens. (1968). *Environ. Sci. Technol.* **2:** 33.

Tosine, H. M. (1978). Capillary GC analysis of aqueous B(a)P photodecomposition. M.Sc. Thesis, York University.

Tunek, A. and F. Oesch. (1979). *Biochem. Pharmacol.* **28:** 3425.

Vandermeulen, J. H. and W. R. Penrose. (1978). *J. Fish. Res. Board Can.* **35:** 643.

Vodicnik, M. J., C. R. Elcombe, and J. J. Lech. (1981). *Toxicol. Appl. Pharmacol.* **59:** 364.

Walker, C. H. (1978). *Drug Metabol. Rev.* **7**(2): 295.

Willis, D. E., M. E. Zinck, D. C. Darrow, and R. F. Addison. (1978). In: B. R. Afghan and D. Mackay, Eds., *Hydrocarbons and halogenated hydrocarbons in the aquatic environment*. Plenum Press, New York, p. 53.

Winchester, J. W. (1980). In: O. Hutzinger, Ed. *The handbook of environmental chemistry*, Vol. 2A, p. 21 Springer-Verlag, Heidelberg.

Wydner, E. and B. Hoffman. (1962). *Cancer* **15:** 103–108.

Yang, S. (1983). In: M. Cooke and A. Dennis, Eds. Proceedings of the Sixth International Symposium on PAH. Battelle Press, Columbus, Ohio.

Zepp, R. G. and P. F. Schlotzhauer. (1969). In: P. W. Jones and P. Lever, Eds., *Polynuclear aromatic hydrocarbons*. Ann Arbor Sci. Publishers, Ann Arbor, Michigan, p. 141.

Zobell, C. E. (1971). Proceedings of a joint conference of the American Petroleum Institute, Environmental Protection Agency, and U.S. Coast Guard on prevention and control of oil pollution. American Petroleum Institute, Washington, D.C., p. 441.

11

ORGANIC POLLUTANTS IN LAKE ONTARIO

W. M. J. Strachan

National Water Research Institute
Canada Centre for Inland Waters
Burlington, Ontario

C. J. Edwards

International Joint Commission Regional Office
Windsor, Ontario

1. INTRODUCTION

The story of toxic substances in Lake Ontario is really that of industrialization in the region and particularly in the Niagara River area. In the late

1800s, the availability of cheap electricity attracted chemical industry to the area and today the list of industrial chemical companies there reads like a *Who's Who* of the corporate chemical world—Dupont, Allied, Olin, FMC, Hooker. These and other smaller chemical companies discharge toxic substances directly to the river in approximately $10^6 m^3$ of wastewater daily plus an equally unknown quantity of chemicals via municipal sewage treatment plants (Hang and Salvo, 1981). In addition, chemicals enter the river from 200 active and inactive landfill sites (DEC and EPA, 1979). There are, of course, other locations and mechanisms for the input of toxic chemicals to the system. They tend to be more diffuse and less localized, however, and have not received the attention that the Niagara River has although their collective input may be important.

Two international agreements related to toxic chemicals and Lake Ontario have been signed by governments of Canada and the United States. The first of the Great Lakes Water Quality Agreements, signed in 1972, focused predominantly on problems of eutrophication and control of nutrient loading, phosphorous in particular. There was, however, a requirement to develop water quality objectives for a number of specified and unspecified toxic chemicals. This was done and these were incorporated in the subsequent agreement signed in 1978. During the interval between these two signings, a great deal of information was developed concerning toxic substances in the Great Lakes system, especially in Lake Ontario. As a consequence of the concerns arising from this new information, the focus of the second agreement was on the control of toxic substances and on the identification of their presence, trends and effects. However, while objectives continue to be developed, new substances identified and established contaminants surveyed, progress in controlling these materials since the 1978 signing appears to have been slow. The Great Lakes community is faced with a growing list of identified contaminants (brought about at least in part by our increasing ability to detect them) without an apparent corresponding improvement in the effective control of their release to the environment.

It is the intent of this chapter to identify some of the major observations of these substances in Lake Ontario. Such identification is a prelude to assessing the substances, which in turn is a necessary step in determining if and which controls may be required. Assessment of the hazard posed by these chemicals toward the system depends on two general factors, exposure and effects. Effects is not the subject of this chapter; exposure is. Exposure can be predicted, at least qualitatively, from physicochemical properties, including transformations, or it can be measured. Both approaches are necessary but, given the number of substances, it is unlikely that sufficient measurements can be undertaken to be acceptable as a description of the exposure problem represented by each chemical. This chapter proposes to provide a description of some of the surveillance results for those chemicals which have been observed, some of which are now included in the routine programs of the Great Lakes institutions. Lakewide efforts are few and have been undertaken largely under the auspices of the Great Lakes Water Quality Agreements. The Great Lakes International Surveillance Program (GLISP) is the vehicle for reporting results from the various agencies involved. Other information, usually of a more localized nature, is provided by researchers in the several universities of the Lake Ontario basin. Some, but not all, of the governmental data gets reported in the Annual Reports of the Water Quality Board, particularly in the Appendices by the Surveillance Committee and these reports are recommended to interested readers.

Where possible, the exposure information for this chapter has been organized by chemical under subparagraphs on water, surface runoff, air, and biota. The Niagara River, as the major input (approximately 90%) of water to the lake and a known source of contamination, receives extra attention. It will become obvious, however, that for most substances, adequate information to allow a lakewide evaluation of exposure does not exist in all of these categories. It should also be of great concern that the concentration or loading of a single chemical may not adequately represent its hazard exposure since the significance of synergism and antagonism is not generally taken account of in the effects side of hazard evaluation.

2. DDT RESIDUES

1,1-Di (*p*-chlorophenyl)-2,2,2-trichloroethane (DDT) and its environmental derivatives are probably the oldest and most studied of the organic contaminants in Lake Ontario (Henderson et al., 1969; Reinert, 1970; Lichtenberg et al., 1970). Its methoxy analog, methoxychlor, is a more recent (and less studied) contaminant. Use of DDT has been essentially banned from the basin since 1971 although it continues to be reported as its environmentally oxidized (DDE), reduced (DDD or TDE), or unchanged forms. The collective total is referred to as ΣDDT or t-DDT.

The Niagara River has been shown to be a major source of organic contaminants entering Lake Ontario. The report by the Canada–Ontario Review Board (CORB, 1981) summarized a number of the studies on this river up to 1981. In a 1980 study, Environment Canada noted that approximately 10% of filtered water samples from the lower Niagara River area had trace contaminations (less than 0.005 μg/liter) of one or another of the DDT residues. A similar investigation by the Ontario Ministry of Environment at virtually the same location gave similar results. In contrast to the water results, all suspended sediments from the Environment Canada study contained DDE and additionally had other related residues in 30–75% of the samples. Concentrations in these samples had a mean t-DDT level of 0.04 μg/g. A more recent study (Kuntz and Warry, 1983) using 1981 samples from Niagara-on-the-Lake and more sensitive methodology, observed overall means for total residues in the dissolved and suspended fractions of 0.0003 μg/liter and 0.04 μg/g, respectively. These values translate into annual lake loadings of approximately 60 kg from each of these compartments.

Atmospheric input (Strachan and Huneault, 1979; Strachan et al., 1980) of DDT residues to Lake Ontario during 1976–1977 occurred at mean rain concentrations of 0.005 μg/liter and snow levels of 0.001 μg/liter. These values, coupled with mean precipitation levels, indicate direct loadings to the lake by rain and snow to be approximately 80 kg/yr. Loading by gaseous exchange is unknown although Eisenreich et al. (1980, 1981) estimates comparable inputs from wetfall and from vapor exchange plus dryfall; air concentrations were reported at 0.01–0.05 ng/m^3 for the region. As a consequence, a reasonable estimate for total atmospheric loading of DDT residues is 150 kg/annum, a figure comparable to loadings from the Niagara River. The insecticide methoxychlor, also reported by Eisenreich et al. (1980), has a single value of 1 ng/m^3 for an atmospheric concentration, which would indicate an estimated direct lake loading of 1.9×10^3 kg/annum. This is probably high and requires confirmation.

In the open waters of Lake Ontario, "dissolved" DDT residues have been looked for but seldom found. Solubility of DDT reported variously as 0.2 to 100 μg/liter (Brooks, 1974) is much above detection capability and hence it must be concluded that removal to suspended and settled sediment or volatalization is responsible for the low or nondetectable levels. At detection levels of 0.005 μg/liter, none of the DDT related chemicals were found in a 1975 offshore study (WQB, 1976). Similar samples collected during the 1972 International Field Year on the Great lakes, IFYGL (Haile et al., 1975), on the other hand, recorded 0.016–0.057 μg/liter t-DDt residues (75% DDE, 16% unchanged DDT, approx. 10% DDD). It is probable that this difference arises from the presence of particular matter in the latter study. Data in STORET (1982), a U.S. environmental data base containing information from a variety of sources, is also conflicting. "Whole" water samples from 1974 to 1977 at both inshore and offshore locations had essentially zero contamination by DDT, DDE, and DDD (0.0001 μg/liter means) while smaller numbers of filtered water samples from 1971 to 1975 showed equal inshore and offshore concentrations at means of 0.005–0.01 μg/liter. Methoxychlor, in a substantial number of determinations also during 1974–1977, had offshore levels (mean 0.013 μg/liter exceeding those at inshore locations (mean 0.0015 μg/liter) by a factor of almost 10. A possible combination of atmospheric deposition and nearshore adsorption to suspended matter may account for this spatial difference.

Sediments have received considerable attention, especially since their successful use in defining the Niagara and Oswego Rivers input of mirex to Lake Ontario (Holdrinet et al., 1978). Among the first extensive surveys for DDT residues in sediments was the study by Frank et al. (1974) in the Bay of Quinte region. A concentration of 0.025 μg/g was reported as a mean for t-DDT in surficial sediments there. Considerably higher values were found in tributaries. In a subsequent study of surface sediment samples from the open lake, Frank et al. (1979) found 1968 t-DDT residues (approximately equal contributions by DDT, DDE, and DDD) averaged 0.043 μg/g with a high of 0.218 μg/g. In the same study, these workers also reported data for three cores—one from off the Niagara River and two from the eastern or Rochester depositional basin. The DDT residue profiles were determined and, aside from the observations that DDT residues did not appear before the 1960s, the authors estimated a lakewide annual loading rate of 246 kg. Atmospheric loadings (150 kg) plus that via the Niagara River (120 kg) would therefore appear to be the major sources. Data in STORET (1982) from mixed sources indicate approximately 3–14 times higher inshore sediment concentrations for DDT residue forms compared with offshore ones. Results are difficult to compare due to different sample treatments, ways of reporting data and, in some cases, the small number of data points; indications are, however, that there is a marginal dominance of DDT and DDD forms over DDE in nearshore environments. Each of these forms was observed at mean concentrations of 0.04–0.06 μg/g (dry weight). Methoxychlor, which was only found in nearshore environments, was at 0.03 μg/g.

Since the signing of the 1978 agreement, and indeed somewhat before, a surveillance program for the contaminants in biota has been in place. Details of some of the fish aspects of the Canadian part of this are represented in the chapter by H. Shear. Generally, there are few studies in which both the samples and the analytical methodology can be compared from area to area or over a significant time period. In Canada, the young-of-the-year spottail shiner program is one such program addressing both these concerns; another is that carried out by the Great Lakes Fisheries Research Branch at the Canada Centre for Inland Waters. Results from other studies appear in STORET (1982) but for most biota, insufficient information on the samples is presented in the summaries to allow necessary comparisons. In the United States, the Toxic Substances Monitoring Program of the New York Department of Environmental Conservation (e.g., DEC, 1983) and the National Pesticides Monitoring Program of the Fish and Wildlife Service are also included to address these problems. In addition to the preceding fish programs, annual sampling of herring gull eggs is carried out by the Canadian Wildlife Service. Results of many of these programs are presented annually in the reports of the Surveillance Work Group of the International Joint Commission (e.g., WQB, 1980, 1981). An attempt to summarize some of the data from these is presented in the attached Table 11.1. The general consensus from those studies with comparable data is that levels of DDT residues in biota (and presumably in the basin generally) have decreased, although at a slower rate than in other parts of the Great Lakes ecosystem. It should, however, be noted that conclusions about lakewide averages and trends may not be applicable at all locations in the basin. Indeed there is some concern that the improving trend is being reversed in the lower Niagara River.

3. MIREX

Mirex, along with PCBs, DDT residues, and possibly dieldrin, is one of the more extensively monitored toxic substances found in Lake Ontario. Pro-

Table 11.1. DDT Residues (μg/g) in Selected Lake Ontario Biota

Sample	Lake Trout		Smelt			Coho Salmon			Spottail Shiner	Herring Gull Eggs
Year	DEC[a]	DFO	DEC	DFO	EPA	OME	DEC	DFO	OME	CWS
1972				1.8e				0.9e		34
1973				1.5e				1.7e		
1974	7.7e[b]			0.4e						23
1975	1.3e				1.4w		0.93e		0.17w	22
1976	0.91e			0.25e			0.93e	0.69f	0.28w	18
1977		2.3e		0.6e		1.4w		1.6e	0.21w	
		2.7w		0.6w						
1978		1.3w		0.44w		0.64w			0.18w	
1979		1.6w		0.39w		0.81w			0.06w	
1980		0.62w		0.25w		0.74w			0.05w	7.7

[a] CWS = Canadian Wildlife Service (Canada); DEC = Department of Environmental Conservation (New York State); DFO = Department of Fisheries and Oceans (Canada); EPA = Environmental Protection Agency (United States); OME = Ontario Ministry of the Environment (Province of Ontario).

[b] e = edible portion; f = fillet; w = whole fish.

duced originally as a pesticide for use in the southeastern United States and subsequently used also as a flame retardant, it escaped to the Lake Ontario ecosystem from the Buffalo, N.Y., production plant (via the Niagara River) and from an industrial accident (via the Oswego River) (Kaiser, 1978). It was also processed at locations on the Credit and Speed/Grand Rivers in Ontario although no significant input was observed at these locations (Task Force on Mirex, 1977).

Soluble mirex has been looked for but even in the waters of the Niagara, the major established source of this compound, it has not been found at a detection level of 0.001 μg/liter. (One raw water sample from Niagara-on-the-Lake (CORB, 1981) was reported to contain 0.0001 μg/liter but this can not be considered statistically significant.) Mirex in the suspended matter of the Niagara has been reported for the period 1979–1981. The Canada-Ontario report (CORB, 1981) also includes a figure which indicates that 1980 levels in this medium increased from 0.002 μg/g in January/February to 0.01 μg/g in July after which levels dropped off sharply. Possibly this reflects increased loading due to spring and early summer runoff. In another report on 1979–1980 samples, Kuntz and Warry (1983) found a mean level of 0.012 μg/g in 76% of the samples collected at Niagara-on-the-Lake. Warry and Chan (1981) reported 0.019 μg/g for similar samples. These authors estimate a loading via this mechanism of 13–20 kg/annum. The levels of this compound in the suspended sediments from this major source are down from those of the most contaminated lake sediments. Hopefully this indicates decreased loadings.

The possibility of atmospheric deposition of mirex has been investigated (Strachan et al., 1980) but to date, this chemical has not been found in rain at detection levels of 0.001 μg/liter.

The chemical mirex has never been observed "dissolved" in the waters of Lake Ontario. Its chemical structure correctly suggests that it is among the most insoluble (<1 μg/liter; Alley, 1973) of the persistent organic substances encountered in the system. STORET (1982) records a 1977 sampling survey in which whole water samples from 38 nearshore and 6 offshore locations were examined. In none of them was any mirex detected.

Sediments have been examined and identified as the major lake compartment containing mirex. The study by Holdrinet et al. (1978) of 1968 surficial sediment samples from a 229-site grid demonstrated the usefulness of this medium in locating and identifying sources. The mean level observed in the third of the samples which were found to contain mirex was 0.069 μg/g; they were located largely along the southern shore of the lake and in Mexico Bay in the eastern basin. This pattern reflects the inputs, the water currents and the anticipated deposition of suspended matter. Confirmation of some of these results with additional samples from the Oswego River area in 1976 was also reported. STORET (1982) includes a smaller number of samples from nine nearshore stations collected during 1981. These samples were found to contain an average of 0.22 μg/g, a value somewhat higher than that reported by Holdrinet et al. (1978). In 1980, the Water Quality Board of the International Joint Commission indicated Oswego Harbour as a problem area as a consequence of these levels; the Niagara River was similarly identified.

By far the compartment which has received the most attention with respect to mirex has been the fishes. An extensive compilation of pre-1976 data is found in the Report of the (Canadian) Task Force on Mirex (1977) and is further summarized by Kaiser (1978). Salmonid and trout species were reported at 0.19 μg/g and bass, perch, and carp at 0.09 μg/g. Since 1976, further surveillance has been undertaken. Data from STORET (1982) is extensive—380 open lake and 85 nearshore samples with mean levels of 0.18 μg/g and 0.34 μg/g, respectively, covering the period 1971–1978. More recently, the Ontario Ministry of the Environment young-of-the-year spottail shiner program (WQB, 1980) does not indicate any significant trend during the 1976–1979 period; annual mean levels varied randomly from undetectable to 0.032 μg/g. Similar comments can be made for lake trout (WQB, 1980) samples over the same period with means averaging 0.23 μg/g for this species. Coho salmon and smelt did, however, indicate a downward trend between 1977 and 1979 with mean levels dropping from 0.16 to 0.05 μg/g and from 0.11 to 0.06 μg/g, respectively. Recent results for 1981 samples (OME, 1982a) would indicate that unfortunately little improvement has occurred since 1979. Lake trout, coho salmon and smelt from similar Canadian sites as used in the earlier studies were reported to contain mean concentrations of 0.14, 0.04, and 0.06 μg/g, respectively. The number of fish, however, was reduced relative to previous years. The study periods for some of the species is reaching a length such that significant changes should be discernible—presuming that they have occurred. Mirex is a very persistent substance, however, and the sedimentation rate is low in Lake Ontario. It may be years before this substance disappears entirely from Lake Ontario fishes.

Herring gulls are the organism for which the most consistent data set exists. The Canadian Wildlife Service has collected and preserved sample extracts from eggs of this species since 1972 (Hallett et al., 1976; WQB, 1976). Levels of 280 μg/g (lipid) for mirex plus photomirex were found at that time for adult birds. A ratio of mirex: photomirex = 2 was consistently observed. Concentrations in eggs (wet weight) of 11.0 μg/g were found for 1972 samples of this matrix. Since then, levels appear to have declined. The 1980 egg extracts contained only 1.7 μg/g mirex (WQB, 1981) and a figure in the 1980 Report (WQB, 1980) indicates a steady decline with a half-life of approximately 2.5 yr.

4. POLYCHLORINATED BIPHENYLS

Polychlorinated biphenyls (PCBs) are industrial chemicals used primarily in the electrical industry as dielectrics in large transformers and capacitors.

The term PCB applies to a wide range of congeners, those with a high degree of chlorination being of greatest environmental concern. PCBs were one of the earliest compounds for which an objective was developed under the Canada-U.S. Great Lakes Water Quality Agreement (1972). In the 1974 Report of the Water Quality Board (WQB, 1975), the proposal was made and agreed to subsequently, to limit water concentrations to 0.001 μg/liter (the detection level at the time) and biota tissue levels to 0.1 μg/g. This, with the Ontario Ministry of the Environment guideline for the disposal of dredged material (0.05 μg/g), are the yardsticks which are available to indicate the presence of unsafe levels of PCBs.

The Niagara River has been strongly implicated as a source of PCBs and has been extensively examined for these chemicals. Water concentrations (dissolved) in the lower Niagara River during 1980 were reported by the Water Quality Board (WQB, 1981) to be nil at a detection limit of 0.01 μg/liter. A different set of samples, collected from the same area during 1980 (CORB, 1981) indicated that 93% contained PCBs at a mean level of 0.010 μg/liter. Whether this difference was due to filtration or not is unclear from the CORB report. A study by Kuntz and Warry (1983) of 1981 raw water samples from Niagara-on-the-Lake reported 97% of the samples had mean concentrations of 0.010 μg/liter. Yet another, recent investigation (Fox et al, 1983) of 1981 water samples from the lower Niagara found 0.011 μg/liter PCB for raw water. Suspended matter in the lower Niagara River universally has been reported to contain substantial levels of PCBs. For 1979–1980 samples, Warry and Chan (1981) found a mean concentration of 0.96 μg/g (0.53 μg/g, if three high values were omitted) and Kuntz and Warry (1983), extending the same data base to 1981, reported a new mean of 0.72 μg/g. The CORB report (1981) contains data with a mean of 0.64 μg/g for 1980 samples and in a study by Fox et al. (1983) on 1981 collections, means of 0.6–6.0 μg/g were observed for several size fractions in the range 75μm to 700μm. Only the fraction retained by a plankton net was investigated, however. Considering these reports, conservative means for the water and the suspended fractions appear to be 0.005 μg/liter and 0.6 μg/g PCBs, respectively. These values, together with a flow of 6400 m^3/s and a suspended load of 8.41 μg/liter (Kuntz and Warry, 1983), lead to rough estimates for annual loadings of 1000 kg PCBs via each of water and suspended matter.

Atmospheric levels of PCBs in rain and snow during 1976–1977 were reported to be 0.02–0.03 μg/liter (Strachan et al., 1980). Air levels for the Great Lakes region have been estimated to be at 1 ng/m^3 (Eisenreich et al., 1981). These data, together with precipitation information for the area, have been used (Eisenreich et al., 1981) to predict total loadings for Lake Ontario via wet and dry precipitation mechanisms of 2300 kg/annum.

Other possible input mechanisms for Lake Ontario include wastewater discharges, landfill leachate and tributary input. Samples collected during 1975 (OME, 1976) contained volume weighted average concentrations of 0.3 μg/liter leachate from five landfill sites and 0.3 mg/liter discharge from 16 municipalities including Toronto and Hamilton. The wastewater loading to the lake was estimated to be 73 kg during 1975 (1974 data for a similar set of municipalities indicated 200 kg); that from landfill cannot be determined without the flows. Similar information from the U.S. side of the lake is needed. Canadian river and creek mouths emptying into Lake Ontario were examined for PCBs during 1979 (OME, 1980). Of the reported samples, 55% were at less than the detection limit of 0.010 μg/liter while the remainder had a mean level of 0.11 μg/liter. It is not certain whether this is a representative value since sufficient seasonal samples have not been investigated. Loadings via this mechanism have not been evaluated either but could represent a significant input (>500 kg/annum) if the annual flows exceed $5 \times 10^9 m^3$ (2.5% of Niagara) at this concentration.

In the open waters of Lake Ontario, there are few examples of dissolved PCBs. STORET (1982) includes data on a 1972 study at eight offshore locations with a mean concentration of 0.06 μg/liter. Another report from this data base pertains to 152 samples collected during 1974–1977; levels of 0.023 and 0.017 μg/liter were observed, respectively, at 30 offshore and 122 nearshore locations. An open lake survey during 1975 in which water was definitely filtered (0.45μm), however, reported no detectable PCBs at a limit of 0.005 μg/liter. Levels for dissolved PCBs must therefore be considered suspect.

Lake Ontario surficial sediment concentrations for PCBs have been reported for samples as far back as 1968 (Frank et al., 1979). This study reported data for 229 sites and included a lakewide mean of 0.059 μg/g quantifiable levels at 91% of the locations. Other early data include an IFYGL study during 1972 (WQB, 1976) at eight offshore lake sites, largely near urban centers. The mean for these analyses was 0.11 μg/g (range nd–0.25 μg/g). STORET (1982) data for samples collected during 1972–1981 showed that largely nearshore samples contained PCBs at concentrations of approximately 0.2 μg/g with highs to 1.6 μg/g. Elevated levels for surfical sediments were found off the Niagara River and reflect the high concentrations found in the suspended materials of that river. Fox et al. (1983) found levels of 0.6 μg/g in surficial samples taken from this area in 1981 and Durham and Oliver (1983), using radiodating techniques on cores from the same area, found approximately 1.6 μg/g down to 14 cm core depths after which the levels decreased considerably.

Due to their relatively high environmental levels, PCBs have received more attention, at least recently, than any other contaminant in Lake Ontario. Surveillance of biota has been high among the compartments examined and among biota, greatest preference has been shown for fishes. Despite this, there have been several other biota subcompartments investigated. The 1972 IFYGL study (Haile et al. 1975) determined levels in net plankton from eight offshore sites near urban areas and found mean levels of 6.1 μg/g (dry weight); means for cladophora were 0.52 μg/g, and for benthic fauna from near three harbors were 0.47 μg/g. Another net plankton study undertaken in 1975 (WQB, 1976) at 11 offshore stations, had a mean PCB level of 1.9

Table 11.2. PCB Residues (μg/g) in Selected Lake Ontario Biota

Sample	Lake Trout			Smelt			Coho Salmon			Spottail Shiner	Herring Gull Eggs
Year	DEC[a]	DFO	OME	DEC	DFO	OME	DEC	DFO	OME	OME	CWS
1970				2.2f			7.9				
1971							6.7				
1972					5.0e		4.7	4.1e			204
1973					7.3e			6.7e			
1974	7.7				1.7e		6.3				155
1975	9.4			2.1f			8.4			0.69w	145
1976	7.1				1.2e	2.6	6.1	3.7e	11.6f	1.3w	138
1977		5.0w	4.9		1.4e	1.5w		3.2e	3.0w	1.5w	104
					1.5w			3.0w			
1978		7.1w	6.1		1.8w	1.6		3.0w		1.1w	61
1979		3.8w	3.8		0.8w	0.75		1.2w	2.8w	0.46w	
1980		4.8w			1.1w			2.3w		0.31w	41

[a] CWS = Canadian Wildlife Service (Canada); DEC = Department of Environmental Conservation (New York State); DFO = Department of Fisheries and Oceans (Canada); OME = Ontario Ministry of the Environment (Province of Ontario).

[b] e = edible portion; f = fillet; w = whole fish.

μg/g. Fox et al. (1983) in a study of benthic fauna from six sites offshore from the Niagara River/Welland Canal area, found levels in oligochaetes at an average of 1.9 μg/g and in amphipods, of 9.0 μg/g. Such concentrations, to the extent that these organisms play a role in bioaccumulation in the Lake Ontario food webs, are of great concern.

Fishes are the primary focus for surveillance activities. There appears to be a great deal of difficulty obtaining the right sort of data, however, in order to establish trends. As with DDT, problems include laboratory intercomparisons, selection of sample sites and species and analytical matrix (fillet, whole fish, etc.). These, together with the need to carry out the program over a protracted period of time, have meant that the necessary data has often been missing. Some of the more suitable data are presented in Table 11.2. In the case of PCBs, the trend picture is similar to that for DDT—general lakewide improvement.

STORET (1982) also reported many studies involving fish "tissues" collected during the period 1971–1980. Generally, only small numbers of specimens were sampled. Two reports, however, were for substantial ($n > 500$ each) numbers. One of these, collected during 1972–1978, reported nearshore and offshore mean levels of 6.1 and 17.0 μg/g, respectively; the other, for the period 1975–1977, reported corresponding values of 4.9 and 8.2 μg/g. These, and many of the reported values from Table 11.2, are far above the IJC Objective of 0.1 μg/g (for the protection of fish-consuming birds and animals) and also above the various food guidelines of the two countries. PCB levels may be decreasing in Lake Ontario, but they still have a long way to go.

The studies with herring gull eggs done by the Canadian Wildlife Service show a half-life, $T\frac{1}{2}$ for the clearance of PCBs, of approximately 3.5 yr (WQB, 1980); that for spottail shiners over the last 4 yr appears to be roughly 1.5 yr. Lakewide comparisons of annual means may not be a particularly useful form of trend analysis since some species (e.g., spottail shiners) are local in habit and should be compared locally. When this was done in the shiner program for the Niagara region (WQB, 1980), a downward trend was observed which, unfortunately, showed a statistically significant upturn in levels for the 1980 samples. Other nearshore areas were generally down.

5. DIELDRIN AND ALDRIN

Dieldrin, like DDT, has been studied for a long time, at least in Lake Ontario. Dieldrin (and aldrin from which it is readily converted environmentally) has been used as a soil and foliage pesticide since the 1950s. The desired properties of these chemicals, i.e. the high stability toward alkali and mild acids, and the high toxicity to insects, have also made them an environmental concern.

The Niagara River report (CORB, 1981) has several notations of dieldrin water concentrations. While one reported no dieldrin or aldrin at levels above 0.001 μg/liter for Niagara-on-the-Lake in 1980, another reported 67% of water samples for the same time and place at 0.001 μg/liter dieldrin, a level equal to the detection level of that study. Dieldrin in samples from the raw water intake (OME, 1982b) at Niagara-on-the-Lake during 1979–1980 were at the detection level. Yet another study (Kuntz and Warry, 1983) observed 0.0006 μg/liter dieldrin in 93% of 1980–1981 raw water samples. Suspended matter from the mouth of the Niagara during 1980 contained 0.004 μg/g dieldrin; there were no reports of the presence of aldrin. Warry and Chan (1981) reporting on 1980 samples also found 0.005 μg/g in the suspended load. It appears, therefore, that while the dissolved fraction may be responsible for approximately 100 kg/annum, the suspended load adds something less than 10 kg.

Atmospheric inputs of these two compounds were investigated by Strachan and Huneault (1979). Trace levels of 0.001–0.002 μg/liter were found in rainfall and comparable levels were observed for accumulated snowfalls. Eisenreich et al. (1980, 1981) used regional air levels of 0.05 ng/m^3 and the foregoing concentrations in precipitation to estimate a total direct loading to the lake of 130 kg/annum, most of which is attributable to gaseous phase transfer.

Early reports (WQB, 1976) included mean concentrations in water at 0.0048 μg/liter for various 1971 locations around the lake; this contrasts with no detections at a 1975 limit of 0.005 μg/liter in filtered water in the same report. Since that report, water concentrations have been reported for the lake itself only in STORET (1982) where mean, nearshore levels from approximately 170 samples (1972–1977) were 0.00011 and 0.00037 μg/liter for aldrin and dieldrin, respectively. Coresponding, offshore values (approximately 30 samples) were 5–10 times these at 0.0012 and 0.0017 μg/liter respectively.

Offshore sediment concentrations in 1968 were observed at 0.0018 μg/g in 19% of samples and at 0.0012 μg/g in 1971 (WQB, 1976). Sediment levels are also reported in STORET (1982). No aldrin was observed in 1972–1979 nearshore samples but analyses for dieldrin were positive—0.012 μg/g for 34 nearshore and 0.0013 μg/g for seven offshore sites during the period 1972–1981. Another report which included sediment levels of dieldrin was that of Frank et al. (1974) who reported a mean of 0.005 μg/g in the Bay of Quinte following two years in which approximately 2000 kg was applied in that watershed which has an approximate area of 1500 km^2. In other samples, the same authors found somewhat higher levels off the Welland Canal, the Niagara River and at a midlake site.

First reports of dieldrin in the Lake Ontario ecosystem were for fishes collected during 1967–1968. Mean levels, depending on species, ranged from 0.01 to 0.50 μg/g (average 0.06 μg/g) in the study by Henderson et al. (1969)

and 0.05 to 0.10 μg/g (average 0.06) for that by Reinert (1970). The report on contaminants (WQB, 1976) also provides early data on levels in biota. Plankton, cladophora, and benthos were found at mean concentrations of 0.12, 13, and 16 μg/g, respectively. For fishes (alewife, smelt, shiny sculpin, bass, perch, and spottail shiners) from various locations, mean concentrations in the several species varied from 0.016 to 0.060 μg/g for the period 1971–1975. The report of Frank et al. (1974) included levels in a wide variety from the Bay of Quinte over the period 1968–1971. Alewives, smelt, yellow perch, coho salmon, and lake trout contained, respectively, 0.042, 0.043, 0.001, 0.064, and 0.026 μg/g in edible tissue. Results from other studies (WQB, 1980, 1981) related to the lake include alewives and smelt collected in 1971 at mean levels of 0.035 and 0.044 μg/g; alewives and smelt collected in 1975 were reported at an overall mean of 0.02 μg/g. Various perch and bass from the Mexico Bay area were reported at 0.05 μg/g and coho muscle at 0.087 μg/g. It would appear that levels of dieldrin have not declined as apparently have those for PCBs and DDT. Aldrin has not been reported in these samples.

The herring gull has also been demonstrated to contain dieldrin. Egg samples from 1975 (Norstrom et al., 1978) were reported to have 0.37 μg/g of this compound and an earlier sample of herring gull lipid had 5.6 μg/g. Since then, mean egg concentrations of 0.2 μg/g have been reported for 1980 samples (WQB, 1981) and a table in the earlier report (WQB, 1980) would seem to indicate a half-life of roughly 6 yr in the gull eggs.

6. LINDANE AND THE BENZENE HEXACHLORIDES

Lindane is the gamma (γ) isomer of the benzene hexachlorides (BHCs), more properly called the hexachlorocyclohexanes. It is the most biologically active of the BHCs and is converted photochemically under environmental conditions to the α and β isomers (Malaiyandi et al., 1982). Consequently, the observation of any of the BHCs may be evidence of an earlier presence of the γ isomer. These compounds have been extensively used as fungicides and for the treatment of seed (Martin and Worthing, 1974).

Considerable data are available for Niagara River levels of BHCs at the entrance to Lake Ontario. The Canada-Ontario Agreement report (CORB, 1981) indicates that BHCs were frequently found in 1980 water samples- the α and γ isomers in 100% of samples at 0.011 and 0.004 μg/liter, respectively. Kuntz and Warry (1983) reported total BHC levels of 0.0115 μg/liter for raw water samples of 1980–1981, a figure in keeping with the preceding. These concentrations indicated a loading to the lake of 2300 kg/yr and would appear to be continuing. In raw water studies at Niagara-on-the-Lake, almost every sample reported (OME, 1982b) since 1974 has contained 0.001-0.008 μg/liter of α- plus γ-BHC with no observable trend. The suspended material from Niagara has also been examined for BHCs. The CORB report (1981) includes references to the presence of α-BHC and lindane in more than half the collected 1980 samples; mean levels were 0.011 and 0.001 μg/g, respectively. Warry and Chan (1981), reporting on 1979–1980 samples, found 0.003 μg lindane/g, whereas Kuntz and Warry (1983), for an additional year, found 0.012 and 0.002 for the α and γ isomers, respectively. They also estimated annual loadings of BHCs to Lake Ontario via suspended matter to be only 9 kg. It is apparent, therefore, that the major river loading is via the dissolved fraction of BHCs.

Atmospheric concentrations have also been determined. Strachan and Huneault (1979) reported values for α-BHC plus lindane in rainfall which showed them to be the second contaminant after PCBs. Total levels of 0.024–0.031 μg/liter for the Lake Ontario region in 1976–1977 were comprised of about 75% α-BHC and 25% lindane. Accumulated snowfall showed only trace (0.001 μg/liter) amounts. Eisenreich (1980, 1981) reported mean air levels for α-BHC (0.3 ng/m^3) and lindane (2 ng/m^3). These, together with the rainfall data, were estimated to produce loadings of approximately 4500 kg/annum to Lake Ontario largely from vapor phase transfer because wetfall levels could account for only 4–500 kg to the lake itself. The indications are, therefore, that this is the major input to the system.

The earliest report on BHC substances in waters of the basin was by Lichtenberg et al. (1970). This report presumes, without stated evidence, that the waters of the St. Lawrence River at Massena, N.Y., contained BHC. In the contaminants report (WQB, 1976), a 1975 open lake cruise is noted in which two of 11 filtered water samples contained 0.006 μg lindane/liter. Subsequent to this report, and possibly stimulated by the incorporation of a lindane objective in the 1978 Great Lakes Water Quality Agreement, additional data has become available. In the open waters themselves, STORET (1982) reports lindane in a large number (166) of nearshore samples (mean 0.004 μg/liter) and a smaller number (22) of offshore ones (mean 0.0014 μg/liter).

The sediments have only a few references for whole lake compilations. STORET (1982) includes data for 1976 nearshore samples (35) with a mean level for lindane of 0.0037 μg/g; a more recent study of 10 nearshore samples listed in this data base is reported at 0.056 μg/g.

The earliest report of BHC chemicals in biota (Henderson et al., 1969) records levels of lindane in several fish species at 0.01–0.36 μg/g, collected in 1967–1968. In 1975, an open lake cruise noted lindane in net plankton at 0.012 μg/g (WQB, 1976). However, only 11 samples were collected and positive observations occurred only in three of these. In the same report, β-BHC was found in herring gull lipid (35 μg/g) and eggs (0.078 μg/g); it was also reported in whole alewives and smelt at 0.002 μg/g. Few further data are available for lake biota. STORET (1982) includes reference to values for a small number of fish in each of 1969 and 1977. In the earlier, two offshore samples (mean 0.34 μg/g) and eight nearshore (mean 0.013 μg/g) were reported for lindane; in the later report, six nearshore samples had a mean of

0.0067 μg/g α-BHC. These, together with the data in the 1976 contaminant reports (WQB, 1976) would appear to constitute the major reports on biota levels for this system. A recent report by Hesselberg and Seeley (1982) on twenty 1977 lake trout from the Oswego area also included BHCs at unspecified but apparently low levels. Current samplings for these compounds are underway for several fish species and should eventually be reported in the IJC surveillance reports.

7. CHLOROBENZENES

Chlorobenzenes (CBs) are industrial compounds used mainly as intermediates or solvents in the preparation of other compounds (Kao and Poffenberger, 1979). They may also be formed inadvertently by chloride–carbon electrode processes such as are found in the manufacture of chlorine. Hexachlorobenzene has also been used as a pesticide (Martin and Worthing, 1974). There are many possible congeners of the CBs which makes their analysis in environmental samples difficult. High resolution capillary gas chromatography is needed to adequately separate isomers.

Data for input via the waters of the Niagara River are available. The Canada-Ontario report (CORB, 1981) notes no HCB in one set of samples collected at Niagara-on-the-Lake during 1980. The same report includes a conflicting report about HCB in 54% of other samples from the same location, at the 0.001 μg/liter level. It is not clear, however, whether these pertain to filtered or to whole water. It is probable that this level is close to the detection limit and this report of HCB should therefore be viewed as qualitative. It should further be noted that Oliver and Nicol (1982) reported a mean of 0.054 μg/liter total CBs at the same location for six sampling occasions during 1980. This level consisted mainly of di- and trichloro- congeners; HCB was reported at a mean concentration of 0.0001 μg/liter. Even at 0.001 μg HCB/liter, the flow of this river (6400 m^3/s; Kuntz and Warry, 1983) is such that a loading of 200 kg/annum would result—a quantity of concern. The real loading is undoubtedly higher via mechanisms other than "dissolved" and the contribution of lower chlorinated CBs is many times that of HCB. There are numerous reports of CBs in the sediments and suspended matter from this river (CORB, 1981) and also reports of industrial discharges (Hang and Salvo, 1981). The report by Kuntz and Warry (1983) indicates extensive contamination of suspended sediments at Niagara-on-the-Lake. Mean levels of total CBs (0.57 μg/g dry weight) together with mean concentrations of suspended matter in the Niagara (8.4 μg/liter) suggest a loading of nearly 1000 kg/annum.

Reports on atmospheric inputs of CBs are restricted to HCB. Trace concentrations (mean 0.0008 μg/liter) were reported (Strachan and Huneault, 1979; Strachan et al., 1980) in 8% of the Lake Ontario basin rain samples. Eisenreich et al. (1980, 1981), in their extensive reviews of airborne contaminants in the Great Lakes system, used mean air and rain levels of HCB of 0.1 ng/m^3 and 0.002 μg/liter, respectively, to estimate an annual atmospheric loading of 390 kg HCB to Lake Ontario.

Environmental CBs were reported in the Lake Ontario ecosystem in the 1976 contaminants report (WQB, 1976). At that time, most of the observations pertained to the presence of hexachloro- congeners in fishes. Subsequent to that report, chlorobenzene examinations of the waters of Lake Ontario have been few. Data from STORET (1982) indicate that CBs were looked for unsuccessfully at ten 1980 inshore stations and that there were no reports for offshore locations. No extensive open lake surveys appear to have been undertaken on the Canadian side although one study by Oliver and Nicol (1982) reported a number of these compounds (mainly 1,2- and 1,4-dichlorobenzenes) at five locations near the Niagara River at a total concentration level of 0.051 μg/liter.

Sediment levels are more extensively reported, at least those relating to the Niagara River. In the lake itself, STORET (1982) provides data for 16 nearshore, U.S. sediment locations which had a mean HCB level of 0.011 μg/g; other congeners are not reported. The report by Oliver and Nicol (1982) included Lake Ontario surface (0–3 cm) sediment levels of 4.5 μg/g from near the Niagara River mouth and mean total levels from 11 sites throughout the lake of 0.5 μg/g. Again, the di- and trichloro- congeners were prominent although tetra- through hexa- comprised 40%; the mean HCB level was 17% of the total.

A number of contaminant reports for biota from the Lake Ontario ecosystem have included chlorobenzenes. STORET (1982) summarizes considerable fish data from the surveillance agencies, from 222 nearshore and 84 offshore locations. Mean concentrations in fish tissues of 0.37 and 0.080 μg/g, respectively, are recorded for HCB; other congeners are not reported. Results from two lake trout samples (Oliver and Nicol, 1982) show low levels (0.001–0.01 μg/g) for all congeners except HCB for which the concentrations observed were 0.06–0.12 μg/g in reasonable agreement with the STORET levels and those cited in the 1976 contaminants report (WQB, 1976). HCBs in salmonids (Niimi, 1979) and in alewives plus smelt (Norstrom et al., 1978) from Lake Ontario have been reported at levels of 0.030–0.060 and at 0.024 μg/g, respectively.

Data for herring gull eggs provide one of the few trend indications for these contaminants. Qualitative results for the years 1973–1979 are presented in the Appendix to the Water Quality Board of 1980 (WQB, 1980). The apparent trend to 1979 is downward, with a half-life for the decrease of 4.5 yr; mean levels of 0.1–0.5 μg/g are indicated for Snake Island in Toronto Harbour. Another program with promise for trend data is the young-of-the-year spottail shiner sampling carried out by the Ontario Ministry of the Environment. Data for these samples are collected from a number of Lake Ontario river mouths but, in general, chlorobenzenes are not reported.

8. TOXAPHENE

Toxaphene is a trade name for chlorinated camphene. Commercial toxaphene has a maximum pesticidal activity when the chlorine content ranges from 67 to 69%, resulting in a typical empirical formula of $C_{10}H_{10}Cl_8$ and an average molecular weight of 414. The technical grade consists of approximately 180 components of which 20–40 have been identified in Great Lake fishes. Use in the basin is minimal but according to Seiber et al. (1979), quantities in excess of 50% can be lost to the atmosphere through volatilization and entrainment of dust. These aspects implicate long-range transport as a probable vehicle for contamination of aqueous systems, including the Great Lakes. Although initial use of toxaphene dates from the early 1950s, it is only recently that sample preparation procedures coupled with capillary GC and electron capture detection have allowed the quantitation of toxaphene in biota. Earlier attempts to isolate and quantify ambient environmental levels were hampered because conventional packed GC methods yielded chromatograms containing peaks that overlapped other contaminants, most notably DDT and its homologs, PCBs, and certain chlordane components (Ribick et al., 1982).

To date, toxaphene values for atmospheric, ambient water and sediment compartments have yet to be reported for Lake Ontario or for the Niagara River. Swain et al. (1982) has indicated concentrations in the range of 0.0005–0.0021 μg/L liter for the waters of Lake Huron and indicated the possibility that this group of substances may have resulted in erroneous PCB results for Great Lakes biota.

A meager amount of data is becoming available for biota (actually only for fishes). Stalling et al. (1982) reported a mean of 0.7 μg/g wet weight (range 0.3–1.4) from a very limited Lake Ontario sample ($n = 3$) of yellow perch and brown trout. Similar results reported by the Department of Fisheries and Oceans to the IJC's Surveillance Group (M. Whittle, personal communication, 1982) for five 1981–1982 brown and lake trout indicated a mean concentration of 0.40 μg/g excluding a high value of 0.66 for a very old lake trout. It is apparent from these preliminary results that toxaphene could be a major contaminant in the Lake Ontario system and steps should be taken, at least to describe the extent of this particular stress.

9. DIOXINS

Dioxins are formed as by-products in the manufacture of herbicides and preservatives such as trichlorophenol, pentachlorophenol (PCP), and 2,4,5-trichlorophenoxyacetic acid (245T). These chemicals have been produced and used in the Great lakes basin. In addition, dioxins are formed by the combustion of certain chemical compounds by a process not fully understood. Dioxins, particularly the 2,3,7,8-tetrachloro isomer (TCDD), are relatively recently observed contaminants for Lake Ontario and the Niagara river. First reports were in the form of a press release by the State of New York in April, 1979. This noted that two 1978 fish samples contained approximately 0.005 μg/kg of TCDD. This stimulated a great deal of activity, not the least of which was the development of ultrasensitive methods for the detection of the chemical.

Identified mechanisms of dioxin input to the lake are combustion of solid wastes and leachate from industrial landfill sites along the Niagara River. The Hyde Park–Love Canal–Bloody Run Creek system of disposal sites has seen TCDD identified at levels of 7.5 μg/liter (*Toxic Material News*, p. 5, Jan. 3, 1979) and at 0.0016–0.0046 μg/liter (O'Keefe, 1980). The significance and quantitation of these loadings to the Niagara River and thence to Lake Ontario is unknown at present. In addition to the leaching of dioxins from these dumps, incineration of municipal refuse produces a range of dioxins and reports exist of this route for the area (Elceman et al., 1979). Among the dioxins produced and found in the fly ash, TCDDs were noted at 0.002–0.01 μg/g; higher amounts were reported for some of the higher-chlorinated congeners.

Fillets of smallmouth bass, yellow perch, and rockbass averaged 0.013, 0.0066, and 0.0024 μg/kg, respectively, from the Niagara River near Queenston, (OME, 1982c) whereas TCDD concentrations in young-of-the-year spottail shiner composites from the same locale were less than the 0.001 μg/kg detection limit (Suns et al., 1983). However, additional spottail shiner composites from other areas along the Niagara River had values of 0.007, 0.0075, and 0.015 μg/kg and a value of 0.059 μg/kg was found for samples from the mouth of the Cayuga Creek. This latter, high value is supported by the result of a study of similarly located samples collected in 1980 in which concentrations of 0.16, 0.091, and a mean of 0.11 μg/kg which were observed for goldfish, goldfish-carp hybrids, and three fillets of carp, respectively.

Reports of dioxins in the waters of Lake Ontario are nonexistent. For sediments, Onuska et al. (1983), have found trace levels (0.003–0.013 ng/kg) in a few samples from the area of the Niagara River. Fishes and herring gulls are the parts of the biotic compartment which have mainly demonstrated the presence of these chemicals. Fillets of 11 lake trout from the Port Credit area had TCDD values ranging from 0.017–0.057 μg/kg and smaller specimens from the Humber Bay area had concentrations in the range 0.006–0.028 μk/kg (OME, 1982c). The N.Y. Department of Environmental Conservation (DEC, 1983) has reported 0.107 μg/kg from a three-fish composite of lake trout taken near Galloo Island in the fall of 1979. Single fish fillets taken in November 1980 from Burlington, Pt. Petrie, and Niagara-on-the-Lake had concentrations of 0.072, 0.042, and 0.041 μg/kg, respectively. Stalling et al. (1982) reported a TCDD concentration of 0.033 μg/kg from a composited whole fish sample of five brown trout taken from the vicinity of Roosevelt Beach. OME (1982c) reported somewhat lower levels in the range of 0.007–0.011 μg/kg for rainbow trout taken in the Port Hope area and specimens of the same species taken in the fall of 1979 from the Salmon River had values in the range of 0.009–0.021 μg/kg (DEC, 1983). Fillets of chinook and coho salmon from the same river were found to have concentrations from a low of 0.019–0.039 μg/kg while three five-fish composites of coho salmon recorded a mean of 0.020 μg/kg. The DEC study further demonstrated that warm water species invariably had values below 0.02 μg/kg, an observation supported by the OME study results for similar species.

The very earliest report of dioxin in Lake Ontario biota was suggestive rather than definitive. Chick edema disease among herring gulls and terns (Gilbertson and Fox, 1977) during 1973–1974 was attributed to TCDD or the analogous tetrachlorodibenzofuran. Confirmation of the presence of these chemicals did not occur until 1980. Analysis of herring gull eggs collected at this time from colonies along the north shore of the lake produced an average range of TCDD from 0.044–0.068 μg/kg wet weight basis (WQB, 1981). The latter value, from the Scotch Bonnet Island Colony, is down from a 1972 high of 1.2 μg/kg performed retroactively on stored samples.

It is noted that the levels of contamination for TCDD in the tissues of biota from Lake Ontario are 10^{-3} to 10^{-6} times those of other traditional contaminants such as mirex and PCBs. It is also true that this compound is very much more toxic (acutely) than most of these other substances. The significance of TCDD in the biota at these levels is unknown but its presence is anthropogenic and there are ample indications if not proof for concern. A recently proposed objective (SAB, 1980) indicates that the limit for TCDD and the other dioxins in any part of the system should be zero.

10. VOLATILE ORGANIC CHEMICALS

This group of chemicals does not readily lend itself to a discussion in this chapter, partly because there is no precise definition of which compounds should be included, and partly because no routine surveillance activities are carried out. This latter, undoubtably, arises because there are no objectives or guidelines for the Great Lakes, a situation which is due to a lack of data on exposure and information on toxicity of probable chemicals to aquatic organisms. Chlorobenzenes, at least the lower homologs, could readily be included in this category. They are, however, noted separately in this chapter. Other compounds, such as DDT and PCBs, are not normally considered as volatile (vapor pressures 10^{-7}–10^{-5} tor; Eisenreich et al., 1980) and yet have been observed at substantial levels in atmospheric precipitation (Strachan et al., 1980). Such substances have also been noted elsewhere in this chapter. In this chapter, the term volatile is restricted to chlorinated compounds of one or two carbon atoms and to the simpler, single-ring aromatics.

The 1976 Report of the Water Quality Board (WQB, 1976) contained several references to substances described as volatile. The major source of

Table 11.3. Volatile Organic Chemicals [μg/liter (% samples)] in Lake Ontario and the Niagara River

Chemical	Lake Ontario pre-1976[a]	Niagara R. Discharges 1977–1981[b]	N.O.T.L. Intake 1978–1981[c]	Lower Niagara R. 1981[d]	Lake Ontario 1981[d]
Methanes					
dichloro		+	1.5(61)	0.021(29)	1.1(36)
trichloro	+	+	0.37(73)	0.027(76)	0.018(39)[e]
tetrachloro	+	+		0.003(100)	0.002(76)
trichloroflouro				0.011(100)	0.24(100)
bromochloro					0.010(1)
bromodichloro		+	0.13(24)	0.007(88)	0.005(36)
dibromochloro		+	0.20(6)	0.008(41)	0.003(14)[e]
bromotrichloro				0.007(6)	0.005(6)
dibromo				0.004(24)	0.004(13)
tribromo	+	+		0.005(12)	0.004(4)
Trichloroethane		+	0.20(9)	0.008(88)	0.004(92)[e]
Ethylenes					
chloro	+				
1,1-dichloro		+			0.14(5)[e]
trichloro		+	0.12(21)	0.008(71)	0.013(19)
tetrachloro	+	+	0.080(15)	0.037(94)	0.003(43)[e]
Aromatics					
benzene	+	+	0.29(52)		
toluene	+	+	0.14(33)		
xylenes		+	0.16(21)		

[a] WQB, 1976.
[b] Hang and Salvo, 1981.
[c] OME, 1982; 33 samples.
[d] Kaiser, 1983; Niagara, 17 samples; L. Ontario, 83 samples.
[e] single high value was not included.

this information was the N.Y. Department of Environmental Conservation and, although estimates were given for losses, presumably to the atmosphere, no information on concentrations in water were given. Many of the same compounds were also identified as being discharged to the Niagara River (Hang and Salvo, 1981), but again, concentrations were not determined for the river water itself. The Province of Ontario (OME, 1982a) provides an annual update of river-related water quality. Raw (and finished) water samples are collected at approximately monthly intervals for several communities; data for Niagara-on-the-Lake (NOTL) are presented in Table 11.3, along with the foregoing indications and data from a report by Kaiser et al. (1983). This latter study, carried out during the summer of 1981, is the most extensive (indeed the only) one on the lake itself. There are disagreements between the OME data and that from the Niagara portion of Kaiser's study.

Analytical detection limits are considerably lower in the latter case and the procedure is headspace analysis rather than purging.

It is apparent from Table 11.3 that there are considerable quantities of some volatiles, both entering and currently present in Lake Ontario. Their significance is unknown but the hazard is more likely to the organisms residing there than to man. Compounds such as the aromatics listed undoubtedly degrade rapidly and are also unlikely to be a hazard.

11. OTHER COMPOUNDS

There are many other chemicals which have been observed in one or another types of samples from Lake Ontario. The Water Quality Board Report (WQB, 1976) contains those which were specifically identified at that time for this lake. A similar report of the Water Quality Board in 1977 (WQB, 1978) contains a much longer list, of substances observed in the other Great Lakes; many of them, however, are probably also present in Lake Ontario. Discovery of such compounds seems to be determined by the nature and sensitivity of the analytical instrumentation applied to the problem as well as the inclination of the individual researchers. Most of the substances noted to date are either organochlorine or are of a natural and abundant nature. The organochlorines are observed because of the availability of the electron capture detector. Gas chromatography (capillary)-mass spectrometry needs to be applied more often to integrating samples of persistent compounds—sediments and top predators. Water-soluble substances are largely ignored, the assumption being that they will degrade readily; means should be developed to detect and analyze for these. The early warning of both soluble and insoluble new contaminants cannot be totally planned because this is basically a research problem; it can, however, be anticipated more often than at present.

A class of substances that should logically be included in a chapter on organic pollutants is the polycyclic aromatic hydrocarbons (PAHs). These compounds are both anthropogenic and natural in origin; they are found in the atmosphere and sediments and a number of them are potent mutagens and carcinogens. They are generally degradable at least by higher organisms but it may be that any concern should more properly focus on the metabolites than on the parent compounds themselves. This group of chemicals is currently being considered for the development of an ecosystem objective under the Great Lakes Water Quality Agreement and is also the subject of another chapter in this volume. They are consequently not dealt with further in this chapter.

12. CONCLUSION

From the foregoing observations, it is apparent that atmospheric and groundwater modes of entry of these substances need to be investigated—the former because of its established loading significance, the latter because it is a quantitatively unknown but indicated significant route. The Science Advisory Board, in its 1982 report to the International Joint Commission and the governments of the United States and Canada (SAB, 1982), highlights these two and seven other research needs. Two of these additional aspects related to exposure information, are epidemiology studies and validation of models. In the former case, little data exists to connect actual field effects to the presence or levels of any particular chemical. Efforts are under way with PCB and human mental development, but they are several years away from completion. The need for data to validate models is obvious and, presuming the successful establishment of a chemical distribution model for Lake Ontario, would do much to alleviate the need for extensive and expensive surveillance programs.

All of the foregoing activities are reactions to problems as they become identified and the greatest need is for preventive action. Waste disposal, of either liquids or solids, has been established as a major mechanism for the release of contaminants to the Niagara; the river, in turn, is a proven major source for Lake Ontario of the substances discussed here. It is also suspect in many other cases. The Great Lakes research community is aware of many other chemicals which are and/or may leach into the system unless efforts are taken to remove them from the ecosystem and dispose of them in a more adequate fashion than at present. Also required is a reexamination of the systems that permit such environmentally unsafe practices to occur.

REFERENCES

Alley, E. G. (1973). The use of mirex in control of the imported fire ant. *J. Environ. Quality* **2**:52–61.

Brooks, G. T. (1974) *Chlorinated insecticides*. Vol. 1, technology and application. C.R.C. Press, Cleveland, p. 55.

Canada-Ontario Review Board. (1981). Environmental Baseline Report of the Niagara River: November 1981 Update, 127 pp.

DEC. (1983). Report on toxic substances in fish and wildlife. New York Department of Environmental Conservation, Albany (in press).

DEC and EPA. (1979). Department of Environmental Conservation (N.Y.) and Environmental Protection Agency (U.S.). Report of Task Force on Hazardous Waste Sites: Potentially hazardous dump sites.

Durham, R. W. and B. G. Oliver. (1983) History of Lake Ontario contamination from the Niagara River by sediment radiodating and chlorinated hydrocarbon analysis. In Proceedings Niagara River Symposium (IAGLR, 1982). *J. Great Lakes Res.* **9**:160–168.

Eisenreich, S. J., B. B. Looney, and J. D. Thornton. (1980). Assessment of airborne organic contaminants in the Great Lakes ecosystem. Science Advisory Board Report: Appendix: Background Reports, April, p. 150.

Eisenreich, S. J., B. B. Looney, and J. D. Thornton. (1981). Airborne organic contaminants in the Great Lakes ecosystem. *Environ. Sci. Technol.* **51:** 30–38.

Elceman, G. A., R. E. Clement, and F. W. Karasek. (1979). Analysis of fly ash from municipal incinerators for trace organic compounds. *Anal. Chem.* **51:** 2343–2350.

Fox, M. E., J. H. Carey, and B. G. Oliver. (1983). Compartmental distribution of organochlorine contaminants in the Niagara River and the western basin of Lake Ontario. In Proc. Niagara River Symposium (IAGLR, 1982). *J. Great Lakes Res.* **9:** 287–294.

Frank, R., A. E. Armstrong, R. G. Boelens, H. E. Braun, and C. W. Douglas. (1974). Organochlorine insecticide residues in sediment and fish tissues, Ontario, Canada. *Pest. Monit. J.* **7:** 165–180.

Frank, R., R. L. Thomas, M. V. H. Holdrinet, A. L. W. Kemp, and H. E. Braun. (1979). Organochlorine insecticides and PCB in surficial sediments (1968) and sediment cores (1976) from Lake Ontario. *J. Great Lakes Res.* **5:** 18–17.

Gilbertson, M. and G. A. Fox. (1977). Pollutant-associated embryonic mortality of Great Lakes herring gulls. *Environ. Pollut.* **12:** 211–216.

Haile, C. L., G. D. Veith, G. F. Lee, and W. C. Boyle. (1975). Chlorinated hydrocarbons in the Lake Ontario ecosystem. EPA-660/3-75-022 (PB-243364/7GI). June, 28 pp.

Hallett, D. J., R. J. Norstrom, F. I. Onuska, M. E. Comba, and R. Sampson. (1976). Mass spectral confirmation and analysis by the Hall detector of mirex and photomirex in herring gulls from Lake Ontario. *J. Agric. Food Chem.* **24:**1189–1193.

Hang, W. L. T. and J. P. Salvo. (1981). *The ravaged river.* The New York Public Interest Research Group, New York, 210 pp.

Henderson, C., W. L. Johnson, and A. Inglis. (1969). Organochlorine insecticide residues in fish (National Pesticide Monitoring Program). *Pest. Monit. J.* **3:**145–171.

Hesselberg, R. J. and J. G. Seelye. (1982). Identification of organic compounds in Great Lakes fishes by gas chromatography/mass spectrometry, 1977. U.S. Fish and Wildlife Service, Great Lakes Fishery Laboratory Administrative Report 82-1, Ann Arbor, January, 49 pp.

Holdrinet, M. V. H., R. Frank, R. L. Thomas, and L. J. Hetling. (1978). Mirex in the sediments of Lake Ontario. *J. Great Lakes Res.* **4:**69–74.

Kaiser, K. L. E. (1978). The rise and fall of mirex. *Environ. Sci. Technol.* **12:** 520–528.

Kaiser, K. L. E., M. E. Comba, and H. Huneault. (1983). Volatile hydrocarbon contaminants in the Niagara River and Lake Ontario. In Proc. Niagara River Symposium (IAGLR, 1982). *J. Great Lakes Res.* **9:** 212–223.

Kao, Che-I and N. Poffenberger (1979). Chlorinated benzenes. In: M. Grayson, ed., Kirk-Othmer *Encyclopedia of chemical technology.* Vol. 5, 3rd ed. editor, J. Wiley, New York, pp. 797–808.

Kuntz, K. W. and N. D. Warry. (1983). Chlorinated organic contaminants in water and suspended sediments of the lower Niagara River. In Proc. Niagara River Symposium (IAGLR, 1982). *J. Great Lakes Res.* **9:**241–248.

Lichtenberg, J. J., J. W. Eichelberger, R. C. Dressman, and J. E. Longbottom. (1970). Pesticides in surface waters of the United States—a five-year summary 1964–68. *Pest. Monit. J.* **4:**71–86.

Malaiyandi, M., K. Muzika, and F. M. Benoit. (1982). Isomerization of γ-Hexachlorocyclohexane to its α-isomer by ultra-violet light irradiation. *J. Envir. Sci. Health* **A17:** 299–311.

Martin, J. and C. R. Worthing, (eds). (1974). *Pesticide manual.* 4th ed. British Crop Protection Council.

Niimi, A. J. (1979). Hexachlorobenzene (HCB) levels in Lake Ontario salmonids. *Bull. Environ. Contam. Toxicol.* **23:**20–24.

Norstrom, R. J., D. J. Hallett, and R. A. Sonstegrad. (1978). Coho salmon (*Oncorhynchus kisutch*) and herring gulls (*Larus argentatus*) as indicators of organochlorine contamination in Lake Ontario. *J. Fish Res. Board. Can.* **35:**1401–1409.

O'Keefe, P. (1980). Technical communication of the Toxicology Institute. N.Y. Department of Health, Albany.

Oliver, B. G. and K. D. Nicol. (1982). Chlorobenzenes in sediments, water and selected fish from Lakes Superior, Huron, Erie and Ontario. *Environ. Sci. Technol.* **16**:532–536.

Ontario Ministry of the Environment. (1976). Polychlorinated biphenyls in the Ontario environment. OME, Toronto, July.

Ontario Ministry of the Environment. (1980). Water Quality Data for Ontario Lakes and Streams. 1979. Vol. XV. Water Resources Branch, OME, 657 pp.

Ontario Ministry of the Environment. (1982a). Niagara River Water Quality Update. OME, Toronto, June 23, 34 pp.

Ontario Ministry of the Environment. (1982b). Data submitted to the Surveillance Working Group of the Water Quality Board, OME.

Ontario Ministry of the Environment. (1982c). New evidence of chemical contaminants in fish: Attachment to prepared statement by the Hon. K. Norton. OME, Toronto, April 4 pp. plus 2 attachments.

Onuska, F. I., A. Mudroch, and K. A. Terry. (1983). Identification and determination of trace organic substances in sediment cores from the western basin of Lake Ontario. In Proc. Niagara River Symposium (IAGLR, 1982). *J. Great Lakes Res.* **9**:169–182.

Reinert, R. E. (1970). Pesticide concentrations in Great Lakes fish. *Pest. Monit. J.* **3**:233–240.

Ribick, M. A., G. R. Dubay, J. D. Petty, D. L. Stalling, and C. J. Schmitt. (1982). Toxaphene residues in fish: Identification, quantification and confirmation at parts per billion levels. *Environ. Sci. Technol.* **16**:310–318.

Science Advisory Board. (1980). Report of the Aquatic Ecosystem Objectives Committee. International Joint Commission, Windsor, Ontario, p. 127.

Science Advisory Board. (1982). 1982 Annual Report: Great Lakes research review. International Joint Commission, Windsor, November, 66 pp.

Seiber, J. W., S. C. Madden, M. M. McChesney, and W. L. Winterlin. (1979). Toxaphene dissipation from treated cotton field environments. *J. Agric. Food Chem.* **27**:284–292.

STORET, (1982). STOrage and RETrieval of Water Quality Data: A computerized information system. U.S. Environmental Protection Agency, Washington, June.

Stalling, D. L., M. A. Ribick, S. Wold, and E. Johansson. (1982). Characterization of toxaphene residues in fish samples analyzed by capillary gas chromatography using SIMCA, multivariate analysis. Proc. Env. Chem. Sect., Amer. Chem. Soc. Meeting, Kansas City.

Strachan, W. M. J. and H. Huneault. (1979). Polychlorinated biphenyls and organochlorines in the Great Lakes region. J. Great Lakes Res. **5**:61–68.

Strachan, W. M. J., H. Huneault, W. M. Schertzer, and F. C. Elder. (1980). Organochlorines in precipitation in the Great Lakes region. In: *Hydrocarbons and halogenated hydrocarbons in the aquatic environment*. B. K. Afghan and D. Mackay, Eds., Plenum Press, New York, pp. 387–396.

Suns, K., G. R. Craig, G. Crawford, G. A. Rees, H. Tosine, and J. Osburne. (1983) Organochlorine countaminant residues in spottail shiners (Notropis hudsonius) from the Niagara River. In Proc. Niagara River Symposium (IAGLR, 1982). *J. Great Lakes Res.* **9**:335–340.

Swain, W. R., M. D. Mullin and J. C. Filkins. (1982). Refined analysis of residue forming organic substances in lake trout from the vicinity of Isle Royale, Lake Superior. Presentation at the 25th Conference of the International Association for Great Lakes Research, Sault Ste. Marie, May 4–6.

Task Force on Mirex. (1977). Mirex in Canada. Departments of Fisheries and Environment and of Health and Welfare Technical Report 77-1. Ottawa, April, 153 pp.

Warry, N. D. and C. H. Chan. (1981). Organic contamination in the suspended sediments of the Niagara River. *J. Great Lakes Res.* **7**:394–403.

Water Quality Board. (1975). Great Lakes Water Quality Third Annual Report. Appendix A:

Annual Report of the Water Quality Objectives Subcommittee. International Joint Commission, Windsor, June, pp. 47–54.

Water Quality Board. (1976). Status report on the persistent toxic pollutants in the Lake Ontario basin. International Joint Commission, Windsor, December, pp. 95.

Water Quality Board. (1978). Status report on organic and heavy metal contaminants in the Lakes Erie, Michigan, Huron, and Superior basins. International Joint Commission, Windsor, July, pp. 373.

Water Quality Board. (1980). 1980 Report on Great Lakes Water Quality: Appendix. International Joint Commission, Windsor, November, 82 pp.

Water Quality Board. (1981). 1981 Report on Great Lakes Water Quality: Appendix: Great Lakes Surveillance. International Joint Commission, Windsor, November, 173 pp.

12

Microcontaminants in Wisconsin's Coastal Zone

Mary E. Pariso, James R. St.Amant, and Thomas B. Sheffy

Bureau of Water Quality
Wisconsin Department of Natural Resources
Madison, Wisconsin 53707

1. INTRODUCTION

Records of Microcontaminants in Wisconsin's coastal zone date to almost 20 years ago. University of Wisconsin researchers (Hickey et al., 1965) found total DDT levels in fish collected from Lake Michigan off Door County to average 3.40 parts per million (ppm) in alewife* regurgitated by gulls, 2.28–7.87 ppm in whole fish samples of bloater chubs, 5.05–7.49 ppm in samples of lake whitefish muscle, and 3.23 ppm in lake whitefish entrails. The State Department of Agriculture sampled market fish taken from Lake Michigan in 1965 (reported in Kleinert et al., 1968), and found 62.36–99.50 ppm total DDT and 0.31–0.58 ppm dieldrin in the fat of raw unbrined bloater chubs. In 1966, DDT was first reported in Wisconsin Coastal Zone sediments (Hickey et al., 1966). A mean DDT level of 14 parts per billion (ppb) was reported in Green Bay sediments.

In 1965, 1966, and 1967, DDT was identified by the U.S. Bureau of Commercial Fisheries in whole fish samples from Lake Michigan (reported in Kleinert et al., 1968). Levels of total DDT in ppm were 2.41–4.99 in alewife, 0.099–7.35 in rainbow smelt, 0.90–0.98 in trout-perch, 1.25 for a single sample of lake herring, 5.29–15.00 for bloater chub, 0.20–0.74 for fingerling lake trout, 0.35–0.50 for suckers, 3.58 for a single sample of carp, and 0.39–4.72 for yellow perch.

A survey of fish by the Wisconsin Department of Natural Resources in 1965, 1966, and 1967 (Kleinert et al., 1968) included five whole-fish samples from Wisconsin's coastal zone. Levels of total DDT ranged from 2.22 ppm in a composite sample of carp and sucker from the Milwaukee River to 6.57 ppm in a Green Bay rainbow trout. Dieldrin ranged from a trace in this same Green Bay trout to 10.00 ppm in a Milwaukee River sucker.

In response to Federal Food and Drug Administration seizures of coho salmon from Lake Michigan in interstate commerce, a survey of pesticide residues in commercial and sport fish of Lakes Michigan and Superior was initiated in 1968 (Poff and Degurse, 1970). Five hundred sixty-three (563) fish samples from Wisconsin's Coastal Zone were analyzed for DDT and dieldrin. The larger individuals of most lake trout and coho salmon from Lake Michigan exhibited residue levels in excess of the 5.00 ppm and 0.30 ppm tolerance levels established by the U.S. Food and Drug Administration for DDT and dieldrin respectively. Only the largest lake trout from Lake Superior were found to contain even moderately high residue levels.

Polychlorinated biphenyls (PCBs) were first identified to be accumulating in the environment in 1966 (Jensen, 1966) in Sweden. PCBs were first identified in Wisconsin's Coastal Zone in 1969 by Veith (1970). A 1971 survey of Lake Michigan fish by Veith (1975) analyzed 13 species of fish for PCBs and DDT. Mean PCB concentrations in the populations of fish averaged over the entire lake ranged from 2.70 ppm in rainbow smelt to 15.00 ppm in lake trout. Most trout and salmon greater than 12 in. in length were found to contain PCBs at concentrations greater than 5.00 ppm. Concentrations of DDT ranged from less than 1.00 ppm in suckers to approximately 16.00 ppm in large lake trout. Large lake trout and white suckers contained 7.00–

* Scientific names of all fish species mentioned in this chapter are given in Table 12.6.

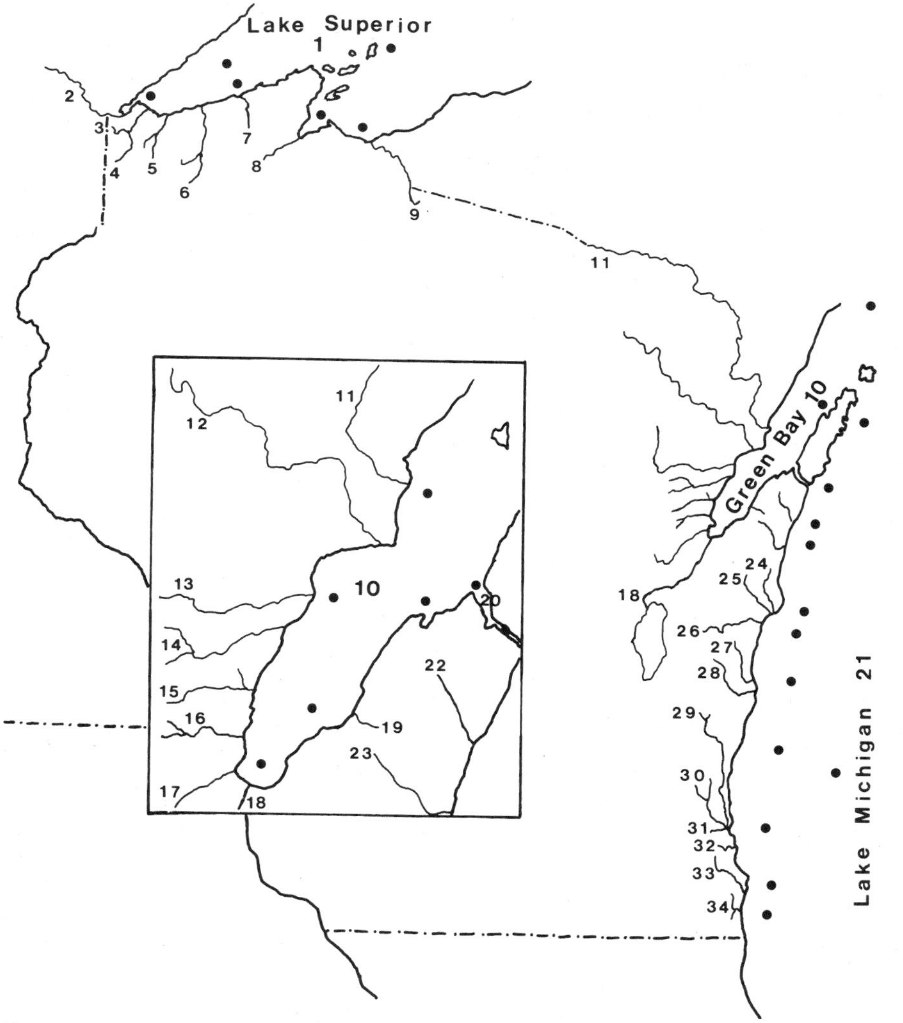

Figure 12.1 Sampling locations in Wisconsin's coastal zone with close-up of Green Bay. Numbers correspond to sampling locations listed in Table 12.1. Dots identify sampling sites within the larger sampling locations.

12.00 ppm DDT. The concentration of PCBs was greatest in fish from southern Lake Michigan. Analysis of southern Lake Michigan sediments in 1969–1970 (Leland et al., 1973) identified the wide distribution and large reservoir of stable organochlorines (DDT and dieldrin) in this area.

Eleven species of fish from Wisconsin's nearshore waters of Lake Superior were sampled in 1974 (Kleinert, 1975) and analyzed for nine metals,

Table 12.1. Sampling Locations and Types of Samples Collected During Each of Three Surveys

		1979 Survey[a]		1980 Survey[a]					1981 Survey[a]			
Map Number	Sampling Location[b]	Fish	Sites Tested	Fish	Sed[c]	Eff	MS Screen	No. of Sites Tested	Fish	Sed[c]	Eff	No. of Sites Tested
1	Lake Superior, nearshore and offshore	24	6	11	1	3	3	3			4	
2	St. Louis River, mouth to Oliver, WI	5	1	10	3		1	3	1		1	2
3	Nemadji River	3	1			1						
4	Black River	3	1									
5	Amnicon River	5	1									
6	Brule River	4	1									
7	Flag River	6	1									
8	Fish Creek	3	1									
9	Montreal River	2	1									
10	Green Bay, nearshore and offshore	40	8						15			3
11	Menominee River	5	1									
12	Peshtigo River, mouth to Peshtigo, WI	6	2									
13	Oconto River, mouth to Hwy. 32	11	3									
14	Pensaukee River, mouth to Hwy. 32	5	1	9	4		1	3				
15	Little Suamico River	5	1									
16	Suamico River	5	1									
17	Duck Creek	5	1	7	3	1	1	2				
18	Fox River, from mouth to Neenah, WI	11	3	12	5	19	3	4	4	3	5	3

19	Red River	5	1	2	2			1				
20	Sturgeon Bay, harbor and ship canal	15	2	12	4		2	4	11	2	1	3
21	Lake Michigan, nearshore and offshore	47	12			3			18	5		4
22	Ahnapee River	5	1									
23	Kewaunee River	6	1									
24	East Twin River	5	1									
25	West Twin River	10	1									
26	Manitowoc River	5	1									
27	Pigeon River	5	1									
28	Sheboygan River	7	1	2				1				
29	Milwaukee River, mouth to Kewaskum, WI	5	1	37	22	14	2	13	19	9	12	6
30	Menomonee River, mouth to Menomonee Falls, WI	3	1		5	3						
31	Kinnickinnic River, mouth to West Allis, WI	3	1	2	4	1	1	1	2	2	12	1
32	Oak Creek, mouth to Oak Creek, WI	4	1		2	1						
33	Root River, mouth to Co. Hwy. G	8	1	11	5		2	5	9	4	1	3
34	Pike River, mouth to Hwy. 1	3	1	8	6	4	1	3	5	2	1	2
	TOTAL	284	63	123	66	50	17	43	90	22	42	27

[a] Column values are the total number of samples from all sites at each location.
[b] Unless otherwise stated, sampling locations began at river mouth and extended 1–2 mi upstream.
[c] Each sediment or effluent sample was collected at a different site for each sampling location.

seven pesticides, and PCB. Although most of the compounds could be detected, the low amounts present indicated high water quality.

At this same time, Lake Michigan fish continued to show high levels of microcontaminants. Degurse and Duter (1975) found that large lake trout and chinook salmon from the waters surrounding Door County contained total DDT at levels greater than 5.00 ppm (9.15 and 5.33 ppm average, respectively). Also high were average PCB concentrations in carp (22.3 ppm), large lake trout (21.80 ppm), chinook salmon (10.13 ppm), large coho salmon (6.62 ppm), and large lake whitefish (5.12 ppm). Dieldrin remained a significant residue in large lake whitefish (0.264 ppm average).

Haile (1977) analyzed samples of eight species of southern Lake Michigan fish (including Wisconsin's coastal zone) collected in 1974. Lake trout contained elevated levels of dieldrin (0.31 ppm average), total DDT (3.60 ppm average), and PCB (6.00 ppm average). Elevated levels of dieldrin (0.31 ppm average) were also found in bloater chub. Both DDT and PCB were detected in all analyzed sediments in the 12–131 ppb range.

PCB (10.00 ppm) was found by Veith et al. (1981) in a composite sample of carp, channel catfish, and white sucker taken from the Fox River below the DePere dam in 1978. Low levels of total DDT, hexachlorobenzene, and chlordane were also found.

Investigators have shown that significant levels of microcontaminants have entered the Great Lakes via streams (Miles and Harris, 1971). In 1978, extremely high levels of PCBs were found in fish from the Sheboygan River (Kleinert et al., 1978). The average PCB concentration of 40 fish samples taken from the mouth to the Sheboygan Falls dam was 115.00 ppm. The source of this problem was traced to an industry that had improperly disposed of PCB-containing wastes in the floodplain of the Sheboygan River.

A limited number of samples from Wisconsin's coastal zone were collected by Kuehl et al. (unpublished manuscript) in 1979. Results indicated that rough fish from the mouths of the Kinnickinnic, Milwaukee, and Pike rivers contained high levels of PCB (16.36, 34.90, and 12.65 ppm, respectively). A sample of lake trout taken offshore at Kenosha had 12.05 ppm PCB and 5.83 ppm DDT.

Despite the efforts of the aforementioned investigators, many gaps existed in Wisconsin's contaminant record. Particularly lacking was a comprehensive record of toxic substances in Wisconsin's coastal zone and tributary streams. To compile a complete and up-to-date inventory of toxic substances in Wisconsin's coastal zone, the Department of Natural Resources received financial support from the Wisconsin Coastal Zone Management Program. The three-year grant was used to estimate the magnitude and identify sources of contaminants to the Great Lakes ecosystem.

The primary objective during the first year of this project was to conduct a systematic survey of toxic substances in fish collected at approximately 60 locations in Wisconsin's coastal zone. Problem drainage systems were identified using fish as a biological indicator. Data gathered from the first year of the study was used to design a more detailed follow-up survey of problem areas for the second year. The second year survey monitored fish, and included sediment and effluent samples to identify point sources of contaminants and more precisely define the geographic extent of contamination. In the third year of the study, fish, sediment, and effluent samples were collected for the purpose of evaluating trends in contaminant levels, and continuing the effort to identify and eliminate point sources of toxic substances in Wisconsin's coastal zone.

Areas that were sampled during the three years of the study are shown in Fig. 12.1. Table 12.1 gives a description of sampling locations and types of samples collected from each station during each year of the survey.

2. WISCONSIN DEPARTMENT OF NATURAL RESOURCES COASTAL ZONE SURVEY

2.1. 1979 Survey

In 1979, 284 composite fish samples were tested for six heavy metals and 10 chlororganic compounds. Table 12.2 summarizes the results of the 1979 survey.

Only one sample exceeded an existing federal tolerance level for heavy metal concentration. A sample of burbot taken near Outer Island, Lake Superior, contained 1.35 ppm mercury, just over the U.S. FDA guideline of 1 ppm. Three other samples of walleye and lake trout from the Lake Superior basin showed elevated levels of mercury (0.85–0.99 ppm). Although these samples are below tolerance and considered safe for human consumption, they exceed the typical levels found elsewhere in the state. The occurrence of elevated mercury burdens in fish from this basin probably reflects an abundance of naturally occurring mercury deposits rather than an industrial discharge.

In the analysis for chlororganic compounds, PCB was found to be the worst contaminant statewide. Ninety-seven samples exceeded the federal tolerance level for one or more of the compounds PCB, DDT, Dieldrin, and chlordane. Excessive PCB levels occurred in 90 of the aforementioned samples.

The most serious PCB problem for Wisconsin's coastal zone was found in the lower Sheboygan River. All fish samples at this location exceeded the PCB tolerance level with a range of 10.50–900.00 ppm. Two samples from the Sheboygan River exceeded the tolerance for chlordane.

Contaminants, particularly PCB, were also found in southeastern Wisconsin. In the Milwaukee–Kinnickinnic–Menomonee River drainage system, all 11 fish samples exceeded the PCB tolerance with a range of 8.60–88.00 ppm. One sample from the Kinnickinnic exceeded the chlordane tolerance. All samples from the Pike River and four of the eight samples from

Table 12.2. Summary of Toxic Substances in Wisconsin Coastal Zone Fish Samples, 1979

Parameter (detection limit ppm) (tolerance ppm)	Number of Samples			Mean Concentration[a] (ppm)	Range[a] (ppm)
	Below Detection Limit	Above Detection Limit	Above Tolerance		
Arsenic (2.00) (5.00 Canadian)	284	0	0	NA	NA
Cadmium (0.20) (0.50)	284	0	0	NA	NA
Chromium (0.50) (NA)	276	8	NA	0.76	0.50–1.20
Copper (0.50) (10.00 Canadian)	8	276	0	1.44	0.50–4.80
Lead (5.00) (10.00 Canadian)	284	0	0	NA	NA
Mercury (0.01) (1.00)	0	284	1	0.15	0.01–1.35
PCB (0.20) (5.00)	22	262	90	30.60	0.20–900.00
DDT (0.05) (5.00)	21	262	2	0.88	0.05–11.37
Dieldrin (0.02) (0.30)	145	138	5	0.10	0.02–0.50
Chlordane (0.05) (0.30)	163	120	21	0.21	0.05–1.05
Aldrin (0.02) (NA)	282	0	NA	NA	NA
Endrin (0.02) (NA)	268	15	NA	0.03	0.02–0.06
Methoxychlor (0.05) (NA)	281	3	NA	0.07	0.06–0.08
Hexachlorobenzene (0.01) (NA)	250	34	NA	0.01	0.01–0.05
Pentachlorophenol (0.05) (NA)	258	4	NA	0.07	0.06–0.09
Alpha BHC (0.01) (NA)	213	69	NA	0.02	0.01–0.09
Gamma BHC (0.01) (NA)	249	29	NA	0.02	0.01–0.05

[a] For samples above detection limit.

the Root River were found to exceed the PCB tolerance level. Respective ranges were 6.60–13.00 and 0.30–16.00 ppm.

Nine of 11 fish samples from the lower Fox River between Neenah and Green Bay exceeded the PCB tolerance level. Previously identified PCB contamination in this area (Sheffy and Aten, 1979) was found to be due primarily to paper recycling mills. Other tributary streams to Green Bay and Lake Michigan where occasional samples exceeded the PCB tolerance included the Manitowoc, West Twin, Kewaunee, Red, Ahnapee, Duck, Suamico, Little Suamico, Pensaukee, Oconto, and Peshtigo rivers. Because of the isolated occurrences of excessive PCB levels, it is thought that these streams do not represent PCB sources, but rather that some fish collected from these rivers were migrants from Green Bay and Lake Michigan. Such migrants would be expected to show elevated PCB levels.

The fact that Lake Michigan and Green Bay act as sinks for chemical pollutants (Armstrong and Weininger, 1980) is demonstrated by offshore sampling results. Forty of 102 fish samples taken from nearshore and offshore stations in Green Bay, Lake Michigan and Sturgeon Bay exceeded the PCB tolerance level. Four samples exceeded the tolerance for dieldrin. Thirteen samples exceeded the tolerance for chlordane and two samples exceeded the tolerance for DDT.

None of the fish samples from the Lake Superior Basin were found to contain excessive amounts of chlororganic compounds. PCB levels were slightly higher in the Superior Entry area when compared to other sites in the basin, reflecting the industrial activity in the Duluth–Superior area, but these concentrations do not represent a problem.

In general, the fish that exceeded tolerance levels were very large (old) individuals. Fish which did not exceed tolerances were generally the smaller, younger fish. For example, a sample of large lake trout (10 lb average) contained 22.00 ppm PCB, 11.40 ppm DDT, 0.40 ppm dieldrin, and 0.60 ppm chlordane, all values in excess of existing standards. In the Lake Michigan basin, lake trout that exceeded the tolerance for PCB averaged more than 24 in. in length whereas those below tolerance averaged only 18.50 in. This supports the concept of bioaccumulation and biomagnification of toxic chemicals in fish through time. The longer fish are exposed to traces of pollutants in a system, the more of these substances they accumulate. It also suggests that the large older fish which lived in a more polluted environment 5–10 yr ago still show residual amounts of pollutants. The younger fish may reflect a cleaner environment due in part to recent laws which have greatly reduced the discharge of toxic substances.

2.2. 1980 Survey

In 1980, 123 composite fish, 66 sediment, and 50 effluent samples were analyzed for contaminants identified in the 1979 survey: PCB, DDT, chlordane, and dieldrin. Table 12.3 summarizes the results of the 1980 survey.

Table 12.3. Summary of Toxic Substances in Wisconsin Coastal Zone Fish, Sediment, and Effluent Samples, 1980

Parameter	Number of Samples: Below Detection Limit	Number of Samples: Above Detection Limit	Number of Samples: Above Tolerance	Mean Concentration[a]	Range[a]
		Fish			
(detection limit ppm) (tolerance ppm)				(ppm)	(ppm)
PCB (0.20) (5.00)	19	104	38	7.71	0.21–79.00
DDT (0.05) (5.00)	29	92	1	0.60	0.05–9.71
Dieldrin (0.02) (0.30)	85	36	2	0.11	0.02–1.60
Chlordane (0.05) (0.30)	100	18	3	0.13	0.05–0.46
		Sediment			
(detection limit ppm)				(ppm)	(ppm)
PCB (0.05)	25	41	NA	2.13	0.06–12.00
DDT (0.01)	32	34	NA	0.16	0.01–2.28
Dieldrin (0.01)	63	3	NA	0.04	0.01–0.09
Chlordane (0.01)	61	5	NA	0.03	0.01–0.07
		Effluent			
(detection limit ppb)				(ppb)	(ppb)
PCB (0.50)	48	2	NA	2.50	1.00–4.00
DDT (0.02)	47	3	NA	0.80	0.66–0.89
Dieldrin (0.01)	48	2	NA	0.85	0.07–0.10
Chlordane (0.01)	50	0	NA	NA	NA

[a] For samples above detection limit.

2.2.1. *Fish*

Of the 123 fish samples, 38 exceeded the PCB tolerance, two exceeded the dieldrin tolerance, one exceeded the tolerance levels for DDT, and three exceeded the chlordane tolerance.

The PCB problem in the lower Sheboygan River remained serious as all fish samples exceeded the PCB tolerance with a mean of 59.00 ppm and a range of 39.00–79.00 ppm.

Elevated levels of PCB were again detected in fish from southeastern Wisconsin. The extensive sampling of the Milwaukee River revealed a problem area that extended from the mouth to Grafton, Wisconsin. Fifteen of the 22 samples in this area exceeded the PCB tolerance with a range of 5.00–49.00 ppm. Fish from the Kinnickinnic River remained above acceptable levels with one sample from the mouth containing 34.00 ppm PCB. Sampling of the Pike River showed that fish commonly exceed the PCB tolerance in the Kenosha area where the mean concentration was 16.70 ppm. Above Kenosha, PCB levels drop to less than the detection limit. Of the eleven samples from the Root River, four samples were above detection limits, and one sample from the Racine area exceeded the tolerance for PCBs.

The lower Fox River, from the DePere Dam to the mouth, remained a rather severe PCB problem area. Eight of nine fish samples from this area exceeded the PCB tolerance level with a mean of 16.00 ppm. Carp sampled in 1980 contained PCB levels twice as high as the levels found in 1979. PCB levels decrease above the DePere Dam with only one sample exceeding the tolerance.

Samples from the Pensaukee River, Red River, and Duck Creek showed that these streams are unlikely sources of PCB. Levels of PCB in fish collected at the mouths of these streams are a reflection of PCB contamination in southern Green Bay. Contaminated fish from the bay probably move in and out of these tributary streams at various times of the year.

1980 sampling of the Sturgeon Bay ship canal confirmed PCB contamination in this area. Resident walleye and carp collected from the canal showed PCB levels between 9.00 and 13.00 ppm. In a related monitoring effort by the Department of Natural Resources, 17 chinook salmon returning from Lake Michigan to Strawberry Creek (tributary to the ship canal) for spawning were tested for PCB. None of these fish exceeded the PCB tolerance (mean 2.90 ppm, range 1.70–4.80 ppm). This suggests that the ship canal remains contaminated with PCB, but the waters of Lake Michigan are improving, since the levels in salmon from Lake Michigan proper are declining.

Fish samples from the St. Louis River and Superior Harbor support the previous claim that no contamination problem exists in this basin. One sample of large carp collected from the St. Louis River at the Wisconsin–Minnesota state line showed 7.40 ppm PCB. This suggests a possible Minnesota source somewhere upstream from the state line.

2.2.2. *Sediment*

Of the 66 sediment samples analyzed, 41 displayed measurable amounts of PCB. Eight of these samples were above 5.00 ppm. DDT was identified in 34 samples (range 0.01–2.28 ppm). Dieldrin was found in three samples (range 0.01–0.09 ppm) and chlordane was found in five samples (range 0.01–0.07 ppm).

With respect to contamination of sediments, the Milwaukee River can be divided into three reaches. The first, between the mouth and Silver Spring Drive, shows an average PCB level of 9.60 ppm. The second, from Silver Spring Drive to County Highway C below Grafton, shows an average PCB level of 0.28 ppm. A sediment sample from Cedar Creek, which flows into the Milwaukee River below County Hwy C, showed a PCB level of 0.73 ppm below the Cedarburg sewage treatment plant. In the third reach, above Grafton, PCB levels were below detection limits. Detectable levels of DDT (0.19 ppm average) were confined to the reach from the mouth to Silver Spring Drive. Four sediment samples from the Woolen Mills impoundment at West Bend show this area to be a low-level source of PCB, DDT, and chlordane. Average values for these parameters were 0.28, 0.13, and 0.04 ppm respectively. No samples displayed detectable levels of other microcontaminants.

The other two rivers draining the Milwaukee metropolitan basin also displayed measurable amounts of sediment contamination. PCB was identified in the Menominee River Sediment from Highway 100 downstream. Three sediment samples were taken from the Kinnickinnic River between Kinnickinnic Avenue and Jackson Park. Elevated levels of PCB were found in all three, with the highest levels near the mouth. Chlordane (0.02 ppm) was found at Kinnickinnic Avenue.

Elevated levels of PCB were found in the Pike River at St. George Cemetery and may also exist upstream in the Carthage College area. PCB was below detection limits from Berryville Road to Petrifying Springs Park but 0.06 ppm was found below Rexnord, Inc., approximately 5 mi upstream. DDT was detected in all samples with the highest level at St. George Cemetery (0.31 ppm). Dieldrin was present in samples at Petrifying Springs Park (0.02 ppm) and below Rexnord, Inc. (0.09 ppm).

The highest levels of PCB in the Root River were found at and below the C&NW Railroad trestle in Racine. Levels decline rapidly above this point and are undetectable at Five-Mile Road. DDT is most abundant at Highway 38 near the mouth, but is present in all samples. No dieldrin or chlordane were detected.

High PCB levels in Lower Fox River fish are reflected in the sediment from this area. All samples showed elevated PCB levels with the highest at the Highway 29 bridge, down stream from the Fort Howard Paper Company. DDT was also found at this location and near the Pulliam Power Plant.

Sediment samples from the Pensaukee River, Red River, and Duck Creek support the assertions made earlier that these are unlikely sources of microcontaminants. No detectable levels of any microcontaminants were found in any of these tributaries.

PCB was detected at three of the four sampling locations in Sturgeon Bay. The average concentration was 0.11 ppm. No other contaminants were detected.

Sediment samples from Superior Harbor and the St. Louis River showed a small amount of PCB contamination (mean 0.12, range 0.06–0.29 ppm) and correlated with detectable but low levels of PCB found in fish from this area.

2.2.3. *Effluent*

Of the 50 effluent samples analyzed in 1980, five showed levels of microcontaminants above detection limits.

A final effluent sample from Fort Howard Paper Company on the Fox River in Green Bay contained 4.00 ppb PCB. This could mean an annual discharge of more than 190 lb of PCB. This industry is a significant source of PCB to the Lower Fox River, and is probably responsible for continued high levels of PCB in fish and sediment in this reach of the river.

PCB (1.00 ppb) was also detected in the final effluent to the Milwaukee River at the Saukville sewage treatment plant. Dieldrin (0.10 ppb) and DDT (0.89 ppb) were detected in the Butler storm sewer discharge to the Menomonee River at 124th Street and Villard Avenue. More intensive sampling is required to determine the exact sources of these microcontaminants. Dieldrin and DDT were also detected in the leachate from the Woolen Mills landfill at West Bend. Two samples were taken, one of which showed dieldrin (0.07 ppb) and both of which showed DDT (0.73 ppb average).

2.2.4. *Gas Chromatography/Mass Spectrometry*

In a review of chemical residues found in fish tissue, Kuehl (1981) discusses the great number of xenobiotic chemical residues present in fish from the environment that are not routinely monitored. In order to identify the presence of new contaminants, seventeen fish samples from the 1980 survey were screened for 39 compounds using gas chromatography and mass spectrometry (GC/MS). Table 12.4 summarizes the results of the analysis.

Samples were cleaned up by using Florisil chromatography, and separated into three fractions by silica gel chromatography before GC/MS analysis. Silica gel was used to isolate high concentrations of PCBs and hydrocarbons from lower concentrations of polychlorinated terephenyls (PCTs) and polycyclic aromatic hydrocarbons (PAHs). The analyses of PCBs, DDT, chlordanes, and dieldrin, reported in Section 3.2.1. and Table 12.3 were confirmed by GC/MS. Thus, GC/MS results for these compounds are not shown in Table 12.4. The Root River B sample was analyzed without silica gel frac-

Table 12.4. Results[a] of GC/MS Analysis in Wisconsin Coastal Zone Fish Samples, 1980

Compounds	Detection Limit by GC/MS (μg/kg)	Sturgeon Bay A	Sturgeon Bay B	Fox River A	Fox River B	Fox River C	Lake Superior A	Lake Superior B	Lake Superior C	St. Louis River	Milwaukee River A	Milwaukee River B	Kinnic-kinnic River	Root River A	Root River B	Pike River	Duck Creek	Pensaukee River
PAHs																		
Methyl Naphthalenes	100	ID	ID	PD	ID	ND	PD	ID	ID	PD	ND	ID	PD	ID	ID	ID	ND	ND
C_2-Naphthalanes	100	ID	ND	ID	ID	PD	PD	ID	ID	PD	ID	ID	PD	ID	ID	ID	ND	ID
C_3-Naphthalenes	100	ID	PD	PD	ND	ND	ND	ID	ID	ND	ID	PD	PD	ID	ID	PD	ND	ID
Methyl biphenyl	100	ND	PD	ND	PD	ND	ND	PD	PD	ND	ID	PD	PD	ID	ND	PD	ND	PD
Fluorene	50	ND	ND	ND	ND	ND	ND	ND	ND	ND	ND	ND	ND	ID	ND	ND	ND	ND
Anthracene/Phenanthrene	50	ID	ID	ND	ND	ND	ND	ID	ID	ND	ID	ID	ID	ID	ND	ND	ND	ID
Fluorathene	50	ND	ND	ND	ND	ND	ND	ND	ND	ND	ID	ND	ID	ID	ND	ND	ND	ND
Pyrene	50	ND	ND	ND	ND	ND	ND	ND	ND	ID	ND	ID	ID	ND	ND	ND	ND	ND
Chrysene	50	ND	ND	ND	ND	ND	ND	ND	ND	ND	ND	ND	ND	ND	ND	ND	ND	ND
Other PAHs		ND	ND	ND	ND	ND	ND	ND	ND	ND	ND	ND	ND	PD	ND	ND	ND	ND
PCB substitutes																		
$C_{15}H_{16}$ (Santosols)	100	ND	ID	ID	PD	ID	ND	ND	ND	ND	ND	ND	ND	ND	ND	ND	ID	ID
$C_{22}H_{22}$ (Santosols)	100	NA	NA	NA	NA	NA	NA	NA	NA	NA	NA	NA	NA	NA	NA	NA	ID	PD
Isopropyl Biphenyl	100	ND	PD	PD	PD	ND	ND	ND	ND	ND	ND	ND	ND	ND	ND	ND	PD	ND
Di-Isopropyl Biphenyl	100	ND	ND	ID	ID	ND	ND	ND	ND	ND	ND	ND	ND	ND	ND	ND	ND	ND
CL'D $C_{15}H_{16}$	100	ND	ND	ND	ND	ND	ND	ND	ND	ND	ND	ND	ND	ND	ND	ND	ND	ND
Phenyl decane	100	NA	NA	ID	NA	NA	NA	NA	NA	NA	NA	NA	NA	NA	NA	NA	NA	NA
Phenyl undecane	100	NA	PD	ID	NA	NA	NA	NA	NA	NA	NA	NA	NA	NA	NA	NA	NA	NA
Phenyl dodecane	100	NA	NA	ID	NA	NA	NA	NA	NA	NA	NA	NA	NA	NA	NA	NA	NA	NA
Other compounds:																		
Polychlorinated terphenyls	400	ND	ID	ND	ND	ND	ND	ND	ND	ND	ID	ND	ND	ND	ND	ID	ND	ND
Diethyl Phthalate	100	ID	ID	ID	ID	ID	ID	ID	ID	ID	PD	PD	PD	ND	ND	ND	ID	ID
Dibutyl Phthalate	100	ID	ID	ND	ID	ID	PD	ID	ID	ID	ID	ID	ID	PD	ND	ND	ID	ID
Dioctyl Phthalate	100	ID	ID	ID	ID	ID	PD	ID	ID	ID	ID	ID	ID	ID	ND	ID	ID	ID
Pentachloroanisole	50	NA	NA	NA	PD	NA	NA	NA	ID	NA	NA	NA	NA	NA	NA	NA	NA	NA
Biphenyl	50	NA	NA	NA	ID	NA	ND	NA	NA	NA	NA	NA	NA	NA	NA	NA	NA	NA
Toxaphene	2000		ID			ND		ID	ID								ND	ND
BHC	50		ID			ND		ND										ND

[a] ID = Identified by GC/MS through spectral interpretation techniques. Quantitated samples have assigned values. PD = Possible detection by GC/MS, compound was present at trace levels or was not reduced from another compound. ND = Not detected by GC/MS. NA = Not detected, but usually not present in this silica gel fraction.

tionation. Results of the Florisil cleanup for this sample indicated no contamination by PCBs or hydrocarbons.

Silica gel fractions one and three (not shown in Table 12.4) were analyzed from Fox River-A, Pike River, Root River-A, and Sturgeon Bay samples. The first gel fraction in each of these samples revealed the presence of hydrocarbon compounds: odd-numbered alkanes, even-numbered alkanes, and specifically heptadecane (C_{17}). Heptadecane and other odd-numbered alkanes are known to be by-products of algal growth and their presence is probably not an indicator of anthropogenic pollutants. Even-numbered alkanes could indicate anthropogenic input. In the third silica gel fraction of Fox River-A, diethyl phthalate and a PCB substitute, $C_{22}H_{22}$, were identified. Nonyl phenol/BHT was identified in fraction three from Sturgeon Bay and the Root River. Dacthal was identified in silica gel three from the Pike and Root Rivers. During the past two years, dacthal has been quantified by GC/MS up to 1.00 ppm.

Dacthal is a pre-emergent herbicide that is used on onion, cabbage, and cotton crops. It has been reported in water samples from the Imperial Valley in California that were analyzed by GC/MS (Picker et. al., 1979). Approximately 100,000 lb. of this herbicide is used per year in the Imperial Valley. It apparently finds its way into irrigation runoff water and eventually into streams and lakes. It is not unlikely that dacthal would be present in Wisconsin rivers and streams considering its common use as a yard and garden herbicide.

With the exception of the Root River B sample, silica gel fraction 2 was analyzed from all samples. PCB substitute compounds were identified in samples from the Fox River, Sturgeon Bay, Duck Creek, and the Pensaukee River. These compounds are suspected to originate from several paper recycling mills along the Fox River (Peterman, 1982).

Polycyclic aromatic hydrocarbons (PAHs) were identified in all samples except from Duck Creek and one sample from Lake Superior. Accumulation and uptake of PAH compounds (methyl napthalene, C_2-napthalene, C_3-napthalene, phenanthrene) by marine bivalves has been related to fossil fuel oil sources (Farrington et al., 1982). The PAH compounds identified in fish tissue from this survey may be fuel oil related as well.

Polychlorinated terphenyls (PCTs) having 3–4 chlorine atoms, were identified in fish from the Milwaukee River, Sturgeon Bay, and Pike River. PCTs have physical properties similar to PCBs although they are more limited in industrial applications (Stratton and Sosebee, 1976). They are desirable in industry because of their high heat capacity, chemical stability, and excellent dielectric properties. Although little information exists concerning PCTs in the environment, their behavior is expected to be similar to that of PCBs in that the rate of degradation is slow and significant accumulation of the compound can occur in the environment. PCTs have been detected in sediments as high as 12 ppm in the vicinity of a manufacturing facility and two users of the investment casting industry (Stratton and Sosebee, 1976). That PCTs were also detected in fish from the Milwaukee River, Sturgeon Bay, and Pike River may be attributed to the abundance of industry in these basins.

In all samples except Root River B, phthalate compounds were identified. Other compounds that were identified by GC/MS included pentachloroanisole, biphenyl, toxaphene, and BHC.

2.3. 1981 Survey

Fifty-two fish, 22 sediment, and 42 effluent samples were collected. Laboratory analysis is not yet complete for the five fish samples selected for GC/MS.

2.3.1. Fish

Results of the 1981 fish survey are summarized in Table 12.5. Twenty-nine samples exceeded the FDA tolerance for PCB and one sample equaled the

Table 12.5. Summary of Toxic Substances in Wisconsin Coastal Zone Fish Samples, 1981

	Number of Samples				
Parameter	Below Detection Limit	Above Detection Limit	Above Tolerance	Mean Concentration[a]	Range[a]
		Fish			
(detection limit ppm)					
(tolerance ppm)				(ppm)	(ppm)
PCB (0.20) (5.00)	5	85	29	6.00	0.21–55.00
DDT (0.05) (5.00)	14	38	0	0.55	0.05–3.16
Dieldrin (0.02) (.30)	41	11	1	0.07	0.02–0.30
Chlordane (0.05) (.30)	48	4	0	0.15	0.10–0.20
		Sediment			
(detection limit ppm)				(ppm)	(ppm)
PCB (0.05)	2	20	NA	3.04	0.06–22.00
DDT (0.01)	5	17	NA	0.15	0.01–0.65
Dieldrin (0.01)	19	3	NA	0.04	0.01–0.05
Chlordane (0.01)	19	3	NA	0.14	0.01–0.34
		Effluent			
(detection limit ppb)				(ppb)	(ppb)
PCB (0.50)	39	3	NA	39.43	1.00–110.00
DDT (0.02)	42	0	NA	NA	NA
Dieldrin (0.01)	42	0	NA	NA	NA
Chlordane (0.01)	42	0	NA	NA	NA

[a] For samples above the detection limit.

dieldrin tolerance level. All samples were below tolerance for DDT and chlordane.

Elevated PCB levels were again found in fish from the Milwaukee River, especially in the reach between the mouth and above the Thiensville dam (12.80 ppm average). Carp and goldfish in this reach averaged 25.50 ppm. Above this point levels drop to an average of 0.50 ppm. Thus the area of contamination corresponds almost exactly to that found in 1980. Low but detectable levels of DDT were found in 11 of the 19 samples. Two samples contained detectable levels of dieldrin while chlordane was not found in any of the samples.

Fish from the other rivers surveyed in southeastern Wisconsin showed lower concentrations of PCBs when compared to 1979 and 1980 samples. Two samples of rough fish from the Kinnickinnic averaged 20.00 ppm compared to 34.00 ppm for all 1980 samples. Mean PCB concentration from the Pike River at Kenosha was 6.14 ppm. In 1980, mean concentration in this area was 16.70 ppm. None of the nine samples from the Root River displayed PCB levels above tolerance, although elevated levels of PCB were found in three samples near the river mouth. Low but detectable levels of DDT were found in all samples from the Pike, Kinnickinnic, and Root Rivers. Detectable levels of dieldrin were found in two samples from the Kinnickinnic River, one sample from the Root River, and three samples from the Pike River (one of which equaled the tolerance level of 0.30 ppm). Elevated levels of chlordane were found in one sample from the Kinnickinnic River and one sample from the Pike River.

Four fish samples were collected from the lower Fox River and three of these samples were above tolerance for PCB. Mean PCB concentration in 1981 was 6.05 ppm, significantly lower than the mean concentration of 16.00 ppm found in 1980. Low levels of DDT were found in all samples.

Fish from Green Bay showed high levels of PCB. Five single carp samples were collected in each of three locations in the bay; offshore at Green Bay, Little Sturgeon Bay, and offshore at Marinette. Average PCB levels for the locations were 9.66, 9.58, and 5.52 ppm respectively. Thus Green Bay remains a problem PCB area due in large part to the influence of the Fox River. None of the Green Bay samples were analyzed for other parameters.

Of the 11 samples taken in Sturgeon Bay, the six carp samples from the ship canal contained an average of 17.02 ppm PCB. Samples at the shipyards and Sawyer Harbor contained an average of only 0.96 ppm PCB. This may be due to the fact that the latter samples consisted of smaller yellow perch, rock bass, brown bullheads, and walleye (representing younger fish that have accumulated less PCB). One Sawyer Harbor sample contained 0.13 ppm dieldrin and 0.18 ppm chlordane.

In Lake Michigan, 12 samples of chinook and coho salmon, and lake trout collected offshore at Milwaukee and Manitowoc showed an average of 2.20 ppm PCB. This is a great improvement over past years. However, four of five large chinook salmon from offshore at Sheboygan were over tolerance

Table 12.6. List of Common and Scientific Names of Fish

Common Name	Scientific Name
Alewife	*Alosa pseudoharengus*
Bloater chub	*Coregonus hoyi*
Brown bullhead	*Ictalurus nebulosus*
Burbot	*Lota lota*
Carp	*Cyprinus carpio*
Channel catfish	*Ictalurus punctatus*
Chinook salmon	*Oncorhyncus tschawytscha*
Coho salmon	*Oncorhyncus kisutch*
Goldfish	*Carassius auratus*
Lake herring	*Coregonus artedii*
Lake trout	*Salvelinus namaycush*
Lake whitefish	*Coregonus clupeaformis*
Rainbow smelt	*Osmerus mordax*
Rainbow trout	*Salmo gairdneri*
Rock bass	*Ambloplites rupestris*
Sucker	*Catostomus* spp.
Trout-perch	*Percopsis omiscomaycus*
Walleye	*Stizostedion vitrium*
White sucker	*Catostomus commersoni*
Yellow perch	*Perca flavescens*

(7.40 ppm PCB average), and one large lake trout collected offshore at Manitowoc was above tolerance (10.60 ppm PCB).

Fish sampled from the St. Louis River confirmed the fact that this region of Wisconsin's coastal zone is very healthy with respect to PCB, DDT, dieldrin, and chlordane. None of the samples from this area contained levels above tolerance for any parameters. Although detectable levels of PCB did occur in six of seven samples, the mean concentration was only 0.58 ppm. Only one sample contained detectable levels of dieldrin, DDT, and chlordane.

2.3.2. Sediment

Of the 22 sediment samples analyzed from 1981, 20 contained detectable levels of PCB (mean 3.04 ppm), three contained dieldrin (0.04 ppm), three contained chlordane (mean 0.14 ppm), and 17 contained DDT (mean 0.15 ppm).

The Milwaukee River from the mouth to Silver Spring Drive remained a problem area. The four sediment samples in this reach contained an average of 8.38 ppm PCB and 0.24 ppm DDT. A single sample from an abandoned landfill below Silver Spring Drive contained 22.00 ppm and 0.66 ppm PCB and DDT, respectively. This sample also showed 0.34 ppm chlordane. The landfill is thus a significant source of these microcontaminants. The four samples from above Silver Spring Drive to Grafton contained lower levels of microcontaminants than samples from below Silver Spring Drive, but higher levels of microcontaminants when compared to 1980 results. Average PCB was 0.18 ppm ($n = 4$), and average DDT was 0.02 ppm ($n = 2$). No chlordane was detected. One sample from Cedar Creek (tributary to the Milwaukee River below Grafton) contained 3.70 ppm PCB, 0.31 ppm DDT, and 0.08 ppm chlordane.

Two sediment samples from the Kinnickinnic River contained an average of 2.30 ppm PCB and 0.12 ppm DDT. One of the samples contained 0.01 ppm chlordane. A sample from the Pike River at St. George Cemetery again showed elevated levels of PCB, although the concentration (3.20 ppm) was much reduced from 1980. This problem does not extend into the Carthage college area as was suspected. Both samples showed low levels of DDT (0.23 ppm average). Neither sample contained dieldrin or chlordane.

Low levels of PCB were found in the Root River at Highway 38 in Racine and at Highway 36 in Milwaukee County (0.28 pm average). The sample from Highway 36 also contained 0.05 ppm dieldrin. DDT was detected in low amounts from all Root River sediment samples (0.15 ppm average). No chlordane was detected.

Sediment samples from the Fox River revealed a stable but elevated concentration of PCBs at the city of Green Bay. The average PCB concentration from this area was 3.70 ppm in 1980 and 3.20 ppm in 1981. A sediment sample taken above DePere contained 7.60 ppm PCB and low levels of DDT. In 1980, the PCB concentration above DePere was 4.50 ppm.

The two sediment samples from Sturgeon Bay showed an average PCB concentration of 0.14 ppm. In 1980, average PCB concentration was 0.11 ppm.

2.3.3. Effluent

Of the 42 effluent samples analyzed from 1981, three samples contained measurable amounts of PCBs. No other microcontaminants were detected.

A sample of noncontact cooling water discharged by the Badger Die Casting Corporation to the Kinnickinnic River contained 110.00 ppb PCB. This is an extremely high value, and represents a significant PCB contribution to the river.

A sample of leachate from the previously mentioned abandoned Milwaukee City landfill on the Milwaukee River contained 7.30 ppb PCB. This is in accord with the high level of PCB found in the sediment at this location (22.00 ppm), and confirms this area as a PCB source.

PCB (1.00 ppb) was also detected in a sample of cooling water effluent to the Kinnickinnic River at the Construction Machinery Division of Rexnord, Inc.

Effluent from Fort Howard Paper Mill on the Fox River was below detection limits for all Parameters in 1981. In 1980, effluent from this paper mill contained 4.00 ppb PCB.

3. SUMMARY

Microcontaminants were identified in Wisconsin's coastal zone almost 20 years ago. Fourteen studies conducted between 1965 and 1979 provide a partial record of these microcontaminants. A subsequent three-year study conducted by the Wisconsin Department of Natural Resources, which compiles a complete and up-to-date inventory of microcontaminants in Wisconsin's coastal zone, is discussed in detail.

In 1979, problem drainage systems were identified using fish as a biological indicator. The results of this survey showed that of the 16 toxic substances monitored, only four represented problems for Wisconsin's coastal zone: PCB, DDT, chlordane, and dieldrin. PCB was found to be the worst contaminant statewide. Fish tested from Lake Superior and its tributary streams showed that the quality of fish from this basin is high. Fish from Lake Michigan and its tributary streams were found to contain varying levels of PCB, chlordane, DDT, and dieldrin. The tributary streams to northern Lake Michigan did not appear to be sources of microcontaminants to the lake. In contrast, the tributary streams to southern Lake Michigan, which drain the more industrialized area of southeastern Wisconsin, are suspected to be sources of various toxic substances.

In 1980, intensive sampling of problem areas identified in the first year survey defined more precisely the limits and potential sources of contamination. Fish and sediment data indicated that a PCB problem persists in specific areas of the following drainage systems: Sheboygan River, Milwaukee River, Kinnickinnic River, Menomonee River, Pike River, Root River, and Fox River. Mass spectrometry analysis of selected fish tissue samples identified the presence of a variety of compounds that had not been monitored before. Compounds that were identified included polychlorinated terphenyls (PCTs), tetrachloroterephthalate (Dacthal), PCB substitute compounds, polycyclic aromatic hydrocarbons (PAHs), phthalate compounds, pentachloroanisole, biphenyl, toxaphene, alkanes, and BHC.

In 1981, fish, sediment, and effluent monitoring continued for the purpose of evaluating trends in contaminant levels and continuing the effort to identify and eliminate point sources of toxic substances. PCB problem areas were again found in the Fox, Milwaukee, Kinnickinnic and Pike rivers. Fish sampled over the three-year period showed decreases in yearly mean concentrations (for samples above detection limit) of PCB, DDT, dieldrin and chlordane: PCB decreased from 30.60 to 4.97 ppm, DDT decreased from 0.88 to 0.58 ppm, dieldrin decreased from 0.10 to 0.07 ppm, and chlordane decreased from 0.21 to 0.15 ppm.

REFERENCES

Armstrong, D. E. and D. Weininger. (1980). Organic contaminants in the Great Lakes. International Symposium for Inland Water and Lake Restoration, September 8–12, 1980, Portland, Maine.

Degurse, P. and V. Duter. (1975). Chlorinated hydrocarbon residues in fish from major waters of Wisconsin. Wisconsin Department of Natural Resources, Bureau of Fish and Wildlife Management, Fish Management Section Rept. No. 79, 21 pp.

Farrington, J. W., A. C. Davis, N. W. Frew, and K. S. Rabin. (1982). No. 2 fuel oil compounds in *Mytilus edulis*. *Mar. Biol.* **66:** 15–26.

Haile, C. L. (1977). Chlorinated hydrocarbons in the Lake Ontario and Lake Michigan ecosystems. Ph.D. dissertation. University of Wisconsin—Madison, 115 pp.

Hickey, J. J., J. A. Kieth, and F. B. Coon. (1965). An exploration of pesticides in a Lake Michigan ecosystem. Completion Rept. (Part 1) for contract 14-16-0008-659 submitted to Bureau of Sport Fish. and Wildlife, Fish and Wildlife Service, U.S. Dept. of the Interior, 30 p.

Hickey, J. J., J. A. Kieth, and F. B. Coon. (1966). An exploration of pesticides in a Lake Michigan ecosystem. *J. Appl. Ecol.* **3**(Suppl.)**:** 141–154.

Jensen, S. (1966). Report of a new chemical hazard. *New Scientist* **32:** 612.

Kleinert, S. J. (1975). Concentrations of metals, pesticides, PCBs, and radioactivity in fish from Wisconsin's nearshore waters of Lake Superior. Wisconsin DNR Technical Report, 1975.

Kleinert, S. J., P. E. Degurse, and T. L. Wirth. (1968). Occurrence and significance of DDT and dieldrin residues in Wisconsin fish. Wisconsin DNR Technical Report, No. 41.

Kleinert, S. J., T. Sheffy, J. Addis, J. Bode, P. Schultz, J. Delfino, and L. Lueschow. (1978). Final report on the investigation of PCBs in the Sheboygan River system. Wisconsin DNR Technical Report, July 12, 1978.

Kuehl, D. W. (1981). Unusual polyhalogenated chemical residues identified in fish tissue from the environment. *Chemosphere* **10:** 231–242.

Kuehl, D. W., E. N. Leonard, B. C. Butterworth, and K. L. Johnson. (Unpublished manuscript). Polychlorinated chemical residues in fishes from the major watersheds near the Great Lakes, 1979. U.S. Environmental Protection Agency, Environmental Research Laboratory—Duluth, 23 pp.

Leland, H. V., W. N. Bruce, and N. F. Shimp. (1973). Chlorinated hydrocarbon insecticides in sediments of southern Lake Michigan. *Environ. Sci. Technol.* **7:** 833–838.

Miles, J. R. and C. R. Harris. (1971). Insecticides residues in a stream and a controlled drainage system in agricultural areas of southwestern Ontario, 1970. *Pest. Monit. J.* **5:** 289–294.

Peterman, P. (1982). Identification of PCB substitute compounds in the Fox River. Masters Thesis, Water Chemistry Program, University of Wisconsin—Madison. (in preparation).

Picker, J. E., M. L. Yates, and W. E. Pereira. (1979). Isolation and characterization of the herbicide dimethyl tetrachloroterephthalate by gas chromoatography–mass spectrometer–computer techniques. *Bull. Environ. Contam. Toxicol.* **21:** 612–617.

Poff, R. J. and P. E. Degurse (1970). Survey of pesticide residues in Great Lakes fish. Wisconsin Department of Natural Resources, Bureau of Fish Management, Technical Report No. 34, May 1970.

Sheffy, T. B. and T. M. Aten. (1979). 1979 Annual summary of PCB levels in Wisconsin fish. Wisconsin DNR Technical Report, July 2, 1979.

Stratton, C. L. and J. B. Sosebee, Jr. (1976). PCB and PCT contamination of the environment near sites of manufacture and use. *Environ. Sci. Technol.* **10:** 1229–1233.

Veith, G. D. (1970). Environmental chemistry of the chlorobiphenyls in the Milwaukee River. Ph.D. dissertation. University of Wisconsin—Madison, 180 pp.

Veith, G. D. (1975). Baseline concentrations of polychlorinated biphenyls and DDT in Lake Michigan Fish, 1971. *Pest. Monit. J.* **9**(1)**:** 21–29.

Veith, G. D., D. W. Kuehl, E. N. Leonard, K. Welch, and G. Pratt. (1981). Fish, wildlife, and estuarties: Polychlorinated biphenyls and other organic chemical residues in fish from major United States watersheds near the Great Lakes, 1978. *Pest. Monit. J.* **15**(1)**:** 1–8.

13

PCB CONTAMINATION IN THE GREAT LAKES

Milagros S. Simmons

Department of Environmental and Industrial Health
The University of Michigan
Ann Arbor, Michigan 48109

1. INTRODUCTION

The Great Lakes have been a vulnerable target sensitive to chemical contamination because of their large size, geographic and demographic characteristics and unique limnological properties. (Sonzogni and Swain, 1980). About 20% of the total U.S. population depends on these water resources with possible increasing dependence in the future, thus the problem of chemical contamination in the Great Lakes has been of great concern.

Polychlorinated biphenyls (PCBs) are a class of organic compounds which, because of certain physical and chemical properties such as low vapor pressure, low water solubility, high dielectric constant, and overall resistance to chemical, heat and biological changes, have been used in a wide variety of ways (Hutzinger et al., 1974). Production of these compounds was begun on a large scale in 1929 by the Swan Corporation which later became part of a larger organization, Monsanto Chemical Company. Monsanto, the sole manufacturer of PCB in North America voluntarily reduced their sales in September 1970 based on consideration of possible contamination of food products and their inability to monitor possible losses in the environment. In 1978, production of PCBs was suspended.

PCB was first detected in Great Lakes fish in 1969 in coho salmon (Price and Welch, 1972). Subsequent environmental monitoring has identified PCBs in some sport fish, especially lake trout, rainbow trout, and various salmon, often reaching levels two to four times the 5 μg/g level set by the U.S. Food and Drug Administration (U.S. FDA, 1976; Willford et al., 1975). It is expected that the inputs of PCB into the lakes from solid waste dumps and other diffuse sources will continue for some time.

A comparison of analytical data shows that the Great Lakes contain relatively high PCB concentrations, with Lake Michigan as the most highly contaminated of all (IJC, 1974, 1976). Recent studies of the environmental impact of PCBs in the Great Lakes have included: monitoring in water, sediments, fish, water fowl, wild life, human blood and tissue; toxicity studies on aquatic and mammalian organisms; transport, cycling, and transformation studies in the lake environment. As a result, a voluminous amount of data exists on studies of the environmental levels and hazard of PCBs in the Great Lakes area. This review is an attempt to (1) present a state-of-the-art view of PCB contamination in the Great Lakes region, (2) indicate problem areas where controversies exist, and (3) identify gaps in existing knowledge on the fate and effects of PCBs in the Great Lakes.

2. PHYSICAL AND CHEMICAL PROPERTIES OF PCBs

Direct chlorination of biphenyl in the presence of an appropriate catalyst produces a class of compounds called polychlorinated biphenyls with varying degrees of chlorination and isomeric substitution. A four-digit number following the general trade name AROCLOR is used to signify the type of PCB. The first two digits (1, 2) identify the molecule as a biphenyl, and the last two digits give the percent weight of chlorine in the compound. These latter two digits thus indicate the degree of chlorination of the biphenyl. The mean percentage of chlorine in the mixture is selected to provide the desired technical properties. The physical and chemical properties of PCBs (solubility, partitioning, degradation, persistence, adsorption) vary considerably depending on the degree and position of chlorine substitution. Thus the distribution and behavior of a particular PCB in the environment will be influenced greatly by its degree of chlorine substitution. Because PCBs characteristically have very low water solubilities they sorb readily to particulate organic matter and aqueous sediments. Their high lipid solubility, on the other hand, plays an important role in their accumulation or concentration in fatty tissue. Adsorption and solubility factors result in the distribution of variable amounts of PCBs in various physical and biological components of the lake. These components, often referred to in the literature as compartments, include: atmosphere, suspended solids, water, sediments, and biota. This distribution is called compartmentalization. A thorough understanding of PCB compartmentalization is a desired but is still a distant goal in the evaluation of PCB contamination of water bodies, particularly the Great Lakes.

3. SOURCES OF PCBs IN THE GREAT LAKES

From an environmental viewpoint the commercial applications of PCBs can be divided into three categories: *controllable closed systems,* in which the life of the PCBs equals the life of the equipment and usually no leakage is observed; *uncontrollable closed systems,* which present problems associated with PCB collection for disposal and which may also lose PCBs through leakage; and *dissipative uses* such as lubricating and cutting oils, pesticides, plasticizers in paints, copying paper, adhesives, sealants, printing ink, rubber manufacturing, water proofings, fungicidal insulations, stabilization of polymers, anticorrosion, nail coatings, and so on, which are in direct contact with, and hence, contaminate the environment.

In 1974 approximately 40 million pounds of PCBs were manufactured in the U.S. and approximately 450,000 lb were imported. Between 65 and 70% were used in capacitors, 29–34% in transformers, and 1% for miscellaneous purposes. As of 1977, Aroclor 1016 was the principal PCB mixture used in 90–95% of all impregnated capacitors made in the United States. PCBs are still in use as coolants in many capacitors and transformers, especially in the utility industry. Capacitors are also used in air conditioners, microwave ovens, and fluorescent light ballasts made before 1977. It has been estimated that between 1930 and 1970, the total loss of PCBs in North America alone was about 500,000 tons with over 30,000 tons dispersed in the air, more than

60,000 tons discharged into bodies of waters and underlying sediments, and 30,000 tons located in dumps and landfills. The ultimate reservoirs of PCBs that enter the environment are mainly sediments of rivers and coastal waters. Because of the specific uses and chemical properties of the various Aroclor mixtures, most of the PCBs lost to the air, water, and land is dominated by one or two mixtures. Hence, most of the PCBs lost to the atmosphere consist of 1248 and 1260, into the water mainly 1242 and 1260 with 1242 dominant in dumpsites. The major routes of PCB contamination in 1978 were incineration of PCB-containing materials; disposal of waste oils, waxes, and paints; landfill leaching, insecticide runoff of PCBs, recycling of PCB-containing paper; and leakage of heat-transfer fluid, or oils.

PCBs enter the Great Lakes from nonpoint sources as well as point sources. Nonpoint sources include atmospheric deposition, dredging activities, runoff from agriculture, silviculture, septic tanks, urban areas, mining operations, and other diffuse sources. Point sources include wastewater treatment plants and industrial waste discharges, tributaries, and seepage from dumpsites and landfills.

PCBs enter the atmospheric environment through incomplete combustion processes, through volatilization from treated surfaces and from aquatic systems, from entrainment of dust in treated soils (Cohen and Pinkerton, 1966), and from spraying programs (Spencer and Cliath, 1975; Peakall, 1976). Aerial transport and deposition of PCBs has been shown to be the most important contributor of PCBs, accounting for 60–90% input to the Upper Great Lakes (Eisenreich et al. 1981; Murphy et al., 1981). Total deposition of airborne PCBs to the Great Lakes has been estimated in metric tons per year as follows: Lake Superior, 9.8; Lake Michigan, 6.9; Lake Huron, 7.2; Lake Erie, 3.1; Lake Ontario, 2.3 (Eisenreich et al., 1980). Swain (1978) found that mean concentrations of PCBs observed in precipitation samples from Siskiwit Lake was 230.0 ng/liter and 50.0 ng/liter for the Duluth–Superior metropolitan region. These data suggest that atmospheric inputs via precipitation may account for the high levels of PCBs in the Isle Royal area, and specifically in fish of Siskiwit Lake. Doskey and Andren (1981a, b), Murphy et al. (1981), and Murphy and Rzuszutko (1977) have studied the atmospheric inputs of PCBs into Lake Michigan. Doskey and Andren (1981a) found that the PCB content in both filterable and nonfilterable air fractions over Lake Michigan had a narrower range as well as lower values than air samples collected over urban areas. They reported an average of 11 times more PCBs contained in the vapor phase as in the particulate, with the average ratios of vapor to particulate PCBs of 9 : 1, 26 : 1, 37 : 1 for Lake Michigan, Chicago, and Madison, WI, respectively. Mixtures of PCBs resembled only Aroclors 1242 and 1254 in air samples collected over the waters of Lake Michigan, whereas PCBs associated with particulate matter tended to contain more of the higher-chlorinated isomers, including Aroclor 1260 type. Doskey and Andren (1981b) estimate that approximately 90% of the PCBs in the atmosphere exist in the vapor state. They suggest that such factors as the suspended solids load, dissolved organic carbon (DOC) content and particulate organic carbon (POC) content of a lake may alter mass transfer and air–water partition coefficients estimated for a lake. This could then result in different volatilization and vapor deposition rates for different lakes.

Surface microlayers of Lake Michigan have been found to be enriched in PCBs (Doskey and Andren, 1981b; Rice and Meyers, 1982); this may have a significant role in the transfer of PCBs back into the atmosphere by bubble ejection at the air–water interface. This concept of PCB transfer via volatilization is a subject of much debate at the present.

In Lake Michigan, Murphy et al. (1981) reported that there is a higher PCB concentration in precipitation at the south end of the lake due to close proximity to industrialized regions, and the greatest overall contribution of PCBs to the lake comes from the atmosphere. Bulk atmospheric loadings to southern Lake Michigan amount to 115 $g/km^2 \cdot yr$, with 77 $g/km^2 \cdot yr$ and 44$g/km^2 \cdot yr$ attributed to wet and dry deposition, respectively.

In Lake Superior, Eisenreich et al. (1981) also claimed atmospheric factors as the major source of PCB input into the lake. In analyzing sediments from different sections of the lake, they found maximum concentrations in surficial sediments between Keweenaw Peninsula and Thunder Bay in the downwind direction of Thunder Bay Ontario. An estimated atmospheric input of 3000–8000 kg/yr, using a sedimentation rate of 0.3–0.4 $\mu g/m^2 \cdot yr$ was reported. They also indicated that dry deposition of pollutants to be 1.5–5.0 times as great as wet deposition, although a theoretical model predicts a 2.5:1 wet:dry deposition ratio.

The northern end of Lake Michigan and Lake Huron receive considerably less PCBs via atmospheric deposition according to the results reported by Murphy et al. (1981). Wet deposition of these regions was 12 $g/km^2 \cdot yr$; dry deposition, 6.4 $g/km^2 \cdot yr$; and bulk deposition, 18 $g/km^2 \cdot yr$. Total mean atmospheric PCB loadings to the Lake Huron basin are estimated to be 1100 kg/yr, and to the Lake Michigan basin, 2500 kg/yr.

Another important source of PCB polution is from wastewater treatment plants, especially in the Great Lakes area. For Lake Erie alone, it has been estimated that around 250 kg/yr of PCBs are lost from municipal wastewater treatment plants to the lake compared to about 1100 kg of PCB which are used annually by 21 industries all over the Great Lakes (IJC, 1976). This accounts for almost one fourth of the amount of PCB added to Lake Erie yearly.

Several studies show that the amount of PCB present in both water and underlying sediment was directly proportional to the degree of urbanization and industrialization of nearby areas. Weininger and Armstrong (1982) have estimated losses from major industrial point sources of PCB in Waukegan Harbor to be 500,000 kg/yr. Sediments of Waukegan Harbor, Il, have been found to contain PCB resembling Aroclor 1242 and 1248. (Armstrong, 1981) with concentrations up to 500,000 $\mu g/g$ (U.S. EPA, 1981). It is suggested that dissolution and resuspension of PCBs in contaminated sediments may increase the PCB concentration of water entering the harbor from Lake Michigan. Armstrong (1981) hypothesizes that dredging activity may contribute to sediment PCB concentrations in the lower Waukegan Harbor. He reports sediment PCB concentrations ranging from 8 to 3600 μg/g in the Waukegan Harbor area, with predominance of Aroclor 1248 in near-harbor samples.

The PCBs disposed of in dumpsites and landfills, exposed to the inevitable rain runoff and leaching, could become one of the major sources of PCBs into the Great Lakes in the future. Likewise, as more point sources of PCBs are curtailed, nonpoint sources will become increasingly important.

Neely (1977) disputes the atmospheric deposition model of PCB influx into Lake Michigan. He studied the possible input sources needed to maintain observed PCB levels in the lakes. By developing a mathematical model, he was able to predict trends in the distribution of PCBs in the lake environment. He concluded that when the direct input of PCB is curtailed, the persistent high levels of PCB in Lake Michigan will not be due to municipal or industrial waste discharge or fallout from rain or snow, but will result from PCB released from lake bottom sediment. This is currently an area of much research interest as the impact of eliminating discharges of PCBs into the lakes are expected to occur.

4. DISTRIBUTION OF PCBS IN THE GREAT LAKES

4.1. Lakewater

Swain (1980) has summarized the concentrations of PCBs in both nearshore and open waters of the Great Lakes. Typical values for the open waters of Lake Superior are 5.0 ng/liter, whereas nearshore waters range up to 110 ng/liter near Minnesota. In Lake Michigan, nearshore waters range from 12 to 50 ng/liter. Eadie (1979) and Rice (1980) reported open water concentrations of 9.0 ng/liter and 3.0 ng/liter, respectively, in Lakes Michigan and Superior. Nearshore waters of Lake Huron may be very high river mouths. The Michigan Department of Natural Resources (1980) reported values up to 700 ng/liter in the vicinity of Rifle River. PCB concentrations in the outer bay of Saginaw Bay, which more closely approximate open water values, range from 0 to 10 ng/liter. For Lake Erie, a lakewide mean value of 27.0 ng/liter was reported, and nearshore concentration of 20.0 ng/liter was reported in the vicinity of Erie, Pennsylvannia. Values for nearshore waters of Lake Ontario typically range from about 40 to 80 ng/liter.

4.2. Sediments

PCB values in sediments are several orders of magnitude higher than those of water. Sediments from tributary mouths in harbors and from nearshores of the lakes have higher PCB values than those of the open lake sediments.

Table 13.1. Concentrations of PCBs in Great Lakes Sediments

Location	PCB (μg/kg)	Reference
Lake Michigan	3.72–132.6	Schacht, 1974
	38.2 (mean)	PLUARG, 1978
	2–20	Frank et al., 1981
Southern Lake Michigan	1–75	Armstrong, 1981
	Trace–20	Glooschenko et al., 1976
Lake Huron	0.1	Berg et al., 1974
	20.0	Burin and Robbins, 1977
	9–33	Frank et al., 1979
Lake Erie	4–800	Frank et al., 1977
Lake Superior	5–390	Eisenreich et al., 1979
	Trace–12.0	Glooschenko et al., 1976
	7.0 (mean)	Veith et al., 1977
	3.3–8.5	Frank et al., 1980
Lake Ontario	120 (mean)	Haile et al., 1975
	3–20	Glooschenko et al., 1976
	<5–280	Frank et al., 1979

The results of an extensive survey in southern Lake Michigan conducted by Armstrong (1981) also suggest PCB concentrations are highest in areas of recent sedimentation and that the lowest levels occur in areas of scour where faster water currents prevent sediment accumulation. Frank et al. (1977, 1979, 1980) found PCB values two to three times higher in depositional zones of the Great Lakes than in the nondepositional zones, except for Lake Superior (Frank et al., 1980) where no difference between the PCB values in these zones was detected. Table 13.1 lists reported PCB concentrations in open lake sediments.

4.3. Fish

The importance of the Great Lakes as a recreational and commercial fishery resource best explains the abundance of PCB fish data in the Great Lakes. A wide variability in the data of PCB level exists, however, perhaps due to weaknesses in analytical methodology, but most likely due to variability in species, sampling locations, and sample size. Table 13.2 shows PCB concentrations reported in Great Lakes fish (PLUARG, 1978). These data indicate that Lake Michigan is the most highly contaminated of the Great Lakes. Recently, there has been speculation that PCBs will decrease in Lake Michigan fish as PCB inputs to the lakes decrease (Sonzogni, 1978; Delfino, 1979). While not unequivocally established, there is increasing evidence that such a trend is occurring at least for certain fish species. The Great Lakes

Table 13.2. PCB Concentrations in Great Lakes Fish

Lake	Sampling Period	Mean PCB Concentrations (μg/g)	Range (μg/g)
Superior	1968–1975	0.61	<0.1–3.7
Michigan	1972–1974	10.2	2.1–18.9
Huron	1968–1976	0.82	<0.1–7.0
Erie	1968–1976	0.88	<0.1–9.3
Ontario	1972–1977	2.37	<0.1–21.1

Fishery Laboratory (Ann Arbor, Michigan) found a gradual decrease of PCB levels in Lake Michigan bloater chubs, lake trout, and salmon during a monitoring program they conducted between 1972 and 1978 (MDNR, 1980). Furthermore, unedited STORET data show a steady decrease in annual aggregate concentration of total PCB in Lake Michigan fish between 1974 and 1980 (Hall, 1980). A more comprehensive treatment of trends in PCB levels in Lake Michigan fish is presented by Amant et al. in the next chapter of this volume.

5. PCB PROBLEM AREAS IN THE GREAT LAKES

One of the principal objectives of this review is to identify PCB "problem areas" or regions of the Great Lakes which are highly contaminated by PCB. These regions can best be identified through a comparison of sediment PCB concentration as obtained through various sediment monitoring programs. The work of Frank and his co-workers is particularly extensive in this area and much of the following discussion is based on their data. Figure 13.1 shows the distribution of PCB problem areas in the Great Lakes. It is interesting to note that the areas designated by the International Joint Commission Water Quality Board as areas of degraded water quality are also identified with the PCB contamination problem. The exceptions are for some areas in Lake Superior where aerial deposition has been established as the main pathway for PCB transport.

In Lake Superior, the highest PCB levels in sediments reported by Frank et al. (1980) were from the Duluth subbasin (8.6 ng/g), the Marathon basin (6.4 ng/g), and Thunder Bay (5.7 ng/g). Eisenreich et al. (1979) also reported highest concentration in sediments from the extreme western end near Duluth–Superior (230 ng/g) and the central part of the lake between the Keweenaw Peninsula and Thunder Bay (290 ng/g).

In Lake Michigan, the most highly contaminated areas reported also contain some of the highest PCB levels reported anywhere in the Great Lakes. These areas include Waukegan Harbor (500,000 μg/g) and North Ditch (250,000 μg/g), tributary to Lake Michigan. Milwaukee Harbor (6.42 μg/g), the Fox River Basin (190 ng/g), and southern Lake Michigan (188 ng/g) have also been designated as highly contaminated areas (U.S. EPA Region V; Armstrong, 1981; Kleinert, 1976).

In Lake Erie, the Western Basin had maximum concentrations of PCB (660 ng/g), and the Central Basin (330 ng/g) and the Eastern Basin (320 ng/g) had lower values (Frank et al., 1977).

In Lake Huron, Frank et al. (1979) indicated that Thunder Bay (Alpena, MI) and Saginaw Bay are the areas where PCB concentration was maximum. Burin and Robbins (1977) also reported 3–12 μg/g PCB in Saginaw Bay sediments.

In Lake Ontario, several areas have been designated as problem areas. Frank et al. (1980) reported 260 ng/g as maximum PCB concentration in sediments from the Bay of Quinte. They also reported greater than 200 ng/g values along the southern shore of the lake implicating the Niagara River as the major source of PCBs to Lake Ontario. The Genesee and Oswego rivers may be important minor sources.

In a field survey in boundary waters, the following problem areas in the Great Lakes have been designated (MDNR, 1980):

Lake Ontario	St. Lawrence River, PCB in water, 1977
Lake Erie	Detroit River, PCB in sediments, 1976
	Rocky River, PCB in water, 1978
	Ashtabula River, PCB in water, 1978
Lake Huron	Saginaw River, PCB in sediments, 1978

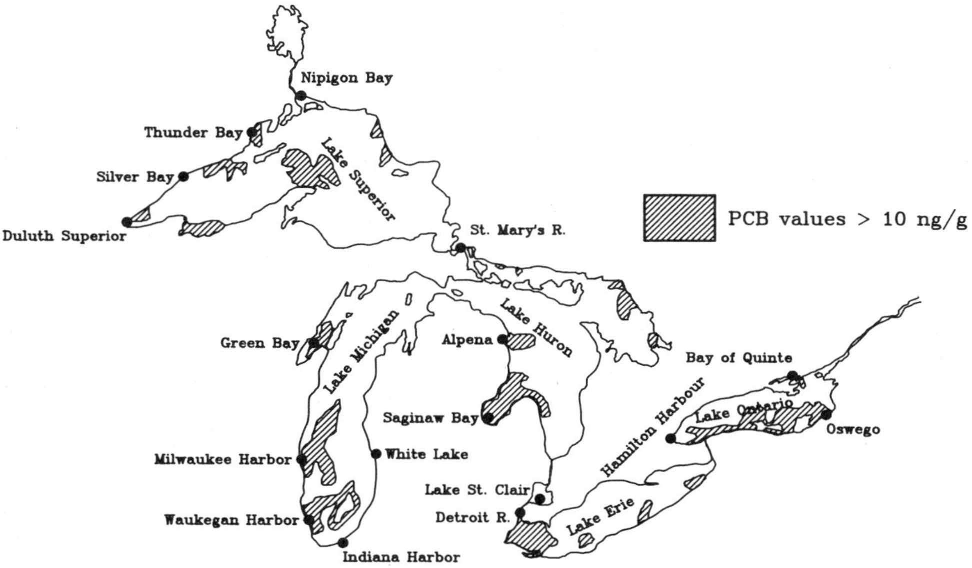

Figure 13.1 Problem areas.

Lake Superior	None
Lake Michigan	Green Bay, PCB in fish and sediments, 1977 Waukegan Harbor, PCB in sediments, 1979 Indiana Harbor, PCB in water, 1979

6. PCB CONTAMINATION IN GREAT LAKES BIOTA

6.1. Fish

The identification and quantification of PCB compounds in Great Lakes fish is widely reported in the literature (Veith, 1975; Willford et al., 1975; Schacht, 1974; Zabik et al., 1978; Passino and Cotant, 1978; Neely, 1977; Veith et al., 1977, Veith, 1980; Guiney and Peterson, 1980; Carr, et al. 1972; Frank et al., 1978; Gessner, 1980; Haile et al., 1975; Spagnoli and Skinner, 1977; Norstrom et al., 1978; Parjeko and Johnston, 1973; Parjeko et al., 1975; Kaiser, 1977). The importance of the Great Lakes as a recreational and commercial fishery resource best explains the wealth of information of PCB residue levels in fish. Swain (1980), in an extensive review, tabulated PCB levels in several species taken from different locations in the Great Lakes.

The first indications of PCB contamination in Great Lakes fishes were determined in museum specimens of six species of Lake Michigan fish collected in 1949 (Neidermeyer and Hickey, 1976). A concentration of 4.85 μg/g was reported for the alewife in 1949 compared to 79.69 μg/g for the 1965 samples. Veith (1975) also reported PCB concentrations for alewife collected in 1971 of 70.80 μg/g.

In 1975, the U.S. FDA confiscated more than 100,000 cans of Lake Michigan coho salmon due to excessive levels of PCBs ranging from 7.6 to 10.9 μg/g PCB (Neely, 1977). In 1971, Veith established baseline concentrations of PCB in Lake Michigan fish in response to the recommendations of the Lake Michigan Interstate Pesticide Committee for predicting trends of PCB in that lake (Veith, 1975). The Great Lakes Environmental Contaminants

Table 13.3. Fish Species for which Consumption Warnings Have Been Established (MDNR, 1980)

Lake	Fish
Lake Michigan	Steelhead, lake trout, salmon
Lake Superior	Lake trout
Lake Huron	Salmon
Lake St. Clair	Carp and catfish
Lake Erie	Carp and catfish

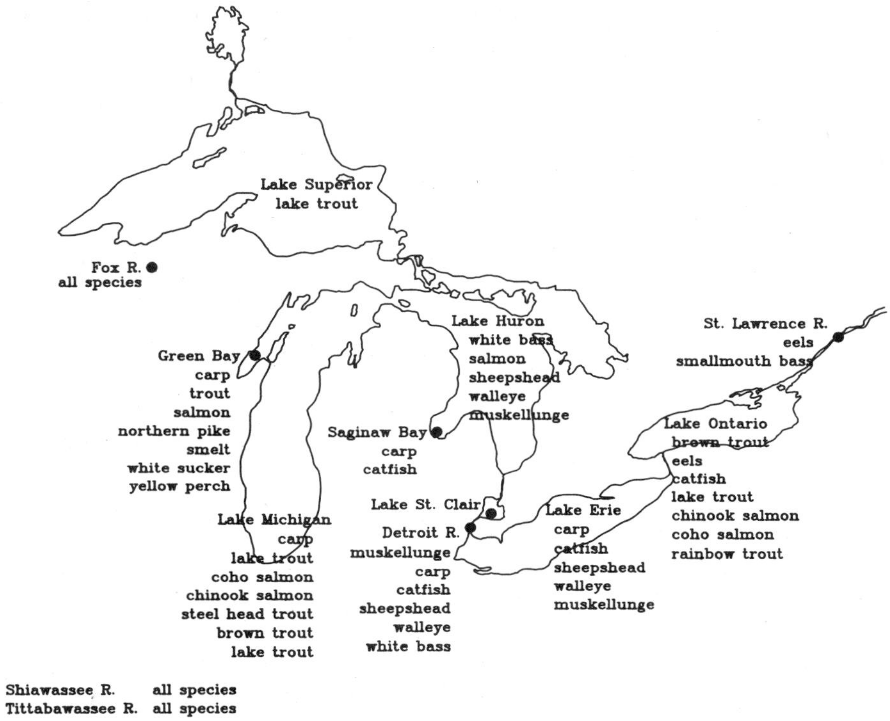

Figure 13.2 PCB contamination in Great Lakes fish.

Survey (GLECS), which has been a cooperative effort between the Michigan Departments of Natural Resources and Agricultural Health, the U.S. Food and Drug Administration's Detroit District, and the U.S. Fish and Wildlife Great Lakes Laboratory, was conducted to determine the contamination levels of PCB and selected toxic materials in the edible portions of fishes found in the Great Lakes. Table 13.3 shows consumption warnings for whole fish found to be highly contaminated by PCB in the designated fishes and remains in effect for all Great Lakes waters within Michigan and its political boundaries.

The PCB levels in fish have been known to vary with feeding habits, age, size, sex, species, and lipid content (Guiney and Peterson, 1980; Gessner, 1980; Rohrer et al., 1980; Kaiser, 1977; Kenaga and Creal, 1979; Swain, 1978). A strong correlation between fat content and total body burden of PCB has been observed in these studies. In a report issued by the Great Lakes Basin Commission (1980) certain fish species in the Great Lakes have been identified as highly contaminated by PCBs. These are indicated in Fig. 13.2.

6.2. Lower Trophic Levels

PCB uptake in algae (Lin, 1981), amphipod (Veith, 1975), benthic algae, and benthic fauna (IJC, 1976) from the Great Lakes have also been reported. In undifferentiated seston samples collected in the Upper Great Lakes in 1974, PCB residues were quantifiable in nearly all samples. Glooschenko et al. (1976) found a maximum concentration of 8.1 μg/g in a sample from the middle of Lake Huron. Two of the highest levels, 6.7 and 5.9 μg/g, were found in samples from Georgian Bay.

Concentrations of PCBs in Lake Superior zooplankton samples of *Mysis relicta* were found to be much less, ranging from 0.05 to 0.12 μg/g in four composited samples. Based on these concentrations and that of Lake Superior water, the biocentration factor is assumed to be on the order of 10^5 (Veith et al., 1977).

6.3. Herring Gulls

Although Vermeer and Reynolds (1970) identified no significant correlation between the concentration of PCBs and shell thickness in great blue heron eggs, suggesting little danger of PCB contamination to aquatic bird natality, other more recent studies conclude otherwise (Vesligh et al., 1979; Gilbertson and Fox, 1977; Sileo et al., 1977). Vesligh et al. (1979) discovered that increased reproductive success in colonies of herring gulls from Lakes Ontario, Erie, Huron, and Superior paralleled a decline in the major organochlorine residues between 1974 and 1978.

Gilbertson and Fox (1977) found high embryonic mortality of herring gulls in two colonies in Lake Ontario and one in Lake Erie. Widespread contamination by PCBs were suspected of causing morphological abnormalities such as enlarged livers, reduced embryo size, porphyna, and accumulations of a subcutaneous mucoservous fluid in some embyros. Post mortem examinations of 138 ring-billed gulls from southern Ontario disclosed concentrations 30–90 times higher in the brain tissues of moribund and dead gulls than in healthy gulls (Sileo et al., 1977). Kaiser (1979) reported mean PCB concentrations in herring gull eggs of 125,000 μg/kg for Lake Ontario specimens and 55,000 μg/kg for birds of Lakes Erie and Superior.

In a study by Frank and Holdrinet (1975), 307 bird eggs were collected from the Niagara Peninsula which represented both terrestrial and aquatic food chain components. Eggs from carnivorous species at the top of the aquatic food chain had the highest average PCB concentration, 3.5–74 μg/g. Eggs from herbivorous and insectivorous birds from both aquatic and terrestrial habitats had lower mean PCB residual levels. Of the aquatic carnivores, herring gulls contained the highest PCB levels, at 74 μg/g. Common tern eggs had 42 μg/g and black-crowned night heron, 27 μg/g. One egg of the red-shouldered hawk had 30 μg/g PCBs. Mean PCB residue levels in the fat of the eggs ranged from 30.2 μg/g in black terns to 851 μg/g in herring gulls. A detailed review of PCB in herring gulls is provided by Mineau et al. in Chapter 19.

6.4. Mink

In 1971, results from a study on the effects on mink reproduction of feeding coho salmon and other Great Lakes fishes were reported (Aulerich et al., 1971). A direct relationship between the degree of reproductive failure and the chlorinated hydrocarbon content of fish fed to the minks was established, but PCBs were not specifically identified as components. In a follow-up study (Aulerich et al., 1973), lesions noted in mink that died while being fed diets containing Lake Michigan coho salmon were very similar to lesions observed in minks receiving 30 μg/g supplemental PCBs. Toxicity studies demonstrated similar PCB residues had accumulated in the brain of the minks. The LD_{50} for 48 hr of three commercial PCB mixtures. Aroclor 1221, 1242, and 1254, when administered as an injection, were 250–500 μg/g, 500–1000 μg/g, and 1000 μg/g, respectively. Other species of freshwater fish from the Great Lakes that were fed to mink as a 30% component of the diet also produced adverse reproductive effects, suggesting that PCB contamination is widespread in the Great Lakes system.

7. TOXICITY EFFECTS IN GREAT LAKES BIOTA

7.1. Algae

In spite of the importance of phytoplankton as the base of aquatic food chains, little research has been conducted to determine the effects of PCBs on this group of aquatic organisms in the Great Lakes ecosystem. In a preliminary investigation on the inhibition of PCBs on photosynthesis and grazing in the Saginaw Bay area, PCBs significantly inhibited carbon fixation and productivity in algae (McNaught, 1977; Rhee, 1982). McNaught (1977) noted that the photosynthetic rate of nanoplankton (<22 μm diam.) was reduced by 52% after only 4–5 hr of incubation in the presence of 2-2′-dichlorobiphenyl. He also noted that the addition of a surfactant resulted in a reduction of only 33% in the photosynthetic rate, indicating that PCBs adsorbed onto cell surfaces have an inhibitory effect on photosynthesis cells which may accentuate bioaccumulation. Glooschenko and Glooschenko (1975) conducted toxicity studies using PCBs on three species of Great Lakes phytoplankton. They found that PCB concentration as low as 1.0 μg/liter resulted in growth inhibition, and a concentration of 50.0 μg/liter was toxic to all cultures. Photosynthetic rates were also depressed by the presence of PCBs. Of the four Aroclor compounds tested, Aroclor 1016 was least toxic to the algae and 1242 most toxic, while 1221 and 1248 initiated intermediate toxic responses. Observations under the light microscope revealed gross changes in cell morphology; closer examination of electron micrographs showed distortion of chloroplast lamellae and increased cytoplasmic vacuole formation in algal cells exposed to 50 μg/liter PCB.

Christensen and Zielski (1980) discovered that PCBs in concentrations of 10–100 μg/liter inhibited chlorophyll production as well as RNA synthesis in *Chlamydomonas,* a species of green alga, which were isolated from Lake Michigan. PCBs depressed growth in this alga when present at concentrations of about 11–111 μg/liter.

7.2. Zooplankton

There is a conspicuous absence of research on PCB toxicity to Great Lakes zooplankton. McNaught (1977) noted that reduced filtering rates in *Chydorus* and reduced grazing rates in copedites resulted when these zooplankton were fed netplankton grown in 100 parts per thousand PCBs. He also found varying accumulation rates among *nauplii, Eubosmina,* and *Daphnia*; the *nauplii* were the most effective ingestors and accumulators of PCBs. The addition of a surfactant significantly reduced the accumulated concentrations of PCBs, suggesting that surface adsorption facilitated bioaccumulation. McNaught found that reduced herbivorous grazing activity and conversion of larger to smaller forms of algae in phytoplankton-based food chains could alter Great Lakes food chains significantly.

7.3. Fish

Some toxicity studies have been conducted using Lake Michigan fish. In a study in Waukegan Harbor in 1979, Lake Michigan species were used to compare rates of bioaccumulation in fish exposed to Waukegan Harbor water as compared to Lake Michigan water, but no PCB data was given (U.S. EPA, 1979). Stauffer (1979) conducted a study to assess the role of DDT and PCBs in reproductive failure of lake trout planted in Lake Michigan as part of a restocking program. Mortality of eggs and fry from Lake Michigan lake trout was significantly higher than that in hatchery lake trout, but no correlation was found between this mortality and the DDT and PCB content.

In a toxicity study conducted by Passino and Kramer (1980) using ciscoe fry, 20% of the fry died in 10 mg PCB/liter and 100% mortality was observed at 1.0–10 mg/liter PCB after 14 days of exposure. Willford (1980) conducted a 6-month study in the winter of 1975–1976 on the effects of chronic exposure of Lake Michigan lake trout to PCBs. Whole body concentrations of 22 μg/g were found in eyed eggs taken from the fish collected, which themselves contained 7.6 μg/g PCBs. One-day-old sac fry hatched from these eggs had concentrations of 3.8 μg/g PCBs measured as Aroclor 1254. Fry exposed for 6 months to levels of PCB contamination approximating resident exposure levels in Lake Michigan of 10 ng/liter in water and 1.0 μg/g in food showed gross deformations after one week. Average total cumulative mortality of fry on the final day of exposure in all treatment concentrations (1–5 and 25 times residence levels in Lake Michigan) was 30.5–46.5 as compared to only 21.7% for controls. These fry mortality results strongly implicate PCBs and chlorinated hydrocarbon contaminants as limiting factors in the reproduction of the lake trout species in Lake Michigan.

In a study using brown trout (*Salvelinus trutta*), Spigarelli et al. (1979) discovered a 6-fold increase in the muscle concentration of PCBs in fish fed whole alewife, containing an average concentration of 2.52 μg/g PCB, compared to Wisconsin pellets containing only 0.18 μg/g PCB at 13°C for 57 days. Analysis of PCBs at three different temperature conditions indicated a significant effect of temperature on bioaccumulation, with increasing accumulation at elevated temperatures. Trophic pathways were demonstrated to be significant to PCB accumulation and transfer. A comparison between fish that consumed alewife and fish that were starved at ambient temperatures suggests that direct uptake of PCBs from water accounts for less than 15% of the total uptake of PCBs by brown trout in Lake Michigan (Spigarelli et al., 1979). For a rigorous treatment on the biological criteria for assessing the effects of PCBs in organisms, the reader is referred to an extensive coverage by Roberts et al. (1978).

8. CHEMICAL DYNAMICS AND TRANSPORT

Literature concerning chemical and transitory processes of PCBs in the Great Lakes is limited. Most of the studies concerning the fate and effects of PCBs are model studies conducted in the laboratory. Transport studies particularly involving atmospheric loadings of PCBs have been extensively reviewed and the reader is referred to the work of Eisenreich et al. (1980) and Murphy (Chapter 3, this volume) Types of Aroclor identified varied in rain, air, and lakewater samples. The predominant Aroclors were 1242 in rain (60%) and air (75%) and 1254 in lakewater. Higher-chlorinated isomers were found to be associated with the particulate fraction (i.e., 29% Aroclor 1242, 15% Aroclor 1260), suggesting that the less-volatile, higher-molecular-weight compounds have a greater tendency to be sorbed onto particulates. Air samples contained a high proportion of Aroclor 1242 (about 84%) while precipitation samples had a much lower 1242 content of approximately 42% (Murphy and Rzeszutko, 1977). Support for the idea that PCBs in rain are associated with the particulate fraction is given by the fact that PCBs in rain are mainly of the Aroclor 1254 type, while those predominating in the atmosphere most closely resemble Aroclor 1242.

Transport within the lakes is generally through sediment and biota. The upper layers of sediment contain the highest quantities of PCBs. Eisenreich et al. (1979) found a concentration of 0.17 ± 0.13 μg/g of PCB in the upper 0.5 cm of sediment. No detectable PCBs were found below 3 cm and most were found in the upper 1 cm. This leads to the conclusion that the depth of sediment mixing by biotic or physical processes is 1 cm. Sorption of PCBs on sediments has been studied extensively. The extent of sorption has been correlated to sediment characteristics such as organic carbon content (Choi and Chen, 1976; Hiraizumi et al., 1979; Simmons et al., 1980; Steen et al., 1978) particle size (Di Toro and Horzempa 1982) as well as the nature and organic carbon content of the overlying water. Most of the PCBs that have entered the lakes are buried in the sediments (Eisenreich et al., 1979; Frank et al., 1979; Smith et al., 1980) which act as a huge reservoir of PCBs for resuspension (Di Toro, 1974; Hollod, 1979; Eisenreich et al., 1980).

The role of aquatic biota in the cycling of PCBs in the lake environment has been a subject of much interest recently. The movement of PCBs in these waters can be mediated through food chain transfers, grazing, and bioturbation. Aquatic organisms have been shown to take up PCB either directly from the waters, usually termed "bioconcentration," or through food chain transfer, indicating "bioaccumulation." The contribution of each of these uptake mechanisms to the concentration levels of PCBs in the Great Lakes are variable (Weininger, 1978; Thomann, 1978). There is strong evidence however, that a larger percentage (75–95) is due to bioaccumulation and a lower percentage (5–25) due to bioconcentration. This suggests therefore that primary producers and primary consumers to some extent take up PCBs into their cell systems.

There are two methods of recirculating the sediment-bound contaminant: (1) by direct uptake and digestion of sediment by benthic organisms (Eadie et al., 1982) and subsequent uptake and accumulation by higher predators, and (2) by resuspension and/or leaching of the sediments into the water column, making the contaminants available to fish via respiration or injestion (Kaiser, 1979). Two primary pathways for PCB transport via aquatic food chains have been outlined by Weininger and Armstrong (1982). The pelagic pathway: water → phytoplankton and suspended particulates → zooplankton → macroinvertebrates → forage fish → pisciverous fish; the benthic pathway: water → particulate matter → sediment → benthic invertebrates → forage fish → pisciverous fish. In the latter case PCBs in sediments remain biologically available. It is obvious that the release of PCBs by bioturbation and other physical processes will lead to increased concentration of PCBs in the water column even when all inputs into the lake have been eliminated, resulting in continued high levels in the fish.

9. DEGRADATION OF PCBs

Because of their chemical inertness, PCBs are resistant to thermal and chemical degradation, hence they are persistent in the environment. A limited amount of research has been conducted on bacterial or microbial degradation of PCBs in the Great Lakes. Laboratory studies have been conducted to determine metabolic products and degradation rates of Aroclors by bacteria (Kaiser and Wong, 1974, Wong and Kaiser, 1975; Liu, 1976). In the degradation studies, Kaiser and Wong (1974) found that a 2-month incubation of the PCB mixture, Aroclor 1242, and bacterial cultures resulted in the degradation of the biphenyl, and none of the metabolites identified contained any chlorine. In the rate studies, it was discovered that bacteria used Aroclor 1221 and 1242, but not 1254, as sole carbon and energy sources for growth. The degradation of Aroclor 1221 into several low-molecular-weight compounds was complete after only one month of incubation. Unchlorinated biphenyl was degraded faster than di- or tetrachlorobiphenyl, indicating bacterial preference for the lower-chlorinated isomers.

PCBs absorb ultraviolet radiation because of their aromatic moities, hence are susceptible to photodegradation. Direct photolysis rates of PCBs in water have been estimated to be slow (Simmons et al., 1981). However, in natural systems, the presence of photosensitizing agents such as humic acids, other organics and algae, can enhance their photolytic degradation.

10. CONTROVERSIES IN RESEARCH FINDINGS

One of the areas of PCB research that has attracted much attention among investigators in the Great Lakes is the question of sources and rates of inputs of PCBs into the lakes. Much debate currently centers around the relative importance of various sources of PCB. Several investigators support aerial transport and deposition as the main pathway for PCB into the lakes, especially for the Upper Great Lakes, while others support river mouth inputs as the most important source of PCB, as has been shown for Lakes Ontario and Erie. Another subject of much debate centers on the volatilization of PCB from lake surfaces as a decontaminating mechanism for PCB in lakes. Most investigators, however, agree that lake sediments can act as reservoirs for PCB and could possibly be the major source to the lakes and biota especially when direct discharges of PCBs into the lakes have been turned off. It is not clear at this point if the mechanisms and "speciation" of PCBs during input to the lakes will necessarily be the same as when PCB is transported from one compartment to another. There is reason to believe that the forms of PCB being transported into the different compartments, and the associated mechanisms and kinetic of transport could be very different.

11. INFORMATION GAPS ON PCB RESEARCH IN THE GREAT LAKES

Most of the existing PCB studies focus on either PCB monitoring or transport processes (atmospheric, in-lake transport, sedimentation, bioaccumulation, bioturbation) and modeling on the cycling of PCB and mass balance in the Great Lakes compartments. There is conspicuous absence of *in situ* transformation studies because most of the work in this area has been conducted in the laboratory. Research on toxicity effects is also lacking and often the limited data can be applied only to toxic effects of PCB to a particular species under specific conditions. Another area of PCB research that needs more work is the question of human toxicity and exposure assessment. Allen et al., (1974) Allen (1975), Allen and Barsotti (1976) documented deleterious effects of PCBs on rhesus monkeys when fed PCBs at levels approaching FDA guidelines for humans. Humphrey (1975) examined the correlation between Great Lakes fish consumption and blood PCB levels in humans. These are the only studies to date that attempt to evaluate human health effects of PCBs in the Great Lakes region. Sufficient data have not been collected to draw any conclusions concerning human health effects from Great Lakes contamination by PCBs. In Chapter 1 of this volume, Sonzogni and Swain discuss in detail some of the health risks associated with PCB contamination in the Great Lakes. Likewise, a recent survey made by Miller (1983) unveils some of the questions surrounding health effects of PCBs.

Finally, there is a great need for studies on human risk and exposure assessment for PCBs. Is there indeed a risk to man from current exposure levels? How do we quantify that risk? These simple fundamental questions have not been answered or even come close to being answered by existing studies.

ACKNOWLEDGMENT

This study is part of a project supported by the Michigan Sea Grant Program No. NA 80 AA-D-00072. Acknowledgment is given to Kim Jones who helped in the literature research and the preparation of this manuscript.

REFERENCES

Allen, J. R. (1975). Response of the nonhuman primate to polychlorinated biphenyl exposure. *Federat. Proc.* **34:** 1675–1679.

Allen, J. R. and D. A. Barsotti. (1976). The effects of transplacental and mammary movement of PCBs on infant rhesus monkeys. *Toxicology* **6:** 331–340.

Allen, J. R., L. A. Carstens, and D. A. Barsotti. (1974a). Residual effects of short-term low level exposure of nonhuman primates to polychlorinated biphenyls. *Toxicol. Appl. Pharmacol.* **30:** 440–451.

Allen, J. R., D. H. Norback, and I. C. Hsu. (1974b). Tissue modification in monkeys as related to absorption, distribution, and excretion of polychlorinated biphenyls. *Arch. Environ. Contam. Toxicol.* **2:** 86–92.

Armstrong, D. (1981). Report on Polychlorinated Biphenyls in Lake Michigan Tributaries, Water and Sediments. Submitted to the U.S. EPA Large Lakes Research Station, Grosse Ile, Michigan, 20 pp.

Aulerich, R. J., R. K. Ringer, H. L. Seagran, W. G. Youaltt. (1971). Effects of feeding coho salmon and other Great Lakes fish on mink reproduction. *Can. J. Zool.* **49:** 611–616.

Aulerich, R. J., R. K. Ringer, S. Iwanoto. (1973) Reproductive failure and mortality in mink fed on Great Lakes fish. *J. Reprod. Fertil.* **19:** (Suppl):365–376.

Burin, G. and J. A. Robbins. (1977). Polychlorinated biphenyls (PCBs) in dated sediment cores from Southern Lake Huron and Saginaw Bay. Abstract, in 20th Conference Great Lakes Research. International Association for Great Lakes Research.

Carr, R., C. E. Finsterwalder, and M. J. Schibi. (1972). Chemical residues in Lake Erie fish, 1970–1971. *Pest. Monit. J.* **6**(1): 23–26.

Choi, W. W. and K. Y. Chen. (1976). Associations of chlorinated hydrocarbons with fine particulates and humic substances in near shore surficial sediments. *Environ. Sci. Technol.* **10:** 782–786.

Cohen, J. M. and C. Pinkerton. (1966). Widespread translocation of pesticides by air transport and rainout. In: R. L. Gould, Ed., *Organic pesticides in the environment.* American Chemical Society, Washington, D.C.

Christensen, E. and P. A. Zielski. (1980). Toxicity of arsenic and PCB to a green alga (*Chlamydomonas*). *Bull. Environ. Contam. Toxicol.* **25:** 43–48.

Delfino, J. J. (1979). Toxic substances in the Great Lakes. *Environ. Sci. Technol.* **13:** 1462–1469.

Di Toro, D. M. (1974). Vertical interactions in phytoplankton models—an eigenvalue analysis. International Association for Great Lakes Research, 17th Conference on Great Lakes Research, Abstract, p. 16.

Di Toro, D. M. and L. M. Horzempa. (1982). Reversible and resistant components of PCB adsorption-desorption isotherms. *Environ. Sci. Technol.* **16:** 594–602.

Doskey, P. V. and A. W. Andren. (1981a) Concentrations of airborne PCBs over Lake Michigan. *J. Great Lakes Res.* **7:** 15–20.

Doskey, P. V. and A. W. Andren. (1981b). Modeling the flux of atmospheric polychlorinated biphenyls across the air/water interface. *Environ. Sci. Technol.* **15:** 705–711.

Eadie, B. (1979). (as cited in the National Research Council Report prepared by the Committee on the Assessment of Polychlorinated Biphenyl in the Environment) Washington, D.C. 182 pp.

Eadie, B. J., C. P. Rice, and W. A. Frez. (1982). The role of the benthic boundary in the cycling of PCBs in the Great Lakes. Proceedings of the Workshop on Physical Behavior of PCBs in the Great Lakes, December 1981, Toronto, Ontario, Canada, Ann Arbor Science Publishers.

Eisenreich, S. J., T. C. Hollod, and T. C. Johnson. (1979). Accumulation of polychlorinated biphenyls (PCBs) in surficial Lake Superior sediments. Atmospheric deposition. *Environ. Sci. Technol.* **13:** 569–573.

Eisenreich, S. J., B. B. Looney, and J. D. Thornton. (1980). Assessment of airborne organic contaminants in the Great Lakes ecosystem. Annual Report, Appendix A, Great Lakes Advisory Board, International Joint Commission, 48 pp.

Eisenreich, S. J., B. B. Looney, M. Holdrinet, D. P. Dodge, S. J. Nepszzy. (1981). Airborne organic contaminants in the Great Lakes ecosystem. *Environ. Sci. Technol.* **15:** 30–38.

Frank, R. and M. Holdrinet. (1975). Residues of organochlorine compounds and mercury in birds' eggs from the Niagara Peninsula, Ontario. *Archiv. Environ. Contam. Toxicol.* **3**(2): 205–218.

Frank, R., M. Holdrinet, H. E. Braun, R. L. Thomas, A. L. W. Kemp. (1977). Organochlorine insecticides and PCBs in sediments of Lake St. Clair (1970 and 1974) and Lake Erie (1971). *Sci. Total Environ.* **8:** 205–227.

Frank, R., M. Holdrinet, H. E. Braun, D. P. Dodge, and G. E. Sprangler. (1978). Residue of organochlorine insecticides and polychlorinated bipenyls in fish from Lakes Huron and Superior, Canada, 1968–1976. *Pest. Monit. J.* **12:** 60–68.

Frank, R., R. L. Thomas, M. Holdrinet, A. L. W. Kemp, H. E. Braun, and R. Dawson. (1979). Organochlorine insecticides and PCB in the sediments of Lake Huron (1969) and Georgian Bay and North Channel (1973). *Sci. Total Environ.* **13:** 101–117.

Frank, R., R. L. Thomas, M. V. H. Holdrinet, and V. Damiani. (1980). PCB residues in bottom sediments collected from the Bay of Quinte, Lake Ontario, 1972–1972. *J. Great Lakes Res.* **6:** 371–376.

Gessner, M. L. (1980). Pesticide and PCB levels in fillet and whole body portions of five Lake Erie fish species. Master Thesis. Ohio State University, Columbus, Ohio, 84 pp.

Gilbertson, M. and G. A. Fox. (1977). Pollutant-associated embryonic mortality of Great Lakes herring gulls. *Environ. Pollut.* **12:** 211–216.

Glooschenko, V. and W. A. Glooschenko. (1975). Effect of polychlorinated biphenyl compounds on growth of Great Lakes phytoplankton. *Can. J. Botany* **53**(7)**:** 653–659.

Glooschenko, W. A., W. M. Strachan, and R. C. J. Sampson. (1976). Residues in water distribution of pesticides and polychlorinated biphenyls in water, sediments and session of the Upper Great Lakes, 1974. *Pest. Monit. J.* **10**(2)**:** 61–67.

Great Lakes Basin Commission, Great Lakes Communicator, *11*(3), December 1980, 10 pp.

Guiney, P. D., M. J. Melancon, L. L. Lech, and R. E. Peterson. (1979). Effects of egg and sperm maturation and spawning on the distribution and elimination of a polychlorinated biphenyl in rainbow trout (*Salmo gairdneri*). *Toxicol. Appl. Pharmacol.* **47:** 261–272.

Guiney, P. D. and R. E. Peterson (1980). Distribution and elimination of a polychlorinated biphenyl after dietary exposure in yellow perch and rainbow trout. *Arch. Environm. Contam. Toxicol* **9:** 667–674.

Haile, C. L., G. D. Veith, G. F. Lee, and W. C. Boyle. (1975). Chlorinated hydrocarbons in the Lake Ontario ecosystem. EPA 660/3-75-022. U.S. Environmental Protection Agency, 45 pp.

Hall, J. (1980). Personal communication, Great Lakes Basin Commission, Ann Arbor, Michigan.

Hiraizumi, Y., M. Takahashi, and H. Nishimura, (1979). Adsorption of polychlorinated biphenyl onto sea bed sediment, marine plankton and other absorbing agents. *Environ. Sci. Technol.* **13:** 580–583.

Hollod, G. J. (1979). Polychlorinated biphenyls (PCBs) in the Lake Superior ecosystem: Atmospheric deposition and accumulation in the botton sediments. Ph.D. dissertation, University of Minnesota, 247 pp.

Humphrey, H. E. B. (1975). Evaluation of changes of the level of polychlorinated biphenyls (PCB) in human tissue. Final Report on FDA Contract 223-73-2209. Michigan Department of Public Health, Lansing, Michigan. 591 pp. Appendices.

Hutzinger, O., S. Safe, and V. Zitko. (1974). *The chemistry of PCBs*. CRC Press, Cleveland Ohio, pp. 1–10.

International Joint Commission. (1974). Great Lakes Water Quality, 1974. Appendix B, Surveillance Subcommittee Report, 161 pp.

International Joint Commission. (1976). Great Lakes Water Quality, 1976. Appendix B, Surveillance Subcommittee Report, p. 36 and p. 79.

International Joint Commission. (1980). Report on Great Lakes Water Quality, Appendix, 82 pp.

Kaiser, K. L. E. (1977). Organic contaminant residues in fishes from Nipigon Bay, Lake Superior. *J. Fish. Res. Board Can.* **34:** 850–855.

Kaiser, K. L. E. (1979). Organochlorine contaminants in the Great Lakes. *Geosci. Can.* **6**(1)**:** 16–19.

Kaiser, K. L. E. and I. Valdmanis. (1978). Organochlorine contaminants in a sea lamprey (*Petromyzon marinus*) from Lake Ontario. *J. Great Lakes Res.* **4**(2): 234–236.

Kaiser, K. L. E. and P. T. S. Wong. (1974). Bacterial degradation of polychlorinated biphenyls. I. Identification of some metabolic products from Aroclor 1242. *Bull. Environ. Contam. Toxicol.* **11**: 291–296.

Kauss, P. B., K. Suns, and A. F. Johnson, (Undated). Monitoring of PCBs in water, sediments and biota of the Great Lakes—some recent examples. (Manuscript) 10 pp.

Kenaga, D. E. and W. S. Creal, (1979). Concentrations of selected contaminants in fish from Lakes Superior and Huron, 1974–1978. GLECS 1974–1978.

Kleinert, S. J. (1976). Sources of polychlorinated biphenyls in Wisconsin. National Conference on PCBs, November 19–21, 1975 Chicago, IL., Dept of Commerce NTIS PB-253, 248 pp.

Lin, C. K. (1981). Uptake of polychlorinated biphenyls by Great Lakes algae. Presented at the 3rd Midwest Water Chemistry Conference, October, 1981, Ann Arbor, Michigan.

Liu, D. L. S. (1976). Biodegradation, an environmental solution to some toxic organic compounds. *Environ. Conserv.* **3**: 137–138.

Liu, D. and V. K. Chawla. (1976). Polychlorinated biphenyls (PCBs) in sewage sludges. Proceedings on Symposium on Trace Substances in Environmental Health, pp. 247–250.

McNaught, D. C. (1977). Uptakes rates and effects of PCBs on phytoplankton and zooplanton. (Manuscript).

Michigan Department of Natural Resources. (1980). Annual Water Quality Report, pp. 210.

Miller, S. (1983). The PCB imbroglio. *Environ. Sci. Technol.* **17**: 11A–14A.

Murphy, T. J. and C. P. Rzeszutko. (1977). Precipitation inputs of PCBs to Lake Michigan. *J. Great Lakes Res.* **3**(3–4): 205–312.

Murphy, T. J. and C. P. Rzeszutko. (1978). PCBs in Precipitation in the Lake Michigan Basin. U.S. EPA Report EPA-600/3-78-071.

Murphy, T. J., A. Schinsky, G. Paolucci, and C. P. Rzeszutko. (1981). *Atmospheric inputs of pollutants to natural waters,* S. J. Eisenreich, Ed. Ann Arbor Science Publishers.

Neely, W. B. (1977). A material balance study of polychlorinated biphenyls in Lake Michigan. *Sci. Total Environ.* **7**: 117–129.

Neidermeyer, W. J. and J. J. Hickey. (1976). Chronology of organochlorine compounds in Lake Michigan fish, 1929–1966. *Pest. Monit. J.* **10**: 92–95.

Norstrom, R. J., D. J. Hallett, and R. A. Sonstegard. (1978). Coho salmon (*Oncorhynchus kisutch*) and herring gulls (*Larus argentatus*) as indicators of organochlorine contamination in Lake Ontario. *J. Fish. Res. Board Can.* **35**: 1401–1409.

Parjeko, R. and R. Johnston. (1973). Uptake of toxic water pollutants (PCB's) by lake trout. Michigan Institute of Water Research, Project Completion Report, Jan. 1973. 16 pp.

Parjeko, R., R. Johnston, and R. Keller. (1975). Chlorohydrocarbons in Lake Superior lake trout (*Salvelinus namaycush*) *Bull. Environ. Contam. Toxicol.* **14**(4): 480–488.

Passino, D. R. M. and C. A. Cotant. (1978). Allantoinase in lake trout (*Salvelinus namaycush*): In vitro effects of PCBs, DDT and metals. *Comp. Biochem. Physiol.* **62C**: 71–75.

Passino, D. R. M. and J. M. Kramer. (1980). Toxicity of arsenic and PCBs to fry of deepwater ciscoes (*Coregonus*). *Bull. Environ. Contam. Toxicol.* **24**(5): 527–534.

Peakall, D. B. (1976). DDT in rainwater in New York following application in the Pacific Northwest. *Atmos. Environ.* **10**: 899–900.

Peterson, R. E. and P. D. Guiney. (1979). Disposition of polychlorinated biphenyls in fish in pesticide and xenobiotic metabolism in aquatic organisms. In: M. A. Q. Khan, J. J. Lech and J. J. Men, Eds., ACS Symposium Series 99. pp. 21–36. American Chemical Society. Washington, D.C.

PLUARG Report. 1978. Final Report to the International Joint Commission, Windsor, Ontario, Canada, July, 1978.

Price, H. A. and R. L. Welch. (1972). Occurrence of polychlorinated biphenyls in humans. *Environ. Health Persp.* **1:** 73–78.

Rhee, G. Y. (1982). Effects of polychlorinated biphenyls on Great Lakes phytoplankton. Personal communication. New York State Dept. of Health, Albany, New York.

Rice, C. R. (1980). Personal communication, The University of Michigan, Ann Arbor, Michigan.

Rice, C. P. and P. A. Meyers, (1982). Contribution of surface microlayer to air/water exchange of organic pollutants. Michigan Sea Grant Project Presentation. The University of Michigan, Ann Arbor, Michigan.

Rohrer, T. K., J. H. Hartig, and J. C. Forney, (1980). Xenobiotic substances on coho and chinook salmon of the Great Lakes, 1980. (manuscript).

Roberts, J. R., D. W. Rodgers, J. R. Bailey, and M. A. Rorke. (1978). Polychlorinated biphenyls: Biological criteria for an assessment of their effects on environmental quality. National Research Council of Canada, NRC Associate Committee on Scientific Criteria for Environmental Quality, 172 pp.

Schacht, R. (1974). Pesticides in the Illinois waters of Lake Michigan. U.S. EPA Ecological Research Series Report. EPA-600/3-74-002.

Sileo, L., L. Karstadt, R. Frank, M. V. H. Holdrinet, E. Addison, and H. E. Braun. (1977). Organochlorine poisoning of ring-billed gulls in southern Ontario. *J. Wildl. Diseases* **13:** 313–322.

Simmons, M. S., D. Bialosky, and R. Rossmann. (1980). Polychlorinated biphenyl contamination in surficial sediments of northeastern Lake Michigan. *J. Great Lakes Res.* **6**(2): 167–171.

Simmons, M. S., D. Zadelis, and M. Hauberstricker. (1981). Photolysis studies of chlorinated biphenyls in aqueous systems. Presented at the 3rd Midwest Water Chemistry Workshop, Ann Arbor, Michigan, October, 1981.

Smith, S. E., W. C. Sonzogni, M. S. Simmons, and C. P. Rice. (1980). Organic and heavy metal contaminants data review—a case study, Lake Michigan. 23rd Conference. *Intern Assoc. Great Lakes Res. Abst.* p. 49.

Sonzogni, W. C. (1978). PCBs and toxics on the decrease? Great Lakes Communicator, Great Lakes Basin Commission, Ann Arbor, Michigan, **9:** 2–3.

Sonzogni, W. C. and W. R. Swain. (1980). Perspectives on U.S. Great Lakes chemical toxic substances research. *J. Great Lakes Res.* **6:** 265–274.

Spagnoli, J. J. and C. Skinner, Lawrence. (1977). PCB's in fish from selected waters of New York State. *Pest. Monit. J.* **11**(2): 69–87.

Spencer, W. F. and M. M. Cliath. (1975). Vaporization of chemicals. In: R. Haque and V. H. Freed, Eds., *Environmental dynamics of pesticides*. Plenum Publishing Co., p. 61.

Spigarelli, S. A., M. M. Thommes, W. Prepejchal, and D. A. Warner. (1979). Accumulation of toxic pollutants by brown trout exposed to various temperatures in Lake Michigan. Argonne National Laboratory. Radiological and Environmental Research. Division Annual Report, January–December 1979. ANL-79-65-PtIII.

Stauffer, T. M. (1979). Effects of DDT and PCB's on survival of lake trout eggs and fry in a hatchery and in Lake Michigan. *Trans. Am. Fish. Soc.* **108:** 178–186.

Steen, W. C., D. F. Paris, and G. L. Baughman. (1978). Partitioning of selected polychlorinated biphenyls to natural sediments. *Water Res.* **12:** 655–657.

Swain, W. R. (1978). Chlorinated organic residues in fish, water and precipitation from the vicinity of Isle Royale, Lake Superior. *J. Great Lakes Res.* **4**(3–4): 398–407.

Swain, W. R. (1980). An ecosystem approach to the toxicology of residue-forming xenobiotic organic substances in the Great Lakes. Environmental Studies Board, National Research Council, National Academy of Sciences, Washington, D.C., 64 pp.

Thomann, R. V. (1978). Size-dependent model of hazardous substances in aquatic food chains. U.S. EPA Ecological Research Series. EPA/600/3-78-036. 39 pp.

U.S. Environmental Protection Agency. (1979). Waukegan Harbor bioconcentration and depuration study. June 19, 1979–October 10, 1979 (Manuscript).

U.S. Environmental Protection Agency. (1981). The PCB Contamination problem in Waukegan, Illinois. EPA Region V, Chicago, 58 pp.

U.S. Food and Drug Administration. (1976) Polychlorinated biphenyls (PCB's) in certain freshwater fish. *Federal Register* **41:** 8409–8410.

Veith, G. D. (1975). Baseline concentrations of polychlorinated biphenyls and DDT in Lake Michigan fish, 1971. *Pest. Monit. J.* **9**(1): 21–29.

Veith, G. D. (1980). Uptake and elimination of PCBs in fish contaminated by the Waukegan Harbor. U.S. EPA (unpubl.).

Veith, G. D., D. W. Kuehl, F. A. Puglisi, G. E. Glass, and J. G. Eaton. (1977). Residues of PCBs and DDT in the western Lake Superior ecosystem. *Arch. Environ. Contam. Toxicol.* **5:** 487–499.

Vermeer, K. and L. M. Reynolds. (1970). Organochlorine residues in aquatic birds in Canadian prairie provinces. *Can. Field-Nat.* **84:** 117–119.

Vesligh, D. V., P. Minzao, and D. J. Hallett, (1979) Organochlorine contaminants and trends in reproduction in Great Lakes herring gulls, 1974–1978. Transactions of the 44th North American Wildlife and Natural Resources Conferences, pp 543–557.

Weininger, D. (1978). Accumulation of PCBs by lake trout in Lake Michigan. Ph.D. dissertation, University of Wisconsin—Madison. 232 pp.

Weininger, D. and D. Armstrong. (1982). Role of microcontaminants in restoration of Great Lakes ecosystems. (Manuscript) 11 pp.

Willford, W. A., R. J. Hesselberg, and L. W. Nicholson. (1975). Trends of polychlorinated biphenyls in three Lake Michigan fishes. Proceedings of National Conference on PCB. EPA-550/6-75-004. March 26, 1975, pp. 177–181.

Willford, W. A. (1980). Chlorinated hydrocarbons as a limiting factor in the reproduction of lake trout in Lake Michigan. Proceedings Third USA-USSR Symposium on Effluent Pollution Aquatic Ecosystems, W. R. Swain and V. R. Shannon, Eds. EPA-660/9-80-034, pp. 75–83.

Wong, P. T. S. and K. L. E. Kaiser (1975). Bacterial degradation of polychlorinated biphenyls II. Rate Studies. *Bull. Environ. Contam. and Technol.* **13:** 249–256.

Zabik, M. E., B. Olsen, and T. M. Johnson. (1978). Dieldrin, DDT, PCBs and mercury levels in freshwater mullet from the Upper Great Lakes. *Pest. Monit. J.* **12**(1): 36–39.

14

POLYCHLORINATED BIPHENYLS IN SEVEN SPECIES OF LAKE MICHIGAN FISH, 1971–1981

James R. St.Amant, Mary E. Pariso, and Thomas B. Sheffy

Bureau of Water Resources Management
Wisconsin Department of Natural Resources
Madison, Wisconsin 53707

1. INTRODUCTION

Polychlorinated biphenyls (PCBs) have been manufactured and sold since 1929. It was not until 1966, however, that these chemicals were discovered bioaccumulating in fish and wildlife (Jensen, 1966). Bioaccumulation undoubtedly occurred before this date. Jensen (1966) obtained eagle feathers preserved at the Swedish National Museum of Natural History from 1880 to 1966 and detected PCB first in an eagle from 1944. Holden and Marsden (1967) discovered compounds that proved to be PCBs (Holmes et al., 1967) in seals and porpoises from Scotland and Canada, far from any PCB sources. Thus by 1967, PCBs had become a ubiquitous environmental contaminant. In this same year, PCBs were discovered to be a problem in Lake Michigan. Aulerich et al. (1971) reported reproductive failure and high kit mortality in mink fed Lake Michigan coho solmon (*Oncorhyncus kisutch*). The cause was later determined to be PCB contamination of these salmon (Ringer et al., 1972).

In 1979, the Environmental Protection Agency prohibited the manufacture, processing, distribution in commerce, and use (except in closed systems) of PCBs. Although PCBs have been restricted, it has been estimated that 500 million pounds still remain in use in closed systems such as electric capacitors and transformers (U.S. EPA, 1979).

Relatively little is known about the processes controlling the distribution, movement, and fate of PCBs in the environment. PCBs are lipophillic and resistant to degradation. Therefore, PCBs tend to persist in the environment and accumulate at the top of the food chain. Fish accumulate PCBs from their food and directly from the water column through their gills. Armstrong and Weininger (1980) have noted that predatory fish accumulate PCBs predominantly through their food via two pathways. One pathway is pelagic: water → phytoplankton and suspended particles → zooplankton → macroinvertebrates → forage fish → predatory fish. The second pathway is benthic: water → particulate matter → sediment → benthic invertebrates → forage fish → predatory fish.

PCBs have been identified in all of the Great Lakes, but the most significant PCB contamination exists in Lake Michigan (Delfino, 1979). The fact that PCBs are present in Lake Michigan sediments, water, and fish poses a threat to the health and stability of the ecosystem and ultimately to man. Although toxicological studies have not directly linked the consumption of fish containing PCBs with human health problems, monkeys fed diets containing PCBs have developed reproductive, skin, and behavioral disorders (Allen, 1975; Allen et al., 1979; Bowman et al., 1978). Perhaps equally important is the damage to Lake Michigan as a natural resource. PCB contamination has led to curtailments of commerical fishing, and warnings have been issued about the hazards of consuming fish obtained by recreational activities. This has dealt a blow to Lake Michigan's sport and commerical fisheries, together valued at more than $28 million for the state of Wisconsin alone (Gould, 1979).

Information on trends of PCB levels in Lake Michigan fish is sparse. Veith (1975) reported 1971 baseline concentrations ranging from 2.7 parts per million (ppm) in rainbow smelt (*Osmerus mordax*) to 15 ppm in lake trout (*Salvelinus namaycush*). Most lake trout, coho salmon, and chinook salmon (*Oncorhyncus tschawytscha*) longer than twelve inches contained PCBs at concentrations greater than the tolerance level of 5.00 ppm established by the U.S. Food and Drug Administration (U.S. FDA). Samples collected in 1974 and 1975 from the waters surrounding Wisconsin's Door county by

Degurse and Duter (1975) revealed high average PCB concentrations in carp (*Cyprinus carpio*) (22.30 ppm), large lake trout (21.80 ppm), chinook salmon (10.13 ppm), large coho salmon (6.62 ppm), and large lake whitefish (*Coregonus clupeaformis*) (5.12 ppm). Comparisons between these two studies are difficult since different areas of Lake Michigan were sampled and different laboratories analyzed the samples. Willford et al. (1976) performed trend analysis on eastern Lake Michigan fish collected from 1972 through 1974. There was no evidence of a decline in average PCB residues and PCB levels in lake trout taken off Saugatuk, Michigan, appeared to increase.

More recently, the Wisconsin Department of Natural Resources monitored PCB levels in seven economically important species of fish: lake trout, chinook salmon, coho salmon, walleye (*Stizostedion vitreum*), lake whitefish, alewife (*Alosa pseudoharengus*), and bloater chub (*Coregonus hoyi*). Samples were collected from Green Bay and Lake Michigan over an 11-year period, 1971–1981. The purpose of this survey was to investigate trends in PCB levels and assess the quality of fish from commercial and sport fishing grounds.

2. DEPARTMENT OF NATURAL RESOURCES STUDY

Fish collection sites are shown in Fig. 14.1. These locations represent major commercial and sport fishing areas as well as areas with known or suspected PCB problems. Results of the 11-year study are shown in Table 14.1 and graphed in Fig. 14.2.

Of the seven species tested, lake trout, chinook salmon, and coho salmon have historically shown levels of PCBs greater than the U.S. FDA tolerance level. As a result, the Wisconsin Division of Health has advised that fish consumers avoid eating more than one meal or ½ pound per week of these fish. It is also advised that lactating mothers, expectant mothers, any women who anticipate bearing children, and children ages six and under not eat any of these fish.

The levels of PCB found in lake trout were particularly high (22.40 ppm) in the early 1970s. This can be attributed to several factors. First, lake trout are predators, feeding on alewives which contain significant levels of PCB. Second, they possess a large amount of fat (average of 12.0%). Roberts et al. (1978) have shown that due to the high solubility of PCB in fat, a positive correlation exists between PCB burdens and fat concentration in living organisms. Third, lake trout have a long life span (average of 8–10 years) and are therefore exposed to PCBs for a longer time than other fish. Fourth, the slow growth rate of lake trout leads to a higher PCB burden (Jensen et al., 1982).

Chinook and coho salmon are also predatory fish, but when compared to lake trout contain less fat, grow faster, and do not live as long. Thus PCB levels in these salmon are elevated, but not to the same degree seen in lake

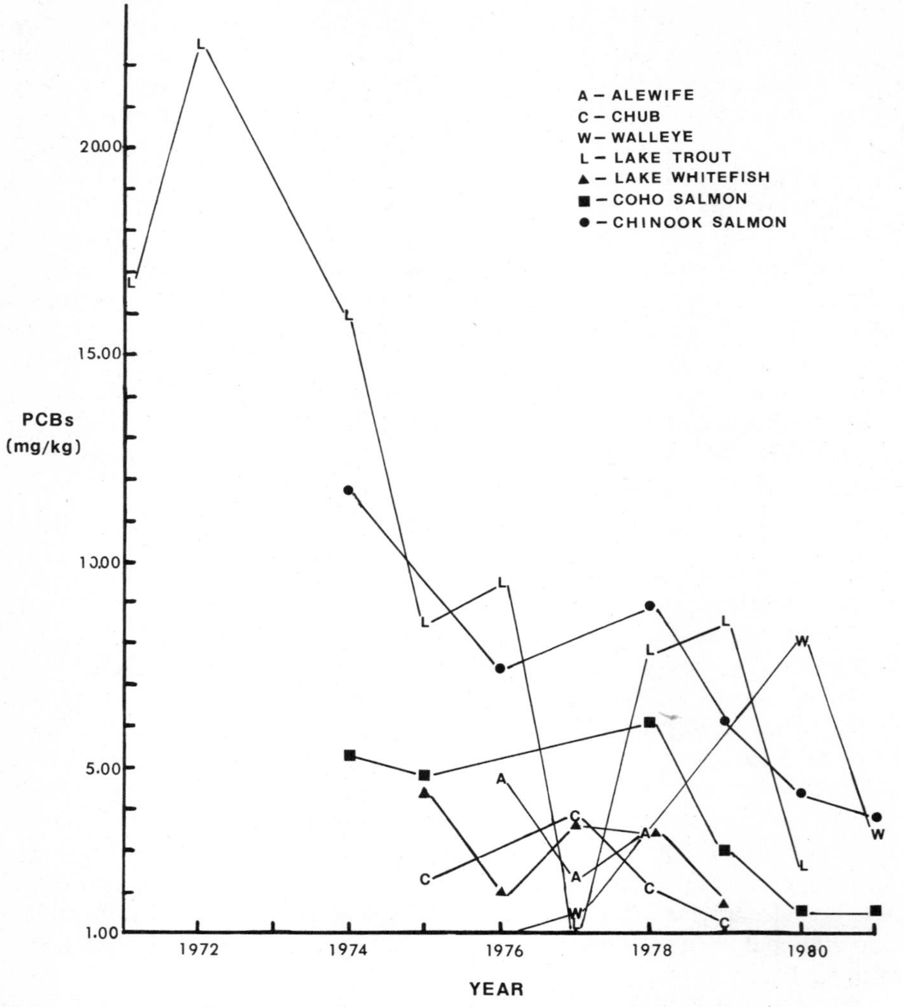

Figure 14.1 Concentration of PCBs found in Lake Michigan fish fillets from 1971 to 1981.

trout. PCB concentrations in chinook are higher than concentrations in coho because coho are exposed for a shorter period of time (about 18 months as opposed to approximately 4 years for chinook) and attain smaller sizes than chinooks (Jensen et al., 1982).

Both lake whitefish and bloater chubs contain high percentages of fat (11.5% and 14.8% respectively) and whitefish have an average life span of 10 years. However, both species show relatively low concentrations of PCB. This is explained by the fact that these are forage fish, relying primarily on a zooplankton diet. Zooplankton contain relatively low levels of PCB and these low levels are reflected in their consumers.

Table 14.1. **Results of PCB Analysis from 1971 to 1981 in Seven Species of Fish from Lake Michigan. PCB and percent fat were determined from fillet portions.**

	Lake Trout[a]				Chinook Salmon				Coho Salmon				Walleye				Lake Whitefish				Chub				Alewife			
Year	n	l	f[b]	PCB[b]	n	l	f	PCB	n	l	f	PCB	n	l	f	PCB	n	l	f	PCB	n	l	f	PCB	n	l	f	PCB
1971	29	23.2	—	16.7																								
1972	10	21.4	—	22.4																								
1973																												
1974	30	24.5	15.2	15.9	8	33.1	2.3	11.7	18	23.5	3.4	5.3										11.6						
1975	54	21.9	10.3	8.5					2	20.1	5.8	4.8					18	19.7	16.9	4.4	2		9.0	2.3				
1976	26	23.9	11.7	9.4	7	30.8	2.0	7.4					4	14.6	0.9	0.6	3	15.9	3.3	2.0		10.6			2	7.0	4.8	4.7
1977	3	13.7	4.3	1.2									8	17.7	2.4	1.4	31	19.7	13.4	3.6	26		17.6	3.8	1	7.0	3.0	2.4
1978	30	22.7	12.3	7.8	24	30.9	4.1	8.9	5	21.9	5.3	6.1					15	19.9	11.2	3.4	5	10.2	15.9	2.1	4	6.1	5.3	3.4
1979	3	23.9	16.3	8.5	10	33.2	2.7	6.1	10	21.6	3.4	3.1					11	18.2	12.9	1.7	13	8.3	16.6	1.2				
1980	7	21.4	13.8	2.6	21	35.4	3.5	4.4	10	23.5	3.8	1.7	1	15.9	9.0	8.1												
1981					30	30.7	4.5	3.8	1	26.0	4.1	1.6	7	21.3	5.2	3.4												
Total	192				97				46				20				78				46				7			

[a] n = number of samples; l = ave. length in inches; f = ave. % fat; PCB = ave. PCB in mg/kg (ppm).

[b] PCB and percent fat were determined from fillet portions.

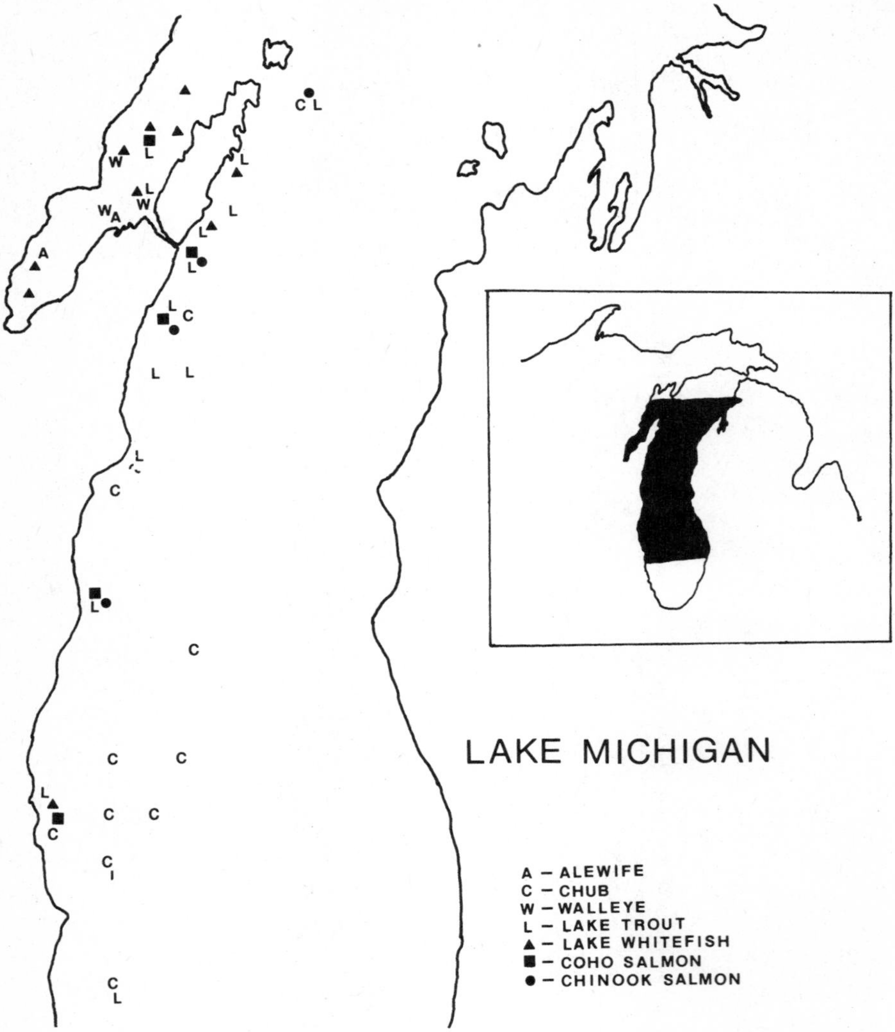

Figure 14.2 Location of sampling sites for PCB analysis in Lake Michigan fish.

The alewife, an exotic species, has been the dominant forage fish in Lake Michigan in the last 18 years (80–85% of the lake's biomass) due to the decimation of predator fish populations by the sea lamprey. Alewives contain relatively low concentrations of fat, but significant levels of PCB. This is due to its exceptional inefficiency at converting food to growth. Over 8 years of life, alewife convert an average of 3.3% of biomass consumed to growth (Stewart, 1980). This is poor compared to lake trout (11% of biomass) or salmon (25% of biomass). This inefficiency results in the consumption of vast quantities of zooplankton and the subsequent bioaccumulation of more PCB than other grazers. This rich and abundant food source serves as the main link between PCBs in the water column/plankton and predatory salmonids.

A decreasing trend was observed in the concentration of PCBs in Lake Michigan fish from 1971 to 1981. The average concentration of PCBs decreased by 84% in lake trout, 70% in coho salmon, 66% in chinook salmon, 61% in lake whitefish, 48% in bloater chub, and 28% in alewife.

An exception to the overall decrease in PCB levels was observed in Green Bay walleyes, where the concentration of PCB increased over the years monitored. This can be explained in part by the yearly walleye stocking effort that began in 1973. The walleye population before 1973 was very low in the shallow, productive waters of Green Bay. In an effort to increase the population and diversify sport fishing in the bay, large numbers of walleye were stocked. Since walleyes are known to be a relatively low-fat species, it was hoped that the stocking effort would produce a large population of sport fish with low levels of PCBs. What actually happened is that the stocking effort has resulted in an increase in the numbers of older walleyes from year to year. This is reflected in the data (Table 14.1) as an increase in walleye length from 1976 to 1981. These walleyes have probably been feeding on alewives, which were found to have higher concentrations of PCB than Lake Michigan alewives due to continued PCB discharge from recycling paper mills on the lower Fox River. The older walleye, being exposed to the PCB-laden alewives for a longer period of time, have bioaccumulated more PCBs.

The overall decrease in PCBs observed in the Lake Michigan basin reflects a cleaner environment due to strict laws which have reduced input to the system. In 1977, the U.S. Environmental Protection Agency established effluent standards restricting PCBs in certain industrial discharges (Federal Register, 1977). In 1978, regulations were enacted for PCB disposal and marking (Federal Register, 1978). In 1979, the use (except in a totally enclosed system), processing, and distribution in commerce of PCBs were banned (Federal Register, 1979). Major point sources of PCBs have also been eliminated in tributary streams to Lake Michigan (Kleinert et al., 1978).

The major removal mechanism for PCBs that have entered the lake is sedementation. Other removal mechanisms of lesser magnitude include volatilization, lake flushing, biodegradation, and harvesting (Armstrong and Weininger, 1980). However, these natural removal mechanisms have probably had less effect on decreasing PCB burdens in fish than reduced input.

Based on an evaluation of the decline of DDT in Lake Michigan after its ban in 1970, Armstrong and Weininger (1980) have estimated that if all input of PCBs to Lake Michigan were eliminated, PCB levels in coho salmon would decline rapidly by 60–80%. The remaining 20–40% would decline more slowly. They attribute the initial rapid and subsequent slow decline to two phenomena. The rapid decrease represents removal of PCBs from the lake water column, and a disruption of the pelagic food chain transport of

PCB to predators. The slow decline in PCB levels represents the benthic transport route, in which PCBs are only gradually buried in the sediment. Results for two predator species in this survey, lake trout and chinook salmon, show a pattern similar to the aforementioned model. The use of PCBs started to decline in 1971–1972 (Armstrong and Weininger, 1980), and many point source discharges were eliminated as a result of use and disposal rules enacted in 1977. PCB concentrations in lake trout and chinook show a more rapid rate of decrease during the early to mid-1970s and a more gradual rate of decrease in the late 1970s, 1980, and 1981.

3. SUMMARY

PCBs were discovered to be a problem in Lake Michigan fish in 1967. Although PCB concentrations in these fish have an important economic impact, little information has been available on the trends of these levels. In a recent study by the Wisconsin Department of Natural Resources, seven species of fish from Lake Michigan and Green Bay were monitored for PCBs from 1971 to 1981. An overall decrease in PCB concentration was observed for all species except walleye. High levels of PCBs (maximum 22.40 ppm) identified at the beginning of the study decreased steadily to the present condition where all species monitored were below the U.S. Food and Drug Administration tolerance level of 5.00 ppm. These trends demonstrate the effectiveness of legislation which banned production, established strict controls on use and disposal, and ultimately reduced the input of PCBs to Lake Michigan.

REFERENCES

Allen, J. R. (1975). Response of primates to polychlorinated biphenyl exposure. *Fed. Proc.* **34:** 1657–1679.

Allen, J. R., D. A. Barsotti, L. K. Lambrecht, and J. P. VanMiller. (1979). Reproductive effects of halogenated hydrocarbons on nonhuman primates. In: W. J. Nicholson and J. A. Moore, Eds., *Health effects of halogenated aromatic hydrocarbons. N.Y. Acad. Sci.* **320:** 419–427.

Armstrong, D. E. and D. Weininger. (1980). Organic contaminants in the Great Lakes. International Symposium for Inland Water and Lake Restoration. September 8–12, 1980. Portland, Maine.

Aulerich, R. J., R. K. Ringer, H. L. Seagran, and W. G. Youatt. (1971). Effects of feeding coho salmon and other Great Lakes fish on mink reproduction. *Can. J. Zool.* **49**(5): 611–616.

Bowman, R. E., M. P. Heironimus, and J. R. Allen. (1978). Correlation of PCB body burden with behavioral toxicology in monkeys. *Pharmacol. Biochem. Behav.* **9:** 49–56.

Degurse, P., and V. Duter. (1975). Chlorinated hydrocarbon residues in fish from major waters of Wisconsin. Wisconsin Department of Natural Resources, Bureau of Fish and Wildlife Management, Fish Management Section Report No. 79. 29 pp.

Delfino, J. J. (1979). Toxic substances in the Great Lakes. *Environ. Sci. Technol.* **13:** 1462–1468.

Federal Register 42. (1977). pp. 6531–6555.

Federal Register 43. (1978). pp. 7150–7164.

Federal Register 44. (1979). pp. 31514–31568.

Gould, W. (1979). Toxic chemicals in the Great Lakes. *Seagrant* **9:** 6–12.

Holden, A. V. and K. Marsden. (1967). Organochlorine pesticides in seals and porpoises. *Nature* **216:** 1274–1276.

Holmes, D. C., J. H. Simmons, and J. O'G. Tatton. (1967). Chlorinated hydrocarbons in British wildlife. *Nature* **216:** 227–229.

Jensen, A. L., S. A. Spigarelli, and M. M. Thommes. (1982). PCB uptake by five species of fish in Lake Michigan, Green Bay of Lake Michigan, and Cayuga Lake, New York. *Can. J. Fish. Aquat. Sci.* **39:** 700–709.

Jensen, J. (1966). Report of a new chemical hazard. *New Scientist* **15:** 612.

Kleinert, S. J., T. Sheffy, J. Addis, J. Bode, P. Schultz, J. Delfino, and L. Lueschow. (1978). Final report on the investigation of PCBs in the Sheboygan River system. Wisconsin DNR Technical Report, July 12, 1978.

Ringer, R. K., R. J. Aulerich, and M. Zabik. (1972). Effect of dietary polychlorinated biphenyls on growth and reproduction of mink. Preprint of paper presented at 164th Natl. Meeting American Chemical Society, **12**(2): 149–154.

Roberts, J. R., D. W. Rodgers, J. R. Bailey, and M. A. Rorke. (1978). Polychlorinated biphenyls: Biological criteria for an assessment of their effects on environmental quality. No. NRCC 16077, National Research Council of Canada. Ottawa, Canada.

Stewart, D. J. (1980). Salmonid predators and their forage base in Lake Michigan; A bioenergetic-modeling synthesis. Unpublished Ph.D. dissertation. University of Wisconsin—Madison.

U.S. Environmental Protection Agency. (1979). Polychlorinated biphenyls 1929–1979: Final report. EPA Report 560/6-79-004. U.S. EPA, Office of Toxic Substances, Washington, D.C. 20460.

Veith, G. D. 1975. Baseline concentrations of polychlorinated biphenyls and DDT in Lake Michigan fish, 1971. Pest. Monit. J. 9(1): 21–29.

Willford, W. A., R. J. Hesselberg, and L. W. Nicholson. 1976. Trends of polychlorinated biphenyls in three Lake Michigan fishes. In: *Conference Proceedings; National Conference on Polychlorinated Biphenyls*. Prepared for the U.S. Environmental Protection Agency, Office of Toxic Substances. Contract No. 68-01-2928. pp. 177–181.

15

PARTITIONING OF TOXIC TRACE METALS BETWEEN SOLID AND LIQUID PHASES IN THE GREAT LAKES

Kenneth R. Rygwelski

Cranbrook Institute of Science
Great Lakes Research Group
P.O. Box 801
Bloomfield Hills, MI 48013

1. INTRODUCTION

The interaction between particulate matter and trace metal species in the water column plays an important role in metals transport and fate throughout the Great Lakes environment. Where there is strong sorption of a metal by particulates and settling is involved, then the metal should not be regarded as a conservative substance in the water column. Dolan and Bierman (1982) modeled the transport of trace metals in Saginaw Bay, Lake Huron, and found that a large percentage of the cadmium, lead, and zinc were lost to the sediments via metal association with particulates suspended in water. Enrichment of cadmium, lead, zinc, and copper was found in surface sediments of Lakes Superior and Huron by Kemp et al. (1978) and of Lakes Ontario and Erie by Kemp and Thomas (1976). Dredging operations and water turbulence can resuspend these particulates and sorbed metals may be subsequently released again to the water column. Other studies have indicated that trace metal magnification in organisms can occur both through the food chain via particulates (bioaccumulation) and directly from the water via sorption on membrane surfaces (bioconcentration). Phillips and Russo (1978) prepared an extensive review of literature on the biomagnification of 21 trace metals in fishes and aquatic invertebrates. The availability of trace metals to aquatic organisms in the Great Lakes can best be understood, at least in general terms, through the phenomenon of solid–liquid partitioning. The following represents a brief review of partition theory and published data, including results of studies by Cranbrook of trace metal partitioning in Saginaw Bay conducted during 1976–1978.

2. ADSORPTION

Much research has indicated that adsorption processes are important in regulating metals partitioning between suspended particles and water (Gardiner, 1974; O'Conner and Connolly, 1980; Oakley et al., 1981). Chemisorption and physical adsorption processes were defined by Smith (1981). Physical adsorption is characterized by weak attraction of the metal solute to the solid surface, by high reversibility, and by nonspecificity of sorbants. Chemisorption, on the other hand, is highly dependent on the nature of the solids, often occurs irreversibly, and forms relatively strong bonds of the type occurring between atoms in molecules. Weber (1972) also included ion exchange as an adsorptive process. All three types may occur simultaneously in natural waters, and often the mechanisms interact.

2.1. Adsorption and Metal Species

The simple metal ion in the aquatic system is subject to numerous chemical reactions, such as precipitation, redox reactions, and complexation with various ligands. As a result, a wide variety of metal species exists in natural water systems. This complicates the adsorption process considerably since each of these metal species in solution can have different adsorption mechanisms. Davis and Leckie (1978) found that the percent of Cu^{2+} adsorbed to amorphous iron oxide increased rapidly as the pH of the water increased. They theorized that the easily adsorbed $CuOH^{1+}$ species was responsible for the higher adsorption of Cu^{2+} in the higher pH range. Also, they found that Cu^{2+} adsorption increased in the presence of the organic ligands, glutamic acid and 2,3-pyrazinedicarboxylic acid. Picolinic acid was found to inhibit adsorption.

Theoretical adsorption models help to deal with the great complexity of natural systems. Vuceta and Morgan (1978) modeled the equilibrium distribution of trace metals in fresh water. They predicted that when complexing agents or ligands are absent or in low concentration, then Cu^{2+} will be substantially removed from solution by adsorption onto particles. However, when complexing agents are abundant then the adsorbed metal should be totally or partially released from particles. They theorized that Zn^{2+} exists in solution as a free ion, but as the adsorbent surface area or organic ligand concentration increases then some adsorption and complexation does occur. They predicted that the distribution of Pb^{2+} between aqueous and adsorbed species would depend primarily on the total surface area of the adsorbent. Huang et al. (1977) reported the effects of pH on trace metal adsorption. With hydrous oxides and soils, adsorption abruptly increased in the pH ranges of 6–7 (cadmium) and 5–6 (lead, copper, zinc).

Due to the generally low levels of trace metals in natural waters it is often very difficult to determine the concentrations of various metal species present. For this reason, data on trace metal species or even total metals in fresh water systems are scarce, particularly in open waters of the Great Lakes where concentrations are very low.

2.2. Adsorption and Solids Composition

The nature of the adsorbent is also an important factor in adsorptive processes. A number of laboratory studies have been designed to determine the types of solids which act as good trace metal adsorbents. Gardiner (1974) found that humic materials in river mud are largely responsible for adsorption of cadmium. Concentration factors for different muds varied between 5,000 and 50,000. Oakley et al. (1981) used an equilibrium adsorption model to predict adsorption of copper and cadmium onto artificial geochemical materials in seawater. They concluded that the clay fraction was a major sink for copper and cadmium. Adsorption constants we e calculated to be much higher for copper than for cadmium.

3. SORPTION: A COMBINATION OF ADSORPTION AND ABSORPTION

Some of the variables affecting adsorption onto inert particles have been discussed. The accumulation of metals in the living component of suspended solids is further complicated by absorption. The term sorption has been used to describe this combination of adsorption and absorption (Weber, 1972).

The effect of various inorganic and organic chemicals in the water on trace metal sorption by a given algal or zooplankton species has been studied by Poldoski (1979). He found that in high concentrations of humic acid, pyrophosphate, or aminopolycarboxylic acids, complexation was effective in reducing cadmium uptake by *Daphnia magna*. He demonstrated that raising the calcium concentration in Lake Superior water five times above the baseline level resulted in a one-third lower uptake of cadmium during an exposure of two days. In the presence of diethyldithiocarbamate, the biomagnification of total cadmium was greater than uptake of the free cadmium ion alone. Keeney et al. (1976) showed that the bioconcentration factor (ratio of metal content in alga to metal concentration in water) in *Cladophora glomerata* was reasonably constant. Two sites in Lake Ontario were chosen to study the concentration factors in *Cladophora glomerata*. The [bioconcentration factors]/10^3 in Lake Ontario (Deadman Bay) were 2.9, 49, 16, and 2.2; and in Lake Ontario (Main Duck) they were 1.0, 18, 20, and 1.9 for zinc, cadmium, lead, and copper, respectively.

Sorption also seems to be species specific for algae or zooplankton. Conway and Williams (1979) measured the sorption of cadmium by two diatom species of the Great Lakes, *Asterionella formosa* and *Fragilaria crotonensis*. Initially, both species rapidly sorbed cadmium; however, the cellular cadmium content of *A. formosa* was about three times that of *F. crotonensis*. They concluded that the higher concentration in *A. formosa* was due to active uptake of cadmium. The initial rate of cadmium desorption was 10–100 times greater than the final rate. Milne and Dickman (1977) found that both attached and planktonic algae significantly concentrated lead when grown over lead-contaminated sediments in snow dump areas in Ottawa. No differentiation was made here between adsorption and absorption.

4. ADSORPTION ISOTHERMS

A number of adsorption isotherms have been used to characterize and simplify the sorption of dissolved metal species by complex mixtures of solid surfaces. The Langmuir, linear, and Freundlich isotherms have been applied to solid–liquid partitioning in sediments, soils, and suspended solids in water. The important assumptions for the Langmuir isotherm are constant activity of surface sites, no interaction between adsorbed metal species, all adsorption occurring by the same mechanism and confined to one complete monomolecular layer on the surface.

At equilibrium then (Smith, 1981):

$$C = \frac{K_c C_m C_l}{1 + K_c C_l} \tag{1}$$

K_c = equilibrium constant
C_m = concentration of trace metal within a complete monomolecular layer on the surface
C_l = concentration of metal dissolved in water
C = adsorbed concentration

In dilute systems where C_l is small then Eq. 1 becomes a linear adsorption isotherm:

$$C = K_c C_m C_l = \P C_l \tag{2}$$

where ¶ is often defined as the partition coefficient (O'Conner and Connolly, 1980).

The Freundlich isotherm is defined as:

$$C = C_m k C_l^{1/n} \tag{3}$$

where n has a value greater than unity
k is a constant

Combining constants then:

$$C = K C_l^{1/n} \tag{4}$$

These adsorption isotherms have been successfully applied to solid–liquid partitioning of trace metals in natural systems. Even though all the assumptions described for the Langmuir isotherm may not be met, the basic assumptions of a finite adsorption time and dynamic equilibrium between rates of adsorption and desorption (on which the Langmuir isotherm is based) are sound. In a complex mixture of solids it may not be possible to determine K_c and C_m in Eq. 2, but the constant ¶ may be determined from the linear portion of the adsorption isotherm (Weber, 1972). One requirement is that the adsorptive equilibrium be achieved rapidly and maintained. Reasoning that the concentrations of trace metals are usually low relative to the adsorptive capacity of solids, O'Conner and Connolly (1980) applied Eq. 2 to various solutes. For a given solute they found that ¶ can vary depending on the nature and concentration of adsorbent.

The two constants in Eq. 4 make the Freundlich isotherm flexible; often constants can be chosen so that experimental data can be curve-fitted over a reasonable range of concentrations. Despite this curve-fitting capability, Travis and Etnier (1981) advise against extrapolating beyond experimental points in the curve. Gardiner (1974) preferred to characterize adsorption of cadmium onto mud using the Langmuir isotherm, since the composition of solids in mud is complex and it is therefore not possible to determine both constant K and n in Eq. 4.

If conditions allow the use of Eq. 2 then C_l and C must be determined. One method is to measure the total and dissolved metal in the water. Dissolved metal is operationally defined as that portion which passes through a 0.45 μm porosity filter. The partition coefficient then becomes:

$$\P = \frac{\frac{T - C_l}{SS}}{C_l} = \frac{C}{C_l} \qquad (5)$$

where SS is the suspended solids concentration
T is the total metal concentration in water

4.1. Applications

The widely accepted filter pore size (0.45 μm) for obtaining a dissolved metal sample (C_l) may not be appropriate for all metals in all types of water. Poldoski (1979) suggested that the 0.1 μm pore size filter may be more appropriate for measuring C_l in some water types. In the presence of complexing ligands, humic acid and diethyldithiocarbamate, he noted decreasing concentrations of cadmium in C_l as he decreased pore sizes from 0.45 μm to 0.1 μm. Copper concentrations showed no pore size dependence (Poldoski, 1974).

Problems can occur in using Eq. 5 when suspended solids concentrations are low. Under these conditions the measured values of $T - C_l$ are often zero or negative. This results in ¶ being negative, which is not possible. When the population of solids >0.45 μm is low, then the fraction of particulate metal in the water is very small. The precision of the analytical method used then may not be sufficient to properly differentiate between T and C_l.

There are alternative ways to determine ¶ when the suspended solids concentrations are low. The partition coefficient may be determined by concentrating the solids in the sample with a continuous flow centrifuge (Nriagu et al., 1980). Relatively large quantities of suspended solids can be collected by this means because of the huge volumes of water that can be processed. Filtering sufficient water and analyzing the solids collected is another approach.

Assuming Eq. 2 still holds, ¶ then becomes:

$$\P = \frac{\frac{\text{weight of metal in solids/volume filtered}}{SS}}{C_l} \qquad (6)$$

5. PARTITION MEASUREMENTS

Due to the complexities of sorption phenomena and the limitations of sampling and analytical methodology, little information exists on metal partitioning between solid and liquid phases in the Great Lakes. Many laboratories lack instruments sensitive enough to analyze open lake waters for trace metals. Some data do exist, but due to the varied nature of suspended solids and chemistry of lake water, there are problems in comparing partition data from different areas.

5.1. Form of Metal in Natural Water

Several studies have focused on the extent to which various metals associate with suspended solids. Nriagu et al. (1980) found that in Lake Ontario 50–80% of the copper, 20–60% of the cadmium, and 60% of the lead were bound to the suspended particulates. Rossman (1981) studied Lake Huron water and found an average total zinc concentration of 310 ng/liter and 174 ng/liter for filtered water (0.5 μm porosity). Mean total cadmium concentrations were 15.4 ng/liter and, for filtered water, 5.10 ng/liter. For copper the total was 415 ng/liter and filtered (0.5 μm), 299 ng/liter. Rygwelski and Townsend (1982) analyzed 313 water samples from Saginaw Bay, Lake Huron, in 1978 and found that, in general, the percentages of samples in which the total metal concentration in water was significantly greater than that of the filtered fraction (0.45 μm porosity) were 43% (Zn), 61% (Pb), and 33% (Cu). These studies suggested that lead was highly associated with particulate matter. The Lake Huron studies cited here all showed that copper was predominantly in a dissolved form.

5.2. Concentration of Metals in Particulates

Nriagu et al. (1980) found that the concentrations of copper, cadmium, and lead in the suspended solids taken from different basins in Lake Ontario were generally higher than those in surficial sediments. Average concentrations in the suspended solids were 180 μg/g for copper, 410 μg/g for zinc, 10 μg/g for cadmium, and 190 μg/g for lead. They also observed that the October concentrations of copper, zinc, and cadmium at open lake stations exceeded the May and June values by factors of 2–4. At nearshore stations in October, however, there was a noticeable decrease in particulate metal concentrations. They attributed this seasonal variation to changes in the organic character of the suspended solids. Others have noted a high correlation of trace metals with organic particulates (Suzuki et al., 1979; Gardiner, 1974).

Elzerman et al. (1979) found significant enrichment of Zn, Cd, Pb, and Cu in particulates of the surface microlayer in southern Lake Michigan. Concentrations of metals in these particulates were Zn, 50–1800 μg/g; Cd, 2–25 μg/g; Pb, 30–1900 μg/g; and Cu, 50–640 μg/g.

Rygwelski and Townsend (1982) studied particulate trace metals in Saginaw Bay water in 1978. Whole water was separated by filters or screens into dissolved (<0.45 μm) and particulate fractions of the following size ranges: 0.45–10 μm, 10–74 μm, 74–210 μm, and 210–1000 μm. Results

Table 15.1. Concentrations of Metal in Particulate Size Fractions, Saginaw Bay, Lake Huron, 1978

	Concentration (μg/g) in Particles		
Metal	10–74 μm	74–210 μm	210–1000 μm
Copper	3.7–1300[a]	4.8–610	3.5–430
Mean	410	70.0	95
Median	300	22	31
n	95	101	97
Lead	23.0–3300	20.0–210	4.6–540
Mean	240	46.0	100
Median	50	32	53
n	100	101	85
Zinc	6.3–870	95.0–430	120–650
Mean	390	170	220
Median	330	130	160
n	98	101	102

[a] Range of metal concentrations found in a given size fraction.

(weight metal per dry weight particles) were obtained for the last three size fractions. The 10–74 μm fraction contained the highest concentration of metals (Cu, Zn, and Pb) in the solids. Eutrophic waters of the inner bay often had significantly lower concentrations of metal in the 10–74 μm size fraction than did oligotrophic outer bay waters (Lake Huron). The results of this study are summarized in Table 15.1.

5.3. Adsorption Isotherm Results

Several empirical adsorption studies have characterized solid–liquid trace metal partitioning in the Great Lakes. Some researchers have examined trace metal partitioning by means of laboratory adsorption tests using lake sediments. Others have filtered or centrifuged lake water samples to isolate particles. Analysis of the particulate and dissolved concentrations of metals in water enabled them to calculate the percent dissolved and ¶.

Bahnick et al. (1978) studied the effect of eroded clay bluff materials in Lake Superior water on the partitioning of trace metals. A linear adsorption isotherm characterized copper sorption onto clay bluff materials. Interpolating from this adsorption isotherm, they concluded that each metric ton of eroded bluff material could remove 175 g of copper from the lake water if it contained 10 μg/liter of copper.

Rygwelski and Townsend (1982) have calculated partition coefficients for cadmium, copper, lead, and zinc from Saginaw Bay, Lake Huron. This system is interesting in that the inner bay is highly eutrophic while the outer

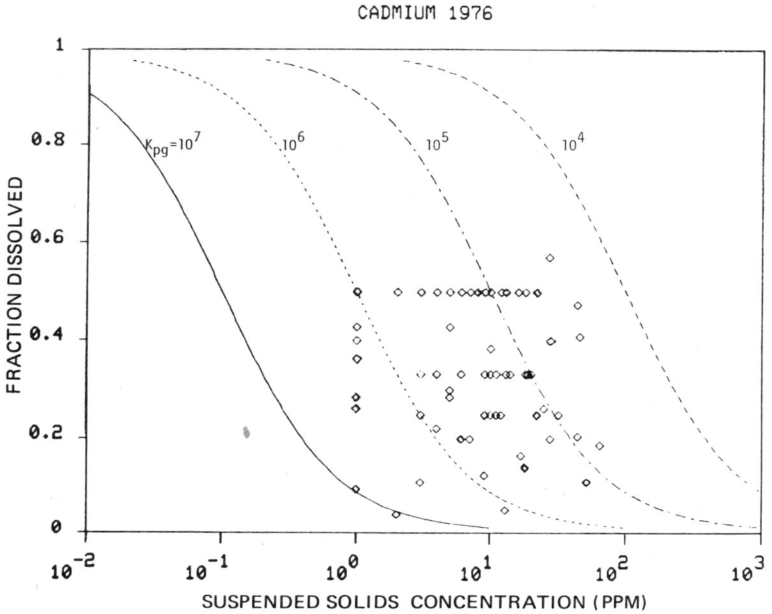

Figure 15.1 Effect of suspended solids concentration (ppm) on partitioning for cadmium, Saginaw Bay, 1976.

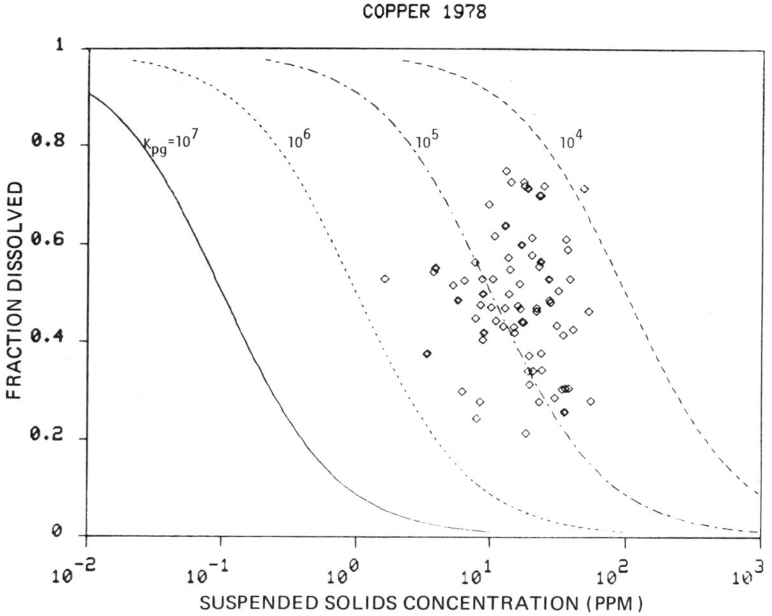

Figure 15.2 Effect of suspended solids concentration (ppm) on partitioning for copper, Saginaw Bay, 1978.

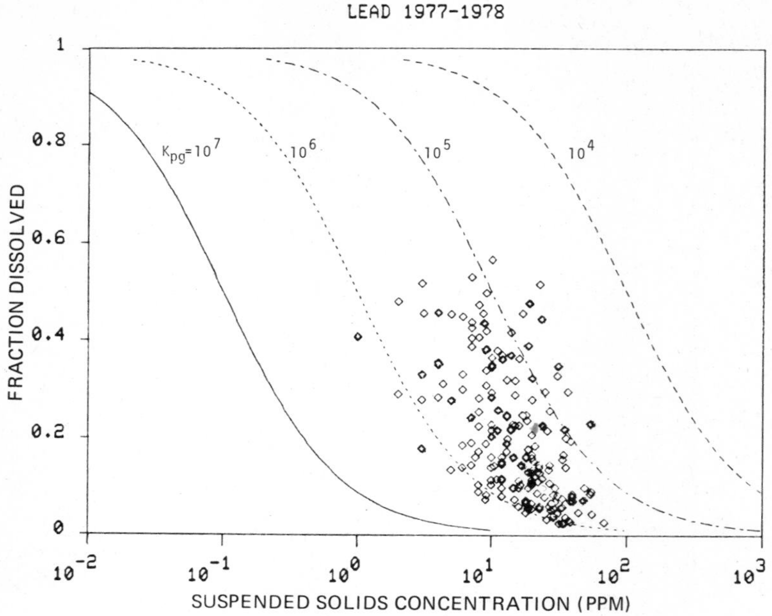

Figure 15.3 Effect of suspended solids concentration (ppm) on partitioning for lead, Saginaw Bay, 1977–1978.

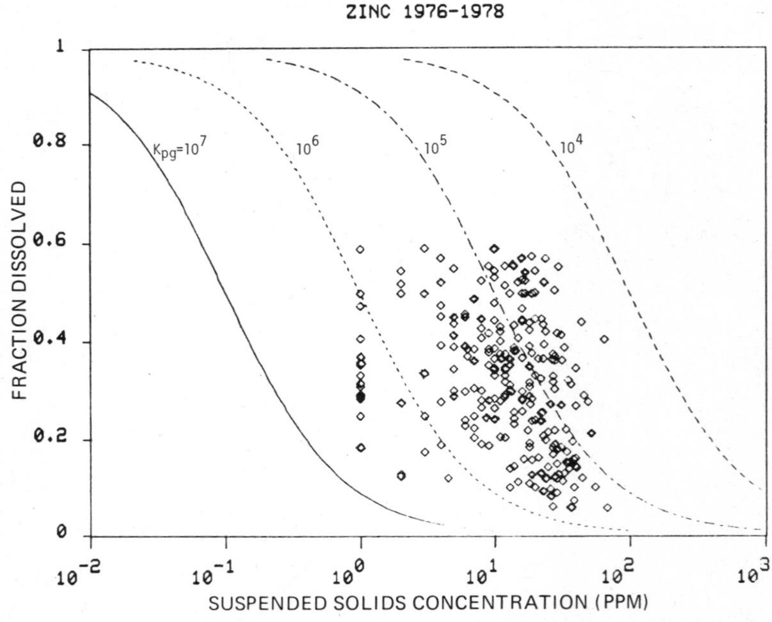

Figure 15.4 Effect of suspended solids concentration (ppm) on partitioning for zinc, Saginaw Bay, 1976–1978.

Table 15.2. General Partition Coefficient (K_{pg}) Data from Saginaw Bay, Lake Huron

Metal	Partition Coefficient Ranges for Inner Bay	Partition Coefficient Ranges for Outer Bay
Cd	$10^{4.5}$–$10^{5.5}$	$10^{5.5}$–$10^{6.5}$
Cu	10^{4}–10^{5}	$10^{5.2}$–$10^{5.7}$
Pb	$10^{4.7}$–10^{6}	$10^{5.5}$–$10^{6.3}$
Zn	$10^{4.5}$–$10^{5.5}$	$10^{5.5}$–$10^{6.5}$

bay water is oligotrophic, much like that of open Lake Huron. General partition coefficients, K_{pg}, (as opposed to K_{ps} for specific fractions) were defined for all suspended solids greater than 0.45 μm.

$$\text{(general partition coefficient) } K_{pg} = \frac{\dfrac{[\text{total metal } (\mu\text{g/liter})] - [\text{dissolved metal } (\mu\text{g/liter})]}{[\text{suspended solids (kg/liter)}]}}{\text{dissolved metal } (\mu\text{g/kg})} \tag{7}$$

This general partition coefficient was found to vary with location and season in the bay. An increasing trend for K_{pg} was noted from the inner to the outer bay (Figs. 15.1–15.4). These figures show data from stations throughout the bay. Values were discarded from the original data set if the dissolved metal sample concentration was not less than the corresponding total metal concentration at the 95% confidence level. Some of the vertical nonrandom alignments of data points in these figures are due to the detection limit for suspended solids. Lines of constant K_{pg} are also plotted on the figures. Partition coefficients for the inner and outer bay are compared in Table 15.2.

On the average, copper had the lowest and lead the highest K_{pg} values. Zinc and cadmium coefficients were intermediate and similar to each other. Dolan and Bierman (1982) were able to apply these partition data in mass balance models for trace metals in Saginaw Bay. They determined that large percentages of cadmium (77%), lead (28–38%), and zinc (15–79%) were lost to the sediment each year. Loss of copper to the sediments was estimated to be minimal.

A specific partition coefficient (K_{ps}) also was calculated for each size fraction in Table 15.1. The fraction which partitioned the highest for all three metals was the 10–74 μm.

6. CONCLUSIONS

The theory of trace metal adsorption processes has been discussed and reviewed. Data were presented which show the variability and extent of trace metals uptake by suspended particles in Great Lakes waters. Models that address the fate of trace metals in water must take into account their partition behavior under various conditions of particle size, organic content, and concentration. Appropriate sampling techniques and analytical instrumentation are available for further empirical studies of metals partitioning, which are both feasible and necessary. Correlations of data on water and particle chemistry with partition data will help to simplify and improve our understanding of partition phenomena in natural waters.

ACKNOWLEDGMENTS

Funding for the Saginaw Bay trace metal study was provided by the U.S. Environmental Protection Agency. Laboratory analyses and data management were conducted at the EPA Large Lakes Research Station, Grosse Ile, Michigan. Field operations were also managed at the Grosse Ile lab. I would like to thank V. Elliott Smith for his careful review of the manuscript.

REFERENCES

Bahnick, D. A., T. P. Markee, C. A. Anderson, and R. K. Roubal. (1978). Chemical loadings to southwestern Lake Superior from red clay erosion and resuspension. *J. Great Lakes Res.* **4**(2): 186–193.

Conway, H. L. and S. C. Williams. (1979). Sorption of cadmium and its effect on growth and the utilization of inorganic carbon and phosphorus of two freshwater diatoms. *J. Fish. Res. Board Can.* **36**: 579–586.

Davis, J. A. and J. O. Leckie. (1978). Effect of adsorbed complexing ligands on trace metal uptake by hydrous oxides. *Environ. Sci. Technol.* **12**(12): 1309–1315.

Dolan, D. M. and V. J. Bierman, Jr. (1982). Mass balance modeling of heavy metals in Saginaw Bay, Lake Huron. *J. Great Lakes Res.* **8**: 676–694.

Elzerman, A. W., D. E. Armstrong, and A. W. Andren. (1979). Particulate zinc, cadmium, lead, and copper in the surface microlayer of southern Lake Michigan. *Environ. Sci. Technol.* **13**(6): 720–725.

Gardiner, J. (1974). The chemistry of cadmium in natural water. II. The adsorption of cadmium on river muds and naturally occurring solids. *Water Res.* **8**: 157–164.

Huang, C. P., H. A. Elliott, and R. M. Ashmead. (1977). Interfacial reactions and the fate of heavy metals in soil–water systems. *J. Water Pollut. Control Fed.* **49**: 745–756.

Keeney, W. L., W. G. Breck, G. W. Vanloon, and J. A. Page. (1976). The determination of trace metals in *Cladophora Glomerata—C. Glomerata* as a potential biological monitor. *Water Res.* **10**: 981–984.

Kemp, A. L. and R. L. Thomas. (1976). Impact of man's activities on the chemical composition in the sediments of Lakes Ontario, Erie, and Huron. *Water Air Soil Pollut.* **5**(4): 469–490.

Kemp, A. L., J. D. William, R. L. Thomas, and M. L. Gregory. (1978). Impact of man's activities on the chemical composition of the sediments of Lakes Superior and Huron. *Water Air Soil Pollut.* **10**: 381–402.

Milne, J. B. and M. Dickman. (1977). Lead concentrations in algae and plants grown over lead contaminated sediments taken from snow dumps in Ottawa, Canada. *J. Environ. Sci. Health* **A12**(4&5): 173–189.

Nriagu, J. O., H. K. T. Wong, and R. D. Coker. (1980). Particulate and dissolved trace metals in Lake Ontario. *Water Res.* **15:** 91–96.

Oakley, S. M., P. O. Nelson, and K. J. Williamson. (1981). Model of trace-metal partitioning in marine sediments. *Environ. Sci. Technol.* **15**(4): 474–480.

O'Connor, D. J. and J. P. Connolly. (1980). The effect of concentration of adsorbing solids on the partition coefficient. *Water Res.* **14:** 1517–1523.

Phillips, G. R. and R. C. Russo. (1978). Metal bioaccumulation in fishes and aquatic invertebrates: A literature review. Ecological Research Series, EPA-600/3-78-103.

Poldoski, J. E. (1979). Cadmium bioaccumulation assays. Their relationship to various ionic equilibria in Lake Superior water. *Environ. Sci. Technol.* **13**(6): 701–706.

Poldoski, J. E. and G. E. Glass. (1974). Methodological considerations in western Lake Superior water–sediment exchange studies of some trace elements. Proceedings of the 7th Materials Research Symposium, Accuracy in Trace Analysis, National Bureau of Standards, p. 10.

Rossman, R. (1981). Trace metal chemistry of Lake Huron's waters. International Association for Great Lakes Research: Twenty-Fourth Conference on Great Lakes Research, Abstracts, p. 34.

Rygwelski, K. R. and J. M. Townsend. (1982). Partitioning of cadmium, copper, lead and zinc among particulate fractions and water in Saginaw Bay, Lake Huron. Paper in preparation, EPA Large Lakes Research Station, Grosse Ile, Mich.

Smith, J. M. (1981). *Chemical engineering kinetics*. 3rd. Ed., McGraw-Hill, New York, pp. 310–322.

Suzuki, M., T. Yamada, T. Miyazaki, and K. Kawazoe. (1979). Sorption and accumulation of cadmium in the sediment of the Tama River. *Water Res.* **13:** 57–63.

Travis, C. C. and E. L. Etnier. (1981). A survey of sorption relationships for reactive solutes in soil. *J. Environ. Quality* **10**(1): 8–17.

Vuceta, J. and J. J. Morgan. (1978). Chemical modeling of trace metals in fresh waters: Role of complexation and adsorption. *Environ. Sci. Technol.,* **12**(12): 1302–1319.

Weber, W. J., Jr. (1972). *Physicochemical processes for water quality control.* Wiley-Interscience, New York, pp. 199–259.

16

LEAD CONTAMINATION OF THE GREAT LAKES AND ITS POTENTIAL EFFECTS ON AQUATIC BIOTA

***Peter V. Hodson,*[1] *D. Michael Whittle,*[1] *Paul T. S. Wong,*[1] *Uwe Borgmann,*[1] *R. L. Thomas,*[1] *Y. K. Chau,*[2] *J. O. Nriagu,*[2] *D. J. Hallett*[3]**

[1] *Great Lakes Fisheries Research Branch and* [2] *National Water Research Institute, Canada Centre for Inland Waters, Burlington, Ontario, L7R 4A6, and* [3] *Environment Canada, Toronto, Ontario, M4T 1M2*

1. INTRODUCTION

Lead is a bluish white metal with a high density, softness, flexibility, malleability, and weldability and with a low melting point and elastic limit. It has an atomic number of 82 and atomic weight of 207.19. Lead may exist as a divalent (plumbous) salt, which is the inorganic form usually found in water,

Table 16.1. Representative Industrial Uses of Lead[a]

Class	Use
Coatings and dyes	Paints, pigments, dyes, metallurgical coatings, ceramic glazes, stained glass, mirrors, wood preservatives, corrosion inhibitors.
Electrical components and electronics	Batteries, electrolytes, electrical insulation, semiconductors, photosensitive electronic components.
Plastics, chemicals	Plastisols, phonograph records, photography processes, plastic stabilizers, catalysts, curing agent for resinous siloxanes, polymer synthesis.
Structural	Printers' type, solder, pipes, containers, fusible alloys, ammunition.
Medical	Veterinary medicines, ointments.
Combustion	Matches, pyrotechnics, explosives, fuel additives.
Miscellaneous	Lubricant, radiation shields.

[a] Information from Rickard and Nriagu, 1978.

and as a tetravalent (plumbic) compound which is the form in most organic lead compounds. Most tetravalent salts are relatively unstable and decompose in water. Lead may also be amphoteric, forming stable plumbates and plumbites with other cations (Nriagu, 1978). The properties of lead allow a variety of uses (Table 16.1). These, plus mining, milling, and smelting of lead and other metal-containing ores, and burning of fossil fuels, cause a loss of lead to aquatic ecosystems at a rate above normal weathering. As an example, Nriagu (1978) has estimated that, of 1360 thousand tonnes of lead used in the United States in 1970, about 56% entered the environment in one form or the other.

The following review briefly outlines the sources of lead inputs to the Great Lakes ecosystem; the forms of lead in air, water, and sediments; the lead concentrations in various ecosystem compartments; and lead toxicity to aquatic biota.

2. INPUTS TO THE GREAT LAKES

The sources of lead in lake water include tributary inflows, industrial and domestic waste discharges, stormwater and coastal runoff, atmospheric deposition (dry particulate and rainfall), shoreline erosion, and dredge spoil disposal. Approximate mass balances for lead are available for Lakes Ontario, Erie, Huron, and Superior (Table 16.2) to indicate the relative con-

Table 16.2. Great Lakes Lead Mass Balances

	Tonnes per Year			
Input/Output	Superior[a]	Huron[a]	Erie[b]	Ontario[b]
Inputs				
Suspended solids and solutes from rivers and interconnecting channels	1110	1200	559	905
Shoreline erosion	?	?	437	50
Dredged spoil	?	?	99	65
Municipal and industrial inputs	4	8	?	?
Atmospheric input	650	780	650–2200	280–440
Total input	>1760	>1990	1745–3295	1300–1460
Outputs				
Accumulating in sediment[c]	362	188	1745	725
Output suspended solids	?	?	319	547
Output solute	?	?	177	207
Total output	>362	>188	2241	1479

[a] Information from Patterson and Kodukula, 1978.

[b] Information from Pollution from Land Use Activities Reference Group, 1977.

[c] Computed from Thomas, 1981.

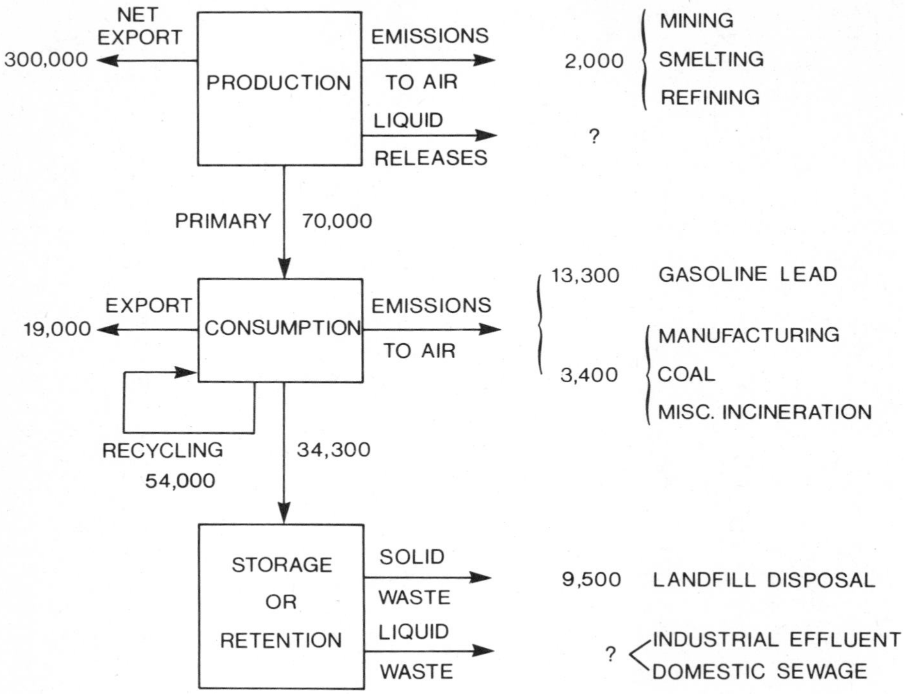

Figure 16.1 Canadian Industrial Lead Budget in 1972 (Tons) derived from Leah (1976). Landfill disposal is calculated from the recycling figure which is estimated to be 87% efficient. Net export equals the export of lead ingots minus the import of lead ingots.

tributions of lead from different sources. Tributary contributions to Lakes Huron and Superior are probably overestimated due to the difficulties of measuring low concentrations of lead (see section on lead in water). Thus, a major fraction (at least 40%) of the lead entering three of the lakes is derived from atmospheric precipitation. Lead in sewage discharges, which often includes storm sewer flows, is partly derived from transportation emissions.

Sediments are one of the major sinks for lead in these lakes. Of the annual load of lead to Lakes Superior, Huron, Erie, and Ontario, the retention in sediments is <20, <10, ≃50, and ≃50%, respectively (Table 16.2).

The 1972 budget for the industrial use of lead in Canada, and associated environmental releases (Fig. 16.1), shows the importance of atmospheric emissions in the environmental lead cycle (Leah, 1976). In particular, gasoline lead accounts for 12,100 out of 17,000 tonnes total lead emissions to the air. In the Canadian portion of the Great Lakes and St. Lawrence drainage basin, gasoline accounts for 7930 tons out of a total of 8152 tons. These figures do not reflect transboundary transport of atmospheric lead or the growing reliance on coal-burning electric power plants with associated lead emissions.

The lead in atmospheric precipitation originates primarily from gasoline additives. On a global basis, about 70% of anthropogenic emissions to the air come from these additives and, in Canada, an additional 23% are from iron, steel, copper, and nickel production. Lead loading from precipitation ranges from 9 to 20 mg/m^2·a, in the Great Lakes Basin, but in highly contaminated areas, it is much greater (e.g., 116, 600, 600, and >6,000 mg/m^2·a near a Sudbury smelter, a Toronto expressway, a Toronto incinerator, and a lead smelter, respectively).

3. FORMS OF LEAD

3.1. Air

The forms of lead in the atmosphere are extremely complex and depend very much on the source of the pollutant (Table 16.3). Atmospheric inorganic lead is generally in amorphous particulate form. The particulates show a broad range of particle size and chemical composition. The variations reflect both the source characteristics and the aging history of the lead aerosols. Organolead compounds are also found. For example, concentrations ranging from ≤6 to 262 μg/m^3 were found at rural and urban sites in Antwerp, Belgium. This was equivalent to 0–24% of total lead, the rest being inorganic (De Jonghe and Adams, 1980). Concentrations of organolead as high as 3570 ng/m^3 have been reported near motorways (Colwill and Hickman, 1973, in De Jonghe and Adams, 1980). The airborne organolead compounds were associated with the use or storage of gasoline. A variety of compounds, including tetramethyllead, trimethylethyllead, dimethyldiethyllead, meth-

Table 16.3. Forms of Atmospheric Inorganic Lead According to Source[a]

Source	Form
Automobile emissions	$PbCl_2$; PbBrCl; Pb(OH)Br; $PbCl_2 \cdot PbBr_2Cl$; $PbO \cdot PbBr_2$; $PbO \cdot PbCl_2$; PbO_x; $PbSO_4$
Mining activities	PbS; $PbCO_3$; $PbSO_4$; $Pb_5(PO_4)_3Cl$; $PbS \cdot Bi_2S_3$
Base metal smelting and refinishing	Pb; $PbCO_3$; $PbSO_4$; PbO_x; $Pb \cdot PbSO_4$; $(PbO)_2$; $PbCO_3$; Pb-silicates
Coal-fired power plants	PbO_x; $PbSO_4$; $Pb(NO_3)_2$; $PbO \cdot PbSO_4$
Cement manufacture	$PbCO_3$; $Pb_5(PO_4)_3Cl$
Fertilizer production	$Pb_5(PO_4)_3Cl$; PbO_x; $PbCO_3$
Rural site (percent distribution)[a]	$PbCO_3$ (30%); $(PbO)_2PbCO_3$ (28%); PbO (21%); $PbCl_2$ (5.4%); $PbO \cdot PbSO_4$ (5%); Pb(OH)Cl (4%); $PbSO_4$ (3.2%); PbBrCl (1.6%); $(PbO)_2 \cdot PbCl_2$ (1.5%); $(PbO)_2PbBrCl$ (1%); $PbBr_2$ (0.1%).

[a] Information from Nriagu, 1978.

yltriethylead and tetraethyllead, could be measured simultaneously, probably due to chemical reactions in the air (De Jonghe et al., 1981).

3.2. Water

In the vast majority of freshwater systems in contact with the atmosphere, lead carbonate complexes ($PbCO_3$) dominate the inorganic chemistry of dissolved lead. Between pH 6 and 8, lead will apparently be entirely complexed as the carbonate species, especially as $Pb(CO_3)_2^{2-}$; $PbCO_3^0$ also occurs at more acidic pH values (Bilinski and Stumm, 1973; Hem, 1976). In oxygenated systems where sulphur-containing organic compounds are not present in measurable quantities, ordinary amino acids will not effectively complex lead because of their low stability constants with lead complexes (log K = 5.5, average for the amino acids complexed with Pb^{2+}). The glycine concentration must approximate the total carbonate concentrations for the lead–glycine complex to dominate. On the other hand, strong, nonspecific chelating agents, such as sodium nitrilotriacetic acid, at concentrations as low at 20 μg/liter, can have a noticeable effect on dissolved lead chemistry.

The forms of lead in Great Lakes waters have not been measured. Information from other natural water systems, although not directly applicable, can be taken as a reference for the Great Lakes.

Lead uptake by aquatic biota is a function of the concentration of biologically available lead and of the ability of biota to actively or passively regulate membrane transport of lead (see sections on lead content of aquatic biota and lead toxicity to aquatic biota). In controlled experimental conditions, lead uptake by fish and sublethal toxicity can be correlated directly to "free" or "ionic" lead and to "total" lead (Davies et al., 1976; Goettl and Davies, 1979; Hodson et al., 1977; 1978a; 1979; Holcombe et al., 1976; Merlini and Pozzi, 1977a; 1977b); "free" or "ionic" lead measurements best describe the direct toxic effects of lead. However, the impacts of lead on aquatic biota may not be restricted to "free" lead. There is evidence suggesting that particulate lead may ultimately be available to herbivorous or detrital feeding fish and to invertebrates (see sections on lead content of aquatic biota). Although the rate of methylation of sediment lead is unknown, alkyl lead compounds have been measured in fish (see section on lead in fish), so that sediment lead conceivably may be transferred to fish in the alkyl form.

3.3. Sediment

Little information exists on species of lead compounds in sediment because of the limited knowledge of the characteristics of organic compounds in sediment. In anaerobic sediments, most of the dissolved lead will be immobilized as the sulfide (PbS). This situation has been exemplified by the Saguenay fjord sediments where 70–80% of the total lead is associated with the sulfide mineral phase (Loring, 1976).

Fulvic and humic acids constitute from 40 to 70% of the organic matter in soils and sediments and can form strong complexes with lead ions (Nissenbaum and Swaine, 1976; Schnitzer, 1971). Lead–fulvic acid complexes will dominate over $Pb(CO_3)_2^{2-}$ at pH 6 and high carbon dioxide partial pressure (0.1 mm) if the fulvic acid concentration is greater than 10^{-4} *M* (Nriagu, 1978).

It has also recently been suggested (Nriagu, 1974) that soluble lead compounds react with phosphates in the soils to form trilead phosphate, $Pb_3(PO_4)_2$, plumbogummite, and pyromorphites.

3.4. Biota

Lead may enter the biomass pool as soluble ions, organolead molecules, or in association with particulate materials. Within biota, lead is likely to be associated with organics and in complexed forms. There is little literature available on the identification of chemical forms of lead in biota, although tetraalkyllead compounds have been reported in fish (Chau et al., 1980; Sirota and Uthe, 1977) (see section on lead in fish).

3.5. Transformation

Incubations of sediments with and without the extraneous addition of organic and inorganic lead compounds produced tetramethyllead (Wong et al., 1975), and this process has been attributed to biological and chemical methylation (Jarvie et al., 1975; Schmidt and Huber, 1976). Several pure bacterial cultures (*Pseudomonas, Alcaligenes, Acinetobacter, Flavobacterium*, and *Aeromonas*) were able to transform trimethyllead acetate to tetramethyllead at various pH's (Wong et al., 1975).

The methylation of Pb(II) does not follow the methyl cobalamin mechanism implicated for mercury methylation, and it has been suggested that the carbonium ion, CH_3^+, is involved in the methylation of Pb(II) (Ahmad et al., 1980).

4. LEAD DISTRIBUTION IN THE GREAT LAKES ECOSYSTEM

4.1. Air

Lead in air is of special concern to human health due to fast respiratory uptake. The following is a verbatim quote from a Canada Department of Health and Welfare (1978) report:

As a result of automobile emissions, elevated lead levels are found in the atmosphere over large cities. Measurements made in the U.S. during 1954–1956 indicated that almost 95 percent of the air samples taken over 22 cities had lead concentrations less than or equal to 0.0039 mg/m^3; variations in the samples correlated with the volume of automobile traffic. In a 1970 study, the major portion of urban air samples contained less than 0.0049 mg/m^3, but levels of some samples were as high as 0.0099 mg/m^3; in contrast, all air samples taken in nonurban localities were in the range 0.000019–0.00099 mg/m^3. In Canada, samples of air taken at peak traffic periods in Vancouver, Toronto, and Montreal contained 0.0082, 0.0084, and 0.004 mg/m^3, respectively. These levels were considered higher than general urban lead levels. Air samples collected from 72 sites across Canada during 1973, had lead concentrations between 0.00013 and 0.00322 mg/m^3. Rainfall cleanses the atmosphere by "scrubbing out" the particles.

Precipitation analyzed in the U.S. had an average lead concentration of 0.034 mg/liter. A study in Toronto showed lead to be present in the precipitation at an average of 0.05 mg/liter.

4.2. Water

An understanding of the true hazards of lead in the Great Lakes creates a need for accurate determinations of the baseline lead concentrations in each lake. Despite numerous surveys, it is unlikely that accurate baseline lead concentrations in any of the Great Lakes have been measured, since reported concentrations in open lake waters range from <0.1 to 100 μg/liter. Such data, obtained by conventional methods, remain suspect until confirmed by measurements in an ultraclean laboratory under exhaustive contamination control. In general, lead levels measured elsewhere in ultraclean facilities tend to be 10- to 100-fold lower than those obtained by conventional methods (Patterson and Settle, 1976). Consequently, the report by Waller and Lee (1979) that 52% of the reporting stations in Lake Ontario had lead concentrations in excess of 25 μg/liter is probably erroneous.

Some recent data on lead concentrations in the Great Lakes suggest that the real lead levels in open lake waters are <1.0 μg/liter (Table 16.4); considering possible sample contamination, the true concentrations are probably closer to 0.1 μg/liter, if not lower.

4.3. Sediments

Lead in Great Lakes sediments was reviewed in the 1977 Annual Progress Report of the Great Lakes Pollution from Land Use Activities Reference Group (1977). Since the arrival of early settlers, lead has been accumulating in sediments at a rate above that due to normal weathering and erosion, and this deposition rate is increasing (Nriagu et al., 1979).

Table 16.4. Recent Data on Lead in Great Lakes Waters (Concentration in μg/liter)[a]

Lake Superior	Lake Huron	Lake Erie	Lake Ontario	Reference
			0.83	Chau et al., 1970
0.4 (nearshore)	0.6 (nearshore)			Poldoski et al., 1978
<1.0	<1.0	1.5(0.2–3.5)	0.7(0.5–1.5)	CCIW, 1979[b]
1.0	1.0	2.0	0.5	Patterson and Kodukula, 1978.
			<0.3	Nriagu et al., 1981
		0.46(0.15–1.40)		Lum, personal communication[c]
<1.0	≤0.5			Water Quality Board, 1977a.
<0.1–10.0	≤5–12			Upper Lakes Reference Group, 1977.

[a] Values preceded by "less than" represent detection limits.

[b] STAR Data File, 1979, Canada Centre for Inland Waters, Burlington, Ontario.

[c] K. Lum, National Water Research Institute, Canada Centre for Inland Waters, Burlington, Ontario, Canada.

Apparent anthropogenic enrichment increases through the lakes in the order Lake Erie < Lake Superior < Lake Huron, Georgian Bay < Lake Ontario. The low enrichment of Lake Erie, which receives the highest total loading, reflects both higher sedimentation rates that dilute anthropogenic inputs, and slightly higher natural lead inputs.

Sediment concentrations tend to be highest in depositional zones (deep, relatively slow water movement) and least in nearshore or shallow zones, where water movements reduce sediment accumulation (Fig. 16.2). Exceptions are "plumes" in Lake St. Clair and Lake Erie due to very high input rates from Sarnia, Detroit, and Toledo and local harbor contamination. There are also elevated sediment lead concentrations near the western shore of Lake St. Clair, perhaps due to lead shot from hunting. The mean (with standard deviation) depositional zone lead concentrations for Lake Superior, Georgian Bay, Lake Huron, Lake St. Clair/Lake Erie, and Lake Ontario were 60 ± 23, 67 ± 27, 66 ± 35, 112 ± 44, and 154 ± 43 mg/kg dry weight, respectively (Pollution from Land Use Activities Reference Group, 1977).

Harbor sediments from Oswego and Rochester had lead concentrations ranging from 17 to 148 and 9 to 39 mg/kg dry weight, respectively. Elutriate tests, however, showed that lead was not released from these sediments to water at concentrations over 58 μg/liter, the detection limit (Water Quality Board, 1977b). More contaminated harbors such as Hamilton may contain up to 930 mg/kg lead (Water Quality Board, 1975). Mean sediment lead concentrations in Canadian harbors on Lake Huron ranged from 4.7 to 162.3

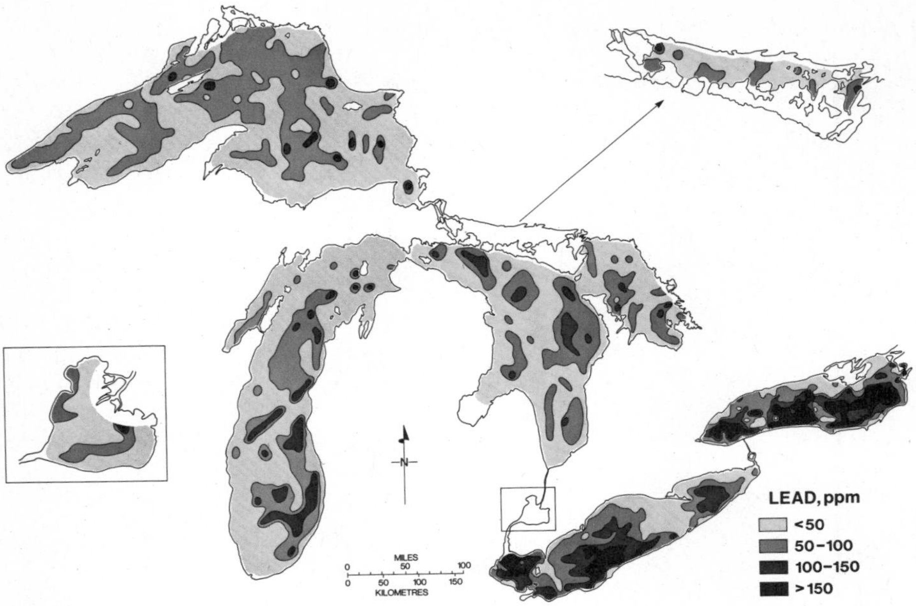

Figure 16.2 Distribution of lead in lake sediments: 0–3 cm sediment thickness. Information from Pollution From Land Use Activities Reference Group, 1977.

mg/kg, with the highest concentration in Collingwood and the next highest in Tobermory and Owen Sound.

While sediments have traditionally been regarded as sinks for lead, the possibility of microbial and chemical lead methylation exists (see section on transformation). Consequently, lead-enriched sediments may produce conditions hazardous to aquatic biota (see sections on toxicity to algae and toxicity to fish). In addition, there are very few data on toxicity of sediment lead to benthos and some indications of transfer to bottom-feeding fish (see section on lead content of fish). Therefore, high lead levels in sediments are important both as indicators of contamination and as potential hazards to the aquatic ecosystem.

4.4. Plankton

Very few data have been published on the lead content of Great Lakes plankton, despite their sensitivity to lead (see section on toxicity). Inventories of the distribution of contaminants in each of the Great Lakes Basins contain no lead data for plankton or benthos (Pollution from Land Use Activities Reference Group, 1978; Water Quality Board, 1977a). A recent unpublished survey by Whittle of the lead content of mixed plankton samples and *Mysis relicta* collected from five offshore sites in Lake Ontario, showed

Table 16.5. Total Lead Content of Lake Ontario Plankton Samples in 1979[a]

Location	Number of Samples	Mean Sample Dry Weight (g)	Mean Concentration (μg/g Dry Weight ± Standard Deviation)	Range
		Net Plankton (>153 μm)		
Eastern Basin	5	1.38 ± 0.06	4.72 ± 2.02	2.8–7.9
Point Traverse	5	1.13 ± 0.02	7.54 ± 3.35	3.5–12.0
Cobourg	5	1.22 ± 0.20	6.84 ± 1.08	6.0–8.7
Port Credit	5	1.13 ± 0.04	5.30 ± 0.52	4.8–6.0
Niagara	5	1.24 ± 0.05	28.6 ± 9.40	17.0–43.0
		Zooplankton (Mysis relicta)		
Eastern Basin	10	1.12 ± 0.12	1.96 ± 0.45	1.3–2.6
Point Traverse	5	1.06 ± 0.5	1.64 ± 0.45	1.2–2.4
Cobourg	5	1.18 ± 0.08	1.20 ± 0.30	0.9–1.6
Port Credit	5	1.08 ± 0.04	5.1 ± 0.53	4.4–5.8
Niagara	4	1.09 ± 0.12	4.5 ± 0.75	3.9–5.6

[a] Whittle, unpublished data.

that the lead content of mixed surface plankton was consistently higher when compared to zooplankton samples collected at the same sites (Table 16.5; see Annex 1 for methodology). The considerable intrasite variability of lead concentrations in plankton may reflect the patchiness of the populations within an area. The zooplankton *Mysis relicta* spends a considerable amount of time near the bottom but also migrates vertically so that lead levels for this species may reflect particulate or sediment lead levels. The total lead concentrations in these samples increase in relation to their proximity to industrialized areas of the Lake Ontario Basin.

Filamentous algae from the littoral zone also reflect this trend. Keeney et al. (1976) reported 9.5–12.2 μg/g total lead in freeze-dried filamentous algae from the eastern basin of Lake Ontario (values on a wet weight basis would probably be one tenth or lower). Samples from the Cobourg area contained 1.22 μg/g wet weight (Wong, unpublished data). Published concentration factors range from 20 for algal samples from the Great Lakes (Keeney et al., 1976) to >100,000 for phytoplankton samples from the English Lake District (Denny and Welsh, 1973).

4.5. Fish

Much of the information on lead in Great Lakes fish is the result of edible portion analyses and reflects a concern for human health. In areas of local contamination, such as Toronto Harbour, tissue lead concentrations were

Table 16.6. Total Lead Concentration of Whole Fish Sampled Offshore in Lakes Huron and Superior in 1976

		Unweighted Lake Average (μg/g Wet Weight)	
Species	Number[a]	Lake Huron	Lake Superior
Bloater chubs (*Coregonus hoyi*)	20	0.08	0.06
Burbot (*Lota lota L.*)	20	0.05	0.04
Lake trout (*Salvelinus namaycush*)	20	—	0.04

[a] Each sample represents a composite containing at least four individual fish.

1.78 μg/g (Brown and Chow, 1975). Elsewhere, muscle lead concentrations were uniformly <0.5 μg/g (Lucas et al., 1970; Upper Lakes Reference Group, 1977; Uthe and Bligh, 1971). Whole body concentrations in large migratory piscivores (e.g., lake trout) in 1975–1976 were <0.08 μg/g in Lakes Huron and Superior (Table 16.6) (Upper Lakes Reference Group, 1977), and similar concentrations were seen in the Lower Lakes (Table 16.7). However, smaller planktivores and omnivores collected at the same sites in the Lower Lakes by Whittle (Unpublished data) had consistently higher concentrations (Tables 16.8 and 16.9; see Annex II for methodology). Recent surveys (Hodson et al., 1980a, 1983) of the blood lead concentrations of Lake Ontario fish demonstrated the following:

1. Lead contamination of lake trout increased from East to West in 1979, but not in 1980.
2. Inshore benthic feeding species such as carp (Cyprinidae) and sucker (Catastomidae) were more highly contaminated than Centrarchidae or Seranidae.
3. Lead contamination was associated with harbors or industrial point sources. The greatest contamination was associated with the effluent from industries producing tetraalkyllead compounds.

These observations are consistent with data presented by Leland and McNurney (1974) for total lead concentrations in biota from urban and rural Illinois streams. Lead concentrations were highest in small planktonic or sestonic organisms with a high surface-to-body weight ratio, suggesting adsorption as a major mechanism of contamination. Herbivorous fish had higher lead concentrations than carnivorous fish, perhaps due to a smaller size, diet composition, or feeding by sifting contaminated sediments. The net effect was an inverse food chain accumulation.

Tissue residues of lead are generally highest in bone, gill, and kidney with lesser amounts in muscle, liver, and remaining tissues (Gire et al., 1974; Goettl and Davies, 1979; Hodson et al., 1978a; Holcombe et al., 1976). Bone lead represents detoxification because it is relatively inactive metabolically,

Table 16.7. Total Whole Body Lead Content of Predatory Fish Samples in Lakes Ontario and Erie in 1978[a]

Location	Species	Number (*N*)	Mean Weight (g)	Mean Concentration[b] (μg/g Wet Weight ± Standard Deviation)	Range (μg/g)	%*N* < Detection Limit
Lake Ontario						
Eastern Basin	Lake trout (*Salvelinus fontinalis*)	50	836	<0.1	—	100
Point Traverse	Lake trout	34	1320	<0.1	—	100
Cobourg	Lake trout	7	1153	<0.1	—	100
Port Hope	Rainbow trout (*Salmo gairdneri*)	39	1752	<0.1	—	100
Port Credit	Lake trout	50	803	<0.1	—	100
Credit River	Coho salmon (*Oncorhynchus kisutch*)	50	3464	0.12 ± 0.01	<0.10–0.13	92.0
Niagara	Coho salmon	49	1101	0.13 ± 0.01	<0.10–0.20	47.9
Lake Erie						
Long Point Bay	Northern pike (*Esox lucius*)	14	2058	0.10 ± 0.00	<0.10–0.10	92.9
Erieau	Walleye (*Stizostedion vitreum*)	11	148	0.18 ± 0.01	<0.10–0.18	81.8
Western Basin	Walleye	44	948	0.13 ± 0.03	<0.10–0.16	95.5

[a] Whittle, unpublished data.

[b] Mean of samples with a lead concentration greater than or equal to the detection limit (0.10 μg/g).

Table 16.8. Total Whole Body Lead Concentrations of Yellow Perch (*Perca flavescens*) Sampled from Lakes Ontario and Erie in 1978[a]

Location	Number (*N*)	Mean Weight (g)	Mean Concentration[b] (μg/g Wet Weight ± Standard Deviation)	Range	%*N* < Detection Limit
Lake Ontario					
Eastern Basin	50	116	0.19 ± 0.01	<0.10–0.61	16
Toronto	28	35	0.40 ± 0.06	0.10–1.0	0
Lake Erie					
Long Point Bay	41	157	0.16 ± 0.01	<0.10–0.38	61
Erieau	36	200	0.20 ± 0.02	<0.10–0.40	33
Wheatley	44	114	0.16 ± 0.01	<0.10–0.28	23
Western Basin	29	86	0.15 ± 0.01	<0.10–0.28	28

[a] Whittle, unpublished data.

[b] Mean of samples with a lead concentration greater than or equal to the detection limit (0.10 μg/g).

Table 16.9. Total Whole Body Lead Concentrations of Rainbow Smelt (*Osmerus mordax*) Sampled from Lakes Ontario and Erie in 1978[a]

Location	Number[b] (*N*)	Mean Weight (g)	Mean Concentration[c] (μg/g Wet Weight ± Standard Deviation)	Range	%*N* < Detection Limit
Lake Ontario					
Eastern Basin	10	14.9	0.12 ± 0.01	<0.10–0.12	80
Point Traverse	10	40.0	0.12 ± 0.01	<0.10–0.17	50
Cobourg	10	29.9	0.22 ± 0.02	<0.10–0.39	30
Port Credit	9	30.2	0.16 ± 0.02	0.11–0.28	0
Niagara	10	30.6	0.09 ± 0.01	<0.10–0.17	40
Lake Erie					
Long Point Bay	11	20.9	0.11 ± 0.01	<0.10–0.12	73
Erieau	10	30.7	0.12 ± 0.01	<0.10–0.16	60
Wheatley	13	33.6	0.13 ± 0.01	<0.10–0.19	39
Western Basin	10	35.7	0.13 ± 0.01	<0.10–0.21	60

[a] Whittle, unpublished data.

[b] Each sample consisted of a composite of five fish.

[c] Mean of samples with a lead concentration greater than or equal to the detection limit (0.10 μg/g).

but tissue lead may be toxic because it is associated with high molecular weight proteins (>55,000) (Reichert et al., 1979). Kidney lead is associated with nephrotoxicity and anemia. Low lead concentrations in brain and neural tissue are surprising in light of lead neurotoxicity (see section on toxicity to fish). This may reflect either high neural sensitivity to lead or difficulties in analyzing small samples. Low concentrations in muscle indicate that "edible portion" or "whole body" analyses (mostly muscle) may provide misleadingly low estimates of lead contamination.

The amount of lead taken up by fish and their resultant concentration factors are dependent upon the fish species, water quality, and waterborne lead concentrations. Log–log plots of tissue lead versus waterborne lead show that concentration factors decrease as waterborne lead concentrations increase, suggesting either accelerated depuration or a saturation of uptake mechanisms. In experiments with brook and rainbow trout in water from Lake Superior and Lake Ontario, respectively, lead in all tissues increased about 6–7 times for each 10-fold increase in waterborne lead (Hodson et al., 1978a; Holcombe et al., 1976); however, lead accumulation rates by various rainbow trout tissues studied by Goettl and Davies (1979) were much higher (16–154 times) than these values. Whole body concentration factors on a wet-weight basis are generally <1000 (Hodson et al., 1978a).

Blood lead equilibrates with waterborne lead very quickly (<1 week), but other tissues require at least 2–4 weeks, and some tissues may require up to 20 weeks (Goettl and Davies, 1979; Hodson et al., 1977; Holcombe et al., 1976). On a whole body basis, the bulk of lead (>90%) is taken up in the first 4 weeks. In contrast, depuration is very slow. Blood lead of rainbow trout has a half-life of at least 4 weeks (Hodson et al., 1977), but this may increase with lead exposure time (Johansson-Sjöbeck and Larsson, 1979). Half-lives in other tissues are not accurately determined, but lead lost from some tissues may be transferred to others instead of being excreted (Varanasi and Gmur, 1978).

Organolead compounds have been identified in both freshwater and marine fish (Chau et al., 1980; Sirota and Uthe, 1977). Chau et al. (1979) demonstrated significant tetramethyllead residues in rainbow trout exposed to a constant concentration of the chemical in the water, and tissue concentrations were 10 times higher than in trout exposed to inorganic lead (Hodson, 1979). Rainbow trout exposed to tetramethyllead at concentrations ranging from 3.5 to 51 μg/liter rapidly accumulated the lead compound, and the highest tissue concentrations were found in the lipid layer of the intestine (Wong et al., 1981). The initial uptake rate was calculated to be 1 μg tetramethyllead per gram fish per day. When contaminated fish were placed in clean water, there was an initial rapid loss of tetramethyllead from their organs followed by a slower release until a base residual level was reached. Thus, levels of alkylleads in wild fish represent a balance between continuous uptake and depuration and suggest a continuous exposure.

Maddock and Taylor (1977) also found uptake of alkyllead compounds by marine fish and invertebrates during laboratory exposures, but concen-

Table 16.10. Total and Hexane Extractable Lead Levels in Several Species of Great Lakes Fish[a]

Lake	Species	Number of Samples	Lead Concentrations (ng/g Wet Weight) Total	Hexane Extractable
Ontario	Lake trout	169	79.0	8.8
	Rainbow trout	10	110.3	9.1
	Coho salmon	25	121.9	9.3
	Rainbow smelt	20[b]	166.3	9.3
Erie	Rainbow trout	9	84.4	12.9
	Coho salmon	24	71.5	4.6
	Walleye	29	74.7	3.9
	Rainbow smelt	22[b]	73.9	4.9
Huron	Splake	50	73.1	6.2
	Rainbow smelt	10[b]	59.5	2.4
Superior	Lake trout	63	66.6	11.5
	Lake whitefish	45	121.6	20.0

[a] Whittle, unpublished data.
[b] Each sample consists of a composite of five fish.

tration factors were 650× or less, which is typical of inorganic lead. Depuration was also observed, with muscle and liver half-lives in the range 40–60 days.

Ionic diethyl- and triethyllead species were found in oysters exposed to tetraethyllead. Analysis of *Macoma* from estuarine locations in the northwest coast of England showed the presence of trialkyl- and dialkyllead compounds (Birnie and Hodges, 1981). Recently, ionic trialkyl- and dialkyllead compounds have been identified in fish samples containing high levels of tetraalkyllead (Chau and Wong, 1983). This finding has a significant bearing on the toxic effects of tetraalkyllead because the trialkyllead species are known to be responsible for tetraalkyllead toxicity in mammals.

A variety of fish species sampled from various rivers and bays around Lake Ontario and from Lake St. Clair all contained measurable quantities of various tetraalkyllead compounds as well as "volatile" and "hexane extractable" lead (Chau et al., 1980). Concentrations were generally in the low μg/kg range. A more intensive study of the occurrence of organolead compounds in Great Lakes fish has shown their widespread occurrence at significant concentrations (Table 16.10).

5. TOXICITY

5.1. Algae

Concentrations of lead toxic to algae are extremely variable, ranging from 10 μg/liter (Lam et al., 1976) to more than 1000 mg/liter (Ruthvens and

Cairns, 1973). This wide range is caused by temperature, growth medium composition, forms of lead tested, complexing capacity of the water, interaction between lead and other metals, and algal species sensitivity (Table 16.11). In general, lead toxicity increases with temperature due, perhaps, to increases in metal solubility, cell metabolism, or membrane permeability to the metal.

Composition of the test medium is also very important. Growth of *Ankistrodesmus falcatus,* a green alga common to the Great Lakes, was reduced 50% by 2 mg/liter lead in a chemically defined medium (CHU-10), but only 10 μg/liter was required for the same effect when the medium was Lake Ontario water (Wong et al., 1978b). Cell growth rate of *Chlorella pyrenoidosa* was unaffected by 10 mg/liter lead, but when the nutrients were diluted a thousand fold, 0.1 mg/liter lead caused a 63% inhibition (Hannan and Patouillet, 1972). Lead toxicity to *Chlamydomonas reinhardtii* was also reduced by phosphate additions to the medium (Schulze and Brand, 1978). As these data suggest, the higher the lead complexing capacity of the medium, the lower the lead toxicity since lead is made less available. When lead (500 μg/liter) was added to water from a variety of inland lakes with different measured complexing capacities, more algal growth was observed in lake waters with higher complexing capacity (Chau and Wong, 1976).

The interactions of several metals affect toxicity. For example, in one experiment, 25 μg/liter lead was not toxic to several species of freshwater algae when tested by itself, but the presence of other metals, also at nontoxic levels, caused growth inhibition in several species of freshwater algae (Wong et al., 1978a) (see also section on invertebrates).

Alkylated leads are more toxic than nonalkylated ones. Triethyl, tributyl, and trimethyllead were more toxic to a green alga, *Scenedesmus quadricauda,* than were lead acetate, bromide, nitrate, and chloride. Volatile methylated lead (tetramethyllead) was twice as toxic as nonvolatile methylated lead and twenty times as lead nitrate (Silverberg et al., 1977; Wong et al., 1975). In contrast, tetraethyllead was not toxic in its original chemical form to the phytoflagellate *Poteriochromonas malhamensis* (Röderer, 1980). However, when this compound was photodegraded into triethyllead, it strongly inhibited cellular growth, mitosis, and cytokinesis (Röderer, 1976). Algae, in contrast to mammals, cannot metabolically convert tetraethyllead to the very toxic triethyllead. Therefore, in algal bioassays, the toxicity of tetraethyllead may be created solely by the influence of light. This hypothesis is supported by the observation that the effects of triethyllead were identical to those observed in tetraethyllead-treated and illuminated cultures of *Poteriochromonas* (Röderer, 1981).

Finally, the sensitivity of different algal species to lead effects should also be considered, since within one experiment, the most resistant species required 37 times more lead for growth inhibition than the most sensitive (Table 16.11). The most sensitive species of Great Lakes algae have not yet been identified in a systematic fashion.

Table 16.11. Examples of Factors Affecting Lead Toxicity to Algae

Factor	Organisms	Lead Concentration Causing 50% Inhibition		Reference
Temperature	*Ankistrodesmus falcatus*	20°C	1–2 mg/L	Wong, Unpublished data
		9°C	>10 mg/L	
		4°C	<10 mg/L	
Medium	*Ankistrodesmus falcatus*	CHU-10 medium	2 mg/L	Wong et al., 1978b
		Lake Ontario water	10 μg/L	Lam et al., 1976
Forms of Lead	*Scenedesmus quadricauda*	$Pb(NO_3)_2$	5 mg/L	Wong et al., 1975
		$(CH_3)_3PbAc$	1.5 mg/L	
		$(CH_3)_4Pb$	<0.5 mg/L	
Complexation	*Scenedesmus quadricauda*	Lake water with <0.75 μmole/L complexing capacity	0.5 mg/L	Chau and Wong, 1976
Algal species variation	*Anabaena* sp.		15–18 mg/L	Malanchuk and Gruendling, 1973
	Cosmarium botrytis		5 mg/L	
	Navicula pelliculosa		15–18 mg/L	
	Chlamydomonas reinhardii		15–18 mg/L	
	Microspora sp.	TIC[a]	1–2 mg/L	Whitton, 1970
	Mougeotia sp.	TIC	3–21 mg/L	
	Ulothrix sp.	TIC	4–9 mg/L	
	Gongrosira sp.	TIC	14 mg/L	
	Stigeoclonium tenue	TIC	18 mg/L	
	Sporotetras pyriformis	TIC	35 mg/L	
	Cladophora glomerata	TIC	37 mg/L	

[a] TIC = Tolerance index concentration: the average of concentrations causing several adverse effects.

5.2. Invertebrates

Biesinger and Christensen (1972) demonstrated a 50% reduction in reproduction of *Daphnia magna* after 3 weeks exposure to 100 μg/liter lead and a 16% reduction at 30 μg/liter (water hardness = 45 mg/liter, alkalinity = 42 mg/liter, T = 18°C). Reproductive impairment as used by Biesinger and Christensen means percentage decrease in young produced relative to controls and probably includes mortality, reduction in growth rate, and true inhibition of reproduction. Borgmann et al. (1978) demonstrated a significant increase in chronic mortality at 19 μg Pb/liter, but not at 12 μg/liter, of the snail *Lymnaea palustris* (hardness = 139 mg/liter, alkalinity = 88 mg/liter, T = 21°C). Growth was not affected by lead. At 48 μg Pb/liter snail mortality equalled growth rates resulting in no net increase in population biomass. Spehar et al. (1978) observed 60% mortality of amphipods *Gammarus pseudolimnaeus* at 32 μg Pb/liter after 28 days exposure, with a calculated LC_{50} of 28.4 μg/liter (hardness = 44–48 mg/liter, alkalinity = 40–44 mg/liter, T = 15°C). Mortality was continuous over time with no incipient LC_{50} observable after 28 days. Spehar et al. (1978), however, found no significant mortality of stoneflies, caddisflies, or snails (*Physa integra*) after 28 days exposure to lead concentrations as high as 565 μg/liter, possibly due to the high turbidity of the incoming water during these tests. Anderson et al. (1980) obtained a 10-day LC_{50} of 258 μg/liter for immature chironomids (*Tanytarsus dissimilis*) (hardness = 47 mg/liter, alkalinity = 44 mg/liter, T = 22°C), a value much lower than most Pb concentrations reported to be toxic to insects, probably because they used immature animals.

Wilson (1982) observed that the behavioral phototactic response of zooplankton was not significantly affected by measured lead levels of up to 3.3 μg/liter in Lake Ontario water at 15°C. Considerable reductions in response were obtained at 3.8 μg/liter for *Diaptomus sicilis* (a 41% reduction), 7.5 μg/liter for *Daphnia galeata mendotae* (32%), and 10.0 μg/liter for *Cyclops bicuspidatus thomasi* (36%). The data of Biesinger and Christensen (1972) and of Anderson et al. (1980) were obtained using static bioassays, whereas the others were obtained using flow-through systems.

Borgmann et al. (1980) using a static system, observed a 20% reduction in the rate of biomass production of copepods in water from the Burlington Canal at anywhere from 240 μg Pb/liter to over 10,000 μg/liter depending on the time of year. In these natural waters, lead toxicity may have been reduced and variable due to high concentrations of particulate matter in the water (Borgmann et al., 1980) [note also suggestion by Spehar et al. (1978) that low lead toxicity to some species may have been related to high turbidity during tests]. Brown (1976), also using static tests, showed that 100 μg/liter lead inhibited the growth of *Asellus meridianus* (isopod) and that animals from lead-polluted areas were more tolerant to lead than those from clean areas.

It is important to note that Borgmann et al. (1978), Spehar et al. (1978), and Wilson (1982) all observed no incipient level. Lead toxicity increased continuously with exposure time. Furthermore, Borgmann (1980) has observed that metal toxicities are additive or slightly synergistic in their effect on copepod production rates so that the presence of other metals at low concentrations may influence the response to lead.

5.3. Fish

Acute lethality of lead to fish might be expected at concentrations above 1000 μg/liter in Great Lakes waters (Wong et al., 1978b). The 48-hr LC_{50} for rainbow trout (*Salmo gairdneri*) in water similar to that of Lake Superior was 1000 μg/liter (Brown, 1968) whereas the 96-hr LC_{50} for rainbow trout in Lake Ontario water was 6500 μg/liter (Lam et al., 1976). At concentrations above 1000 μg/liter in Lake Ontario water, a visible precipitate of lead carbonate is observed. Toxicity of the remaining unprecipitated lead is believed to be a function of the amount of "free lead" (theoretically, lead ions plus weak complexes as measured by pulse polarography) rather than of the amount of lead complexed by anions, such as carbonate, hydroxyl, and phosphate (Davies et al., 1976). It is estimated that lead concentrations above 1000 μg/liter in Lake Ontario water diminish to less than 1000 μg/liter within 24 hr due to the carbonate complexation and lead precipitation noted earlier (Lam et al., 1976). High concentrations of lead titrate the inorganic complexing capacity. When the capacity is completely utilized, residual uncomplexed lead is available for toxicity. However, mixing of this water with fresh water provides more complexing capacity so that toxic ionic lead continues to be removed (Lam et al., 1976). Consequently, if efficient mixing occurs, it is unlikely that acutely lethal concentrations will be maintained long enough to kill fish in Lake Ontario. In Lakes Superior, Michigan, and Huron, the carbonate concentrations are much lower, and lead toxicity is greater in these soft waters. Therefore, lead would be maintained at concentrations greater than 1000 μg/liter for significant periods of time and lethality may be expected in these lakes. Carbonate concentrations in Lake Erie are close to those in Lake Ontario, and it is doubtful that acute lethality would occur. Lingering or chronic mortality of larval fish due to sustained exposure occurs at concentrations as low as 84 μg/liter (Sauter et al., 1976). It is possible that these lower concentrations could be maintained for significant periods of time in Great Lakes water because precipitates are not observed. Experimentally, lead availability to trout is directly proportional to waterborne lead concentrations up to 1000 μg/liter in filtered Lake Ontario water (Hodson et al., 1978b).

Significant sublethal effects of lead on fish include haemotological, neurological, teratogenic, growth, and histological responses that occur at concentrations as low as 13, 8, 119, 22–65, and >1000 μg/liter waterborne lead, respectively (Table 16.12). Other enzymatic and physiological changes have

Table 16.12. Sublethal Effects of Lead on Fish

Class	Specific Responses	Minimum Effective Lead Concentration (μg/liter)	References
Hematological	Inhibition of hemoglobin synthesis, red blood cell stippling, premature mortality of red blood cells (haemolytic anemia), compensatory erythropoiesis.	13	Christenson et al., 1977; D'Amelio et al., 1974; Dawson, 1935; Hodson, 1976; Hodson et al., 1977; 1978a; Johansson-Sjöbeck and Larsson, 1979; Péguignot et al., 1975; Srivastava and Mishra; 1979.
Neurological	Blackening of the caudal area, lordosis and scoliosis, neural degeneration in spinal cord, learning impairment.	8	Davies et al., 1976; Hodson et al., 1979; 1978a; 1980b; Weir and Hine, 1970; Hicks,[a] personal communication.
Teratogenesis	Lordosis and scoliosis.	119	Holcome et al., 1976; Ozoh, 1979.
Growth Impairment		22–65	Christensen, 1975; Hodson et al., 1980b; Sauter et al., 1976.
Histopathology	Heart, intestine, pyloric caeca, liver, kidney, gonads, chemoreceptors.	>1000	Crandall and Goodnight, 1962; 1963; Haider, 1964; 1975; Sastry and Gupta, 1978; Vijaymadhaven and Iwai, 1975.
Miscellaneous	Enzyme inhibition in liver, intestinal tract, kidneys, ovaries, and spleen, increased skin mucus fluidity; increased plasma sodium and chloride.	various, ≥10	Christensen et al., 1977; Johansson-Sjöbeck and Larsson, 1979; Narbonne et al., 1973; 1975; Sastry and Agrawal, 1979; Sastry and Gupta, 1978; Varanasi et al., 1975.

[a] B. Hicks, Department of Pathology, Ontario Veterinary College, University of Geulph, Guelph, Ontario, Canada.

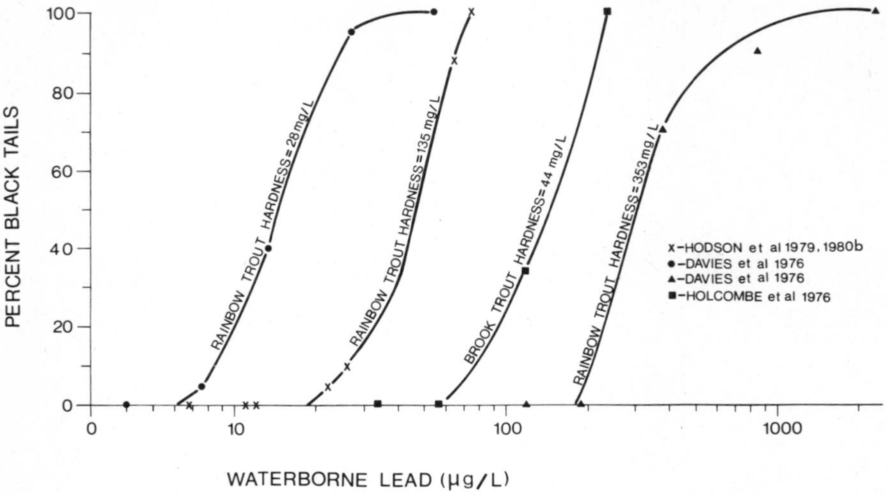

Figure 16.3 The effect of chronic exposure to waterborne lead on the incidence of black tails, a symptom of lead neurotoxicity, in rainbow trout and brook trout exposed to lead in waters of different hardness and alkalinity.

been observed at similar lead concentrations. All of these symptoms increase with increasing exposure time and waterborne lead concentration (e.g., Fig. 16.3).

Lead toxicity to fish may be affected by a variety of physical, chemical, and biological factors that change the availability of lead to fish, the ability of the fish to take up lead, and the response of fish to lead taken up (Hodson, 1979).

5.3.1. Source and Form of Lead

Inorganic lead is very poorly accumulated by fish from their diet and does not appear to contribute significantly to lead toxicity or to lead levels in fish (Hodson et al., 1978a). This supports the conclusion that lead does not bioconcentrate up food chains and corresponds to the observation in humans that only 10% of dietary lead is absorbed.

Chau and Wong (1983) summarized the toxicities of various alkyllead compounds to aquatic biota. In general, tetraalkyllead compounds were more toxic than trialkyllead compounds with tetraethyllead being the most toxic of all. This order of toxicity is the reverse of that seen with mammals (Grandjean and Nielsen, 1979) and may be caused by the rapid accumulation by aquatic biota of tetraalkyllead from water (Harrison, 1980; Hodson, 1979). Fish were the most sensitive species, and bacteria were most resistant to alkyllead toxicity.

A study of alkyllead toxicity to marine fish indicated that toxicity increased as the number of alkyl side chains increased (Maddock and Taylor,

1977). The 96-hr LC_{50}'s for tetramethyl and tetraethyllead were 50 and 230 μg/liter, or about 3600 and 700 times more toxic, respectively, than inorganic lead compounds. Since the exposures were in salt water, which tends to precipitate inorganic lead, the true ratios of toxicity may be considerably less. The 48-hr LC_{50}'s for larvae of the marine fish *Morone labrax* were 100 and 65 μg/liter for tetraethyl and tetramethyllead, respectively (Marchetti, 1978).

5.3.2. Water Quality

Increased water hardness (or alkalinity) reduced chronic lead toxicity to fish (Fig. 16.4) through carbonate complexation (described earlier) and through reduction of gill permeability by calcium (Davies et al., 1976; Hodson, 1979;

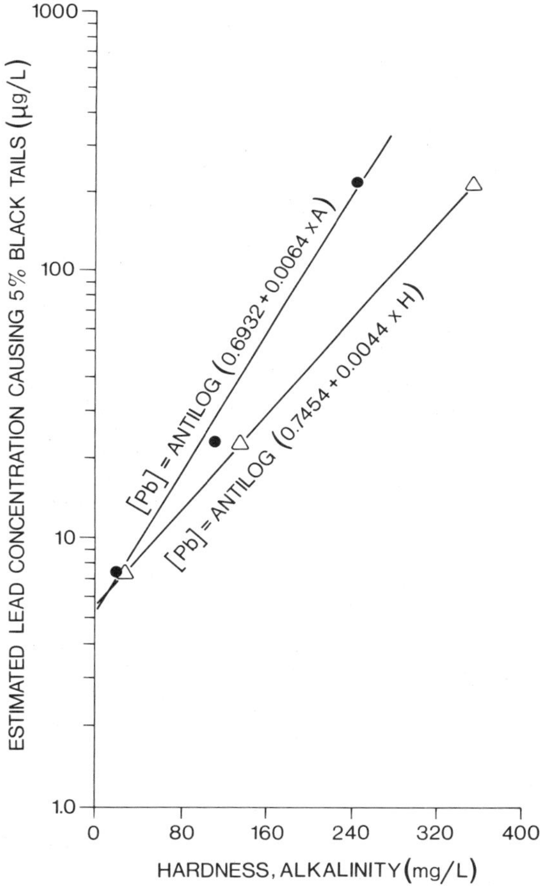

Figure 16.4 The effect of hardness (H) and alkalinity (A) on the waterborne lead concentration required to induce black tails in 5% of a population of rainbow trout chronically exposed to lead. The 5% concentration was estimated from Fig. 16.3.

Hodson et al., 1979; Varanasi and Gmur, 1978). The hardness effect may be described by relating the lead concentration (Pb, μg/liter) estimated to cause 5% black tails in rainbow trout (Fig. 16.3) to water hardness (H, mg/liter):

$$[\text{Pb}]_{5\%} = \text{antilog}_{10}(0.7454 + 0.004406 \times H) \qquad \text{(from Fig. 16.4).}$$

From this equation, predicted threshold lead concentrations for black tails in trout range from 9.0 μg/liter in Lake Superior to 22 μg/liter in Lake Ontario. Analyses based on alkalinity give similar results.

Acidity increases lead uptake by fish through interactions with lead solubility and gill permeability (Hodson et al., 1978b). Lead uptake, on a blood lead or whole body basis, can be related at pH's between 6.0 and 10.0 to a reference water pH by the equation (Hodson et al., 1978b):

$$\text{Relative blood lead} = 100 \times 2.1^{-\Delta\text{pH}}$$

Since toxicity is directly related to blood lead levels, both blood lead and toxicity should double for each decrease in pH by 1.0 unit. Since pH and alkalinity are directly related, toxicity can also be predicted from Fig. 16.4.

Organic particulates reduce lead uptake through complexation. An increase of particulate organic carbon from <0.2 mg/liter (laboratory filtered Lake Ontario water) to 1.8 mg/liter (raw Lake Ontario water) reduced blood lead of lead-exposed rainbow trout by about 1.8 times (Hodson, 1979). In late fall, when particulate organic carbon decreased to <0.2 mg/liter, there was no difference in lead uptake between fish exposed to lead in filtered or unfiltered water. Consequently, toxicity should vary with primary productivity, being highest in winter and least in summer. Increased temperature, however, may counteract these changes by increasing lead uptake, probably as a result of an enhanced metabolic and gill ventilation rate (Hodson, unpublished data).

5.3.3. Fish Species

Brook trout (*Salvelinus fontinalis*) appear more resistant to lead than do rainbow trout, based on studies by two different authors in two different water types (Davies et al., 1976; Holcombe et al., 1976) (see Fig. 16.3). Lead uptake by both species exposed simultaneously to lead under the same conditions demonstrated that rainbow trout accumulate lead at much lower waterborne lead concentrations than do brook trout (Hodson et al., 1977). Therefore, ability to prevent lead accumulation may impart resistance to lead toxicity to brook trout. Based on lead accumulation, pumpkinseed (*Lepomis gibbosus*) should be equally as sensitive to lead as brook trout (Hodson et al., 1977). Short-term studies (60 days) of percent hatch, mortality, growth, and survival of the eggs and larvae of a variety of species of fish demonstrated the following decreasing order of sensitivity: lake trout (*Salvelinus namaycush*) > rainbow trout = channel catfish (*Ictalurus punctatus*) = bluegill

(*Lepomis macrochirus*) > white sucker (*Catastomus commersoni*) (Sauter et al., 1976). Since the effects of lead are fully expressed only over the long term (>100 days or more from spawning), the order of sensitivity might change in longer experiments.

5.3.4. Fish Age and Size

Rainbow trout exposed to lead immediately after hatch develop symptoms faster and at lower concentrations than do fish exposed as fingerlings (Davies et al., 1976). The fish exposed from hatch seem to accumulate lead to a greater extent throughout their life than do those exposed at a later stage (Hodson et al., 1979). This effect is not simply one of size because size has been shown to have little influence on lead accumulation by fish and the chronic toxicity of lead (Hodson et al., 1982). Lead effects on the nervous systems of young rapidly developing fish may be greater than in mature individuals as is the case with mammals (Jaworski, 1979). Young fish growing rapidly develop neurotoxicity faster than do cohorts growing more slowly (Hodson et al., 1982). Therefore both stage of development and growth rate seem to be important determinants of toxicity.

5.3.5. Diet

An increase in dietary intake of calcium from 0 to 8.4 mg reduced uptake of waterborne lead by coho salmon, perhaps due to interactions with gill membrane permeability (Varanasi and Gmur, 1978); the result should be reduced toxicity.

5.4. Waterfowl

Bellrose (1959) wrote an extensive review of lead poisoning in waterfowl due to spent lead shot from hunting. His study showed that records of such lead poisoning dated back to the 19th century and have been associated with major flyways of migrating ducks where hunting is intense. Poisoning is characteristic of late fall and early winter (occasionally spring) and the most affected species are mallards and pintails, followed by geese, swans, and other duck species. Affected species are generally those that consume the bottom material of shallow ponds. The degree of poisoning is a function of the amount of shot on the bottom, bottom firmness, shot size, water depth, and ice cover. Shot tends to sink into soft bottoms and is covered over with annual detrital loads to the sediments. The percentages of wildfowl taken by hunters that had ingested lead in their gizzards were: Canada geese (>1); blue and snow geese (>3); buffleheads, green-winged teals, baldpates, and common goldeneyes (2–5); ruddy ducks, mallards, black ducks and pintails (5–10); and canvas backs, lesser scaups, redheads, and ring-necked ducks (>10). Of the gizzards containing shot, 64.7% contained one pellet, 14.9% two pellets, and 7.4% more than six pellets. Ingested lead produced weakness and fatigue in wild mallards that reduced their ability to migrate and increased their susceptibility to hunting. Mortality of mallards increased by 9%, 23%, 36%, and 50% due to ingestion of 1, 2, 4, and 6 pellets per bird, respectively. It is estimated that one quarter of the wild mallards of North America in any year ingest lead shot and that 4% of those in the Mississippi Flyway die due to lead poisoning. Another 1% are poisoned, but are shot by hunters before dying. For all waterfowl species in North America, 2–3% annual mortality is estimated as a result of lead poisoning.

In the Great Lakes, lead shot poisoning of wildfowl will be characteristic of any marshy area open to hunting, but will be especially important in major waterfowl habitats such as the marshes of Lake St. Clair, Pt. Pelee, and Long Point Bay (Lake Erie) (see section on lead in sediments).

5.5. Humans

Numerous extensive reviews exist on this subject (Boggess and Wixson, 1977; Grandjean and Neilson, 1979; Hilburn, 1979; Jaworski, 1979; National Academy of Science, 1972a; Nriagu, 1978) and the information is not repeated here. The current maximum permissible lead concentration in drinking waters is 50 μg/liter, both in Canada and the United States (Canada Department of National Health and Welfare, 1979; National Academy of Science, 1972b). A Canadian guideline for lead in edible marine products of 10 mg/kg was recently cancelled so there are no regulations in Canada or the United States governing lead levels in fresh fish as human food. England has a guideline for fish, fish paste, and canned fish of 2 mg/kg and for fried and salted fish of 5 mg/kg (Marshall,* personal communication). None of the lead concentrations measured in Great Lakes fish (see section on lead in fish) exceeded these guidelines.

6. SUMMARY AND RESEARCH NEEDS

This review has demonstrated that much is known of lead in the Great Lakes and its potential impacts on aquatic biota; this information is summarized by Fig. 16.5, a conceptual diagram of lead in a lake ecosystem. Overall, lead input to the Great Lakes is considerably above expected "natural" inputs due to human activities, and much of this input is inorganic lead from the atmosphere. However, due to complexation of inorganic lead, it appears that significant concentrations of biologically available "free" lead will be found only near point sources. This should restrict the negative impacts of waterborne lead to relatively small areas of the lakes. However, lead con-

* Keith Marshall, Canadian Wildlife Service, Environment Canada, Ottawa, Ontario, Canada.

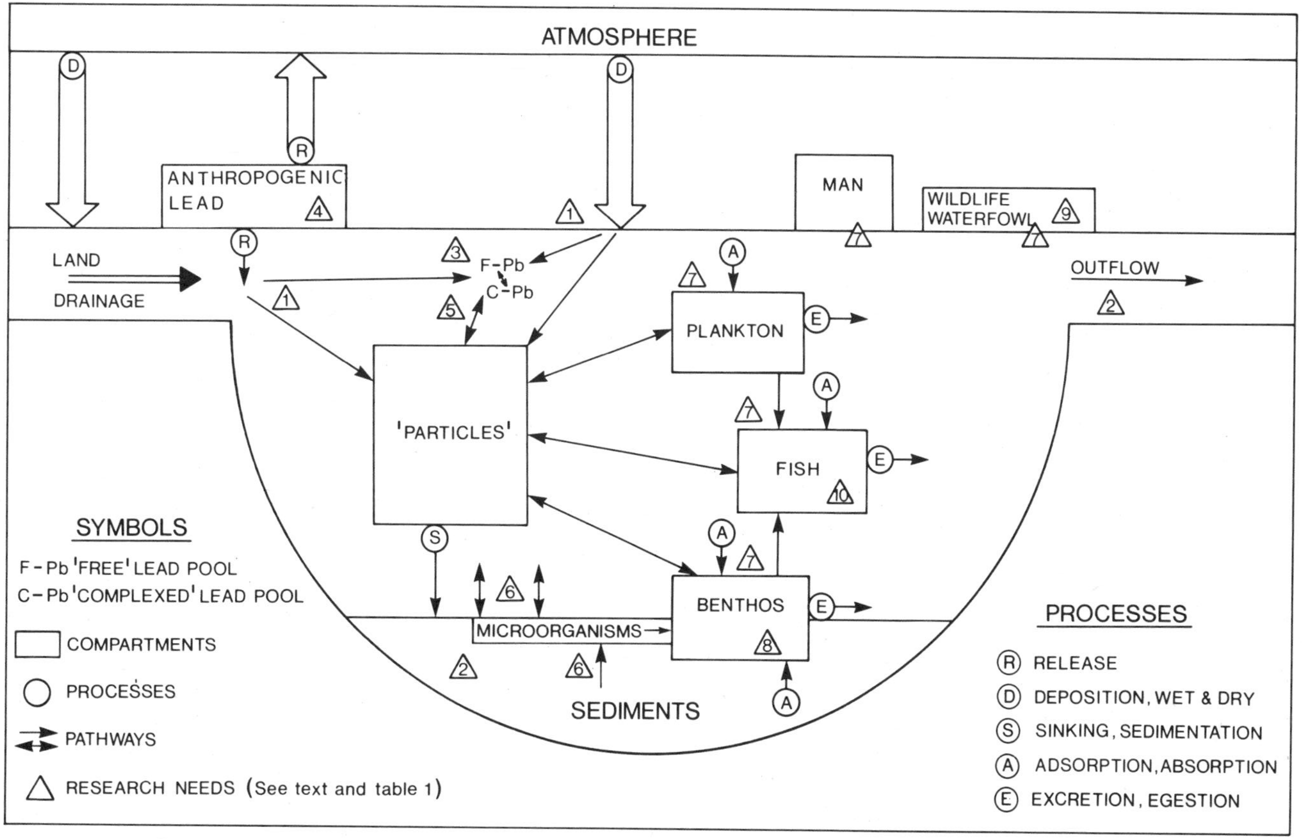

Figure 16.5 Symbolic overview of lead flux in lake ecosystems.

centrations in sediments are elevated over a wide area due to continual atmospheric loadings and the movement and settling of contaminated particulates from point sources. The observations of lead methylation by sediment bacteria, rapid uptake of alkyl lead compounds by aquatic biota, and the greater contamination of benthic-feeding fish relative to pelagic-feeding fish suggests that the impacts of lead may be somewhat more widespread than expected. A true understanding of impact will only arrive from research to address the questions outlined in Table 16.13 and Fig. 16.5.

Table 16.13. Research Needs in the Great Lakes. (Note: The numbers below refer to Fig. 16.5 and not to research priorities)

Quantification

1 Estimate, by source, loading of lead to each lake and identify contamination "hot spots" in relation to known point sources.
2 Estimate the rate of lead accumulation in the sediments of each lake and amount of sediment leaving each lake via resuspension.
3 Determine background lead concentrations in lake and river waters using "ultraclean" methods.
4 Cost strategies that will reduce the primary anthropogenic lead sources, gasoline additives, and production in the Great Lakes Basin.
Compute maximum permissible anthropogenic loadings based on the overall fate and effect of lead in each lake.

Transformation

5 Determine the exchange kinetics between different forms of lead (e.g., between "free" and particulate lead) in lake waters as a function of trophic status, season, hardness, organic content, etc.
6 Estimate the rates of lead alkylation in, and lead releases from, lake sediments and the contribution of lead from this source to biota and other ecosystem compartments.
7 Evaluate the availability of different forms of lead to the various classes of biota, and explore the use of "fingerprinting" techniques (e.g., isotope ratio variations) to show the contribution of different sources of lead to levels found in biota, particularly fish.

Effect

8 Determine the response of benthos to "free" lead in lake water, sediment-bound lead, and lead in interstitial waters, and the role of microbes in lead toxicity to benthos.
9 Examine the impact of alkyllead on aquatic biota and wildlife.
10 Measure the sensitivity of fish to lead as a component of liquid effluents released into the lake, and as a function of their food habits (e.g., planktivores vs. benthic detritivores).

ANNEX I—METHODOLOGY USED TO MEASURE LEAD CONTENT OF NET PLANKTON AND ZOOPLANKTON

The sampling design and, specifically, the station pattern, for plankton collections was derived from the program design of the Water Quality Board's Fish Contaminant Work Group (Water Quality Board, 1977c). All collections of net plankton (>153 μm) and zooplankton (*Mysis relicta*) were conducted during the day with collection nets of steel and aluminum frames and nylon netting. Samples were concentrated on nylon screens and placed in acid-washed and acetone-hexane-rinsed glassware. Samples were held at 4°C for 12 hr and then filtered on nylon screens. All samples were dried at 60°C for 24 hr, ground with a glass mortar and pestle, and subdivided into a minimum of five aliquots per sample. All processing equipment was washed and rinsed in distilled water, acetone, and hexane between each sample. All samples were stored in acid-washed and solvent-rinsed sealed glass containers until analysis. Sample analysis was by the method of Agemian et al. (1980).

ANNEX II—METHODOLOGY USED TO MEASURE LEAD CONTENT OF WHOLE FISH

The sampling design for the fish collections was that of the Water Quality Board's Fish Contaminant Work Group (Water Quality Board, 1977c). After collection, individual whole fish were wrapped in hexane-rinsed aluminum foil and stored at −25°C until homogenization. Individual fish were weighed, measured, and sexed and homogenized a minimum of five times with a commercial meat grinder. The whole fish tissue was further homogenized for five minutes with a large stainless steel food blender. All equipment and processing instruments were washed between samples and rinsed in distilled water, acetone, and hexane. A 50 g aliquot of the homogenate was stored in an acid-washed and acetone- and hexane-rinsed glass container at −25°C until analysis by the method of Agemian et al. (1980).

ACKNOWLEDGMENTS

The material in this review was originally prepared by the authors as the supporting rationale for a water quality objective. We would like to thank the Aquatic Ecosystem Objectives Committee of the Science Advisory Board, International Joint Commission, for their editorial comments. Jo-Anne Mann and Caryl Fawcett typed the manuscript.

REFERENCES

Agemian, H., D. P. Sturtevant, and K. D. Austen. (1980). Simultaneous acid extraction of six trace metals from fish tissues by hot-block digestion and determination by atomic-absorption spectrometry. *Analyst* **105:** 125–130.

Ahmad, I., Y. K. Chau, P. T. S. Wong, A. J. Carty, and L. Taylor. (1980). Chemical alkylation of lead (II) salts to tetraalkyllead (IV) in aqueous solution. *Nature* **287**(5784): 716–717.

Anderson, R. L., C. T. Walbridge, and J. T. Fiandt. (1980). Survival and growth of *Tanytarsus dissimilis* (Chironomidae) exposed to copper, cadmium, zinc, and lead. *Arch. Environ. Contam. Toxicol.* **9:** 329–335.

Bellrose, F. R. (1959). Lead poisoning as a mortality factor in waterfowl populations. *Ill. Nat. History Survey Bull.* **27**(3): 235.

Biesinger, K. E. and G. M. Christensen. (1972). Effects of various metals on survival, growth, reproduction, and metabolism of *Daphnia magna*. *J. Fish. Res. Board Can.* **29:** 1671–1700.

Bilinski, H. and W. Stumm. (1973). Pb(II)—Species in natural waters. *EAWAG News*, No. 1, Jan. 1973.

Birnie, S. E. and D. J. Hodges. (1981). Determination of ionic alkyl lead species in marine fauna. *Environ. Technol. Lett.* **2:** 433–442.

Boggess, W. R. and B. G. Wixson. (1977). *Lead in the environment.* National Science Foundation, p. 272. NSF/RA-770214.

Borgmann, U. (1980). Interactive effects of metals in mixtures and biomass production kinetics of freshwater copepods. *Can. J. Fish. Aquat. Sci.* **37:** 1295–1302.

Borgmann, U., O. Kramar, and C. Loveridge. (1978). Rates of mortality, growth, and biomass production of *Lymnaea palustris* during chronic exposure to lead. *J. Fish. Res. Board Can.* **35:** 1109–1115.

Borgmann, U., R. Cove, and C. Loveridge. (1980). Effects of metals on the biomass production kinetics of freshwater copepods. *Can. J. Fish Aquat. Sci.* **37:** 567–575.

Brown, B. E. (1976). Observations on the tolerance of the isopod *Asellus meridianus* Rac. to copper and lead. *Water Res.* **10:** 555–559.

Brown, J. R. and L. Y. Chow. (1975). The comparison of heavy metals in fish samples taken from Baie du Dore, Lake Huron and Toronto Harbor, Lake Ontario. *Internat. Conf. on Heavy Metals in the Environment*, Toronto, October, 1975.

Brown, V. M. (1968). "The Calculation of the Acute Toxicity of Mixtures of Poisons to Rainbow Trout." *Water Res.* **2:** 723–733.

Canada Department of Health and Welfare. (1978). Guidelines for Canadian Drinking Water Quality 1978. Supporting Documentation. Ottawa. Department of Supply and Services Catalogue No. H48-10/1978/IE.

Canada Department of Health and Welfare. (1979). Guidelines for Canadian Drinking Water Quality. Federal-Provincial Working Group on Drinking Water of the Federal-Provincial Advisory Committee on Environmental and Occupational Health, Ottawa.

Chau, Y. K. and Wong, P. T. S. (1976). Complexation of Metals in Natural Waters. In: R. W. Andrew, P. V. Hodson, and D. E. Konasewich, Eds., *Toxicity to biota of metal forms in natural water.* Proceedings of a workshop held in Duluth, Minnesota, October 7–8, 1975. International Joint Commission, Windsor, Ontario. pp. 187–196.

Chau, Y. K. and P. T. S. Wong. (1983). Determination of molecular and ionic alkyllead species in environmental samples. Presented at the 4th International Conference on the Organometallic and Coordination Chemistry of Germanium, Tin and Lead. McGill University, Montreal, Canada, August 8–12, 1983.

Chau, Y. K., Chawla, V. K., Nicholson, H. F., and Vollenweider, R. A. (1970). "Distribution of Trace Elements and Chlorophyll-*a* in Lake Ontario." Proc. 13th Conf. Great Lakes Res., 659–672.

Chau, Y. K., P. T. S. Wong, G. A. Bengert, and O. Kramar. (1979). Determination of tetraalkyllead compounds in water, sediment and fish samples. *Anal. Chem.* **51:** 186–188.

Chau, Y. K., P. T. S. Wong, O. Kramar, G. A. Bengert, R. D. Cruz, J. O. Kinrade, J. Lye, and J. C. Van Loon. (1980). Occurrence of tetraalkyllead compounds in the aquatic environment. *Bull. Environ. Contam. Toxicol.* **24:**265–269.

Christensen, G. M. (1975). Biochemical effects of methylmercury chloride, cadmium chloride, and lead nitrate on embryos and alevins of the brook trout, *Salvelinus fontinalis*. *Toxicol. Appl. Pharmacol.* **32:** 191–197.

Christensen, G., E. Hunt, and J. Fiandt. (1977). The effects of methylmercuric chloride, cadmium chloride and lead nitrate on six biochemical factors of the brook trout (*Salvelinus fontinalis*). *Toxicol. Appl. Pharmacol.* **42:** 523–530.

Colwill, D. M. and A. J. Hickman. (1973). The concentration of volatile and particulate lead compounds in the atmosphere. Measurement at four road sites. Transport and Road Research Laboratory. Dept. of Environment, Great Britain. TRRL Report LR545.

Crandall, C. A. and C. J. Goodnight. (1962). Effects of sublethal concentrations of several toxicants on growth of the common guppy, *Lebistes reticulatus*. *Limnol. Oceanogr.* **7:** 233–239.

Crandall, C. A., and C. J. Goodnight. (1963). The effects of sublethal concentrations of several toxicants on growth of the common guppy, *Lebistes reticulatus*. *Trans. Am. Microscop. Soc.*, **82:** 59–73.

D'Amelio, V., G. Russo, and D. Ferraro. (1974). The effect of heavy metals on protein synthesis in crustaceans and fish. *Rev. Intern. Oceanogr. Med.* **33:** 111–118.

Davies, P. H., J. P. Goettl, Jr., J. R. Sinley, and N. F. Smith. (1976). Acute and chronic toxicity of lead to rainbow trout, *Salmo gairdneri*, in hard and soft water. *Water Res.* **10:** 119–206.

Dawson, A. B. (1935). The hemopoietic response in the catfish, *Ameiurus nebulosus*. to chronic lead poisoning. *Biol. Bull.* **68:** 335–346.

DeJonghe, W. R. A. and F. C. Adams. (1980). Organic and inorganic lead concentrations in environmental air in Antwerp, Belgium. *Atmos. Environ.* **14:** 1177–1180.

De Jonghe, W. R. A., D. Chakraborti, and F. C. Adams. (1981). Identification and determination of individual tetraalkyllead species in air. *Environ. Sci. Technol.* **15:** 1217–1222.

Denny, P. and R. P. Welsh. (1979). Lead accumulation in plankton blooms from Ullswater, the English Lake District. *Environ. Pollut.* **18:**1–9.

Gire, N. P., J. F. Narbonne, and A. Serfaty. (1974). Intoxication par le nitrate de plomb chez la Carpe (*Cyprinus carpio*). Localisation du polluant dans l'organisme. *Europ. J. Toxicol.* **7:** 98–103.

Goettl, J. P., Jr., and P. H. Davies. (1979). Water pollution studies. Job Progress Report, Colorado Division of Wildlife, Federal Aid Project F-33-R-14. Fort Collins, Colorado.

Grandjean, P. and T. Neilson. (1979). Organolead compounds. Environmental health aspects. *Residue Rev.* **72:** 98–148.

Haider, G. (1964). Heavy metal toxicity to fish. I. Lead poisoning of rainbow trout (*Salmo gairdneri*) and its symptoms. *Zeitschrift für Angewandte Zool.* **51:** 347–368.

Haider, G. (1975). The effect of sublethal lead concentrations on the chemoreceptors of two freshwater fish species. *Hydrobiologia* **47:** 291–300.

Hannan, P. J. and C. Patouillet. (1972). In: M. Ruivo, Ed., *Marine pollution and sea life*. Fishing News Ltd., London.

Harrison, G. F. (1980). The Cavtat Incident. In: M. Branica and Z. Konrad, Eds., *Lead in the marine environment*. Pergamon Press, Oxford, pp. 305–317.

Hem, J. D. (1976). Geochemical controls on lead concentrations in stream water and sediments. *Geochim. Cosmochim. Acta* **40:** 599–610.

Hilburn, M. E. (1979). Environmental lead in perspective. *Chem. Soc. Rev.* **8:** 63–84.

Hodson, P. V. (1976). δ-Amino levulinic acid dehydratase activity of fish blood as an indicator of a harmful exposure to lead. *J. Fish. Res. Board Can.* **33:** 268–271.

Hodson, P. V. (1979). Factors affecting the sublethal toxicity of lead to fish. Presented at the International Conference on Management and Control of Heavy Metals in the Environment, London, England, Sept. 18–21, 1979. Proceedings, pp. 135–138.

Hodson, P. V., B. R. Blunt, D. J. Spry, and K. Austen. (1977). Evaluation of erythrocyte δ-amino levulinic acid dehydratase activity as a short-term indicator in fish of a harmful exposure to lead. *J. Fish. Res. Board Can.* **34:** 501–508.

Hodson, P. V., B. R. Blunt, and D. J. Spry. (1978a). Chronic toxicity of water-borne and dietary lead to rainbow trout (*Salmo gairdneri*) in Lake Ontario water. *Water Res.* **12:** 869–878.

Hodson, P. V., B. R. Blunt, and D. J. Spry. (1978b). pH-Induced changes in blood lead of lead-exposed rainbow trout (*Salmo gairdneri*). *J. Fish. Res. Board Can.* **35:**437–445.

Hodson, P. V., B. R. Blunt, D. Jensen, and S. Morgan. (1979). Effect of fish age on predicted and observed chronic toxicity of lead to rainbow trout in Lake Ontario water. *J. Great Lakes Res.* **5:** 84–89.

Hodson, P. V., B. R. Blunt, and D. M. Whittle. (1980a). Biochemical monitoring of fish blood as an indicator of biologically available lead. *Thalassia Jugoslav.* **16:** 389–396.

Hodson, P. V., J. W. Hilton, B. R. Blunt, and S. J. Slinger. (1980b). Effect of dietary ascorbic acid on chronic lead toxicity to young rainbow trout (*Salmo gairdneri*). *Can. J. Fish. Aquat. Sci.* **37:** 170–176.

Hodson, P. V., D. G. Dixon, D. J. Spry, D. M. Whittle, and J. B. Sprague. (1982). Effect of growth rate and size of fish on rate of intoxication by waterborne lead. *Can. J. Fish. Aquat. Sci.* **39:** 1243–1251.

Hodson, P. V., B. R. Blunt, and D. M. Whittle. (1983). Utility of a biochemical method for assessing exposure and response of feral fish to lead. Accepted for Publication in the 6th Symposium of Aquatic Toxicology, American Society for Testing and Materials, Philadelphia, Pennsylvania.

Holcombe, G. W., D. A. Benoit, E. N. Leonard, and J. M. McKim. (1976). Long-term effects of lead exposure on three generations of brook trout (*Salvelinus fontinalis*). *J. Fish. Res. Board Can.* **33:** 1731–1741.

Jarvie, A. W., R. N. Markall, and H. R. Potter. (1975). Chemical methylation of lead. *Nature* **255:** 217–218.

Jaworski, J. F. (1979). Effects of lead in the environment—1978. Quantitative aspects. National Research Council of Canada, Associate Committee on Scientific Criteria for Environmental Quality, Ottawa. Publication No. NRCC/CNRC 16736.

Johanssen-Sjöbeck, M.-L. and A. Larsson. (1979). Effect of inorganic lead on delta-aminolevulinic acid dehydratase activity and hematological variables in the rainbow trout (*Salmo gairdneri*). *Arch. Environ. Contam. Toxicol.* **8:** 419–431.

Keeney, W. L., W. G. Breck, G. W. Van Loon, and J. A. Pag. (1976). The determination of trace metals in *Cladophora glomerata* as a potential biological monitor. *Water Res.* **10:** 981–984.

Lam, D. C. L., C. K. Minns, P. V. Hodson, T. J. Simons, and P. T. S. Wong. (1976). Computer model for toxicant spills in Lake Ontario. In: J. O. Nriagu, Ed., *Environmental biogeochemistry*, Vol. 2. Ann Arbor Science Publishers, Ann Arbor, Michigan.

Leah, T. D. (1976). The production, use and distribution of lead in Canada. Environmental Contaminants Inventory Study No. 3, Report Series No. 41, Inland Waters Directorate, Department of the Environment, Ottawa.

Leland, H. V. and J. McNurney. (1974). Lead transport in a river ecosystem. International Conference on Transport of Persistent Chemicals in Aquatic Ecosystems, University of Ottawa, Ottawa, May 1–3, 1974, pp. III 17–III 23.

Loring, D. H. (1976). The distribution and partition of Zn, Cu, and Pb in the sediments of the Saguenay Fjord. *Can. J. Earth Sci.* **13:** 960–971.

Lucas, H. F., Jr., D. N. Edgington, and P. J. Colby. (1970). Concentrations of trace elements in Great Lakes Fishes. *J. Fish. Res. Board Can.* **27:** 677–684.

Maddock, B. G. and D. Taylor. (1977). The acute toxicity and bioaccumulation of some lead alkyl compounds in marine animals. Presented at the International Experts Meeting on Lead—Occurrence, Fate and Pollution in the Marine Environment, Rovinj, Yugoslavia, Oct. 18–22, 1977.

Malanchuk, J. L. and G. K. Gruendling. (1973). Toxicity of lead nitrate to algae. *Water Air Soil Pollut.* **2:** 181–190.

Marchetti, R. (1978). Acute toxicity of alkyl leads to some marine organisms. *Mar. Pollut. J.* **7:** 206.

Marshall, K. Personal Communication, Canadian Wildlife Service, Environment Canada, Ottawa, Ontario.

Merlini, M. and G. Pozzi. (1977a). Lead and freshwater fishes. Part I. Lead accumulation and water pH. *Environ. Res.* **12:** 167–172.

Merlini, M. and G. Pozzi. (1977b). Lead and freshwater fishes. Part II. Ionic lead accumulation. *Environ. Pollut.* **13:** 119–126.

Narbonne, J. F., J. C. Murat, and A. Serfaty. (1973). Intoxication by lead nitrate in the carp (*Cyprinus carpio*). Data on modifications of nucleoprotein glucidic metabolism. *Comptes Rendus Séances de l'Academie des Sciences, Soc. Biol.* **167**(3–4): 572–575.

Narbonne, J. F., M.-P. Gire, J. C. Murat, and A. Serfaty. (1975). Intoxication par le nitrate de plomb chez la Carpe (*Cyprinus carpio* L.). *Europ. J. Toxicol.* **8:** 159–164.

National Academy of Science. (1972a). *Biological effects of atmospheric pollutants. Lead. Airborne lead in perspective.* National Academy of Sciences, Washington, D.C., 330 pp.

National Academy of Science and National Academy of Engineering. (1972b). *Water Quality Criteria.* For Ecological Research Series, Environmental Protection Agency, Washington, D.C., EPA-R3-073-033.

Nissenbaum, A. and D. J. Swaine. (1976). Organic matter–metal interactions in recent sediments. The role of humic substances. *Geochim. Cosmochim. Acta* **40:** 809–816.

Nriagu, J. O. (1974). Lead orthophosphates. IV. Formation and stability in the environment. *Geochim. Cosmochim. Acta* **38:** 887–898.

Nriagu, J. O., Ed. (1978). *The biogeochemistry of lead in the environment.* Part A. Elsevier/North-Holland, New York, pp. 1–14 and 137–184.

Nriagu, J. O., A. L. W. Kemp., H. K. T. Wong, and N. Harper. (1979). Sedimentary record of heavy metal pollution in Lake Erie. *Geochim. Cosmochim. Acta* **43:** 247–258.

Nriagu, J. O., H. K. T. Wong and R. D. Coker. (1981) Particulate and dissolved trace metals in Lake Ontario. *Water Res.* **15:** 91–96.

Ozoh, P. T. E. (1979). Malformations and inhibitory tendencies induced to *Brachydanio rerio* (Hamilton-Buchanan) eggs and larvae due to exposure in low concentrations of lead and copper ions. *Bull. Environ. Contam. Toxicol.* **21:** 668–675.

Patterson, C. C. and D. M. Settle. (1976). Reduction of errors in lead analysis. *Proc. 7th IMR Symposium*, pp. 321–351.

Patterson, J. W. and P. Kodukula. (1978). Heavy metals in the Great Lakes. *Water Quality Bull.* **3:** 6–7.

Péguignot, J., M.-P. Gire, and A. Moga. (1975). Intoxication par la nitrate de plomb chez la

Carpe (*Cyprinus carpio* L.). Influence sur la structure l'histologique de la rate et l'hematocrite. *Europ. J. Toxicol.* **8:** 165–168.

Pollution from Land Use Activities Reference Group (1977). Annual progress report of the International Reference Group on Great Lakes Pollution from Land Use Activities. International Joint Commission, Windsor, Ontario.

Pollution from Land Use Activities Reference Group (1978). *Environmental Management Strategy for the Great Lakes System.* Final Report to the International Joint Commission, Windsor, Ontario.

Poldoski, J. E., E. M. Leonard, J. T. Fiandt, L. E. Anderson, G. F. Olson, and G. E. Glass. (1978). Factors in the determination of selected trace elements in nearshore U.S. waters of Lakes Superior and Huron. *J. Great Lakes Res.* **4:** 206–215.

Reichert, W. L., D. A. Federighi, and D. C. Malins. (1979). Uptake and metabolism of lead and cadmium in coho salmon (*Oncorhynchus kisutch*). *Comp. Biochem. Physiol.* **63C:** 229–234.

Rickard, D. T. and J. O. Nriagu. (1978). Aqueous environmental chemistry of lead. In: J. O. Nriagu, Ed., *The biogeochemistry of lead in the environment.* Part A. Elsevier/North-Holland, New York.

Röderer, G. (1976). Induction of giant, multinucleate cells with tetraethyl lead. *Naturwiss.* **63:** 248.

Röderer, G. (1980). On the toxic effects of tetraethyllead and its derivatives on the chrysophyte *Poteriochromonas malhamensis.* I. Tetraethyllead. *Environ. Res.* **23:** 371–384.

Röderer, G. (1981). On the toxic effects of tetraethyllead and its derivatives on the chrysophyte *Poteriochromonas malhamensis.* II. Triethyllead, diethyllead and inorganic lead. *Environ. Res.* **25:**361–371.

Ruthvens, T. A. and J. Cairns, Jr. (1973). Response of freshwater protozoan artificial communities to metals. *J. Protozool.* **20:** 127–135.

Sastry, K. V. and M. K. Agrawal. (1979). Effects of lead nitrate on the activities of a few enzymes in the kidney and ovary of *Heteropneustes fossilis. Bull. Environ. Contam. Toxicol.* **22:** 55–59.

Sastry, K. V. and P. K. Gupta. (1978). Histopathological and enzymological studies on the effects of chronic lead nitrate intoxication in the digestive system of a freshwater teleost *Channa punctatus. Environ. Res.* **17:** 472–479.

Sauter, S., K. S. Buxton, K. J. Macek, and S. T. Petrocelli. (1976). Effects of exposure to heavy metals on selected freshwater fish. U.S. Environmental Protection Agency, Duluth, Minn. EPA 600/3-76-088.

Schmidt, U. and F. Huber. (1976). Methylation of organolead and lead (II) compounds to $(CH_3)_4Pb$ by microorganisms. *Nature* **259:** 157.

Schnitzer, M. (1971). Metal–organic matter interactions in soils and waters. In: S. J. Faust and J. V. Hunter, Eds., *Organic compounds in aquatic environments.* Marcel-Dekker, New York. pp. 297–316.

Schulze, H. and J. J. Brand. (1978). Lead toxicity and phosphate deficiency in *Chlamydomonas. Plant Physiol.* **62:** 727–730.

Silverberg, B. A., P. T. S. Wong, and Y. K. Chau. (1977). Effect of tetramethyllead in freshwater green algae. *Arch. Environ. Toxicol. Contam.* **5:** 305–313.

Sirota, G. R. and J. F. Uthe. (1977). Determination of tetraalkyllead compounds in biological materials. *Anal. Chem.* **49:** 823–825.

Spehar, R. L., R. L. Anderson, and J. T. Fiandt. (1978). Toxicity and bioaccumulation of cadmium and lead in aquatic invertebrates. *Environ. Pollut.* **15:** 195–208.

Srivastava, A. K. and S. Mishra. (1979). Blood dyscrasia in a teleost, *Colisa fasciatus*, after acute exposure to sublethal concentrations of lead. *J. Fish. Biol.* **14:** 199–203.

Ter Haar, G. L. and M. A. Bayard. (1971). Composition of airborne lead particles. *Nature* **232:** 553–554.

Thomas, R. L. (1981). Sediments of the North American Great Lakes. *Verh. Internat. Verein. Limnol.* **21:** 1666–1680.

Upper Lakes Reference Group (1977). *The waters of Lake Huron and Lake Superior*, Vol. III, Parts A and B. Report to the International Joint Commission, Windsor, Ontario.

Uthe, J. F. and E. G. Bligh. (1971) Preliminary survey of heavy metal contamination of Canadian freshwater fish. *J. Fish. Res. Board Can.* **28:** 786–788.

Varanasi, U. and D. J. Gmur. (1978). Influence of water-borne and dietary calcium on uptake and retention of lead by coho salmon (*Oncorhynchus kisutch*). *Toxicol. Appl. Pharmacol.* **46:** 65–75.

Varanasi, Y., P. A. Robisch, and D. C. Malins. (1975). Structural alterations in fish epidermal mucus produced by water-borne lead and mercury. *Nature* **258:** 431–432.

Vijaymadhavan, K. T. and T. Iwai. (1975). Histochemical observations on the permeation of heavy metals into taste buds of goldfish. *Bull. Jap. Soc. Sci. Fish.* **41:** 631–639.

Waller, T. T. and G. F. Lee. (1979). Evaluation of observations of hazardous chemicals in Lake Ontario during the International Field Year for the Great Lakes. *Environ. Sci. Technol.* **13:** 79–85.

Water Quality Board. (1975). *Great Lakes Water Quality 1974. Appendix B. Surveillance Subcommittee Report.* International Joint Commission, Windsor, Ontario.

Water Quality Board. (1977a). *Great Lakes Water Quality 1976.* Appendix E. Status Report on Organic and Heavy Metal Contaminants in the Lakes Erie, Michigan, Huron and Superior Basins. D. Konasewich, W. Traversy, and H. Zar, Eds. International Joint Commission, Windsor, Ontario.

Water Quality Board. (1977b). *Great Lakes Water Quality. Status report on the persistent toxic pollutants in the Lake Ontario basin by the Implementation Committee of the Water Quality Board.* International Joint Commission, Windsor, Ontario.

Water Quality Board. (1977c). Great Lakes International Fish Contaminants Surveillance Program. International Joint Commission, Windsor, Ontario.

Weir, P. A. and C. H. Hine. (1970). Effects of various metals on behavior of conditioned goldfish. *Arch. Environ. Health* **20:** 45–51.

Wilson, J. B. (1982). Phototaxis impairment as a sensitive indicator of lead or cadmium stress in freshwater crustacean zooplankton. Ph.D. dissertation, Dept. of Zoology, University of Guelph, Guelph, Ontario.

Whitton, B. A. (1970). Toxicity of heavy metals to freshwater algae. A review. *Phykos* **9:** 116–125.

Wong, P. T. S., Y. K. Chau, and P. L. Luxon. (1975). Methylation of lead in the environment. *Nature* **253:** 263–264.

Wong, P. T. S., Y. K. Chau, and P. L. Luxon. (1978a). Toxicity of a mixture of metals on freshwater algae. *J. Fish. Res. Board Can.* **35:** 479–487.

Wong, P. T. S., B. A. Silverberg, Y. K. Chau, and P. V. Hodson. (1978b). Lead and the Aquatic Biota. In: J. O. Nriagu, Ed., *The biogeochemistry of lead in the environment.* Elsevier/North-Holland, New York, pp. 279–342.

Wong, P. T. S., Y. K. Chau, O. Kramar, and G. A. Bengert. (1981). Accumulation and depuration of tetramethyllead by rainbow trout. *Water Res.* **15:** 621–625.

17

SELENIUM CONTAMINATION OF THE GREAT LAKES AND ITS POTENTIAL EFFECTS ON AQUATIC BIOTA

Peter V. Hodson[1], D. Michael Whittle[1], Douglas J. Hallett[2]

[1] *Great Lakes Fisheries Research Branch*
Department of Fisheries and Oceans
Canada Centre for Inland Waters
867 Lakeshore Road, Burlington,
Ontario, L7R 4A6

[2] *Environmental Protection Service*
Environment Canada
55 St. Clair Ave. East
Toronto, Ontario M4T 1M2

1. INTRODUCTION

Selenium is a common element that occurs in the earth's crust at concentrations of approximately 0.7μg/g. It is present largely as heavy metal selenides (together with sulphide minerals) but can also occur as selenates and selenites. In soils, excluding seleniferous soils not normally found in the Great Lakes region, it has been found at levels ranging from 0.1 μg/g to less than 2 μg/g (Cooper et al., 1974). Elevated levels of selenium are found in some sedimentary rock formations and their derived soils in central areas of Canada and the United States. There are no known mining activities for selenium and its production is as a by-product of copper and lead refining.

Commercial use of selenium was about 500 metric tons in 1968, mostly in the elemental form as red crystals or grey powder. It is used in electronics for rectifiers, photocells, and xerography, in steel and in pigments for paints, glass, and ceramics (Cooper, 1967; Lymburner and Knoll, 1973).

Selenium is usually present in water as selenate and selenite; the elemental form is insoluble but may be carried in suspension. Weathering of rocks and soil erosion is a major source of selenium in water. On a worldwide basis, approximately 10,000 metric tons yearly is weathered and carried downstream to the sea. Of this, 140 tons is in solution, but only 16 tons remains dissolved in the sea; the remainder goes into sediments (Schroeder, 1974). The burning of fossil fuels is another source of soluble selenium. Analysis of coal, bottom ash, and fly ash from a single burner has turned up levels of 2, 3.4, and 41.3 μg/g, respectively (Lymburner and Knoll, 1973). The use of fossil fuel releases about 450 tons per year of selenium as SeO_2 into the atmosphere, about 4.5% of the amount eroded naturally (Schroeder, 1974).

Liquid wastes may be another source of selenium, although concentrations in effluents seem to be low. Sewage in California (both raw and treated) was found to have only 10–60 μg/liter of selenium, except for a high value of 280 μg/liter in an industrial area (Feldman, 1974). The importance of effluents should not, however, be judged solely on concentrations. Total loadings may determine the impact of selenium on an aquatic ecosystem (see section on field studies).

2. SELENIUM DISTRIBUTION IN AQUATIC ECOSYSTEMS

Selenium loadings to the Great Lakes have been estimated for Lakes Huron and Superior (Table 17.1), but these estimates are very unreliable due to errors in measurement of high flow rates and low selenium concentrations, and no data on other important sources such as atmospheric input. Copeland (1970) demonstrated that zooplankton of Lake Michigan downwind of Chicago were contaminated with selenium. He suggested that the source was the atmosphere due to the combustion of fossil fuels.

Waterborne selenium has a half-life in fresh water of 25–50 days and this

Table 17.1. Selenium Loadings to Lake Huron and Lake Superior (kg/day)[a]

Source	Lake Huron	Lake Superior
Municipal discharges	<0.001	0.013
Industrial discharges	2.09	NM[b]
Tributary inputs	145	184
Atmospheric	NM	NM
Shoreline erosion	—[c]	NM
Dredge spoil disposal	NM	NM

[a] ULRG, 1977, Vols. IIA, IIIA.

[b] Not measured.

[c] All samples less than detection limit of 1 mg/kg.

time may be a function of particulate density and proximity to sediments. Selenium was precipitated to the sediments in enclosures of lake water only when the water was in contact with sediments (Rudd et al., 1980). While contact with sediments appeared to be necessary for precipitation, sediment type did not seem to influence disappearance rates from the water column. Several investigators have suggested that selenium may be bioaccumulated up food chains (Rudd et al., 1980; Sandholm et al., 1973; Cumbie, 1978). Microbial transformations of selenium, particularly methylation to volatile compounds (Chau et al., 1976) or reduction to elemental selenium (Silverberg et al., 1976), may change the availability and toxicity of waterborne and sedimentary selenium to aquatic biota as well as affect the form stored in their tissues.

Selenium concentrations in water are usually low. The literature has been reviewed in several places (e.g., NAS, 1973), but many of the older estimates of concentrations are probably too high because of the limitations of analytical methods. Most uncontaminated surface waters have less than 5 μg/liter of selenium, and most drinking waters contain less than 10 μg/liter (APHA, 1971). For example, surface waters in a province of Germany averaged 4 μg/liter (Heide and Schubert, 1973). The normal concentration in seawater is only 0.4 μg/liter (Chau and Riley, 1965). Seepages from seleniferous soils contain less than 500 μg/liter and the majority is lost in ponds or lakes by co-precipitation with ferric hydroxide (APHA, 1971).

Waterborne selenium concentrations in the Great Lakes range from 0.001 to 5.0 μg/liter (Table 17.2). This wide range probably reflects variation in analytical capability rather than in real concentrations. Rain water contains much higher selenium concentrations than lake water and the concentrations reflect proximity to urban and industrial development (Traversy et al., 1975). Therefore, atmospheric loading of selenium to the Great Lakes could be significant.

Lake sediments seem to act as reservoirs or sinks. In the northern United

Table 17.2. Selenium Concentrations in Great Lakes Waters and Sediments

	Superior	Michigan	Huron—	Georgian Bay	—North Channel	St. Clair	Erie	Ontario	Reference
Rainwater or snow (μg/liter)	<0.1–0.2			0.1–0.4		0.5	0.2–0.8	0.10–0.75	1.
Water (μg/liter)	<0.1	0.083	<0.1–0.2[a]	<0.1	<1.0		<0.1 1–5 filtered 11–36 unfiltered	<0.1	1,2,3,4,5
Sediments—Lake and Harbors (mg/kg dry wt.)	0.63	<0.5[b]	0.90				0.16, 0.56, 0.79	1.00	1, 6

[a] Detection limit = 0.1 μg/liter.

[b] Detection limit.

[c] References: 1. Traversy et al., 1975; 2. Copeland and Ayers, 1972; 3. Warry, 1978; 4. ULRG 1977 vol IIA; 5. Adams and Johnson 1977; 6. Copeland, 1970.

States they contain from 1.0 to 3.5 μg/g dry weight of selenium, considerably more than the usual concentrations in soils (Weirsma and Lee, 1971). Small model ecosystem experiments showed that of the total amount of selenium in rain that fell on soil, 75% stayed in soil and 25% ran off into an aquatic system. Thirty-six percent of the selenium entering the aquatic system ended in the sediments whereas most of the remainder was in the biota (Huckabee and Blaylock, 1974). Great Lakes sediments contain about 0.1–1.0 μg/g dry weight and the concentrations are slightly higher in the lower lakes relative to Lake Superior.

Net plankton selenium concentrations vary both within and between lakes with no obvious trends that could be related to contamination (Table 17.3). Zooplankton, however, showed higher concentrations in Georgian bay compared with Lakes Ontario and Erie (Table 17.4), perhaps due to extensive mining and smelting of copper ores in the Georgian Bay watershed. Copeland (1970) speculated that elevated selenium levels (1–7 μg/g dry weight) in Lake Michigan zooplankton downwind from Chicago were due to fossil fuel combustion.

Concentrations of selenium in fish tissues from a wide range of locations in fresh and marine water vary from 0.16 to about 0.7 μg/g wet weight as follows: Canadian dressed fish from industrial and isolated locations, 0.17–0.38 μg/g (Uthe and Bligh, 1971); freshwater fish from New York, 0.2–0.5 μg/g (Pakkala et al., 1972); ocean and freshwater fish in Finland, 0.2–0.58

Table 17.3. The Selenium Content of Net Plankton (153μ mesh) Sampled from the Great Lakes in 1980[a]

Lake	Site	No. of Samples	Mean Concentration) (μg/g dry wt.)	Standard Deviation	Range
Ontario	Main Duck Is.	5	2.52	0.08	2.4–2.6
	Cobourg	5	2.74	0.09	2.6–2.8
	Port Credit	5	2.04	0.09	2.0–2.2
	Port Dalhousie	5	3.34	0.06	3.3–3.4
Erie	Long Pt. Bay	5	2.26	0.06	2.2–2.3
	Erieau	5	2.44	0.06	2.4–2.5
	Wheatley	5	2.84	0.06	2.8–2.9
	Pigeon Bay	5	1.90	0.00	1.9
	Amherstburg	5	0.93	0.02	0.90–0.95
Huron	Goderich	2	1.70	0.00	1.7
	S. Baymouth	5	0.80	0.03	0.76–0.84
	Cape Rich	2	2.10	0.00	2.1
	Burnt Is.	3	1.43	0.06	1.4–1.5
	French R.	2	1.60	0.00	1.6

[a] D. M. Whittle, unpublished data.

Table 17.4. The Selenium Content of Invertebrates Sampled from the Great Lakes

Lake	Site	No. of Samples	Mean Concentration (μg/g dry wt.)	Standard Deviation	Range
		Mysis relicta—1980[a]			
Ontario	Main Duck Is.	1	2.3	—	2.3
	Cobourg	3	2.77	0.15	2.6–2.9
	Port Credit	5	2.42	0.05	2.4–2.5
	Port Dalhousie	4	1.88	0.32	1.6–2.2
Huron	Goderich	5	3.82	0.05	3.8–3.9
	S. Baymouth	8	2.83	0.67	2.3–3.7
	Burnt Is.	5	3.24	0.09	3.1–3.3
	French R.	4	4.70	0.08	4.6–4.8
		Pontoporeia spp.—1980[a]			
Ontario	Main Duck Is.	5	2.14	0.06	2.1–2.2
Huron	S. Baymouth	5	3.88	0.88	3.8–4.0
		Zooplankton (500μ mesh)—1973[b]			
Erie	Western Basin	5	2.38	0.24	—
		Zooplankton (280μ mesh) 1969–1970[c]			
Michigan	Lakewide	50	0.60	—	0.2–1.7

[a] D. M. Whittle, unpublished data.

[b] Adams and Johnson, 1977.

[c] Copeland and Ayers, 1972.

μg/g (Sandholm et al., 1973); seafoods, about 0.32–0.56 μg/g (Morris and Levander, 1970); edible portion of trout, about 0.28–0.68 μg/g (Arthur, 1972).

In the Great Lakes, concentrations of seleniun in fish from the North Channel of Lake Huron, Georgian Bay, Lake Erie, and Lake Ontario ranged from 0.56 to 1.59, 0.60 to 1.55, 0.15 to 1.51, and 0.10 to 0.57 μg/g, respectively (Table 17.5). Variations in selenium concentrations between fish species exceed the variations within species (Table 17.5); standard deviations generally are less than 13% of the mean of any sample (Adams and Johnson, 1977). Selenium concentrations within any fish species decrease from Georgian Bay/North Channel to Lake Erie to Lake Ontario (Table 17.5). Although sample sizes are small in some cases, variation between years and between authors is remarkably low. Therefore, the trend toward higher selenium concentrations in biota from Georgian Bay relative to other lakes is probably real and may again reflect the influence of mining and smelting activities in the French River drainage basin (Warry, 1978). Fish appear to be good indicators of selenium contamination of water. Native centrarchids and sal-

Table 17.5. **The Selenium Content of Fish from the Great Lakes**

Species	Lake	Year Sampled	Mean Concentration (μg/g wet wt.)	No. of Samples	Range	Author
Catfish (?)	Ontario	1973	0.10	2	0.06–0.14	Beal, 1974
	Erie	1973	0.15	3	0.12–0.17	Beal, 1974
Walleye (*Stizostedion vitreum*)	Ontario	1973	0.25	2	0.14–0.35	Beal, 1974
	Erie	1973	0.31	12	0.24–0.36	Beal, 1974
	Erie-West	1973/1974	0.52	7	SD = 0.52[a]	Adams and Johnson, 1977
		1980	0.37	30	0.27–0.54	Whittle, unpublished data
	North Channel	1973	0.56	1	—	Beal, 1974
	Georgian Bay	1973	0.60	6	0.42–0.79	Beal, 1974
		1980	0.74	50	0.57–1.40	Whittle, unpublished data
Smelt (*Osmerus mordax*)	Ontario	1973	0.32	3	0.26–0.38	Beal, 1974
		1980	0.33	68[b]	0.26–0.79	Whittle, unpublished data
	Erie	1973	0.34	8	0.15–0.45	Beal, 1974
		1980	0.31	35[b]	0.23–0.37	Whittle, unpublished data
	Georgian Bay-South	1979	0.64	12[b]	0.59–0.73	Whittle, unpublished data
	Georgian Bay-North	1980	0.77	12[b]	0.60–0.88	Whittle, unpublished data
Yellow perch (*Perca flavescans*)	Ontario	1973	0.34	6	0.26–0.38	Beal, 1974
	Ontario	1977	0.40	51	0.27–0.74	Whittle, unpublished data
	Erie-West	1973/1974	0.74	79	SD = 0.05	Adams and Johnson, 1977
	Erie	1977	0.46	73	0.26–0.72	Whittle, unpublished data
	North Channel	1973	0.63	3	0.59–0.67	Beal, 1974
	Georgian Bay	1973	0.94	2	0.77–1.11	Beal, 1974
	Lake Huron	1974	0.60	7	SD = 0.04	Adams and Johnson, 1977

Table 17.5. (*Continued*)

Species	Lake	Year Sampled	Mean Concentration (μg/g wet wt.)	No. of Samples	Range	Author
Whitefish (*Coregonus* spp.)	Ontario	1973	0.22	1	—	Beal, 1974
	North Channel	1973	1.55	3	0.87–2.00	Beal, 1974
	Georgian Bay	1973	1.00	3	0.89–1.07	Beal, 1974
Sheepshead (*Aplodinotus grunniens*)	Ontario	1973	0.26	3	0.19–0.35	Beal, 1974
	Erie	1973	0.45	4	0.36–0.50	Beal, 1974
	Erie-West	1973/1974	1.51	13	SD = 0.19	Adams and Johnson, 1977
Rock Bass (*Ambloplites rupestris*)	Ontario	1973	0.38	4	0.35–0.40	Beal, 1974
	Erie	1973	0.25	2	0.10–0.39	Beal, 1974
Pike (*Esox lucius*)	Ontario	1973	0.30	3	0.23–0.39	Beal, 1974
	Georgian Bay	1973	0.78	2	0.51–1.04	Beal, 1974
Coho Salmon (*Oncorhynchus kisutch*)	Ontario	1980	0.43	25	0.32–0.51	Whittle, unpublished data
	Erie	1973	0.44	2	0.42–0.46	Beal, 1974
		1980	0.50	23	0.32–0.80	Whittle, unpublished data
Carp (*Cyprinus carpio*)	Ontario	1973	0.34	4	0.17–0.52	Beal, 1974
	Erie-West	1973/1974	0.82	6	SD = 0.13	Adams and Johnson, 1977

Lake Trout (*Salvelinus namaycush*)	Ontario	1980	0.44	176	0.33–0.66	Whittle, unpublished data
	Huron	1980	0.81	47	0.61–0.95	Whittle, unpublished data
	Superior	1980	0.38	50	0.27–0.61	Whittle, unpublished data
Rainbow Trout (*Salmo gairdneri*)	Ontario	1980	0.57	15	0.43–0.71	Whittle, unpublished data
	Erie	1980	0.65	10	0.41–0.96	Whittle, unpublished data
Slimy sculpin (*Cottus cognatus*)	Superior	1972	0.59	25	SD = 0.07	Korda et al., 1977
Four horn sculpin (*Myoxocephalus quadricornis*)	Superior	1972	0.50	25	SD = 0.07	Korda et al., 1977
Chub (?)	Ontario	1973	0.52	1	—	Beal, 1974
	North Channel	1973	0.59	1	—	Beal, 1974
	Georgian Bay	1973	0.73	2	0.62–0.85	Beal, 1974
Splake[c]	Georgian Bay-South	1979	0.70	47	0.42–1.20	Whittle, unpublished data
		1980	0.75	50	0.42–0.91	Whittle, unpublished data

[a] SD = standard deviation.

[b] Five fish composites.

[c] *Salvelinus fontinalis* × *Salvelinus namaycush*.

Table 17.6. The Selenium Content of Herring Gull Tissue Sampled from Great Lakes Gull Colonies in 1977[a]

Lake	Site	Tissue	No. of Samples	Mean Value (μg/g wet wt.)	Standard Deviation
Erie	Middle Island	Egg	2	1.0	—
	Port Colbourne	Egg	2	1.3	—
Huron	Chantry Island	Egg	2	<0.4	—
	Double Island	Egg	2	<0.4	—
Michigan	Little Sister Is.	Egg	2	1.3	—
	Hat Island	Egg	2	0.5	—
Superior	Mamainse Island	Egg	2	1.0	—
	Granite Island	Egg	2	<0.4	—
Ontario	Muggs Island	Egg	8	0.56	0.15
	Snake Island	Egg	9	0.72	0.14
	Kingston	Adult liver	17	0.311	0.11
		Adult feather	17	2.60	1.22

[a] Source: D. J. Hallett, unpublished data.

monids from selenium-contaminated Belews Lake (Cumbie, 1978; Cumbie and Van Horn, 1978) and Western U.S. lakes (Kaiser et al., 1979) contained elevated concentrations relative to fish from low selenium lakes.

A relationship between selenium levels and fish size has not been conclusively demonstrated. Adams and Johnson (1977) found a relationship of selenium concentration to weight in yellow perch, the only species with sufficient sample numbers for a comparison. An examination of Adam's thesis (Adams, 1976) indicates that the weight relationship was based on two distinct pools of fish of different weights. Within each pool there was no weight effect. Whittle (unpublished data) could not find any relationship between selenium and length using the large sample sizes of smelt, walleye, rainbow trout, lake trout, splake, or coho reported in Table 17.5.

The selenium content of herring gull tissues sampled from Great Lakes colonies shows a considerable geographic variability (Table 17.6). Local contamination is a possible cause of these high levels but the accumulation of selenium from sources other than the Great Lakes during migrations cannot be discounted. The observed range of concentrations encompasses those seen in other biota and their significance to the health of gulls is unknown.

3. METABOLISM OF SELENIUM BY AQUATIC BIOTA

In an experimental system, Sandholm et al. (1973) found that *Scendesmus dimorphus* could actively concentrate selenomethionine but showed no active or passive uptake of inorganic selenium. *Daphnia pulex*, however, could absorb selenium from selenite. Waterborne selenium was taken up rapidly by *Daphnia pulex* with an equilibrium observed within 24 hr (Schultz et al., 1980). Copeland (1970) reported that concentrations of selenium from Lake Michigan zooplankton were highest downwind of industrialized areas, although this was not reflected in the sediments, where concentrations were uniformly less than 0.5 μg/g. Concentrations in zooplankton however, increased from 1 μg/g in uncontaminated areas to 7 μg/g in contaminated waters.

Rainbow trout achieve equilibrium with waterborne selenium in about 30 days (Gissel-Nielsen and Gissel-Nielsen, 1978). Uptake rates by trout are a function of the exposure concentration since the relative uptake at low concentrations is greater than at high concentrations [i.e., concentration factors (tissue/water) are greater at low concentrations] (Hodson et al., 1980). Since this is true for eggs, sac fry, and fry, and is independent of dietary loading, it is suggested that gill membrane permeability limits selenium uptake (Hodson and Hilton, 1982).

Dietary selenium is taken up rapidly but no plateau is evident within four days. The degree of uptake was again inversely proportional to the dietary loading (Hodson and Hilton, 1982). Selenium taken up from the water or diet is found in all tissues, but the highest concentrations are in the liver, kidney and intestines (Hodson et al., 1980; Hodson and Hilton, 1982). For fish exposed to waterborne selenium, gill concentrations are also high. Muscle selenium concentrations are generally lower and equivalent to whole body concentrations.

Daphnia and rainbow trout also excrete selenium. Waterborne selenium is depurated by trout at a fixed logarithmic rate with a half-life of 29 days (Gissel-Nielsen and Gissel-Nielsen, 1978), indicating a passive excretion model relying on simple diffusion kinetics. Dietary selenium is excreted more actively with half-lives inversely proportional to dietary loading (Hodson and Hilton, 1981). It is possible that inorganic selenium taken up from water is transferred from gills to tissues and stored as inorganic selenium, whereas that taken up from the diet is transformed by the liver to an organic form that is both more toxic and more easily excreted. Within tissues of *Daphnia pulex*, selenium is associated with low molecular weight (64%) and protein (25%) components, while lesser amounts are associated with nucleic acids and lipids (10 and 0.1%, respectively) (Schultz et al., 1980). Autoradiography indicated highest concentrations of ^{75}Se in cytoplasm and these results correspond to those observed in mammals. Therefore, selenium may be excreted in a fashion similar to that in humans. A normal human intake of 0.06–0.15 mg/day is balanced by an output of 0.03 mg in feces, 0.05 mg in urine, and 0.08 in sweat, air, and hair (Schroeder et al., 1970).

Selenium is known to be methylated biologically. Chau et al. (1976) demonstrated methylation of sodium selenite, sodium selenate, selenocystine, selenourea and seleno-DL-methionine by microbial action in lake sediments. All sediments that demonstrated microbial action were capable of methylating selenite and/or selenate. Three compounds, mono- and dimethyl selenide, and an unknown were produced. Because bacterial action may have produced an unknown selenium compound of high toxicity to fish (Niimi and LaHam, 1976), the environmental significance of selenium methylation should be assessed.

4. SELENIUM AS A NUTRIENT FOR FISH AND MAMMALS

Fish fed diets of 0.07 mg/kg showed signs of incipient selenium deficiency that included reduced growth rates and low levels of serum glutathione peroxidase activity relative to fish on diets containing 0.35 mg/kg or higher. However, acute symptoms of deficiency, such as muscle pathology, were not evident (Hilton et al., 1980). Similar symptoms, plus elevated mortality rates and muscle pathology, were observed in Atlantic salmon (*Salmo salar*) fed diets deficient in both selenium and vitamin E (Poston et al., 1976). Vitamin E deficiencies were somewhat alleviated by selenium supplementation, but a diet with adequate vitamin E and only 0.04 μg Se/g dry weight was deficient.

Deficiency of selenium in the soil and in grass eaten by livestock leads to "white muscle disease." Dietary needs of livestock are in the vicinity of 0.1–0.2 mg/day (NAS, 1973), whereas the daily selenium requirement of humans has not been accurately determined. It would appear to be in the range of 0.1–0.2 mg/day (Levander, 1972), an amount normally found in an adequate diet (NAS, 1973).

5. ACUTE TOXICITY OF SELENIUM TO AQUATIC BIOTA

Bowen (1966) described selenium as moderately toxic to plants (toxic effects at concentrations between 1 and 100 mg/liter in the nutrient solution). Apparently this applies to freshwater algae as well. The concentrations of selenite causing 95% growth inhibition of *Anabaena variabilis* and *Anacystis nidulans* were 20 and 70 mg/liter, respectively (Kumar and Prakash, 1971). Selanate produced the same results with these species at 30 and 50 mg/liter, respectively. Kumar (1964) showed that growth of *Anacystis nidulans* was also completely inhibited by 20 mg/liter of selenate. However, a culture of this alga at increasing concentrations of selanate over several generations, produced a tolerant strain that could grow in 250 mg/liter of selenate. *Scenedesmus* sp., however, was more sensitive', 2.5 mg/liter was lethal (Bringmann and Kuhn, 1959).

The acute lethality of selenium to invertebrates has been shown for daphnids and *Hyalella azteca*. The 96-hr LC_{50}'s for *Daphnia magna* in hard water (329 mg/liter as $CaCO_3$) and *Daphnia pulex* in soft water (standard test medium) were 0.43 and 0.50 mg/liter, respectively (Halter et al., 1980; Schultz et al., 1980). Lethality to *D. pulex* was a function of age with 96-hr LC_{50}'s being 0.126 mg/liter for fed juveniles and 0.50 mg/liter for fed adults. Feeding reduced toxicity; the 96-hr LC_{50} for unfed *D. pulex* was 0.07 mg/liter (Schulz et al., 1980). The 96-hr LC_{50} for an amphipod *Hyalella azteca* was 0.34 mg/liter, whereas the 336-hr (14 day) LC_{50} was 0.07 mg/liter (Halter et al., 1980).

Niimi and LeHam (1975, 1976) published the most comprehensive reports to date on the toxicity of selenium to fish. Acute studies (Niimi and LaHam, 1976) indicated that lethality of selenium to zebrafish larvae (*Brachydanio rerio*) varied with the selenium salt used. The 96-hr and 10-day LC50's (Table 17.7) showed that selenate salts are less toxic than selenite salts. These salts are the most common forms normally occurring in freshwaters. The selenides, selenomethionine and selenocystine, were also shown to be toxic. Selenocystine was about as toxic as the selenates and selenomethionine was more toxic. Reliable LC_{50}'s for selenides could not be calculated, however, due to a loss of compounds from the solution because of biological action. This action was also a problem in early experiments with inorganic compounds. It was noticed that bacterial slimes in test containers could produce a highly toxic, unidentified organic selenium compound. Daily cleaning alleviated the problem but it suggested that hazardous transformations of inorganic to organic selenium compounds might occur in aquatic systems.

Studies on the toxicity of selenium dioxide to zebrafish embryos showed that they were quite resistant compared to larvae; concentrations up to 10 mg/liter had no effect on hatching (Niimi and LaHam, 1975). This was due probably to the extreme low permeability of the egg membrane. Larvae, by comparison, were quite sensitive and high mortality was observed at concentrations as low as 3 mg/liter after 10 days. No effect was observed at 1 mg/liter.

The acute toxicity to fish of selenium varies with the species tested and the duration of exposure (Table 17.8). Carp was the most resistant species tested and fathead minnow was the least. Within a species (rainbow trout), fish exposed to selenium for 96 hr in soft water (30–36 mg/liter as $CaCO_3$; Goettl and Davies, 1978) were less sensitive than those exposed in a hard water (135 mg/liter as $CaCO_3$; Hodson et al., 1980). The difference in toxicity

Table 17.7. Acute Toxicity of Selenium Salts to Zebrafish Larvae[a]

Selenium Salt	96-hr LC_{50} (mg/liter)	10-day LC_{50} (mg/liter)
Selenium dioxide	20	5
Sodium selenite	23	4
Potassium selenite	15	≃2
Sodium selenate	82	40
Potassium selenate	81	50

[a] Niimi and LaHam, 1976.

Table 17.8. The Acute Lethal Toxicity of Selenium to Several Species of Freshwater Fish

Species	Hours of exposure	LC_{50} (mg/liter)	Author
Carp (*Cyprinus carpio*)	96	35	Sato et al., 1980
Zebrafish larvae (*Brachydanio rerio*)	96	23	Niimi and LaHam, 1976
Goldfish (*Carassius auratus*)	120	10	Ellis et al., 1937; Weir and
	168	12	Hine, 1970
Rainbow trout (*Salmo*	96	8–13	
gairdneri)	216	6.5	Goettl et al., 1976; Hodson
	384	5.0	et al., 1980
Fathead minnows (*Pimephales promelas*)	96	0.6–1.0	Halter et al., 1980

was less than a factor of 2 (Table 17.8) so that the cause may have been experimental error rather than the increased hardness. However, the most sensitive fish, fathead minnows (Table 17.8), were tested in a very hard water (329 mg/liter as $CaCO_3$; Halter et al., 1980).

Injected sodium selenite was lethal to channel catfish (*Ictalurus punctatus*) within 48 hr at a dose of 3 mg Se/kg (Ellis et al., 1937). At lower concentrations, toxicity was not evident for 4–10 days after which there was mortality due to liver, spleen, and kidney damage associated with an apparent loss of osmoregulation (edema) and abnormal erythropoiesis. These effects were caused by a single dose of 0.9 mg Se/kg or daily injections of 0.04 mg/kg (total Se dose = 0.2 mg/kg).

Exposure to selenium has also been shown to reduce the acute toxicity of inorganic mercury to fish but, paradoxically, mercury accumulation increases in survivors (Heisinger et al., 1979).

6. CHRONIC AND SUBLETHAL TOXICITY OF SELENIUM TO AQUATIC BIOTA

Prolonged exposures of fathead minnow eggs and fry to selenium concentrations of 1.0 mg/liter or higher reduced times to hatch, but had no effect on percent hatched (Halter et al., 1980). Survival times were reduced relative to controls at all selenium exposure levels, but even controls exhibited some mortality. There is a possibility of other lethal factors interacting with selenium in this study.

Chronic exposures of rainbow trout to 130 μg/liter of waterborne selenium caused elevated mortality rates and incidence of deformity relative to controls and the next lowest concentration tested (60 μg/liter) (Goettl et al., 1976). Exposure of trout to selenium for 44 weeks resulted in subtle hematological responses at 28 μg/liter or higher (Hodson et al., 1980). These results suggest that waterborne concentrations up to 70 times background levels (<0.4 μg/liter) should not have direct adverse effects on fish. Using a conditioned avoidance response as an index, Weir and Hine (1970) discovered that 250 μg/liter could significantly affect learning behavior as compared to controls. A concentration of 150 μg/liter had no meaningful effect.

Dietary selenium may be more toxic to fish than waterborne selenium. Prolonged feeding of rainbow trout with diets containing 13 mg/kg of selenium caused liver pathology, elevated mortality rates, decreased growth efficiency, and decreased growth rates (Hilton et al., 1980); there were no obvious toxic effects at the next lower dietary concentration (3.7 mg/kg). Research by Goettl and Davies (1978) has also shown effects of dietary selenium on trout growth and mortality rates. Fifty percent mortality occurred at 10 mg/kg (dry weight) over a one-year trial. In comparison, symptoms of selenium poisoning of mammals occur at 4–5 mg/kg (dry weight) (Oldfield et al., 1974).

7. FIELD STUDIES OF SELENIUM TOXICITY TO AQUATIC BIOTA

The high toxicity of dietary selenium has also been suggested by field studies of fish mortality in a reservoir (Belews Lake, North Carolina) receiving effluent containing high concentrations of selenium from a fly-ash settling pond (Cumbie and Van Horn, 1978). Fish populations were severely reduced with evidence of decreased standing stock and a total lack of reproduction in the years following the start of operation of a coal-fired power plant. Studies of conditions in the lake showed that pesticide concentrations, water levels, temperatures, population structure, impingement and entrainment, diseases and parasites could not account for the loss of fish, especially since the loss was not evident in upstream waters or remote parts of the same reservoir. Analysis of the elemental composition of fish tissues in affected and unaffected areas showed that, of 16 elements measured, only selenium was correlated to the condition of the populations.

Selenium concentrations of upstream fish were in the range of 0.5–7 mg/kg (wet weight) whereas those from the affected main lake were consistently higher, with concentrations of 10–50 mg/kg. Selenium in ovaries of ripe females from the affected area was 1- to 3-fold higher than in muscle tissue and this was most pronounced in various sunfish (*Lepomis* spp), among the most affected species. Selenium concentrations in plankton were 4–20 mg/kg (dry weight) upstream in contrast to 40–100 mg/kg in the affected area (Cumbie, 1978). Waterborne concentrations averaged 150–200 μg/liter in the effluent and 5–10 μg/liter in the lake although one peak of 20 μg/liter was recorded. The majority of this selenium in both effluents and lake water passed through a 0.45 μm filter and hence was available for sorption by biota and sediments (see discussion in Section 2). In the sediments, selenium concentrations were 6–8 mg/kg (dry weight) in contrast to 3.4 mg/kg at the control site. However, these values resulted from the mixing during sampling of a thin surface layer of contaminated sediment with underlying uncontaminated sediments. The actual concentrations in surficial sediments at contaminated sites were greater than 20 mg/kg (dry weight) compared to 1–5 mg/kg at control sites (Cumbie, personal communication).

Further studies of stocked bluegill (*Lepomis macrochirus*) showed that fish released in clean areas survived indefinitely, either in cages or in the lake. Fish held in cages in the contaminated area, however, gradually died over a 3–4 month period, while fish released directly to the area died almost immediately. The stomach contents of caged fish showed few benthic organisms whereas those of dead fish outside the cage contained a high proportion of benthic organisms, which corresponds to this species feeding habits (Scott and Crossman, 1973). Dying fish exhibited symptoms typical of acute selenium toxicity (Ellis et al., 1937); that is, the peritoneal cavity was distended with ascites and the fish had "popeye" (Cumbie, personal communication). Since the waterborne selenium concentration by itself was insufficient to cause mortality, the death of fish with symptoms characteristic of acute selenium toxicity indicates that dietary selenium was very high. It is highly probable that selenium was taken up from the sediments during foraging, either from ingestion of sediments directly or from ingestion of contaminated benthos. Benthos from the contaminated area contained 20–70 μg Se/g dry weight while those from the control area contained 4–8 μg Se/g dry weight (Cumbie, personal communication).

Fish mortality in a Colorado reservoir was also caused by selenium from bottom deposits which had passed through the food chain to accumulate to levels of 300 μg/g Barnhart (1958). In a less contaminated aquatic ecosystem, animals were shown to have higher residues than plants, but there was no pattern of continuing accumulation. Also, fish from pond culture, where the artificial food was low in selenium, contained less selenium than those from a natural system (Sandholm et al., 1973).

A study of ash-pit effluents at a Wisconsin power plant showed elevated selenium concentrations in the water of a creek receiving this effluent (Magnuson et al., 1980). However, other metals (e.g., Cr, Fe, Zn) were also elevated so that observed effluent effects on crayfish (change in respiration rate) could not be attributed solely to selenium. Crayfish caged in the ash-pit drain accumulated about 30 mg/kg dry weight of selenium in the hepatopancreas and about 0.4 mg/kg in the muscle.

8. TOXICITY OF SELENIUM TO MAMMALS

Selenium poisoning of livestock has been divided into two classes: the acute type, termed blind staggers, and the chronic, called alkali disease. Acute poisoning is associated with ingestion of highly seleniferous plants containing

1000 μg/g or more of selenium, whereas the chronic type is associated with grains and plants which contain 5–20 μg/g of selenium (Moxon, 1958). The extensive literature on natural poisoning of livestock from selenium in their food agrees, in general, that 5 μg/g or more can cause death in the herbivore, and that such levels in plants result from soil concentrations in the range 0.5–6 μg/g (McKee and Wolf, 1963; NAS, 1973; U.S. Dept. of Interior, 1968). Also, a diet containing 3 μg/g of selenium as selenite killed rats in a lifetime study (Schroeder, 1967). The usual chronic effects in mammals may include weakness; visual impairment; paralysis; damage to heart, liver, and viscera; stiff joints, and loss of hair and hooves. Additional symptoms in humans are marked pallor; red tainting of fingers, teeth, and hair; dental caries; debility; depression and irritation of nose and throat. Acute toxicity in humans may be characterized by nervousness, vomiting, cough, dyspnea, convulsions, abdominal pain, diarrhea, hypotension, and respiratory failure (NAS, 1973; Schroeder, 1974). Selenium effects on human health are reviewed in detail in a number of places including ''Drinking Water and Health'' (U.S. NRC, 1977).

The carcinogenic potential of selenium has been widely investigated (Schroeder, 1974). Recent critical evaluations made of these early studies concluded that insufficient high-quality data exist to assess the carcinogenicity of selenium compounds (Palmer and Olson, 1974; WHO, 1976). No suggestion that selenium is carcinogenic in man can be found in the available data (WHO, 1976).

Antagonism between the toxicity of selenium and other metals has been found. Levander (1972) reviewed the action of arsenic in counteracting selenium toxicity. Several cases in which cadmium poisoning is decreased by selenium were listed in Pakkala et al. (1972) and in ''The Selenium Paradox'' (Anon., 1972). The action against mercury toxicity was mentioned by Koeman et al. (1973). There are other aspects such as the interrelationship with vitamin E and possible teratogenic effects (Anon., 1972).

Toxicity due to selenium in drinking water is not common, probably because concentrations in water are generally low, and cases of toxicity to livestock are usually related to intake with food. However, a level of 9000 μg/liter in well water resulted in human poisoning in 3 months (Beath, 1962). It is not possible to assess from this data the hazards to aquatic mammals of selenium in the Great Lakes.

9. SUMMARY AND RESEARCH NEEDS

Selenium concentrations in the Great Lakes are obviously elevated in some areas due to anthropogenic inputs. Given the potential adverse effects on fish of selenium accumulation in benthic food chains, it is important to better document existing sources, loading rates, and sites of contamination. In particular, the degree of contamination of industrial harbors and Georgian

Bay needs to be described. The cause of selenium accumulation by benthos and the role of bacterial transformations in the accumulation and its subsequent toxicity to bottom-feeding fish needs more research. This would aid in understanding the factors that will foster or inhibit effects on fish, particularly sediment conditions that favor contamination of benthos.

ACKNOWLEDGMENTS

This review is an amalgamation of two rationales written in support of recommendations for water quality objectives for selenium in the Great Lakes. These rationales were reviewed by members of the Water Quality Objectives Subcommittee of the Water Quality Board, and the Aquatic Ecosystem Objectives Committee of the Science Advisory Board of the International Joint Commission. The authors wish to thank the members of both committes for their constructive criticisms. The manuscript was typed by Caryl Fawcett.

REFERENCES

Adams, W. J. (1976). The toxicity and residue dynamics of selenium in fish and aquatic invertebrates. Ph.D. dissertation, Dept. of Fisheries and Wildlife, Michigan State University.

Adams, W. J. and H. E. Johnson. (1977). Survey of selenium content in the aquatic biota of western Lake Erie. *J. Great Lakes Res.* **3:** 10–14.

Anonymous. (1972). The selenium paradox. *Food Cosmet. Toxicol.* **10:** 867–874.

APHA. (1971). Standard methods for the examination of water and wastewater. 13th ed. Prepared by the American Public Health Assoc., American Water Works Assoc., and Water Pollution Control Fed. Washington, D.C.

Arthur, D. (1972). Selenium content of Canadian foods. *Can. Inst. Food Sci. Technol. J.* **5:** 165–169.

Barnhart, R. A. (1958). Chemical factors affecting the survival of game fish in a western Colorado reservoir. Colorado Co-operative Fisheries Research Unit, Quarterly Report *4:* 25–28. Abstract No. 1673, p. 114.

Beal, A. R. (1974). A study of selenium levels in freshwater fishes of Canada's central region. Environment Canada, Fisheries and Marine Service, Tech. Rept. Ser. No. CEN/T-74-6.

Beath, O. A. (1962). Selenium poisons Indians. *Sci. News Lett.* **81:** 254.

Bowen, H. J. M. (1966). *Trace elements in biochemistry*. Academic Press, London.

Bringmann, G. and R. Kuhn. (1959). The toxic effects of waste water on aquatic bacteria, algae, and small crustaceans. *Gesundheits-Ing.* **80:** 115–120.

Chau, Y. K. and J. P. Riley. (1965). The determination of selenium in sea water, silicates and marine organisms. *Anal. Chim. Acta* **33:** 36–49.

Chau, Y. K., P. T. S. Wong, B. A. Silverberg, P. L. Luxon, and G. A. Bengert. (1976). Methylation of selenium in the aquatic environment. *Science* **192:** 1130–1131.

Cooper, W. C. (1967). Selenium toxicity in man. In: O. H. Muth, J. E. Oldfield, and P. H. Weswig, Eds., *Selenium in biomedicine*. Avi Publ. Co., Westport, Connecticut.

Cooper, W. C., K. G. Bennett, and F. B. Croxton. (1974). The history, occurrence, and properties of selenium. In: R. A. Zingar and W. C. Cooper, Eds., *Selenium.* Van Nostrand Reinhold, New York.

Copeland, R. (1970). Selenium: the unknown pollutant. *Limnos* **3:** 7–9.

Copeland, R. A. and J. C. Ayers. (1972). Trace element distributions in water, sediment, phytoplankton, zooplankton and benthos of Lake Michigan: A baseline study with calculations of concentration factors and buildup of radioisotopes in the food web. Special Report No. 1, Environmental Research Group, Inc., Chicago, Illinois.

Cumbie, P. M. (1978). Belews Lake environmental study report: Selenium and arsenic accumulation. Duke Power Company, Charlotte, N.C., Technical Report Series No. 78-04.

Cumbie, P. M. and S. L. Van Horn. (1978). Selenium accumulation associated with fish mortality and reproductive failure. Proc. Ann. Conf. S.E. Assoc., Fish & Wildl. Agencies **32:** 612–624.

Cumbie, P. M. Personal Communication. Duke Power Company, Charlotte, North Carolina.

Ellis, M. M., H. L. Motley, M. D. Ellis and R. O. Jones. (1937). Selenium poisoning in fishes. *Proc. Soc. Exptl. Biol. Med.* **36:** 519–522.

Feldman, M. G. (1974). Trace materials in wastes disposed to coastal waters: fates, mechanisms, ecological guidance and control: 107. National Technical Information Service Report No. PB 202346.

Gissel-Nielsen, G. and M. Gissel-Nielsen. (1978). Ecological effects of selenium application to field crops. *Ambio* **2**(4)**:** 114–117.

Goettl, J. P., Jr., and P. H. Davies. 1978. Water pollution studies. Job Progress Report, Federal Aid Project F-33-R-13, Colorado Division of Wildlife, Fort Collins, Colorado.

Goettl, J. P., Jr., P. H. Davies, and J. R. Singley. (1976). Laboratory studies: Water pollution studies, selenium, p. 74. In: O. B. Cope, Ed., *Colorado fisheries review* 1972–1974, Colorado Division of Wildlife Publication DOW-R-R-F72-F75, Fort Collins, Colorado.

Halter, M. T., W. J. Adams, and H. E. Johnson. (1980). Selenium toxicity to *Daphnia magna, Hyallela azteca*, and the fathead minnow in hard water. *Bull. Environ. Contam. Toxicol.* **24**(1)**:** 102–108.

Heide, F. and P. Schubert. (1960). The content of selenium in Saale water. *Naturwissenschaften* **47:** 176–177.

Heisinger, J. F., C. D. Hansen, and J. H. Kine. (1979). Effect of selenium dioxide on the accumulation and acute toxicity of mercuric chloride in goldfish. *Arch. Environ. Contam. Toxicol.* **8:** 279–283.

Hilton, J. W., P. V. Hodson, and S. J. Slinger. (1980). The requirement and toxicity of selenium in rainbow trout (*Salmo gairdneri*). *J. Nutr.* **110**(12)**:** 2527–2535.

Hodson, P. V. and J. W. Hilton. (1982). The nutritional requirements and toxicity to fish of dietary and waterborne selenium. Proceedings of the 5th International Symposium on Environmental Biogeochemistry, Stockholm, June 1–5, 1981. *Ecol. Bull.* **35:** 335–340.

Hodson, P. V., D. J. Spry, and B. R. Blunt. (1980). Effects on rainbow trout (*Salmo gairdneri*) of a chronic exposure to waterborne selenium. *Can. J. Fish. Aquat. Sci.* **37**(2)**:** 233–240.

Huckabee, J. W. and B. G. Blaylock. (1974). Microcosm studies on the transfer of mercury, cadmium and selenium from terrestrial to aquatic ecosystems. In Proc. 8th Ann. Conf. on Trace Substances in Environmental Health, 1974, pp. 219–222. Abstracted in NSF-Rann Trace Contaminant Abstracts 283: 112.

Kaiser, I. I., P. A. Young, and J. D. Johnson. (1979). Chronic exposure of trout to waters with naturally high selenium levels: Effects on transfer methylation. *J. Fish. Res. Board Can.* **36:** 689–694.

Koeman, J. H., W. H. M. Peeters, C. H. M. Koudstaal-Hol, P. S. Tjioe, and J. J. M. de Goeij. (1973). Mercury-selenium correlation in marine mammals. *Nature* **342** (5425)**:** 385–386.

Korda, R. J., T. E. Henzler, P. A. Helmke, M. M. Jimenez, L. A. Haskin, and E. M. Larsen. (1977). Trace elements in samples of fish, sediment and taconite from Lake Superior. *J. Great Lakes Res.* **3**(1–2): 148–154.

Kumar, H. D. (1964). Adaption of a blue-green alga to sodium selenate and chloramphenicol. *Plant Cell. Physiol.* **5:** 465–472.

Kumar, H. D. and G. Prakash. (1971). Toxicity of selenium to the blue-green algae, *Anacystis nidulans* and *Anabaena variabilis. Ann Bot.* **35:** 697–705.

Levander, O. A. (1972). Metabolic interrelationships and adaptations in selenium toxicity. *Ann. N.Y. Acad. Sci.* **192:** 181–192.

Lymburner, D. B. and H. Knoll. 1973. Selenium: Its production and use in Canada, a background paper. (Unpublished). Canada Centre for Inland Waters, Burlington, Ontario.

Magnuson, J. M., A. M. Forbes, D. M. Harrell, and J. D. Schwarzmeier. (1980). Responses of stream invertebrates to an ash-pit effluent. Wisconsin Power Plant Impact Study. U.S. EPA, Duluth, Minnesota. EPA-600/3-80-081.

McKee, J. E. and H. W. Wolf. (1963). Water quality criteria. 2nd ed. California State Water Quality Control Board Publ. 3A. Sacramento, California, USA.

Morris, V. C. and O. A. Levander, (1970). Selenium content of foods. *J. Nutr.* **100:** 1383–1388.

Moxon, A. L. (1958). *Trace elements.* Academic Press, New York.

National Academy of Sciences. 1973. Water quality criteria 1972. National Academy of Sciences and National Academy of Engineering. U.S. Environmental Protection Agency Ecol. Res. Series R3-73-033.

National Academy of Sciences. (1977). Drinking Water and Health. National Academy of Sciences, Washington, D.C.

Niimi, A. J. and Q. N. LaHam. (1975). Selenium toxicity on the early life stages of zebrafish, *Brachydanio rerio. J. Fish. Res. Board. Can.* **32:** 803–806.

Niimi, A. J. and Q. N. LaHam. (1976). Relative toxicity of organic and inorganic compounds to newly hatched zebrafish (*Brachydanio rerio*). *Can. J. Zool.* **54:** 501–509.

Oldfield, J. E., W. H. Allaway, H. A. Laitinin, H. W. Lakin, and O. H. Muth. (1974). Selenium. In: *Geochemistry and the environment.* Vol. 1. The relation of selected trace elements to health and disease. National Academy of Science, Washington, D.C., pp. 57–63.

Pakkala, I. S., W. H. Gutenmann, D. J. Lisk, G. E. Burdick, and E. J. Harris. (1972). A survey of the selenium content of fish from 49 New York State waters. *Pest. Monit. J.* **6:** 107–114.

Palmer, I. S. and O. E. Olson. (1974). Relative toxicities of selenite and selenate in the drinking water of rats. *J. Nutr.* **104:** 306–314.

Poston, H. A., G. G. Combs, Jr., and L. Leibovitz. (1976). Vitamin E and selenium interrelations in the diet of Atlantic salmon (*Salmo salar*): Gross, histological and biochemical deficiency signs. *J. Nutr.* **106:** 892–904.

Rudd, J. W. M., M. A. Turner, B. E. Townsend, A. Swick, and A. Furutani. (1980). Dynamics of selenium in mercury-contaminated experimental freshwater ecosystems. *Can. J. Fish. Aquat. Sci.* **37**(5): 848–857.

Sandholm, M., H. E. Oksanen, and L. Pesonen. (1973). Uptake of selenium by aquatic organisms. *Limnol. Oceanogr.* **18:** 496–498.

Sato, T., Y. Ose, and T. Sakai. (1980). Toxicological effects of selenium on fish. *Environ. Pollut.* (Series A) **21:** 217–224.

Schroeder, H. A. (1967). Effects of selenate, selenite, and tellurite on the growth and early survival of mice and rats. *J. Nutr.* **92:** 334–338.

Schroeder, H. A. (1974). *The poisons around us: Toxic metals in food, air, and water.* Indiana Univ. Press, Bloomington, Indiana.

Schroeder, H. A., D. V. Frost, and J. J. Balassa. (1970). Essential trace elements in man: Selenium. *J. Chronic Disease* **23:** 227–243.

Schultz, T. W., S. R. Freeman, and J. N. Dumont. (1980). Uptake, depuration, and distribution of selenium in *Daphnia* and its effects on survival and ultrastructure. *Arch. Environ. Contam. Toxicol.* **9:** 23–40.

Scott, W. B. and E. J. Crossman. (1973). Freshwater fishes of Canada. Bulletin 184. Fisheries Research Board of Canada, Ottawa, 966 pp.

Silverberg, B. A., P. T. S. Wong, and Y. K. Chau. (1976). Localization of selenium in bacterial cells using TEM and energy dispersive X-ray analysis. *Arch. Microbiol.* **107:** 1–6.

Traversy, W. J., P. I. Goulden, Y. M. Sheikh, and J. R. Leacock. (1975). Levels of arsenic and selenium in the Great Lakes region. Environment Canada, Inland Waters Directorate, Burlington, Ontario. Scientific Series No. 58, 18 pp.

Upper Lakes Reference Group. (1977). The waters of Lake Huron and Lake Superior. Vols. IIA, IIB, IIIA, IIIB. Report by the Upper Lakes Reference Group to the International Joint Commission, IJC Regional Office, Windsor, Ontario.

U.S. National Research Council. (1977). Drinking water and health. United States National Research Council, Washington, D.C.

U.S. Dept. of the Interior. (1968). Water Quality Criteria: Report to the Secretary of the Interior. National Technical Advisory Committee on Water Quality. U. S. Dept. of the Interior, Washington, D.C.

Uthe, J. F. and E. G. Bligh. (1971). Preliminary survey of heavy metal contamination of Canadian freshwater fish. *J. Fish. Res. Board Can.* **28:** 786–788.

Warry, N. D. (1978). Chemical Limnology of the North Channel, 1974. Fisheries and Environment Canada, Inland Waters Directorate, Water Quality Branch, Burlington, Ont. Scientific Series No. 92.

Weir, P. A. and C. H. Hine. (1970). Effects of various metals on behavior of conditioned goldfish. *Arch. Environ. Health* **20:** 45–51.

Weirsma, J. H. and G. F. Lee. (1971). Selenium in lake sediments—analytical procedure and preliminary results. *Environ. Sci. Technol.* **5:** 1203–1206.

World Health Organization. (1976). Monograph of the Evaluation of Carcinogenic Risk of Chemicals to Man. World Health Organization. International Agency for Research on Cancer. Geneva, Switzerland. 9: 260.

18

INORGANIC CONTAMINANTS IN LAKE MICHIGAN SEDIMENTS

Richard A. Cahill and Neil F. Shimp

Illinois State Geological Survey
Natural Resources Building
615 East Peabody
Champaign, Illinois 61820

1. INTRODUCTION

Lake Michigan, the third largest of the five Great Lakes, has a surface area of 58,000 km^2, a mean depth of 85 m, and a maximum depth of 280 m (Fig. 18.1). It receives most of its water as direct precipitation and as runoff from a drainage area of 176,000 km^2. The lake is divided into two parts by the midlake high. The Southern Basin slopes rapidly and smoothly to the west to a maximum depth of 163 m; the midlake high may be as shallow as 20 m. The northern part of the lake bottom has the deepest basin, is irregular with small ridges and valleys, and contains a great deal of exposed bedrock. The surface circulation is primarily counterclockwise in both the Northern and Southern Basins, and bottom currents are complex and variable. Southern basin surface flows are known to be subject to wind direction.

Sedimentation rates are low because this is a sediment-starved lake; they range from 0.1 to 4 mm/yr (Lineback and Gross, 1972; Robbins and Edgington, 1975). Tributaries to the lake are relatively small, and only eight discharge more than 1000 ft^3/s (Robbins et al., 1972; Fitchko and Huchinson, 1975).

Lake Michigan has the smallest outflow of the Great Lakes (1560 m^3/s) occurring at the Straights of Mackinac and at the Chicago River. The average flow-through time, estimated at 100 years, is long, so Lake Michigan bottom sediments in the deeper basins should reflect the history of man's activities in the surrounding watershed.

Since 1969, the Illinois State Geological Survey (ISGS) has been conducting investigations of the bottom sediments in Lake Michigan. This chapter is an account of these investigations which, for convenience, have been divided into two parts: those conducted before 1975 and those conducted from 1975 to the present. Because this chapter is primarily a review of ISGS investigations, the numerous other investigations of the Great Lakes are not all cited.

2. PRE-1975 INVESTIGATIONS OF LAKE MICHIGAN SEDIMENTS

2.1. Sampling

Our initial studies were done in cooperation with H. V. Leland at the University of Illinois. The studies were conducted on the ship RV *Inland Seas*, which was operated by the University of Michigan Great Lakes Research Division under National Science Foundation sponsorship. In the early phase of our work, sampling was restricted to southern Lake Michigan and was accomplished by several means: the Shipek grab sampler for surficial sediments; the Benthos coring device for cores up to 2 m long; and for cores up to 18 m long, a 550-kg Alpine piston coring device was used for stratigraphic correlations (Gross et al., 1970).

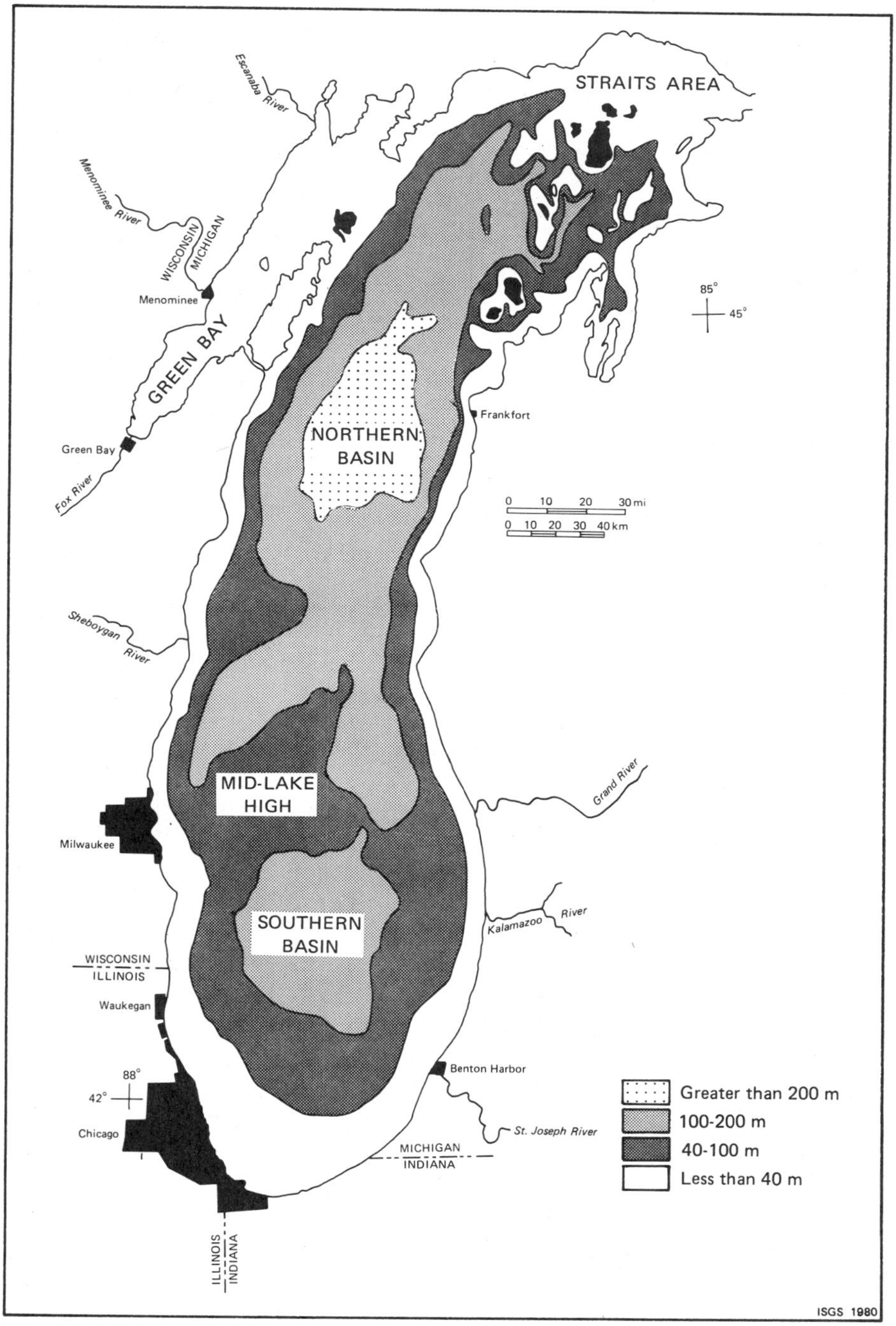

Figure 18.1 Generalized water depth of Lake Michigan.

2.2. Vertical Distribution of Trace Elements in Southern Lake Michigan Sediments

Analysis of the first few core sections showed striking concentrations of trace elements in the surficial sediments; these elements included Pb, Br, Zn, Cr, and organic carbon. Results for the first three cores were published

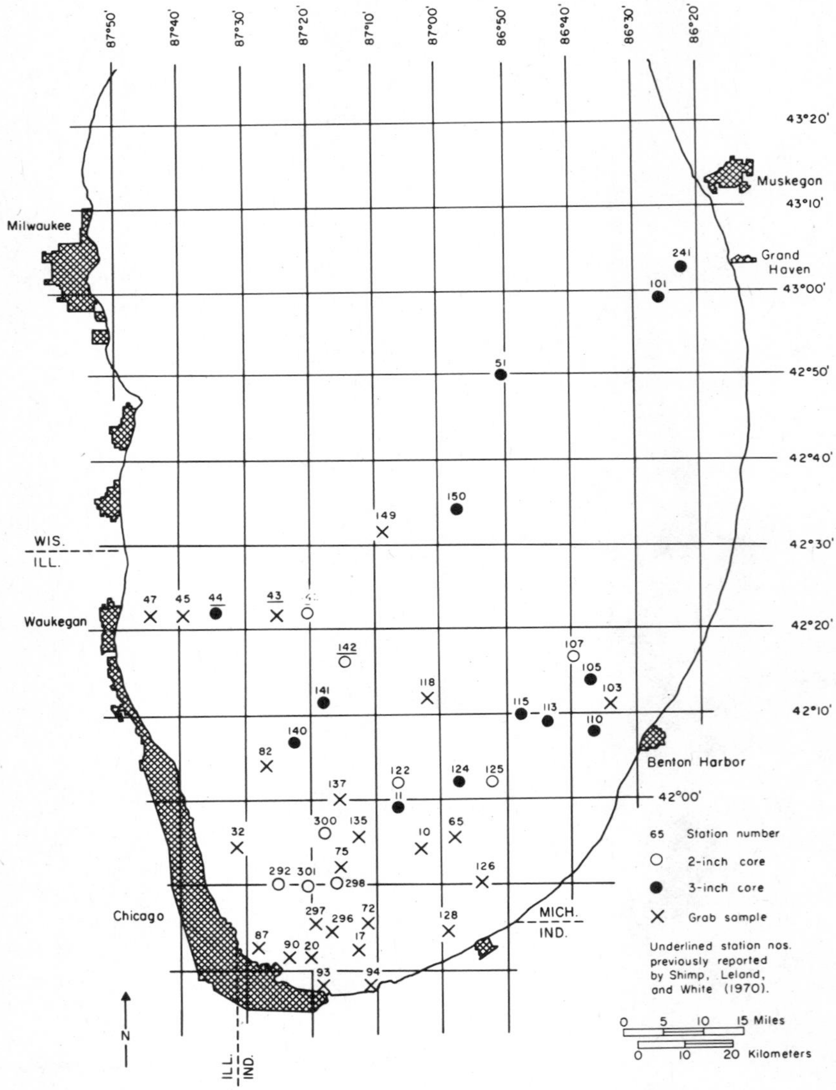

Figure 18.2 Locations of stations in southern Lake Michigan from which cores and grabs were taken. (Shimp, et al, 1971). Underlined samples are locations of first three cores analyzed.

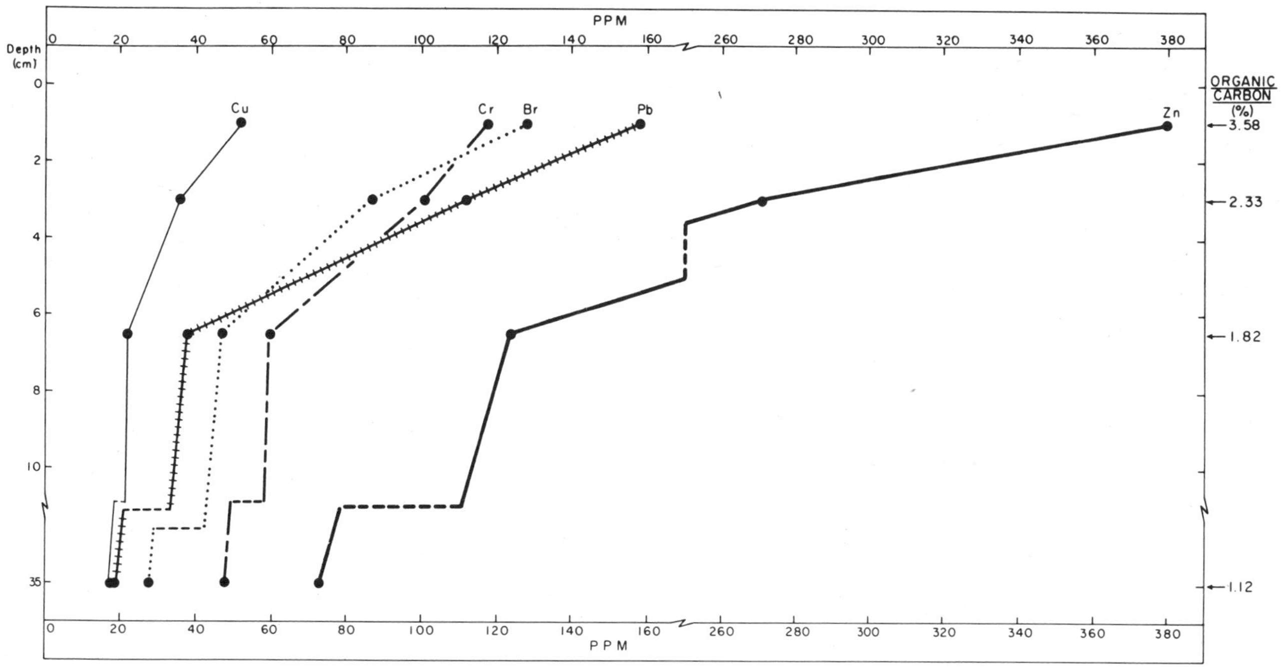

Figure 18.3 Typical distribution of Br, Cr, Cu, Pb, Zn and organic carbon in southern Lake Michigan core 101. The zone of trace element accumulation extends 8 or 10 cm because of relatively high sedimentation rates.

Table 18.1. Correlations of Trace Elements in Southern Lake Michigan Sediments with Organic Carbon and <2μm Clay Material

	Correlation Coefficients	
Element	Organic Carbon	2μm Clay
Br[a]	.86	.77
Pb[a]	.82	.30
Cr[a]	.81	.54
As[a]	.72	.56
Zn[a]	.70	.32
Cu[a]	.65	.45
Hg[a,b]	.71	.24
Be	.65	.64
Sc	.64	.90
V	.63	.80
Co	.61	.79
B	.44	.60
La	.24	.51
Cd	.23	.34
Ni	.65	.75

[a] Accumulating trace elements.
[b] Hg also correlates well with total sulfur (0.71).

in 1970 (Shimp et al., 1970). Later that year results of additional sediment work on arsenic and phosphorus were published (Ruch et al., 1970; Schleircher and Kuhn, 1970). In 1971 the results for 15 trace elements in 21 core and 21 grab samples (Fig. 18.2) were published (Shimp et al., 1971; Kennedy et al., 1971) along with profiles for the vertical distributions of these trace elements (Fig. 18.3). We concluded from these data that at least seven trace elements (Table 18.1) were accumulating in the uppermost, recently deposited portions of the organic-rich, fine-grained sediments of southern Lake Michigan. The concentrations of these elements gave good positive correlations with organic carbon and clay-sized material, although correlations for the accumulating elements were better for organic carbon (Table 18.1). Good correlations with clay-sized material were observed for the nonaccumulating trace elements.

2.3. Sedimentary Patterns

As chemical work progressed, geologists as ISGS were studying sediment depositional patterns in southern Lake Michigan. They had described mem-

Series	Stage	Fm	Member	Description
PLEISTOCENE SERIES	HOLOCENE STAGE	LAKE MICHIGAN	Ravinia	Sand on beaches
			Waukegan	Dark gray to brown, soft sandy silt to silty clay; sand
			Lake Forest	Dark gray silty clay with black beds and mottling; more compact than Waukegan Member
	-?-			
	WISCONSINAN STAGE		Winnetka	Dark brownish gray clay; a few black beds and some black mottling
			Sheboygan	Reddish brown clay
			Wilmette Bed	Dark gray clay with some black beds
				Reddish brown clay
			South Haven	Reddish gray clay
		EQUALITY	Carmi	Gray, sandy, pebbly clay; clay; silt; clay-pebble conglomerate
	Unnamed			
		WEDRON	Unnamed	Reddish brown, silty, clayey till
			Wadsworth	Gray, silty, clayey till

Figure 18.4
Generalized stratigraphic column for unconsolidated late Pleistocene sediments underlying southern Lake Michigan.

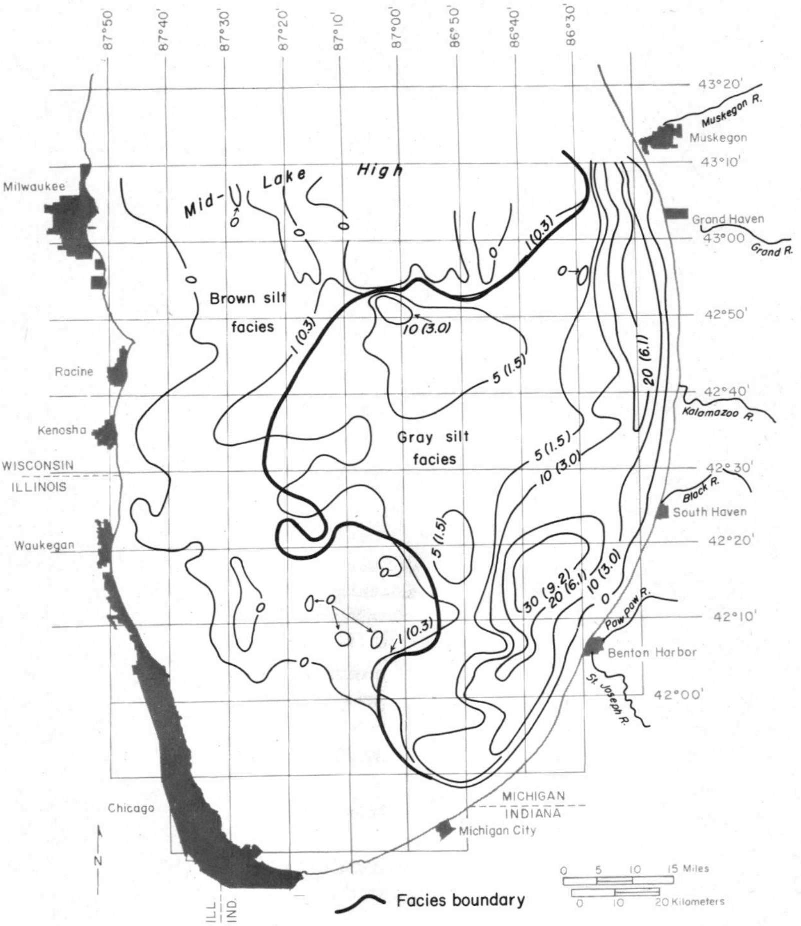

Figure 18.5 Thickness and facies of the Waukegan Member in southern Lake Michigan. Contour interval is 10 ft (3.0 m), but contours at 1 ft (0.3 m) and 5 ft (15 m) have been added. The facies boundary here defined by geologic criteria is virtually identical to one constructed from Eh values.

bers of the late Pleistocene Lake Michigan Formation (Lineback and Gross, 1972) (Fig. 18.4). The Waukegan Member contained the most recent sediments, in which high concentrations of trace elements were found. In this member, two facies were identified (Fig. 18.5): the gray silt facies in the center and along the eastern side of the Southern Basin, and the brown silt facies along the western side of the lake and at the midlake high. A thick belt of Waukegan along the east side of Lake Michigan indicated that the rate of sedimentation has been higher there (Fig. 18.5). The presence of sediments of the gray silt facies along the eastern side of Lake Michigan was thought to be caused by the basin's slope toward the west and the continuing deposition of material from Michigan rivers. These sediments were higher in organic carbon than those of the thinner, brown silt facies to the west. It was not surprising, therefore, that the trace elements that had accumulated in the sediments—As, Br, Cr, Cu, Pb, Hg, and Zn—were higher in the surficial sediments of the gray silt facies than in other Southern Basin surficial sediments.

The sediment of the gray silt facies is derived from erosion of the southeastern and eastern shores of the lake and from rivers that drain southern Michigan. These rivers contribute most of the waterborne trace elements that enter the Southern Basin. The nature of the sediments (fine-grained and carbonaceous) and the rapid sedimentation rate dictate that the greatest abundance of most elements transported to the area would occur in the gray silt facies.

Electrode potentials (Eh values) were almost always lower in the gray silt than in other sediments of the basin (Leland et al., 1973b). In fact, the facies boundary, which was independently plotted from electrode potentials, was identical to the boundary obtained from geologic observations (Fig. 18.5). Consequently, conditions in the gray silt facies are reducing (more negative), and those in the brown silt facies are oxidizing (more positive). The reducing conditions are most likely due to microbial decomposition of organic matter in the surficial sediments. The color of the gray silt facies is caused by the reduction of oxides; the color of the brown silt facies is caused by the presence of hydrated iron oxides.

2.4. Mechanisms for Enrichment of Trace Elements

Several questions arose concerning the mechanism(s) by which trace elements are enriched in surficial sediments: (1) Are they deposited with the sediment particles? (2) Do they diffuse upward from deeper layers of sediments? (3) Is the enrichment a natural phonemenon that occurs elsewhere in the geologic record?

In 1973, the distinct geographic distribution of accumulating trace elements led us to wonder whether recent deposition of insecticides would follow similar patterns. Indeed, this was found to be so. Accumulations of organochlorine insecticides were higher in the gray silt facies than in other regions, their vertical distribution in the sediments was similar to that of the trace elements (Leland et al., 1973a).

Furthermore, we found that suspended particles collected from 1 m above the lake floor in deeper parts of the Southern Basin contained trace element concentrations in suspended matter that equalled or exceeded amounts found in surficial sediments (Leland et al., 1973b). For example, organic carbon ranged up to 47%, Zn was as high as 2000 ppm, and As was over 40 ppm in some samples of suspended matter. These results, along with those for

Table 18.2. First Approximation of Trace- and Minor-Element Baseline Ranges for Fine-Grained Southern Lake Michigan Sediments

Element	Baseline (ppm)	Element	Baseline (ppm)
As	<8	Hg	0.05–0.10
B	30–50	Ni	15–40
Be	1–2	Pb	15–30
Br	<40	V	30–60
Co	10–20	Zn	50–100
Cr	20–40	MnO(%)	0.05–0.10
Cu	15–30	P_2O_5(%)	<0.3

pesticides, were convincing enough for us to conclude that the accumulation of trace elements in the surficial sediments resulted primarily from deposition of fine-grained particulate material with associated trace elements.

The theory that the enrichment of trace elements in southern Lake Michigan was a natural geologic process was tested by studying the sediment record for a glacial lake called Lake Saline, which once existed in southeastern Illinois (Frye and Shimp, 1973). The Lake Saline deposits were about the same age (the youngest being 13,000 radiocarbon years BP) and were derived from the same general region as those deposited in Lake Michigan. Textural composition and mean concentrations of major elements in Lake Saline deposits are similar to those in Lake Michigan sediments. Although concentrations of organic carbon as high as 6% were found in Lake Saline sediments, no consistent correlations of trace elements with organic carbon were found. In fact, no sign of trace-element accumulation was observed, and the mean trace-element values from Lake Saline sediments were similar to those found in the deeper Lake Michigan sediments. These results supported the conclusion that the values for deeper sediments are acceptable as baseline concentrations (Table 18.2) for evaluation of surficial sediment enrichment factors. Thus, over a span of some 20,000 years, no trace-element enrichment similar to that in Lake Michigan sediments occurred in the sediments of Lake Saline. These data indicate that significant amounts of trace elements were being added to southern Lake Michigan bottom sediments as a direct result of man's activities.

3. CURRENT INVESTIGATIONS OF LAKE MICHIGAN BOTTOM SEDIMENTS

3.1. Sampling

The first truly systematic sampling of Lake Michigan sediments was begun in 1975 in cooperation with R. L. Thomas of the Canadian Centre for Inland

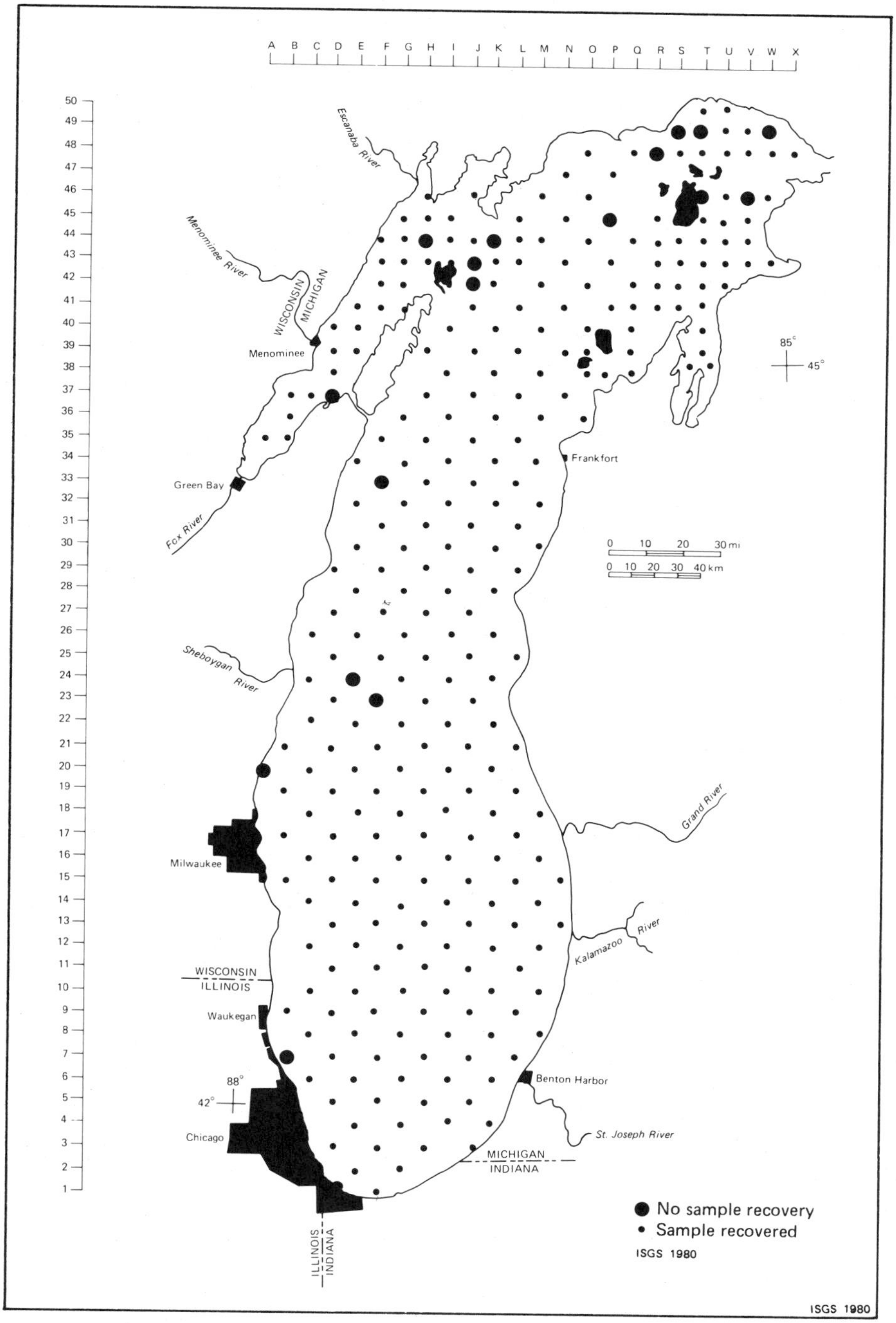

Figure 18.6 Sample location grid for the 1975 cruise of the CSS *Limnos* in Lake Michigan.

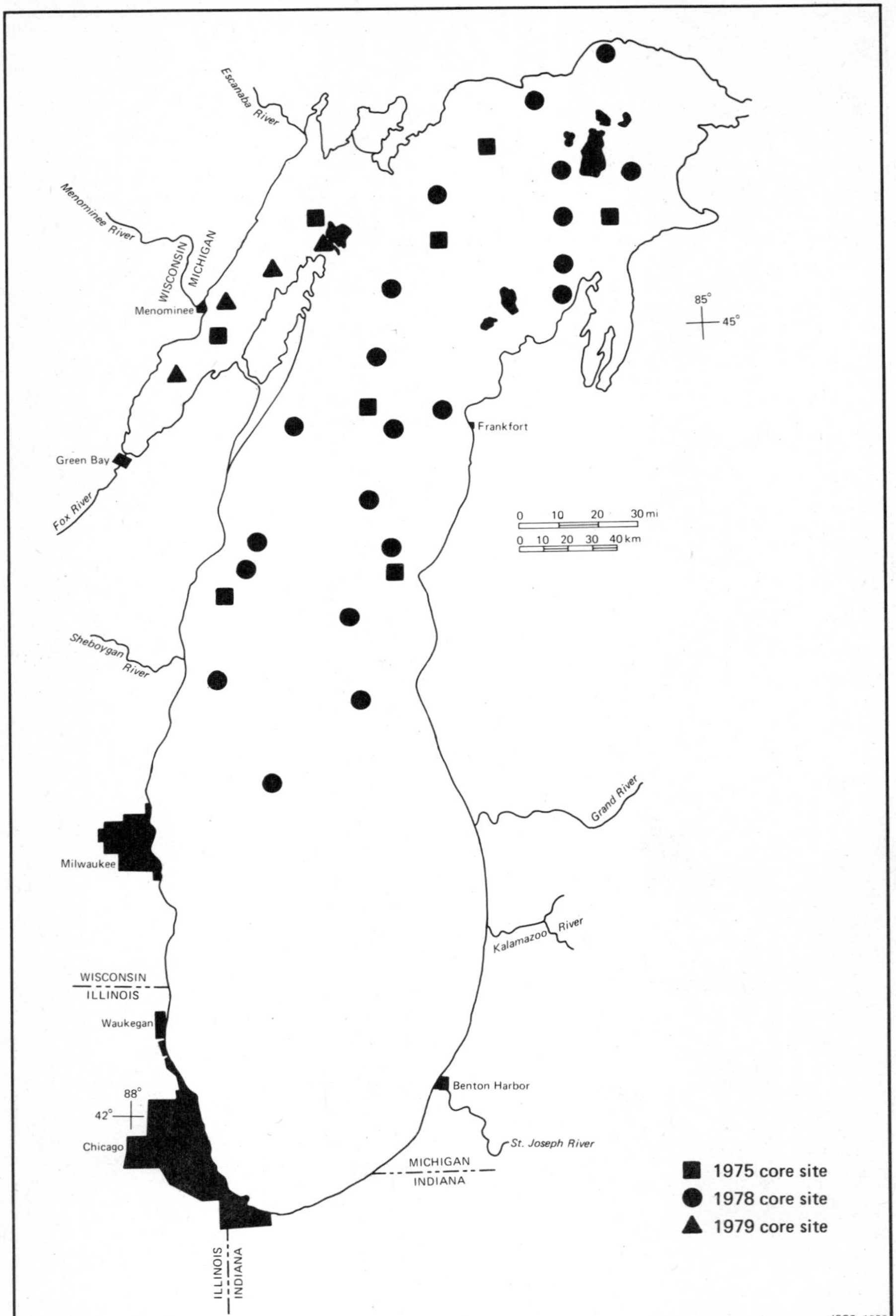

Figure 18.7 Sample locations for Benthos cores from northern Lake Michigan in 1975, 1978, and 1979.

Water (CCIW), Burlington, Ontario. Through his auspices, the work was conducted with the Canadian Centre's research vessel the CSS *Limnos*. Shipek grab samples were collected from a 12 × 12 km grid over the lake bottom, except in Green Bay and Traverse Bay, where a 7 × 7 km grid was used (Fig. 18.6). Grab samples were obtained from 286 stations and a number of Benthos cores (Fig. 18.7) were also obtained, primarily from northern Lake Michigan. These samples were sectioned and processed aboard ship before being returned to the laboratory for analysis by various methods (Cahill, 1981).

3.2. Sedimentary Subbasins in Lake Michigan

A distribution of bottom types for Lake Michigan based on grain-size analyses of the surficial sediments was mapped by Thomas of CCIW (Fig. 18.8). These bottom types may be summarized as follows (Cahill, 1981): (1) nearshore areas are composed of glacial tills, bedrock, or sand; (2) glaciolacustrine sediments occur between nearshore deposits and muds in basins; (3) thick postglacial muds (type A) occur in current areas of continuous sedimentation; and (4) thin postglacial muds (type B) occur in current areas of continuous sedimentation.

Comparison of the bottom sediment map (Fig. 18.8) with one of the thickness of gray clay (Fig. 18.9) shows that the thickest deposits of the gray clay correspond roughly with areas of type A mud (Wickham, et al., 1978). The thick deposits are located mainly along the eastern side of the lake and result from postglacial sediments carried into the lake by western Michigan rivers, as previously noted.

A plot of the areal distribution of clay-sized sediments (Fig. 18.10) illustrates that the higher percentages of clay-sized material occur in the deeper parts of both the Northern and Southern Basins of Lake Michigan. The distribution roughly follows that of Thomas's sediment type distribution.

Thomas (Cahill, 1981) has defined the depositional basins in Lake Michigan using physical descriptions, grain-size information, and echo sounding tracks. Using these same criteria, plus chemical data unavailable to him, we have redefined depositional basins in Lake Michigan (Fig. 18.11). These basins are the Southern, Grand Haven, Milwaukee, Northern, Green Bay, and several other small depositional areas near the Straits.

3.3. Areal Distrubition of Trace Elements

Approximately 15,000 chemical determinations were made in the 286 grab samples (top 3 cm) to ascertain the areal distribution of chemical constituents in Lake Michigan bottom sediments. These data are given in detail in another study (Cahill, 1981), but are summarized here in Table 18.3. The areal dis-

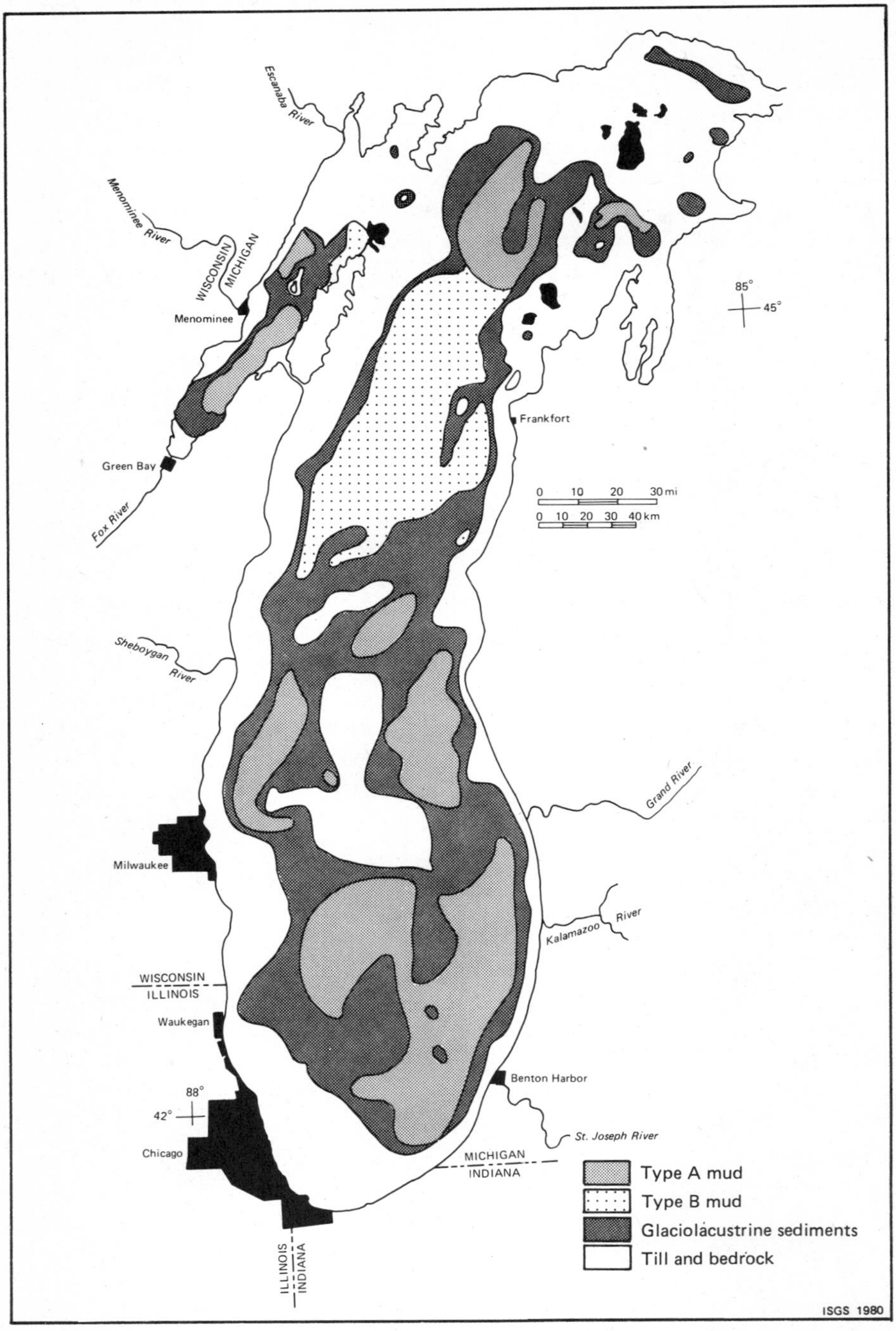

Figure 18.8 Distribution of bottom sediment types in Lake Michigan.

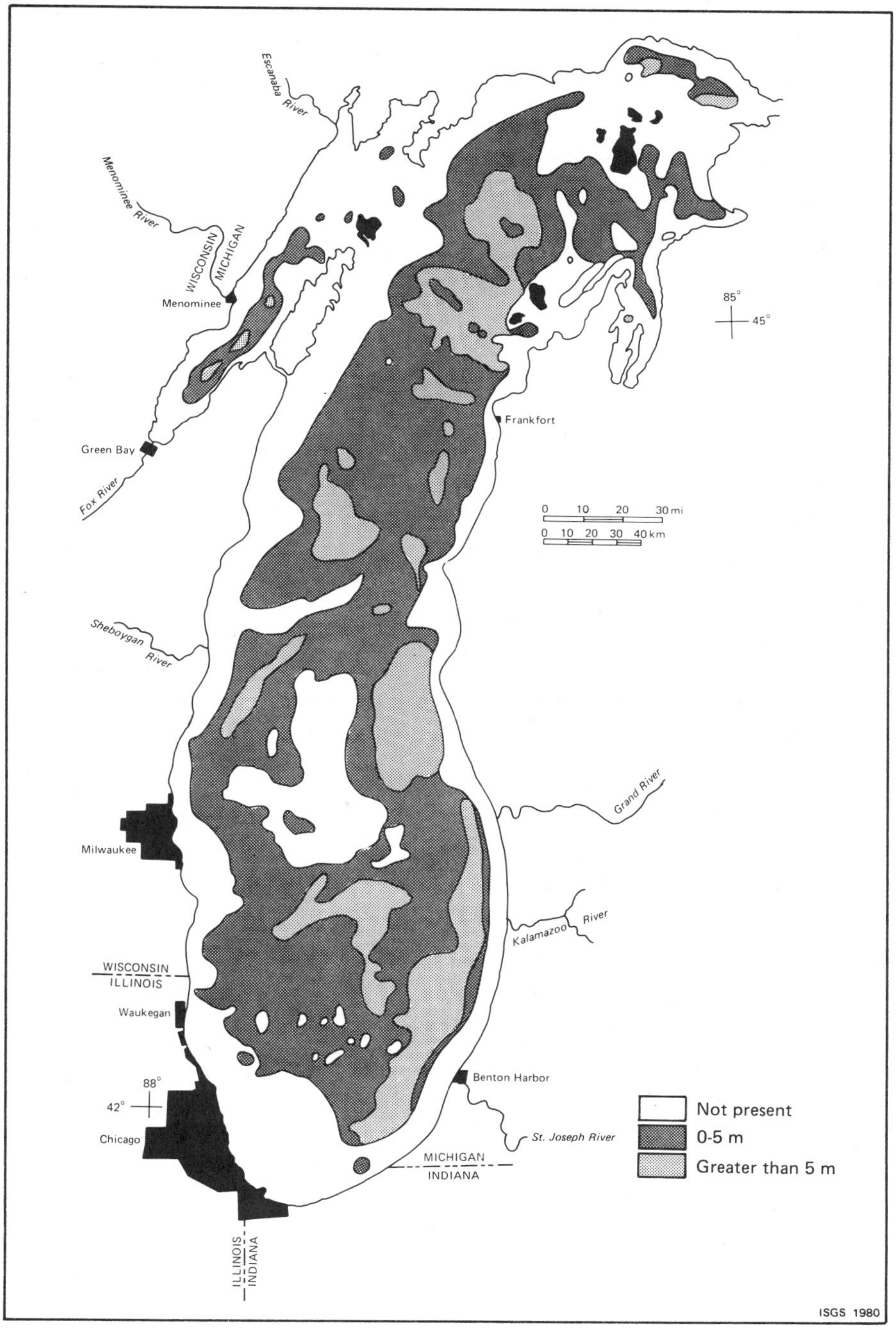

Figure 18.9 Thickness of gray clay, which represents the Winnetka, Lake Forest, and Waukegan Members combined.

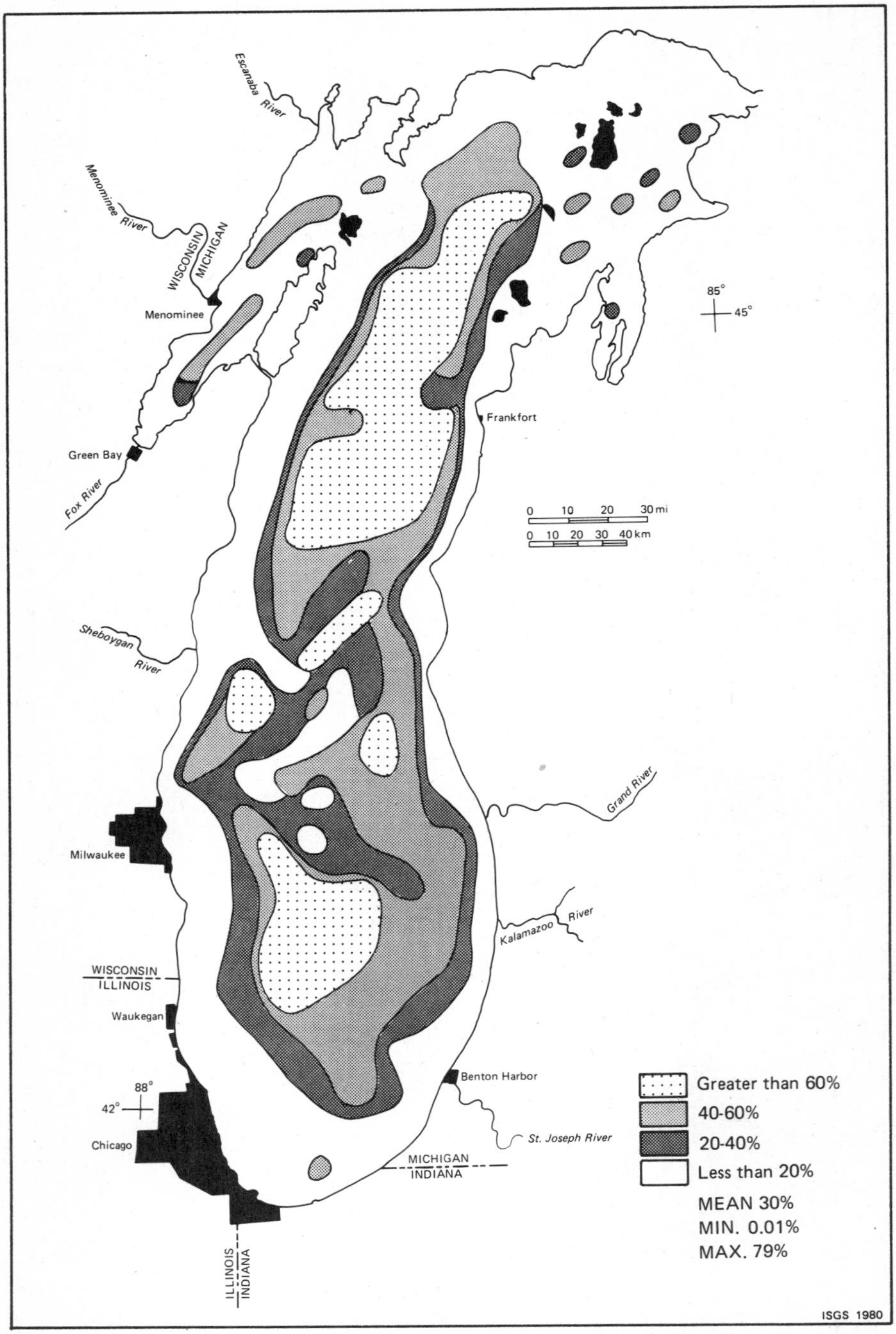

Figure 18.10 Clay-size sediment distribution in the upper 3 cm of Lake Michigan.

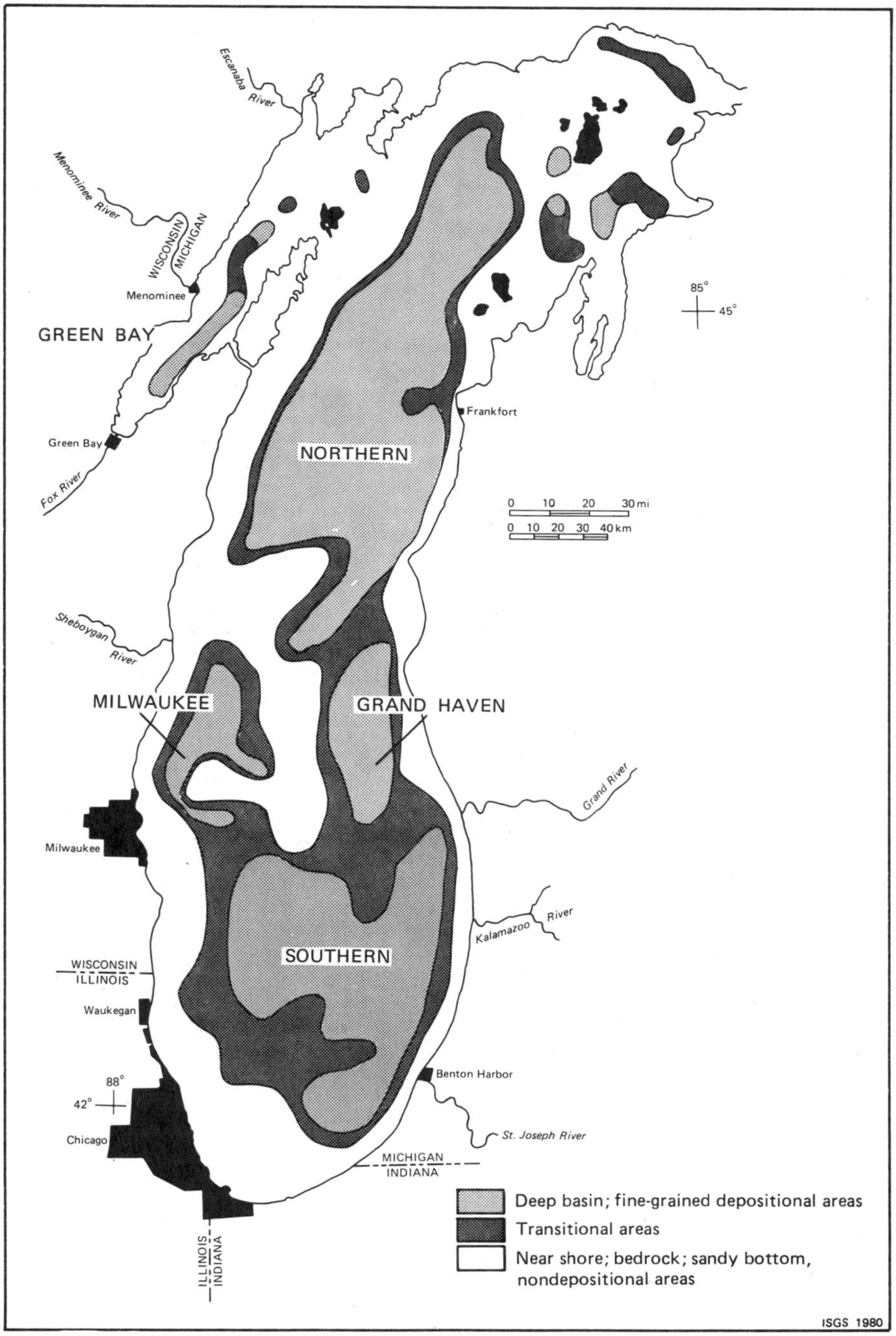

Figure 18.11 Generalized depositional environments and sub-basins of Lake Michigan.

Table 18.3. Mean Values for Trace-Element Concentrations in 286 Lake Michigan Surficial Sediments (Cahill, 1981) (all values in ppm unless noted)

Element	Arithmetic Mean	Geometric Mean	Minimum	Maximum	Standard Deviation	Number Less Than Values
Ag	0.46	0.4	0.1	1.4	0.25	209
As	10.5	6.8	0.8	153	16	2
Ba	494	437	120	7400	497	
Be[a]	1.7	1.7	0.9	2.5	0.4	32
Br	33	18	0.8	175	32	
Cd	0.9	0.9	0.5	2.5	0.4	189
Ce	48	40	5	360	30	
Co	9.0	7.1	0.7	59	6.1	
Cr	46	34	3	176	32	
Cs	2.9	2.1	0.2	8.5	2.1	
Cu	22	13	1.0	84	19	7
Eu	0.8	0.7	0.2	1.9	0.3	
Ga	10	8.4	0.8	32	5.3	2
Hf	5.1	4.8	1.4	18	1.9	2
Hg(ppb)	107	77	20	800	111	2
La	23	21	6.4	76	11	
Lu	0.2	0.2	0.01	0.7	0.1	2
Mo	7	5.4	1	18	5	230
Ni	24	17	1	198	21	15
Pb	40	21	1	153	41	9
Rb	85	77	18	220	37	
Sb	1.1	0.8	0.1	4.7	0.9	
Sc	6.6	5.1	0.3	16.4	4.0	
Se	1.2	1.0	0.1	3.3	0.7	137
Sm	3.7	3.3	1	11	1.8	
Sr	132	122	30	340	54	14
Ta	0.5	0.4	0.1	1.6	0.3	1
Tb	0.5	0.4	0.1	1.4	0.2	
Th	5.8	4.9	0.4	13.6	3.0	
U	2.3	2.1	0.6	9.2	1.2	119
V[a]	53	35	1.4	130	38	7
W	1.1	1.1	0.4	2.7	0.5	140
Yb	1.7	1.5	0.4	6.0	0.8	
Zn	97	58	4	350	90	2
Zr[b]	138	116	15	281	73	

[a] Values were determined on only 93 samples.

[b] Values were determined on only 103 samples.

tributions for As, Hg, and Pb (Figs. 18.12 to 18.14) illustrate the types of distributions observed. High concentrations of lead and mercury correspond to depositional areas, to type A muds, and to silty clay and clayey silt. In contrast, high levels of arsenic are localized in areas of Green Bay. The ratios between mean concentrations of most elements in the depositional

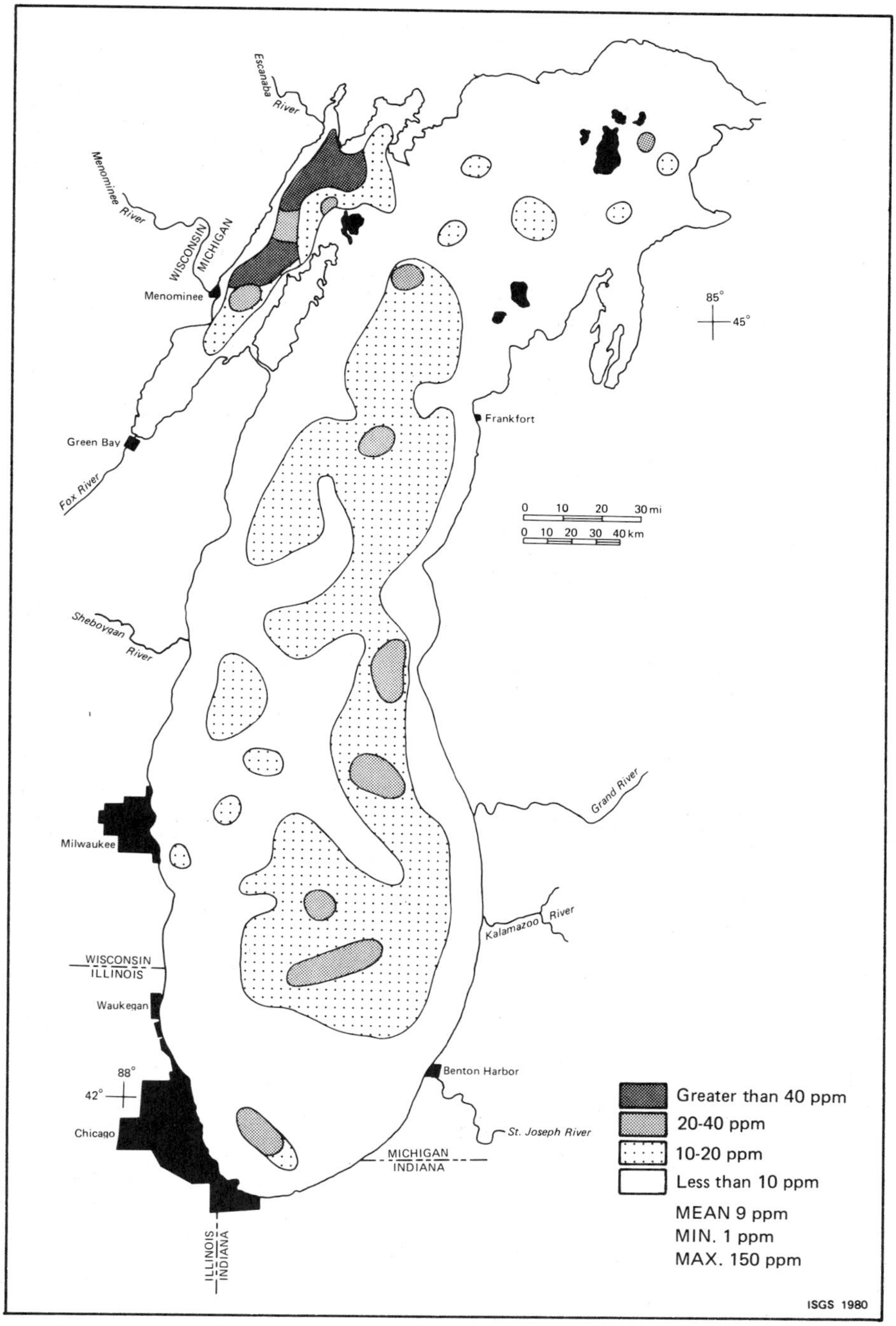

Figure 18.12 Arsenic distribution in the upper 3 cm of Lake Michigan sediments.

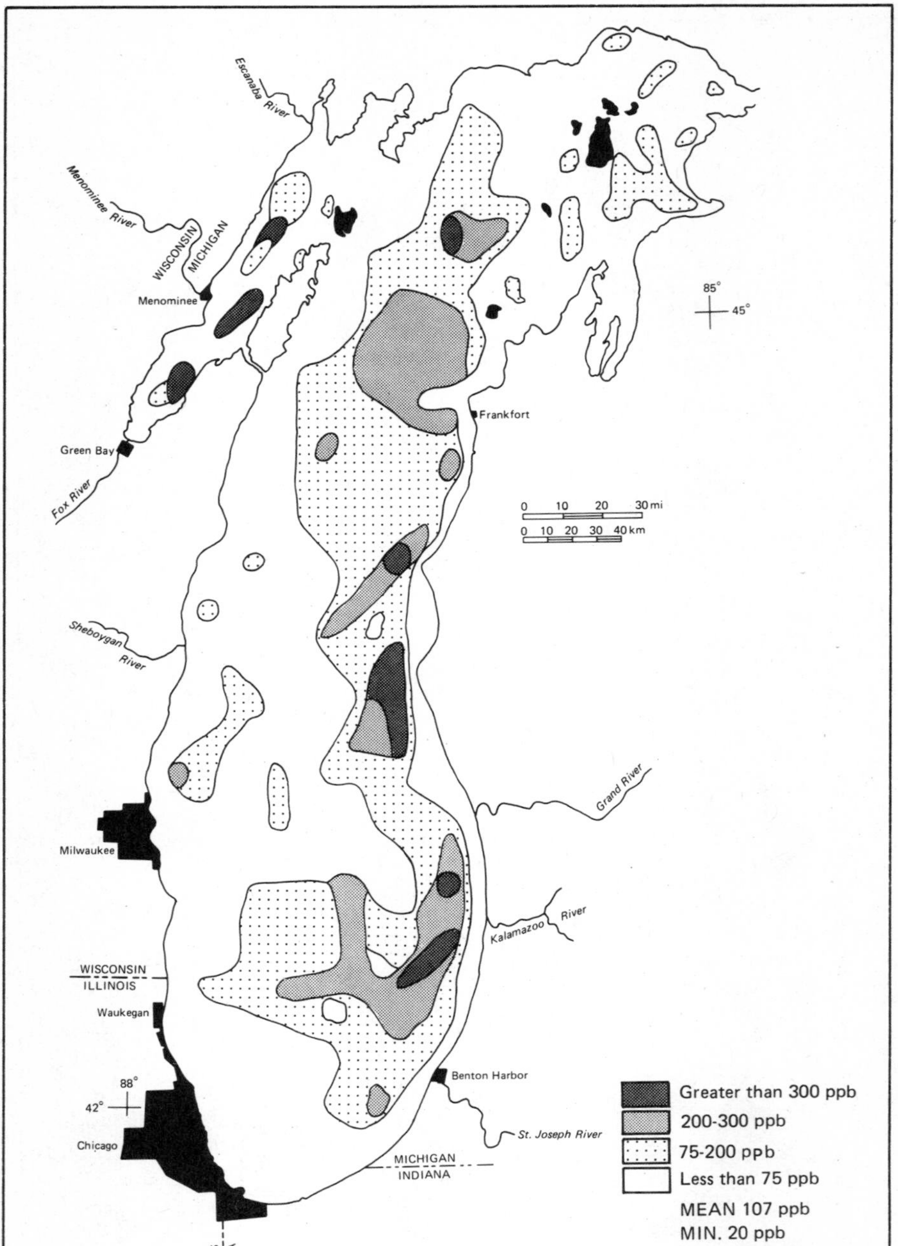

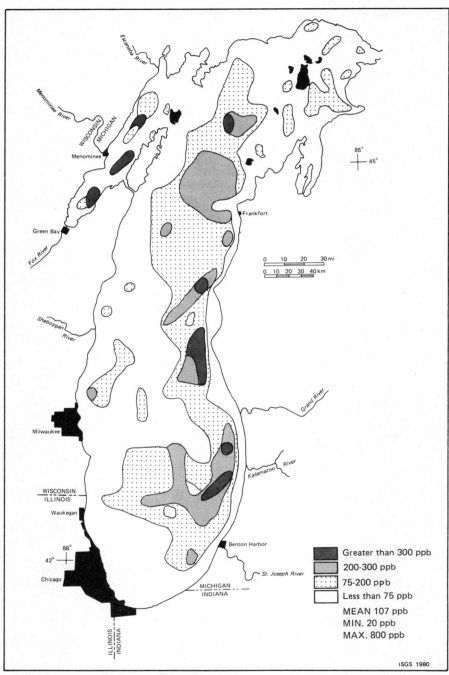

Figure 18.13 Mercury distribution in the upper 3 cm of Lake Michigan sediments.

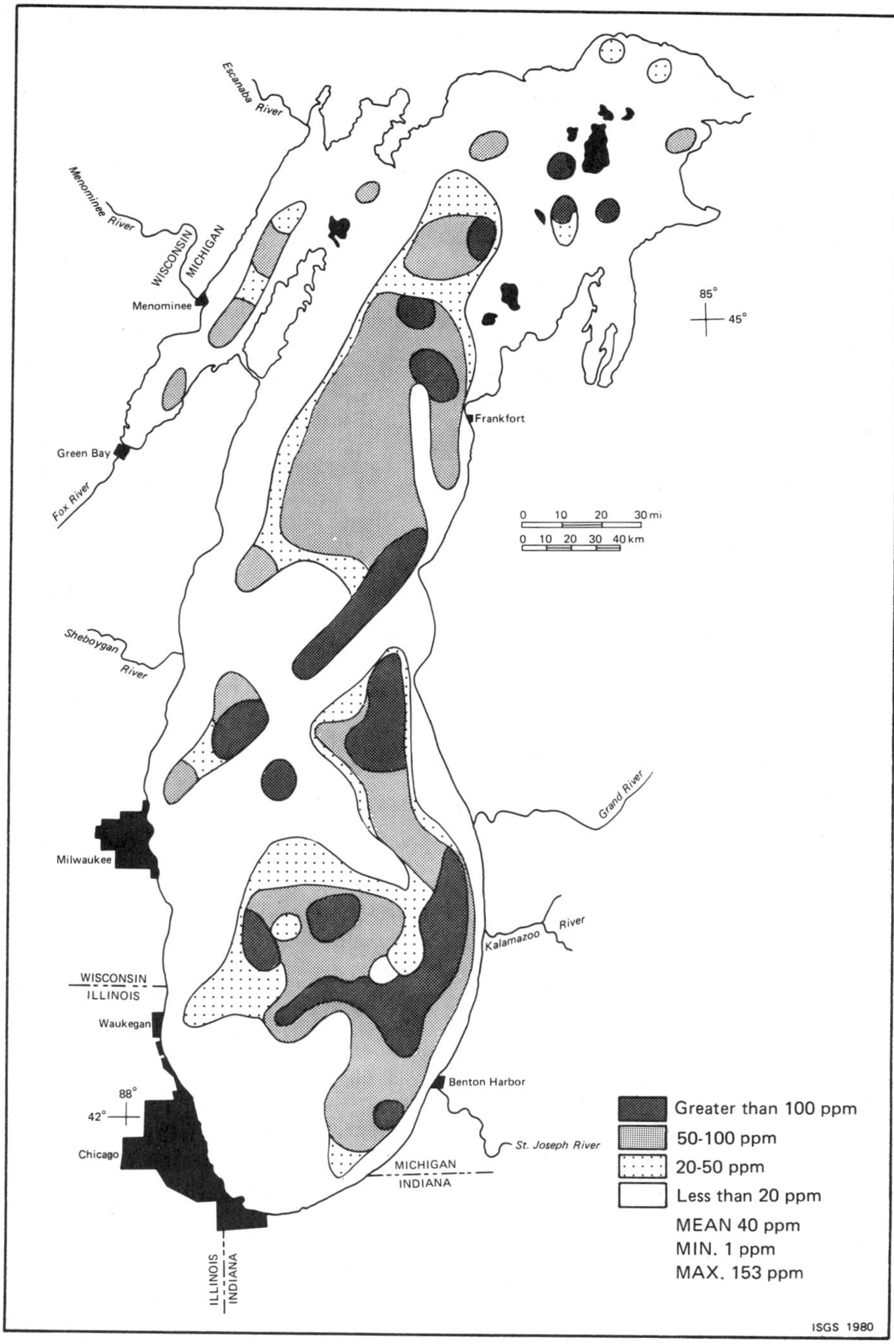

Figure 18.14 Lead distribution in the upper 3 cm of Lake Michigan sediments.

Table 18.4. Selected Mean Concentrations (ppm) for Sediments in Lake Michigan Subbasins (Cahill, 1981)

	Subbasin				
	Southern	Northern	Milwaukee	Grand Haven	Green Bay
As	12	11	11	17	15
Br	54	62	81	51	62
Cr	83	72	70	91	61
Cu	39	41	49	43	31
Hg[a]	168	156	118	205	401
Pb	88	68	77	98	57
Sb	2.0	1.9	2.4	2.3	1.8
Zn	198	173	182	228	119
Organic carbon	3.9%	3.1%	3.8%	3.9%	5.0%

[a] Values in ppb.

and nondepositional areas is 1.7:1.8. For Br, Cr, Cu, Hg, Ni, Pb, Sb, V, and Zn, however, the ratios are much higher; for example, Pb was 7.0, Cu was 6.7, C was 6.3, Br was 6.0, and Zn was 5.9 (Cahill, 1981). These are many of the same elements previously found to be accumulating near the sediment–water interface in the vertical concentration profiles for depositional areas of the Southern Basin (Fig. 18.3).

Table 18.5. Concentrations (ppm) Observed in Northern Lake Michigan Cores Collected in 1978[a]

Core	As		Br		Pb		Zn	
Site	0–1 cm	20–23 cm	0.1 cm	20–23 cm	0–1 cm	20–23 cm	0–1 cm	20–23 cm
F-19	10	5.5	99	62	86	<40	150	63
I-23	34	7	65	32	110	<38	250	64
J-31	6	8	104	43	<40	<40	198	144
M-35	18	8	61	40	75	<38	365	89
K-40	22	5	97	31	110	<38	410	86
M-44	13	5	105	64	73	<38	350	100
T-50	5	2	23	5	<50	<38	68	10
Green Bay								
B-36	8	6	40	40	48	<25	116	87
D-39	114	4	37	3	104	26	308	12
F-41	91	7	14	2	31	<25	42	45

[a] Results are arranged roughly from south to north and include only samples in areas of active sediment accumulation. The high levels of arsenic in the two samples from Green Bay correspond to D-39, the outfall of the Menominee River, and F-41, an area of nodule occurrence.

When mean trace-element concentrations in surficial sediments of the five depositional subbasins were compared (Table 18.4), they showed good general agreement. Evidently, trace-element loading varies little in the depositional subbasins (which is surprising, considering the higher population density surrounding the Southern Basin). This implies either that the sources of these elements are distributed uniformly around the lake or that circulation patterns in Lake Michigan are effective, or both.

3.4. Vertical Distribution of Trace Elements in Northern Lake Michigan Sediments

In 1975 and 1978, Benthos cores were taken in northern Lake Michigan (Fig. 18.7) for study of their vertical concentrations of trace elements. Unlike cores from the 1975 cruise, which were subsampled at 4-cm intervals, the cores taken in 1978 were subsampled by a hydraulic core extruder at 1- and 2-cm intervals. Cores were kept in the vertical position at all times to minimize mixing in the upper few cm. Because of better site selection on the basis of information obtained from the 1975 cruise and because of the improved resolution caused by smaller sampling intervals, trace-element enrichment in the surficial sediments was evident even in areas with low sedimentation rates (as low as 0.1 mm/yr). Table 18.5 compares concentrations in the uppermost sediment interval sampled (0–1 cm) with concentrations observed at a depth of 20–23 cm for arsenic, Br, Pb, and zinc. The results indicate that enrichments are observed even in the northernmost sample listed (T-50), and that "baseline values" in the 20- to 23-cm intervals vary with location. There is good general agreement, however, with the approximate baseline concentration ranges previously obtained for southern Lake Michigan sediments (Table 18.2).

Grab samples and benthos cores were obtained in Green Bay in 1979, in cooperation with the University of Wisconsin, with the research vessel, *Neeskay*. Elevated concentrations of a number of elements occur in surficial sediments—notably As, Ba, Mn, and Fe. One explanation for this is the occurrence of ferromanganese nodules in Green Bay and northwestern Lake Michigan (Edgington and Callender, 1970; Rossman, et al., 1972). Arsenic and Ba are relatively enriched in the nodules, as are, to a lesser extent, other elements such as Zn, Pb, Cu, Ni, and Co. Intense industrial pollution is a possible source of As in Green Bay sediments (Cahill, 1981). Figures 18.15 and 18.16 delineate the arsenic and lead distributions observed in Green Bay. The lead distribution (Fig. 18.15) closely follows areas of silty clay accumulation, which traces anthropogenic sources surrounding Green Bay. The arsenic distribution (Fig. 18.16) indicates an industrial source at the Menominee River and a natural concentration mechanism resulting in nodule occurrences in the northern reaches of the bay.

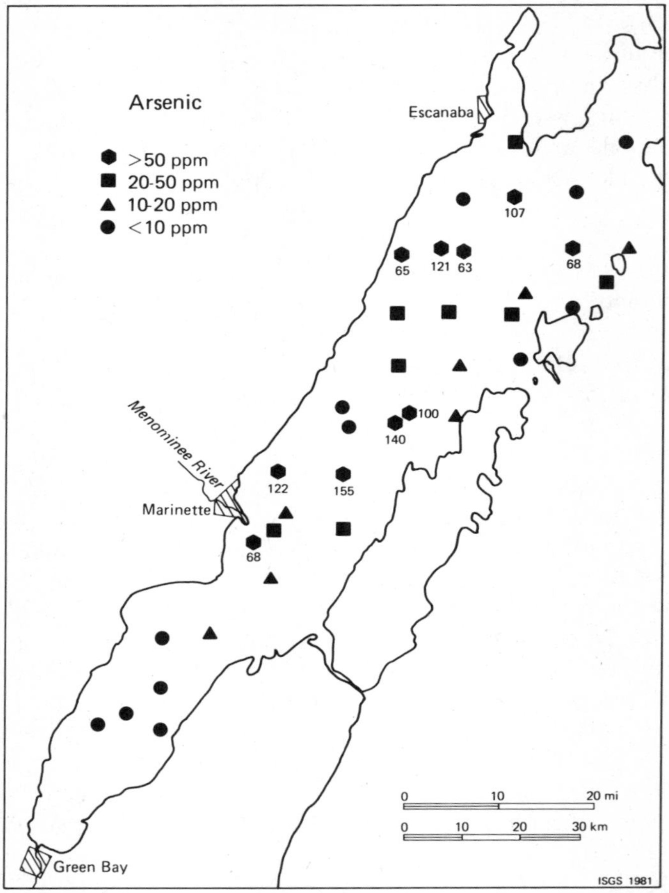

Figure 18.15 Arsenic distribution in the upper 3 cm of Green Bay sediments.

3.5. Controlling Physical Factors

Linear correlations of the Lake Michigan data were made to ascertain which major physical factors might be controlling the variability of the system. Good positive correlations for most elements were found with clay-sized sediments (clay minerals) (Cahill, 1981). Exceptions were Ca, Mg, and Sr, which occur as carbonates, and Si, which is concentrated in sands. Br, Cr, Cu, Pb, and Zn all have high correlations (0.7 or higher) with organic matter and clay-sized material. Organic material coatings on silt- and clay-sized sediments may contribute significantly to the exchange complex, which is responsible for trace-element enrichment in the sediments. Arsenic and Ba

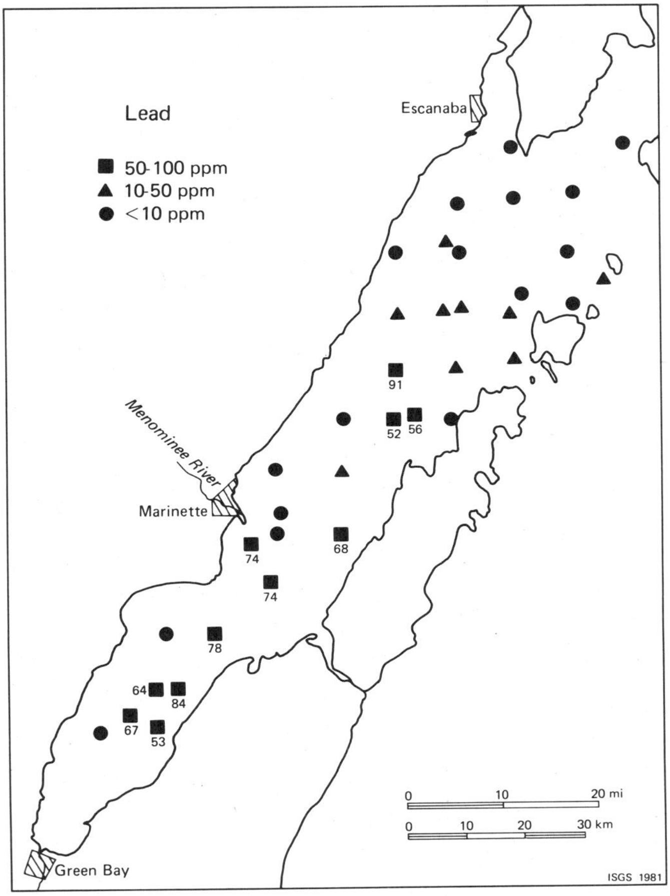

Figure 18.16 Lead distribution in the upper 3 cm of Green Bay sediments.

show good correlations (0.8 and 0.9, respectively) with Mn, probably because of highly selective adsorption by hydrous ferromanganese oxides in localized areas of the lake.

Factor analysis is a statistical tool that summarizes relationships between variables in a matrix of factor (Table 18.6) which then can be attributed to common geochemical or physical processes. The technique has been found to be particularly useful for intrepreting data sets that have a large number of elemental concentrations determined at a number of locations. The loading values that emerge from the ensuing calculation specify what portion of the total concentration of each element can be attributed to each factor. Seven factors were found to account for 87% of the variance of the system (Cahill, 1981).

Factor 1 accounts for 48.6% of the variance of the system and has high loadings for 22 of the chemical elements. It represents an abundance of clay-sized sediment and the elements derived from shoreline erosion and river input. Rare earths have high loading on this factor.

Factor 2 accounts for 16% of the variance and has high loadings for organic matter and for many trace elements with high enrichment factors (accumulating elements) that are probably anthropogenic in origin. Silt and organic carbon also have high loadings on factor 2; therefore, this factor probably represents fine-grained sediments with a high proportion of silt-sized material. Volatile elements S, Hg, Pb, Sb, and Zn associated with this factor

Table 18.6. Orthogonally Rotated Factor Matrix for Lake Michigan Surficial Sediments (Cahill, 1981)

	Factor						
Elements	1	2	3	4	5	6	7
SiO_2	−0.71	−0.41	−0.15	−0.49	−0.13	−0.05	−0.02
Al_2O_3	0.90	0.19	0.07	0.10	0.16	0.00	0.03
Fe_2O_3	0.65	0.28	0.51	0.15	−0.19	0.05	0.02
MgO	0.36	0.30	−0.07	0.80	0.22	0.01	0.05
CaO	0.21	0.23	−0.04	0.88	0.11	−0.04	0.05
Na_2O	0.27	−0.24	−0.07	0.07	0.63	0.42	−0.01
K_2O	0.73	0.16	0.03	−0.04	0.20	0.18	0.02
TiO_2	0.86	0.29	0.10	0.19	0.03	0.13	0.02
P_2O_5	0.70	0.36	0.33	0.10	0.03	0.00	0.00
MnO	−0.07	0.02	0.94	−0.10	0.01	−0.05	−0.01
TOC	0.52	0.61	0.23	−0.24	0.15	−0.12	−0.04
S	0.29	0.62	0.04	0.28	0.17	0.18	−0.09
Cl	−0.23	0.15	−0.19	0.41	0.62	0.01	0.03
As	−0.01	0.08	0.85	0.00	−0.26	0.04	0.02
Ba	−0.01	−0.04	0.90	−0.02	0.03	0.06	0.02
Br	0.66	0.44	0.29	−0.10	0.24	−0.11	−0.01
Co	0.54	0.10	0.78	0.05	0.05	−0.07	0.03
Cr	0.76	0.44	0.07	0.27	−0.08	−0.02	0.06
Cs	0.89	0.26	0.05	0.13	−0.04	−0.05	0.04
Cu	0.75	0.55	0.07	0.06	0.07	−0.15	0.03
Ga	0.85	0.23	0.08	0.07	0.11	0.03	−0.01
Hf	0.18	0.05	0.00	−0.02	0.21	0.88	−0.01
Hg	0.23	0.72	0.04	0.03	−0.04	0.26	−0.01
Ni	0.49	0.28	0.64	−0.06	0.18	−0.09	−0.03
Pb	0.46	0.75	0.10	0.14	0.02	−0.14	0.09
Rb	0.89	0.10	0.08	0.09	0.03	0.09	0.07
Sb	0.58	0.60	0.15	0.12	−0.06	−0.04	0.08
Sc	0.93	0.21	0.09	0.17	0.07	0.01	0.02
Sr	0.32	0.05	0.24	0.18	0.50	0.16	0.12

Table 18.6. (*Continued*)

	Factor						
Elements	1	2	3	4	5	6	7
Ta	0.77	0.33	0.05	0.27	−0.09	0.18	0.06
Th	0.90	0.24	0.10	0.13	0.01	0.08	0.04
Zn	0.54	0.70	0.17	0.15	−0.01	−0.13	0.07
Ce	0.48	0.10	0.78	0.03	0.09	0.01	0.02
Eu	0.77	0.21	0.47	0.10	0.12	0.10	−0.01
La	0.77	0.17	0.54	0.05	0.05	0.03	−0.02
Lu	0.78	0.13	0.15	0.11	−0.04	0.21	0.05
Sm	0.78	0.19	0.50	0.06	0.07	0.05	−0.02
Yb	0.76	0.19	0.22	0.16	0.00	0.22	0.03
Eh	−0.27	−0.71	−0.02	−0.18	0.12	0.06	0.03
Ph	0.09	−0.01	0.02	0.06	0.06	−0.01	0.97
Sand	−0.77	−0.41	−0.05	−0.22	−0.11	−0.4	0.09
Silt	0.49	0.51	0.03	0.43	0.25	0.21	−0.04
Clay	0.88	0.30	0.06	0.05	0.06	−0.04	0.00
Mean	0.83	0.38	0.05	0.13	0.13	0.04	0.01
Pct. Variance	48.6%	16.5%	15.8%	7.6%	4.6%	4.0%	2.9%
Cum. Pct.	48.6%	65.1%	80.9%	88.5%	93.1%	97.1%	100%

probably have an atmospheric source. A strong negative Eh loading indicates this factor is not an oxidizing environment in which organic matter would be unstable. As in southern Lake Michigan (Fig. 18.5), relatively low Eh values are associated with sediment deposition and trace-element enrichment.

Factor 3 is related to the occurrence of hydrous ferromanganese nodules; Fe, Mn, As, Ba, Ni and Co have high loadings on this factor.

Factor 4 has high loadings for Mg and Ca, which reflects the presence of carbonates in the sediment.

Factors 5, 6, and 7 account for only 10% of the variance of the data. No satisfactory explanations have been found for them. Silcon, sand-sized particles, and Eh have negative loadings on nearly all of the factors.

The factor scores for each sample were calculated and then plotted, using a trend surface analysis computer graphics program. The resulting plots indicate which areas of the lake are impacted by, or control, the distribution of elements with a common association.

Elements of anthropogenic orgin (Cd, Hg, Pb, Zn, S, and organic carbon) had a high loading on factor 2. The plot of the factor 2 scores (Fig. 18.17) clearly shows that high loadings of these elements are associated with the Fox River. The influence of the Grand Kalamazoo and St. Joseph Rivers,

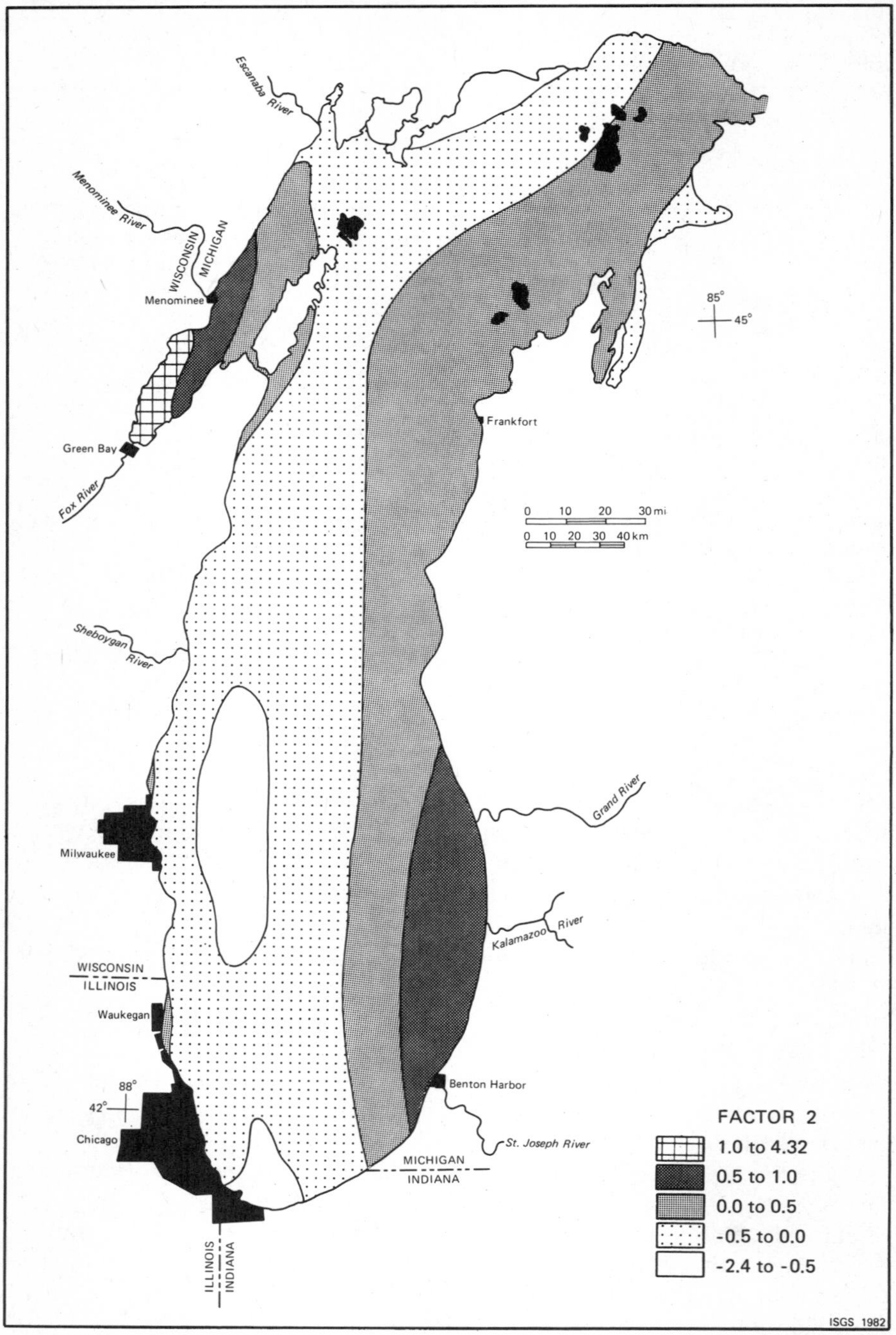

Figure 18.17 Trend surface analysis plot of Factor 2 scores in Lake Michigan. Factor 2 represents total organic carbon, S, Hg, Pb, Sb, Zn, and Cd loadings.

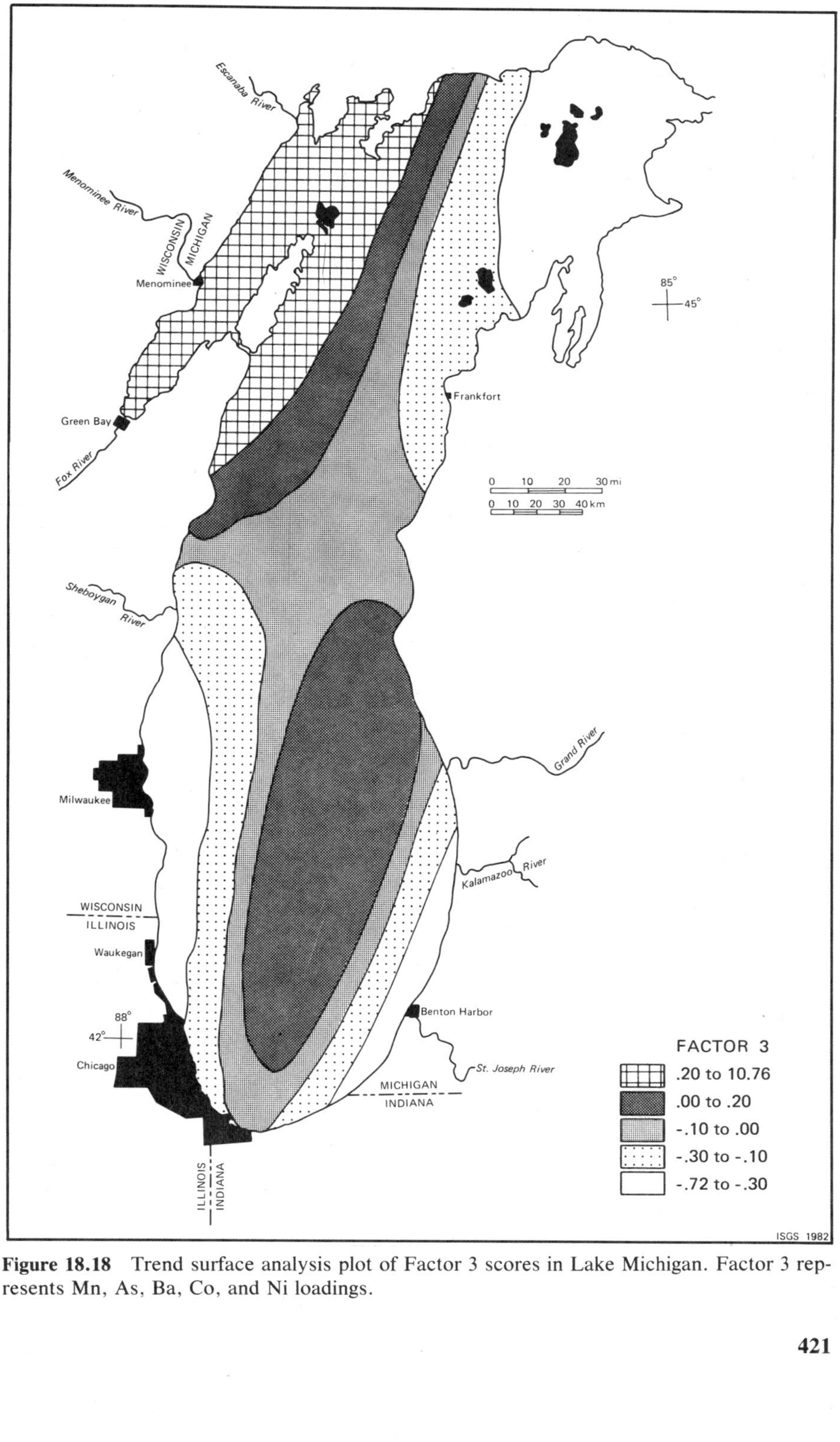

Figure 18.18 Trend surface analysis plot of Factor 3 scores in Lake Michigan. Factor 3 represents Mn, As, Ba, Co, and Ni loadings.

and water circulation patterns control the loadings of these elements in southern Lake Michigan.

Manganese, As, Ba, Ni, and Co were elements associated with factor 3. The results of trend surface analysis (Fig. 18.18) clearly show that occurrences of ferromanganese nodules in Green Bay control the loadings on factor 3.

4. CONCLUSIONS

Variations between depositional and nondepositional areas in Lake Michigan can be delineated from grain- and clay-sized information and from the organic carbon distribution. Trace elements such as Br, Cr, Cu, Pb, and Zn are highly enriched in surficial sediments of depositional areas in the lake. Because of efficient mixing by lake currents or atmospheric transport, differences in concentrations of most trace elements between depositional subbasins in Lake Michigan are not significant. Evidently such mixing averages out local differences in source materials before they become incorporated into the sediments. The accumulating elements are anthropogenetic and are incorporated into organic matter and clay minerals present in the finer-grained sediments.

ACKNOWLEDGEMENTS

The authors gratefully acknowledge the help given by the following persons who have made significant contributions to these investigations of Lake Michigan sediments: Richard L. Thomas, Jerry Lineback, Jerry Wickham, and David L. Gross (Geology); John A. Schleicher, Phil K. Hopke, Brad A. Roscoe, and Jeanne Dunn (Statistics); Harry V. Leland, Rodney R. Ruch, and members of the Analytical Chemistry Section (Chemistry).

REFERENCES

Cahill, R. A. (1981). Ill. State Geol. Survey Cir. 517, 94 pp.

Edgington, D. N., and E. Callender. (1970). *Earth Planet. Sci. Lett.* **8:** 97.

Fitchko, J. and T. Huchinson. (1975). *J. Great Lakes Res.* **1:** 46.

Frye, J. C. and N. F. Shimp. (1973). Ill. State Geol. Survey, *Environ. Geol. Notes* 60, 14 pp.

Gross, D., J. Lineback, W. White, N. Ayer, C. Collinson, and H. Leland. (1970). Ill. State Geol. Survey, *Environ. Geol. Notes* 30, 20 pp.

Kennedy, E. J., R. R. Ruch, and N. F. Shimp. (1971). Ill. State Geol. Survey, *Environ. Geol. Notes* 44, 18 pp.

Leland, H. V., W. N. Bruce, and N. F. Shimp. (1973a) *Environ. Sci. Technol.* **7:** 833.

Leland, H. V., S. S. Shukla, and N. F. Shimp. (1973b). *Trace metals and metal-organic interactions in natural waters.* Ann Arbor Science Publishers, Ann Arbor, pp. 89–129.

Lineback, J. A., and D. L. Gross. (1972). Ill. State Geol. Survey, *Environ. Geol. Notes* 58, 25 pp.

Robbins, J. A. and D. N. Edgington. (1975). *Geochem. Cosmochim. Acta* **39:** 285.

Robbins, J., E. Landstrom, and M. Wahlgren. Proc. 15 Conf. Great Lakes Res., 270.

Rossman, R., E. Callender, and C. J. Bowser. (1972). 24th Int. Conf. Geol. Congr., Montreal, Canada, pp. 336–341.

Ruch, R. R., E. J. Kennedy, and N. F. Shimp. (1970). Ill. State Geol. Survey, *Environ. Geol. Notes* 37, 16 pp.

Schleicher, J. A. and J. K. Kuhn. (1970). Ill. State Geol. Survey, *Environ. Geol. Notes* 39, 15 pp.

Shimp, N. F., H. V. Leland, and W. A. White. (1970). Ill. State Geol. Survey, *Environ. Geol. Notes* 32, 19 pp.

Shimp, N. F., J. A. Schleicher, R. R. Ruch, D. B. Heck, and H. V. Leland. (1971). Ill. State Geol. Surv., *Environ. Geol. Notes* 41, 25 pp.

Wickham, J. T., D. L. Gross, J. A. Lineback, and R. L. Thomas. (1978). Ill. Geol. Survey, *Environ. Geol. Notes* 84, 26 pp.

19

USING THE HERRING GULL TO MONITOR LEVELS AND EFFECTS OF ORGANOCHLORINE CONTAMINATION IN THE CANADIAN GREAT LAKES

P. Mineau[1], *G. A. Fox*[1], *R. J. Norstrom*[1], *D. V. Weseloh*[2], *D. J. Hallett*[1] *and J. A. Ellenton*[1]

[1] *Canadian Wildlife Service*
National Wildlife Research Centre
Environment Canada
Ottawa, Ontario K1A OE7

[2] *Canadian Wildlife Service*
Canada Centre for Inland Waters
Environment Canada
Burlington, Ontario L7R 4A6

---

1. INTRODUCTION

> The study of environmental contamination is justified on the assumption that the environmental media: air, water and soil may be contaminated to such a degree that their use by living organisms or man may constitute a hazard to life. The possible effects of such contamination on one or more components of an ecosystem may lead to a disruption of the normal balance of species, or to a decline or destruction of one or more plant or animal populations. Furthermore, it may give rise to an unacceptable level of contamination of certain species (e.g., fish) which may be used for human food, or to the loss of such species as a significant food resource.(OECD, 1980)

This statement accurately summarizes our concerns about the effects of environmental contaminants on our ecosystem. These concerns date back to the mid-1960s when there was a growing awareness that the global ecosystem was becoming increasingly contaminated with persistent organochlorinated chemicals such as DDT and its metabolites, and polychlorinated biphenyls (PCBs). Some avian populations, particularly of fish-eating species, provided indications that many of these substances were accumulating through food chains. Concomitant observations of eggshell thinning and the resulting breeding failures, especially in raptors, provided the evidence that this contamination had grave biological repercussions.

The Great Lakes were not spared this contamination; indeed, they have borne the brunt of it. It is estimated today that 30,000 man-made chemicals are in use in the Great Lakes Basin, and so far more than 400 have been found in one or more components of the Great Lakes ecosystem (IJC, 1980). Cases of breeding failures in fish-eating birds from the Great Lakes were first published in 1966 (Keith, 1966; Ludwig and Tomoff, 1966 in Lake Michigan) and 1970 (Edwards, 1970 in Lake Ontario). At that time it was logical to consider as prime suspects those organochlorines that could be detected.

Wildlife samples from the Great Lakes were analyzed for known organochlorines with remarkable results: populations of piscivorous birds such as herring gulls (*Larus argentatus*), double-crested cormorants (*Phalacrocorax auritus*), and common terns (*Sterna hirundo*) were found to be among the most contaminated in the world.

It soon became apparent in the Great Lakes and elsewhere that some of these heavily contaminated species could serve as useful monitors of the very contaminants that threatened their continued existence. As man is also a consumer species, the need for information on the effects of contamination was all the more vital. Following the discoveries on Lakes Michigan and Ontario, the Canadian Wildlife Service (CWS) gradually developed a comprehensive program designed to delimit the extent of organochlorine contamination of wildlife in the Great Lakes and to understand its ecotoxicological significance. This program eventually became incorporated as part of the surveillance and monitoring activities conducted in support of the Canada—United States agreement of 1978 on Great Lakes water quality which emphasized toxic chemicals.

Very early in the evolution of the program, the value of focusing attention on a single indicator species, the herring gull, was recognized (Gilbertson, 1974). In Section 2 we shall see why this species was chosen and how it has become a valuable early warning system for the detection of new persistent, lipophilic contaminants. We discuss its value as a monitor of levels of known organochlorines like DDT and PCBs and its usefulness in assessing the effectiveness of efforts to prevent or reduce contamination by toxic chemicals in the Great Lakes basin. Its value as an indicator of biological damage is also discussed. By documenting both the strengths and the weaknesses of the herring gull as an indicator of ecosystem health, we hope that this chapter will serve as a case study of the selection of wildlife species for use as monitors of contamination.

2. WHY THE HERRING GULL?

In 1974 it was proposed that the herring gull be used as a monitor of organochlorine contaminants for the Great Lakes (Gilbertson, 1974). Justifications for this choice have been widely reviewed (Peakall et al., 1978; Gilman et al., 1979; Dauphiné, 1980) and have not changed appreciably since they were originally formulated.

The herring gull is a top predator in the food web of the Great Lakes. Althouth it is opportunistic in its feeding habits and may eat insects, birds, small mammals and refuse, it remains primarily a fish eater. It is probable that herring gulls eat only fish during the winter because they become scarce at shore sites (such as landfills) while large numbers are sighted offshore on floes and ice edges, especially on Lake Erie. A good monitor of organochlorine contaminants in the Great Lakes must obtain its contaminant load from the aquatic food chain. In Lake Ontario at least, ratios of PCBs, mirex, and photomirex to DDE in herring gulls are substantially the same as in coho salmon (*Oncorhynchus kisutch*), and their common prey, rainbow smelt (*Osmerus mordax*) and alewife (*Alosa pseudoharengus*) (Norstrom et al., 1978). The amount of these organochlorine contaminants obtained through terrestrial systems or from refuse therefore appears to be inconsequential to total body burdens.

Perhaps the most important attribute of the herring gull in comparison with other fish-eating bird species in the Great Lakes is that from the time it reaches breeding age it is a year-round resident. Herring gulls only migrate away from the lakes as immatures (Moore, 1976); hence there is no likelihood that organochlorine levels in adult herring gull tissues reflect conditions elsewhere. Interlake movements do occur during the winter, more commonly among birds from the upper lakes (Moore, 1976; Gilman et al., 1977). Between 1978 and 1980, we color-marked breeding adults in southern Lake Huron, Lake Erie, and Lake Ontario to study their winter movements. Preliminary analysis of these data show that most movements away from the breeding colony take place between fall and midwinter. The attachment to the breeding colony and its surrounding area from mid-winter onward might explain why herring gulls sampled on colony during the breeding season show such different contaminant burdens between lakes (Section 3.5). Further analysis of banding data has also revealed that immigration of individuals to the Great Lakes population from other North American populations is minimal.

Herring gulls nest colonially and are therefore easy to locate, count, and sample. Not only does the species have a more extensive breeding range in the Great Lakes than any of the other piscivorous species, but it also has a holarctic distribution, which allows for comparisons over very broad geographical areas. In consequence, the biology of the herring gull has been extensively studied.

3. THE HERRING GULL AS AN INDICATOR OF CONTAMINANT LEVELS

3.1. Sample Collection

As early as 1965, adult herring gulls were collected for analysis in Lakes Ontario and Erie. However, from 1972 onwards the favored tissue for analysis has been eggs, for several reasons. The lipid content of an egg is less variable than that of most other tissues. Eggs can be easily collected with little disturbance to the colony. A number of studies (Herman et al., 1969; Vermeer and Reynolds, 1970; Enderson and Berger, 1970) have shown the similarity between contaminant burdens of eggs and bodies of female birds. Similarly, we have some evidence that in the herring gull, organochlorine residue levels in egg lipids approximate those in the liver lipids of the female that laid the eggs.

In 1974, a sampling protocol was established. It has been followed yearly except for 1976 when little sampling took place. A visit is made early in the laying season and 10 eggs collected from 10 completed clutches chosen along a transect of the entire colony. Eggs are collected at random from these clutches. Usually, fresh eggs are collected to reduce intersample variance in contaminant levels and to facilitate homogenization. Occasionally, however, embryonated eggs have to be collected. For this reason, all residue values are expressed on a fresh-weight basis since this method of calculation is less affected by embryonic development (Peakall and Gilman, 1979). It has recently been shown (Mineau, 1982) that contaminant levels increase systematically in later-laid eggs and that a lower sampling variance could probably be achieved by always taking eggs from the same position in the laying sequence from all clutches.

The routine sampling to establish time trends (Section 3.4) takes place on two colonies per lake. These are: Snake and Muggs Islands in Lake Ontario, Port Colborne Breakwater and Middle Island in Lake Erie, Chantry and Double Islands in Lake Huron, and Mamainse Harbour and Granite Island in Lake Superior. Positions of these colonies are given in Fig. 19.1. Nearby colonies have on occasion been sampled instead. From 1978 to 1981 we collected herring gull eggs from colonies other than those routinely monitored on Lakes Erie, Huron, and Ontario to correspond with intensive year assessments of those lakes in accordance with the Great Lakes International Surveillance Plan. The purpose of these collections was to provide samples from which we could determine the geographical variation in organochlorine levels within each lake and to examine colonies exposed to more localized areas of industrial contamination (Section 3.5). Positions of these colonies as well as of geographical features mentioned in the text are given in Fig. 19.1. In using herring gull eggs to determine the geographical distribution of contaminants in the Great Lakes, the coverage can obviously only be as complete as the distribution of the gulls' nesting colonies within each lake.

3.2. Quantitative Chemical Analysis

After homogenization, sample aliquots are analyzed while the rest of the sample is stored in CWS's National Specimen Bank. The presence of *p,p'*-DDT, *p,p'*-DDD, *p'p*-DDE, and *o'p*-DDT was confirmed in all but a few of the adult herring gulls collected in 1965. However, quantitative information was unreliable because the confounding influence of PCBs had not been removed. When eggs were collected in 1972, the compounds that were looked for included *p,p'*-DDT, *p,p'*-DDD, *p'p*-DDE, heptachlor epoxide, dieldrin, hexachlorobenzene (HCB), and PCBs. The PCBs had been known as ubiquitous environmental contaminants since about 1968. An early method for separating PCB peaks from those of the common pesticides was first worked out on herring gull eggs (Reynolds, 1969). In general, levels of dieldrin, heptachlor epoxide, PCBs (calculated as Aroclor 1260), and HCB,

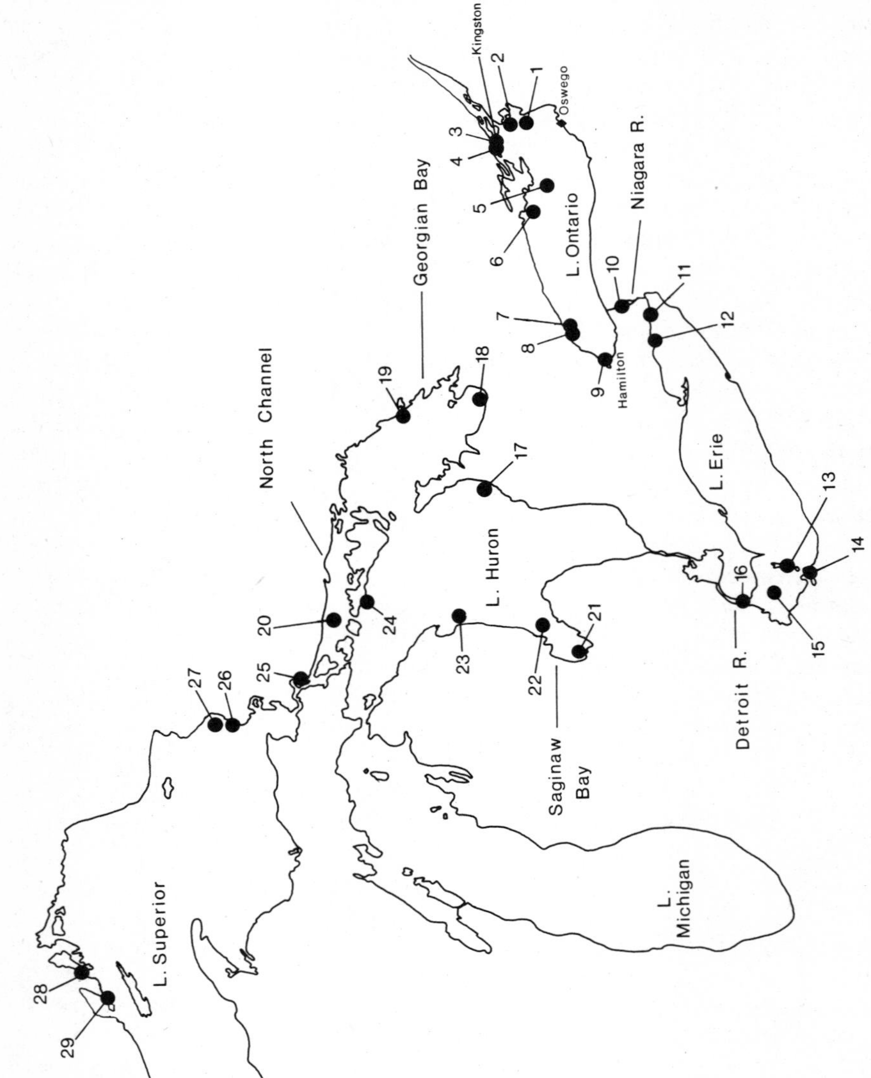
Kingston
Oswego
Hamilton
L. Ontario
Niagara R.
L. Erie
Detroit R.
L. Huron
Saginaw Bay
Georgian Bay
North Channel
L. Michigan
L. Superior
1
2
3
4
5
6
7
8
9
10
11
12
13
14
15
16
17
18
19
20
21
22
23
24
25
26
27
28
29

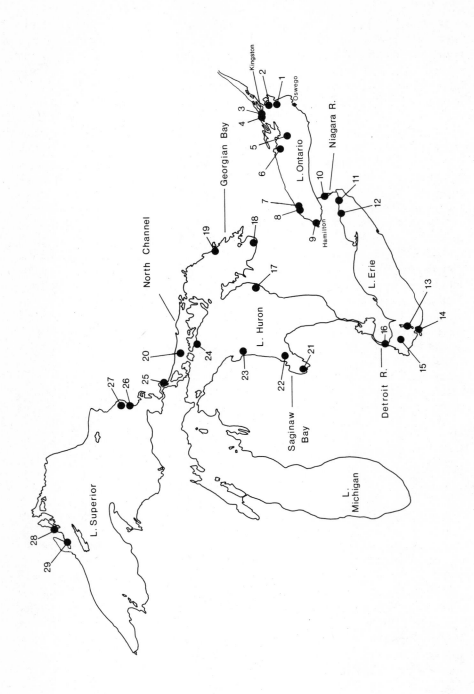

as well as *p'p*-DDT and derivatives, in Lake Erie and Lake Ontario Herring Gulls were as high as those previously found anywhere in the world for any fish-eating bird, and much higher than elsewhere in Canada (Keith and Gruchy, 1972; Gilbertson and Reynolds, 1974). Data from western Lake Superior (Ryder, 1974) proved more encouraging because contaminant levels there were substantially lower.

Samples collected on a routine basis as outlined in Section 3.1 are now analyzed for *p,p'*-DDT, DDE and DDD, dieldrin, heptachlor epoxide, β-BHC, tetrachlorobenzenes, pentachlorobenzene, HCB, chlordane, oxychlordane, mirex, photomirex (8-monohydromirex), and PCBs. The latter are quantified by capillary gas chromatography against an Aroclor 1260 standard or a 1 : 1 mixture of Aroclors 1254 : 1260, which gives a consistent if slightly higher value for residues found in wildlife tissues (Reynolds, 1971; Norstrom et al., 1978; Won and Norstrom, 1980). These routine analyses are carried out at the Ontario Research Foundation, Mississauga, Ontario. Exact procedures are described in Reynolds and Cooper (1975) and Norstrom et al. (1980b). Residue identities have all been confirmed by mass spectrometry (Hallett et al., 1977a). In recent years, DDD, β-BHC, chlordane and, oxychlordane levels have dropped to near or below the detection limit.

In 1978, a review of methodology indicated that losses of DDD, DDT, HCB and oxychlordane of up to 50% were occurring due to the use of vacuum oven drying. The losses were reproducible, so correction factors were applied to pre-1978 data. Fig. 19.2 summarizes level of six major organochlorine contaminants obtained between 1974 and 1979 in the routinely sampled herring gull colonies on the Canadian Great Lakes. An analysis of these data is given in Section 3.4.

3.3. Qualitative Chemical Analysis

Because the level of persistent, readily identifiable organochlorine contaminants is so high in herring gulls, this species is ideal for investigating the bioaccumulation of other persistent chemicals suspected or known to be present in fish, water, sediments or other components of the Great Lakes ecosystem. Therefore herring gulls can function as an early warning system

Figure 19.1 Map of the Great Lakes indicating the locations of herring gull colonies mentioned by name in the text or by number in Figs. 19.2 and 19.3. The colonies are as follows: 1, Little Galloo Island; 2, Pigeon Island; 3, Snake Island; 4, West Brothers Island; 5, Scotch Bonnet Island; 6, Island off 'Prequ'ile'; 7, Leslie Street Spit; 8, Muggs Island; 9, Hamilton Harbour; 10, Niagara River; 11, Port Colborne Breakwater; 12, Mohawk Island; 13, Middle Island; 14, Sanduski Turning Basin; 15, Middle Sister Island; 16, Fighting Island; 17, Chantry Island; 18, Nottawasaga Island; 19, Castle Rock; 20, Double Island; 21, Chanel and Shelter Islands; 22, Little Charity Reef; 23, Black River Island; 24, Manitoba Reef; 25, Island off Pumpkin Point; 26, Mamainse Harbour; 27, Agawa Rocks; 28, Granite Island; and 29, Silver Islet.

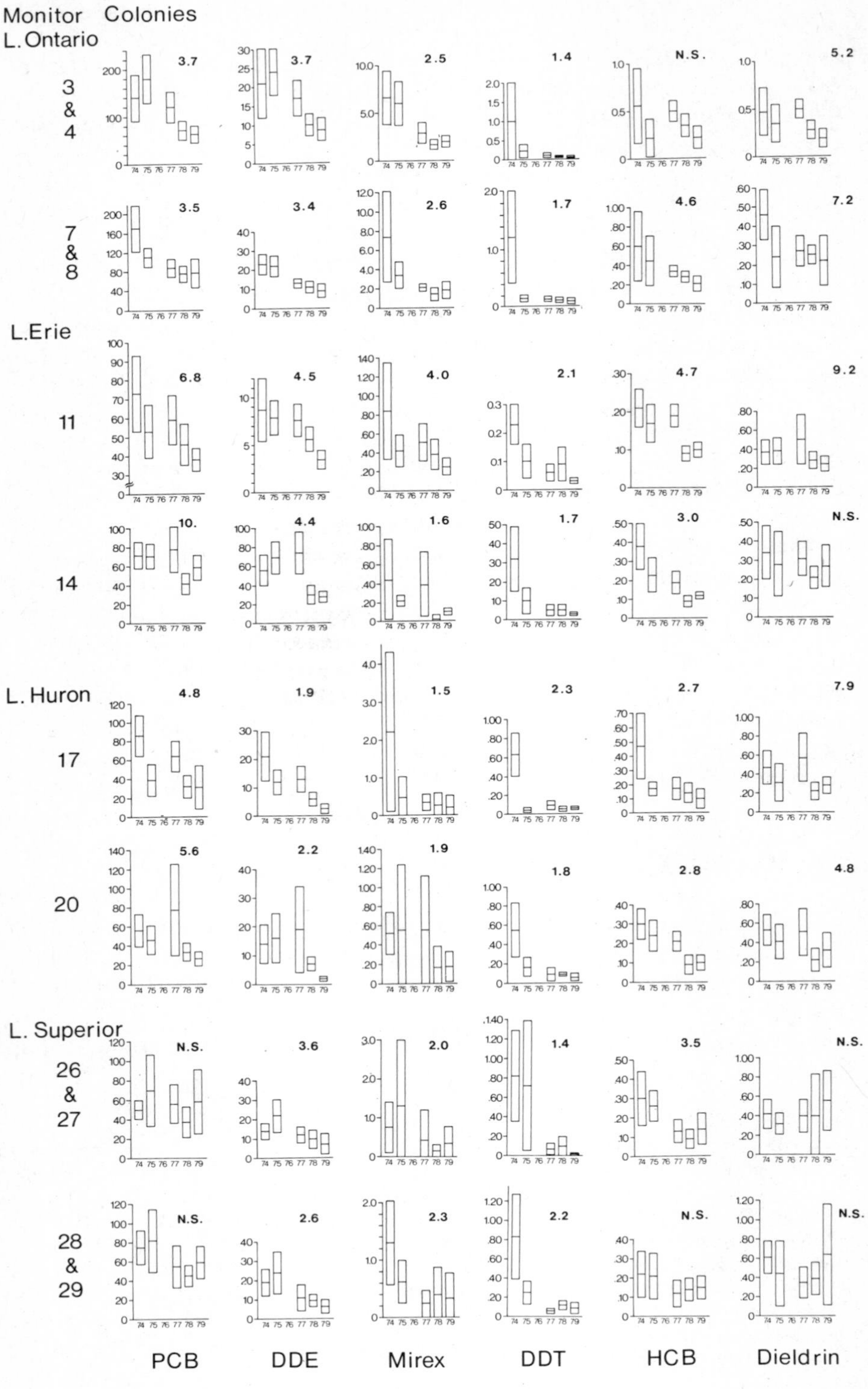

Monitor Colonies
L. Ontario
3 & 4
7 & 8
L. Erie
11
14
L. Huron
17
20
L. Superior
26 & 27
28 & 29
PCB
DDE
Mirex
DDT
HCB
Dieldrin

for contaminants not known to be present in the environment. Many contaminants that are barely detectable in fish are present in sufficiently higher quantities in herring gulls and their eggs to obtain diagnostic mass spectra free from biogenic interferences. With the further help of gas chromatographic retention times, residue identification is usually possible.

In 1975, a 1 kg pool of lipid was extracted from adult herring gulls from a colony on Pigeon Island, Eastern Lake Ontario. Shortly before this, mirex had been identified as a contaminant in Lake Ontario fish (Kaiser, 1974), so the lipid pool was investigated by GC/MS and GC/Hall electrolytic conductivity chromatography (Hallett et al., 1976). Not only was mirex detected (at levels around 200 ppm in the lipid), but a photodegradation product, 8-monohydromirex (labeled as photomirex), was also discovered at levels approximately one half those of mirex. This was the first indication of the presence of photomirex in the Great Lakes ecosystem, and the first report of this compound as an environmental contaminant. A procedure was then developed for routine determination of mirex and photomirex in the presence of large PCB interference (Norstrom et al., 1980b). Five minor mirex-related compounds were also identified, and the distribution of mirex and photomirex in all the routinely monitored Great Lakes herring gull colonies was determined (Norstrom et al., 1980a). The distribution was consistent with the hypothesis that the mirex originated from an industrial source or sources leading to Lake Ontario, as herring gull eggs from that lake had consistently higher levels of mirex than eggs from the other lakes. Only eggs from a colony in eastern Lake Erie, which is within foraging distance of Lake Ontario and near a known source of mirex on the Niagara River, had mean levels of mirex approaching those of Lake Ontario (Section 3.5). These data played an important part in defining the extent of the mirex problem in the Great Lakes (Mirex Task Force, 1977).

In addition to mirex-related residues, the lipid pool from Lake Ontario was found to contain a number of chlorinated organic compounds related to the cyclodiene class of pesticides, including *cis*-chlordane, *trans*-nonachlor, photo *cis*-nonachlor, oxychlordane, heptachlor epoxide, and dieldrin (Hallett et al., 1977a). Chlorobiphenyl-methyl ethers (methoxy-PCBs) containing 4, 5, 6, and 7 chlorines were also identified. These compounds have not subsequently been found in egg lipids, and it is therefore possible they were formed from hydroxy-PCBs or other PCB metabolites during saponification of the whole-bird extracts with methanolic KOH.

Herring gull eggs from Lake Ontario were also investigated by GC/MS in 1975, 1976, and 1977. Over this period the residues identified included

---

Figure 19.2 Residue data in ppm fresh weight obtained from 1974 to 1979 on the routinely sampled herring gull monitor colonies (see text and Fig. 19.1 for names and locations). Mean levels (± 1 standard deviation) are given. N equals 10 in most cases and is always greater than or equal to 8. The boldface number in the right-top quadrant of each plot is the calculated half-life for the plot (see text). N.S. signifies that residue levels do not follow an exponential decline.

tri- and tetrachloroethylene and pentachlorobenzene as well as isomers of di-, tri-, and tetrachlorobenzene (Hallett et al., 1982). Note that the units for all residue levels in that publication should be μg/kg, not ng/kg. Levels of all tetrachlorobenzene isomers combined ranged from 13 to 39 μg/kg fresh weight. Levels of pentachlorobenzene ranged from 25 to 51 μg/kg fresh weight. Three isomers of heptachlorostyrene were identified tentatively by GC/MS. Low levels of polynuclear aromatic hydrocarbons (PAHs) with two to five rings were also found (Table 19.1) in eggs as well as in adults collected (Hallett et al., 1977b). The absence of methylated PAHs indicated a combustion-related source rather than petroleum hydrocarbon contamination.

In 1978 and 1979 the intensive GC/MS analysis was shifted to samples from Lake Erie and the Detroit River. The colony on Fighting Island in the lower Detroit River was found to be more heavily contaminated with PCBs and other industrial chemicals than colonies in Lake Erie. Other chemicals identified by GC/MS in gull eggs from Fighting Island were: octachlorostyrene, three isomers of heptachlorostyrene, polychlorinated terphenyls with four to eight chlorines, and mixed chloro-fluoro compounds with the molecular formula $C_{14}H_7F_6Cl_2$.

Reports of the presence of polychlorinated dibenzodioxins in fish from Lake Ontario and from the Tittibawasi River, which flows into Saginaw Bay, Lake Huron, changed the emphasis in 1980 to a search for this class of compounds, particularly the most toxic isomer 2,3,7,8-tetrachlorodibenzo-*p*-dioxin (TCDD). A survey of tetrachlorodibenzo-*p*-dioxins in pooled samples of eggs from all of the Great Lakes revealed that the 2,3,7,8-isomer was present in each of the samples. There was a "background" level of about 10 ng/kg in eggs from Lakes Superior, Michigan, Erie, and the main body of Lake Huron. Levels in Lake Ontario colonies averaged 59 ± 9 ng/kg ($n = 4$), and were 40–90 ng/kg in Saginaw Bay of Lake Huron (Norstrom et al., 1982). This pattern of distribution, as well as the absence of other TCDD isomers, suggested that this contamination was a result of past production of 2,4,5-trichlorophenol by chemical plants with access to Saginaw Bay and the Niagara River. These results led to an assessment of the levels of 2,3,7,8-TCDD in Lake Ontario commercial and sport fish as well as an intensive bilateral effort to find and control the sources of this compound.

In order to determine the trends of TCDD residue levels, we analyzed pooled samples of herring gull eggs from Scotch Bonnet Island, Lake Ontario, which had been collected over the period 1970–1980 and deposited in the CWS National Specimen Bank. Levels were found to be much higher in the early 1970s (1200 ng/kg in 1972), declining steadily since 1974 with an apparent half-life of 2.5 yr in the herring gull population. The most likely source of this chemical was a manufacturer in the Niagara River that ceased producing trichlorophenol (of which 2,3,7,8-TCDD is a by-product) in the early 1970s. At about the same time, Hyde Park, the last active large landfill site (containing over 70,000 tonnes of chemical waste) was capped, thereby limiting but not necessarily eliminating further inputs.

Table 19.1. List of Polynuclear Aromatic Hydrocarbons (PAHs) Identified from Pools of Herring Gull Eggs (Pigeon Island) and Adults (Kingston) (Levels are given in ppm lipid weight)

Compounds	Concentration (ppm)[a] Egg Lipid	Adult Lipid	Mass Spectral Confirmation
Naphthalene	0.05	0.05	*
2-Methyl naphthalene	0.04	0.06	*
1-Methyl naphthalene	0.04	0.01	*
Biphenyl	0.15	0.02	*
Acenaphthene	0.04	0.01	*
4-Methyl biphenyl	0.06	0.01	*
Fluorene	0.04	0.003	*
Anthracene	0.15	0.02	*
Penanthrene	ND	0.002	*
1-Phenyl naphthalene	0.01	0.008	
2-Methyl penanthrene	0.02	0.007	*
1-Methyl penanthrene	0.01	0.02	*
9-Methyl anthracene	0.01	0.03	*
3,6-Dimethyl phenanthrene	ND	0.01	*
Fluoranthene	0.08	0.02	*
Pyrene	0.08	0.02	*
1-Aza pyrene	a	a	
9-Acetylanthracene	a	a	
1,2-Benzofluorene	a	a	*
2,3-Benzofluorene	a	a	*
1-Methyl pyrene	a	a	*
2-Acetyl phenanthrene	a	a	
1,1-Binaphthyl	a	a	
Chrysene	0.05	a	*
Benz(e)Pyrene	0.03	0.02	*
Benz(a)Pyrene	0.04	0.03	*
Perylene	0.05	0.03	

[a] a = PCB interference; ND = not detected.

Work is now continuing on the search for polychloro-dibenzodioxins, -dibenzofurans, -styrenes, -diphenyl ethers, and toxaphene in herring gulls. It has recently been reported (Swain et al., 1982) that residues of toxaphene, a persistent agricultural organochlorine, are present in fish from Lake Superior. Alternative methods of analysis are also being applied. Neutron activation analysis has recently shown that PCBs and DDE account for 96% of the total weight of organically bound chlorine in egg samples from Lake Ontario, and that organically bound bromine levels range from 0.3 to 2.5 ng/kg in the same samples (Norstrom et al., 1981).

3.4. Trend Analysis

A powerful tool in wildlife residue monitoring is the systematic, periodical sampling of the same population. Regardless of the precise levels of contamination, changes in contaminant levels can be accurately measured and correlated with remedial measures.

Preliminary analysis of the data up to 1978 (Weseloh et al., 1979; Mineau et al., 1979) and reanalysis after the addition of 1979 data revealed that there had been a net decrease in contaminant levels over the years and that this decrease could best be described by a first-order loss rate. The advantage of such a first-order model in studies of residue dynamics is that half-lives can be calculated from year-to-year changes without regard to the initial concentration. These half-lives have been termed 'average population half-lives' (Armstrong and Sloan, 1980). An added benefit of using log-transformed residue values is a reduction of the skewness inherent in most residue data. The calculated "average population half-lives" are given in Fig. 19.2.

It is important to remember that these determinations of average population half-lives represent the integration of a complex set of processes in the herring gull and its prey. DDT and PCB residue levels in caged herring gulls equilibrated; that is, the input and excretion rates were the same after about 1 yr of exposure to Lake Michigan smelt (Anderson and Hickey, 1976). Anderson and Hickey estimated the half-life of DDE clearance in the birds to be 90–100 days. We have recently obtained a direct measurement of the half-life of ^{14}C-DDE in wild herring gulls of approximately 250 days. The clearance half-lives of other highly chlorinated compounds like PCBs may be similar because they are slowly metabolized in a number of species. Compounds such as mirex may have longer half-lives than DDE because they are not known to be metabolized at all.

It is apparent that for compounds such as DDE which reach equilibrium concentration in female gulls in approximately 1 yr, levels in eggs will be proportional to those in the diet integrated throughout the year. For compounds that are cleared quickly from the gull, residue levels in eggs will reflect the diet shortly before yolk deposition. If the rate of decline of residue levels in food is slow compared to the rate of excretion of the compound, the proportionality between food and gull egg will remain constant. If, on the other hand, the half-life of the contaminant decline in food is similar to or faster than the excretion half-life in the gull, the rate of decline in contaminant levels in gulls (and eggs) will be a hybrid of the two half-lives. Computer modeling of the dynamic processes has been initiated to obtain quantitative interpretation of the trend data.

The declining residue levels in herring gulls suggest that control measures have been successful in curbing contamination at the higher trophic levels within a surprisingly short time. Pesticides such as DDT, aldrin, and dieldrin were severely restricted in Canada and the United States between 1969 and

1971. Production of PCBs was voluntarily restricted by Monsanto Corporation in 1971, and a ban on further use of PCBs in new, nonelectrical equipment came into effect in 1977 in both Canada and the United States. Discharge of hexachlorobenzene (HCB), as a by-product in the manufacture of organochlorines was consequently reduced. It is likely that these control measures have resulted in a decreased bioavailability of the contaminants. This reduced bioavailability probably stems from irreversible sedimentation, atmospheric loss, or flushing from the system. If all direct discharges of contamination were stopped we could expect contaminants to be present for much longer in the upper lakes because of the prevailing low sedimentation rates, long flushing times, and large surface areas which receive greater atmospheric inputs. The seemingly contradictory data from Lake Erie, that is, long average population half-life despite known rapid sedimentation and short flushing time, suggest a continued input of material into the aquatic food web through continued discharge (see Section 3.5) or resuspension of contaminants due to the lake's shallow depth. Resuspension of sediments from the western basin of Lake Erie is known to be very extensive (Thomas et al., 1976) although release rates of contaminants from sediments are not known.

3.5. Geographical Distribution of Contaminants: Lakes Erie, Huron, and Ontario

As mentioned in Section 3.1, we extensively sampled herring gull colonies on lakes Erie, Ontario, and Huron to investigate contaminant distribution. Levels of DDE, PCBs, and mirex are presented in Fig. 19.3. These compounds were chosen here as they represent, respectively, the main breakdown product of a widespread agricultural organochlorine insecticide, a ubiquitous industrial organochlorine, and a persistent organochlorine with known sources limited to Lake Ontario at the Oswego River and the Niagara River.

Eighteen herring gull colonies are located at the western and eastern extremities of Lake Erie and in the Detroit River. There are no colonies in the central section of the lake. Eggs from Fighting Island in the Detroit River had significantly higher levels of DDE and PCBs than those from any of the other colonies samples. Eggs from the other four colonies sampled in 1979 showed no significant pattern in DDE distribution but there was a significant regression between decreasing PCB levels and increasing distance from the Detroit River. Similarly, there was a significant regression between decreasing concentration of mirex residues and increasing distance from Lake Ontario and the Niagara River on those same Lake Erie colonies.

Herring gull colonies are widespread throughout Lake Huron. Gulls nest on more than 600 islands and we were able to obtain samples from all major areas of the lake. As in Lake Erie, colonies with elevated levels of DDE

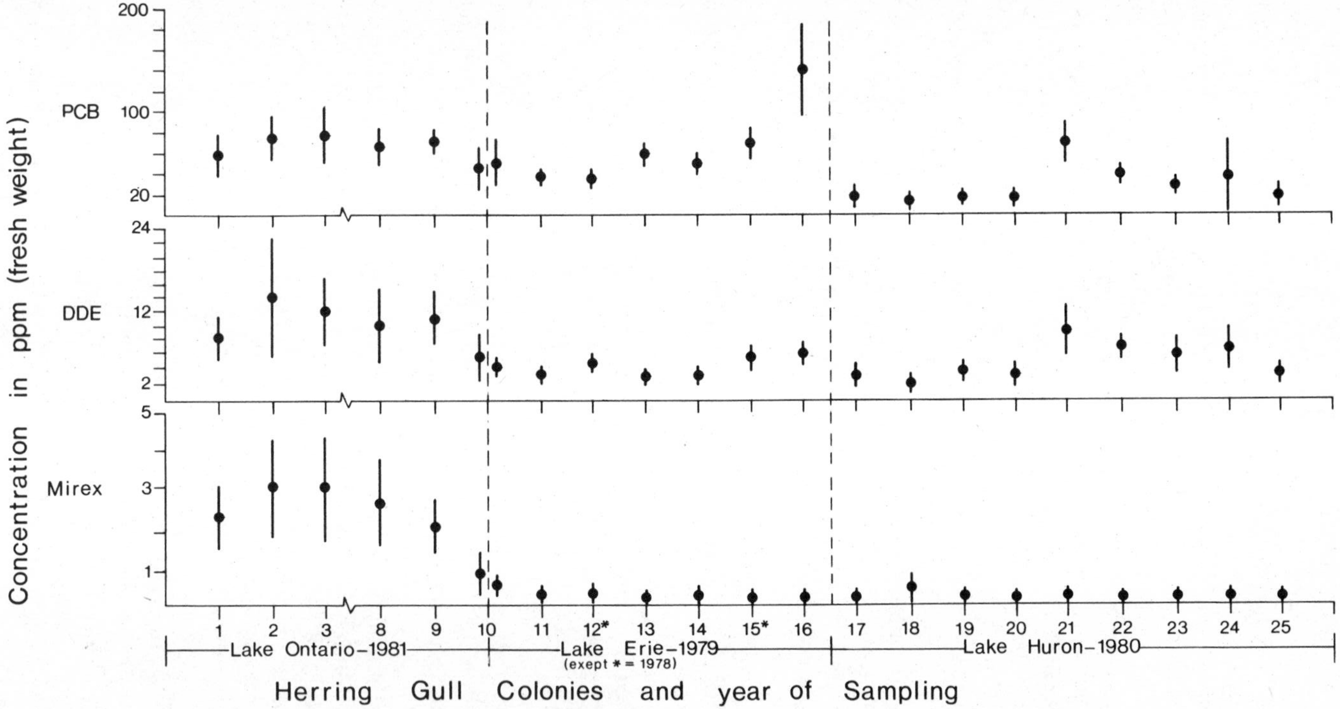

Figure 19.3 Distribution of residue levels (ppm fresh weight) for several herring gull colonies in Lakes Ontario, Erie, and Huron, in numerical sequence from east to west. Mean level ± 1 standard deviation are given. $N = 10$ in all cases. Refer to text and to Fig. 19.1 for names and locations of the colonies.

also had elevated levels of PCBs. For DDE and PCBs, eggs from colonies in Saginaw Bay (Channel and Shelter Island and Little Charity Island) and some of those in the main body of the lake (Manitoba Reef and Black River Island) had levels significantly higher than those from Georgian Bay and North Channel. Levels of DDE and PCBs from colonies in Georgian Bay and North Channel did not differ significantly. Levels of mirex from colonies throughout Lake Huron were low and they showed wide variation and overlapping ranges. There were no significant differences between colonies.

Lake Ontario has the smallest population of herring gulls of any of the Great Lakes. Ten herring gull colonies are located at the extreme western and eastern ends of the lake and along the north shore. There are no colonies on the southern half of the lake from Hamilton to Oswego. Contaminant levels in herring gull eggs from different colonies in Lake Ontario were more homogeneous than in either Lakes Huron or Erie. Of the five colonies sampled, there were no significant differences in the levels of DDE, PCBs, or mirex. However, the eggs from Snake and Pigeon islands had the greatest absolute mean levels of all three compounds. Levels of DDE, PCBs, and mirex in the colony at Niagara Falls were significantly lower than levels in the most contaminated Lake Ontario colony.

The locations of the gull colonies with the greatest mean levels of DDE and PCBs in each lake appear to correspond with the location of major sources of these contaminants as indicated by areas of elevated residue levels in sediments. Such was the case for eggs from Fighting Island in the Detroit River which were heavily contaminated, especially with PCBs. This contamination decreased in sediments (Frank et al., 1977) and herring gulls with increasing distance from the river. Similarly, gulls in Saginaw Bay, Lake Huron had the highest levels of DDE and PCBs of any Lake Huron colony, which was again consistent with the sediment data (Frank et al., 1979b). Frank et al. (1979a) have shown that DDE residues in the sediments from Lake Ontario were uniformly high. This suggested widespread (past) use of DDT with many sources into the lake. High PCB levels in sediments, extending from the mouth of the Niagara River eastward along the south shore of the lake, indicated that the major source of this contaminant in Lake Ontario was the Niagara River. Elsewhere in the lake, PCB levels in sediment were relatively low and uniform. There are no gull colonies along the entire southern shore of Lake Ontario that could reflect the elevated PCB residues found there. Mirex levels in Lake Ontario gulls were more or less uniform as there are no colonies situated in either of the areas of high contamination, namely the mouths of the Niagara and Oswego rivers. It is noteworthy that mirex levels in eggs from the Niagara River were lower than in those from Lake Ontario. Preliminary data suggest that most of the persistent organochlorines entering the Niagara River are carried downstream before they can be incorporated into the local biomass (John Carey, personal communication).

When we obtain a better knowledge of organochlorine dynamics in the herring gull and we know what prey species constitute the bulk of its diet at critical times of the year from the standpoint of contaminant accumulation, we will then more accurately know how these egg residues are reflecting those found in the food chain. Between-lake differences are such that we can make a case for the herring gull being a lakewide monitor. The evidence presented here shows that it may be a good indicator of serious contamination of its food chain in specific areas such as the Detroit River and Saginaw Bay, although we have seen that it is not very useful as an indicator of contamination in the Niagara River.

4. THE HERRING GULL AS AN INDICATOR OF BIOLOGICAL EFFECTS

4.1. Documentation of Reproductive Problems

Reproductive problems in herring gulls from the Great Lakes were first noted on Lake Michigan (Keith, 1966; Ludwig and Tomoff, 1966). Thin, flaking eggshells, high embryonic mortality and poor chick survival were reported, and abnormal parental behavior was suggested. Hickey and Anderson (1968) also concluded that the herring gull was suffering from DDE-induced eggshell thinning in Lake Michigan. The implications of these findings to other species were serious as we now know the herring gull to be relatively resistant to shell thinning.

Effects of contaminants on fish-eating birds of the Canadian Great Lakes were first reported in 1970. In June of that year, members of the Kingston Field Naturalists made their annual visit to Pigeon Island in Eastern Lake Ontario and noted that black-crowned night herons (*Nycticorax nycticorax*) were having poor reproductive success, and herring gull eggs were found with dented or collapsed shells (Edwards, 1970). The next observations were made during the same summer in Hamilton Harbour. A high proportion of common tern (*Sterna hirundo*) embryos were dying in the egg before and at hatching (Gilbertson, 1975). Following the decision to concentrate on the herring gull as a long-term monitor of biological effects (Section 2), colonies from Lakes Ontario, Erie, and Huron were investigated in 1973. Eastern Lake Ontario was singled out for more intensive work because fledging success (the number of young able to leave the breeding colony per breeding pair) there was found to be only 0.12 to 0.21 chicks per pair compared with a normal fledging success of more than 1 per pair (Gilbertson, 1974). Thin eggshells were also found. From more detailed studies in 1973 of one colony from Eastern Lake Ontario (Scotch Bonnet Island), Gilbertson and Hale (1974a, b) concluded that the number of eggs laid was low, that eggs were disappearing at a high rate, that few of the remaining eggs hatched on account of the high rate of embryonic mortality during the first week of incubation, and that survival of hatched young to fledging was also low. In short, reproduction was found to be affected at all possible levels. A similar situation in 1973 was observed on nearby Brothers Island (Teeple, 1977). No direct cause–effect relationship could be established between residues of the common organochlorine contaminants measured and any aspect of the reproductive failure. Infertility was not a problem because all eggs contained embryos at some stage of development. Abnormal parental behavior was thought to be a contributing factor. Studies of a colony at Port Colborne in eastern Lake Erie between 1973 and 1976 revealed values for hatching and fledging success intermediate between those observed in the Lake Ontario colonies and those normally observed (Morris and Haymes, 1977).

From 1974 to 1976 reproductive data were collected from widely separated colonies in all four of the Canadian Great Lakes (Gilman et al., 1977; Morris and Haymes, 1977; Ryder and Carroll, 1978). These data revealed that the reproductive problem was confined to Lake Ontario and eastern Lake Erie. Eggshell thinning was greatest in Lake Ontario colonies but no flaking or denting was recorded. Artificially incubated eggs from Lake Ontario colonies had consistently lower hatching rates than those from colonies in Lake Erie, Lake Huron, or the Atlantic Coast, suggesting that intrinsic factors were in part responsible for the reproductive failures observed (Gilman et al., 1977).

4.2. Search for Intrinsic Factors Causing Hatching Failure

Because levels of PCBs in eggs from Lake Ontario were very high, it was suspected that the high rates of embryonic and chick mortality in the herring gull population might be caused by chick edema disease (Gilbertson, personal communication). Pericardial edema in poultry (chick-edema or hydropericardium factor) had been linked to contamination of PCB mixtures by chlorinated dibenzofurans (Vos et al., 1970). Some chlorinated dibenzo-*p*-dioxins, specifically tetrachloro and hexachloro isomers (Firestone, 1973) and chlorinated azobenzenes (Shrankel et al., 1982) have also been shown to be strong inducers of chick edema. In order to test the hypothesis that chick edema disease was present, eggs collected in 1974 from Scotch Bonnet Island (Eastern Lake Ontario), the Port Colborne colony (Eastern Lake Erie), and control colonies in Northern Alberta and the Bay of Fundy were artificially incubated. Pathophysiological studies of the resulting embryos and newly hatched chicks indicated that symptoms of chick edema disease were present in the Lake Ontario and Lake Erie colonies in 1974 (Gilbertson and Fox, 1977; Gilbertson 1983). Observed pathologies in the embryos included: accumulations of a muco-serous exudate in the pectoral and abdominal regions, ascites, larger volumes of pericardial fluid, shorter tarsal lengths, higher liver weights, and possibly, elevated porphyrin levels. Although not seen in the herring gull, congenital anomolies were abnormally prevalent in chicks of other species of fish-eating birds in the Port Colborne and Lake Ontario colonies in 1971–1973 (Gilbertson et al., 1976). This suggested that a teratogenic agent was also present.

High porphyrin levels, growth reduction and enlarged livers are all consistent with known biological effects of one or more of the organochlorine contaminants known to be present at high levels in the Great Lakes, namely DDE, PCBs, and HCB. HCB is a well-known porphyrogenic agent but is not chick-edema active. An initial search for chlorinated dibenzo-*p*-dioxins and chlorinated dibenzofurans at μg/kg levels was made in eggs collected from Scotch Bonnet Island (Bowes et al., 1973), but it was unsuccessful. However, as seen in Section 3.3, eggs from all lakes have recently been found to contain measurable quantities of 2,3,7,8-TCDD, with the highest levels in eggs from Lake Ontario and Saginaw Bay, Lake Huron (Norstrom et al., 1982). Retrospective analyses of frozen eggs from Scotch Bonnet Island from 1971 to 1980 indicated that levels have declined steadily from a high of 1200 μg/kg in 1972. Based on laboratory studies, 2,3,7,8-TCDD is chick-edema active and the most embryotoxic, teratogenic, hepatotoxic, porphyrinogenic substance tested in chick embryos (J. H. Koeman and J.J.T.W.A. Strik, personal communication). Therefore, 2,3,7,8-TCDD is a prime suspect for the pathology and poor reproductive success observed in the early 1970s. Levels of other chemicals with similar properties (eg. chlorinated dibenzofurans and other chlorinated dibenzo-*p*-dioxins) are currently being investigated.

Chick-edema active and hepatoporphyrinogenic aromatic polyhalogenated hydrocarbons tested to date are strong inducers of the enzyme aryl hydrocarbon hydroxylase (AHH) (based on: Strik et al., 1980; Poland et al., 1976; and Poland et al., 1979). AHH induction was measured in livers of 25-day-old herring gull embryos collected from a series of colonies throughout the Great Lakes in 1981 (Ellenton, in prep.). AHH levels were markedly higher in embryos from colonies in Lake Ontario and Saginaw Bay, Lake Huron, and were significantly correlated with 2,3,7,8-TCDD levels but not with any of the routinely measured organochlorine contaminants as discussed in Section 3.2. Levels of other strong AHH inducers were not measured.

Finally, injection of synthetic mixtures of the common organochlorine pollutants or extracts of 1976 Lake Ontario eggs into fresh, relatively uncontaminated eggs from the Atlantic Coast failed to produce any decrease in hatching success or any of the pathologies observed in the early 1970s (Gilman et al., 1978). This may be interpreted as indicating that the common organochlorine contaminants or their metabolites, as extracted and injected did not include the intrinsic factor(s) or that they were not present at sufficient levels to produce any grossly observable pathologies (2,3,7,8-TCDD levels are now known to have been approximately 200 ng/kg in Lake Ontario eggs in 1976). This is consistent with the fact that 1976 was the first year in which a marked improvement in reproduction was observed in Lake Ontario colonies, and that the rate of congenital anomalies has been normal in fish-eating birds in Lake Ontario since 1976.

4.3. Parental Care: The Extrinsic Factor

Gilbertson (personal communication) and Teeple (1977) both suggested that poor parental care was a factor in the low reproductive success of Lake Ontario herring gulls. Egg exchanges between nests in "clean" maritime colonies and "dirty" Lake Ontario colonies, as well as artificial incubation of eggs from these locations, provided strong evidence that both intrinsic and extrinsic factors were involved in the reduced hatchability experienced in the Lake Ontario colonies (Peakall et al., 1978). The extrinsic factor appeared to be incubation behavior. Anomalous behavioral responses to the investigators (and hence to potential predators) had been reported by Ludwig and Tomoff (1966) and later by Fox et al. (1978), but in neither case was the natural variability of the response to an observer assessed.

The egg exchange experiments were run for three consecutive nesting seasons. In 1975, it was obvious that parental care was deficient and that an intrinsic factor was affecting hatchability. In 1976, parental care appeared to be normal but the intrinsic factor was still present. In 1977, no effects were seen suggesting that perhaps some threshold had been crossed. Failure of the 1977 experiment to detect any problem is consistent with fledging success data presented in Section 4.4.

Telemetered eggs placed in nests of Lake Ontario gulls in 1976 indicated that incubation attentiveness was poorer than that measured in a "clean" coastal colony. Nest air temperatures were also lower on the Lake Ontario colony. The same differences were noted between successful and unsuccessful nests in the Lake Ontario colony. Incubation attentiveness was inversely correlated with organochlorine content of the eggs, suggesting pollutant-induced endocrine dysfunction. A combination of the major organochlorine contaminants found in Lake Ontario herring gull eggs has been shown to induce endocrine dysfunction accompanied by similar anomalies in parental behavior in ring doves (*Streptopelia risoria*) (MacArthur et al., 1983). Telemetered egg studies failed to show any behavioral anomalies in Lake Ontario colonies in 1978. Similarly, attempts in 1978 to compare the behavioral repertoire of gulls from two colonies in Lake Erie and one in Lake Ontario with divergent pollutant burdens failed to show any differences in nest attentiveness.

4.4. Measurement of Reproductive Success

The method used for assessing the reproductive success of herring gull colonies in 1975 and 1976 was very labor-intensive because individual nest histories were maintained and nests were visited at least once a week. These frequent visits introduced the very real confounding factor of human disturbance and restricted the number of colonies that could be assessed in any one year. However, the technique did allow the investigator to identify the major types of hatching failure and to differentiate between low hatching success and increased chick mortality. In 1977, the herring gull monitoring program was incorporated into the larger surveillance program of the Great Lakes Water Quality Board of the International Joint Commission (IJC) which required that monitoring efforts be expanded geographically and that emphasis be placed on trend determination rather than research. From 1978 through 1981, reproductive assessments in the form of a simplified measure of survival to 21 days (Mineau and Weseloh, 1981) were carried out on all the colonies routinely monitored for residue levels (see section 3.1). This assessment of fledging success is conducted rapidly with relatively little disturbances. By aging all chicks (Mineau et al., 1982), one can ensure that the data obtained are comparable between colonies. Weseloh et al. (1979) reviewed available fledging rate data obtained on monitor colonies from 1974 to 1978; these results, as well as unpublished data obtained from 1979 to 1981, are presented in Table 19.2. They illustrate the recovery of colonies from eastern Lake Ontario and eastern Lake Erie. Net reproductive output on Chantry Island in southern Lake Huron seems to have improved

Table 19.2. Average Number of Herring Gull Chicks (to 21 Days of Age) per Nest on the Main Monitor Colonies and Scotch Bonnet Island between 1974 and 1981 (See Fig. 19.1 for colony locations)

Location	1972	1974	1975	1976	1977	1978	1979	1980	1981
Lake Ontario									
Snake Island	0.21[a]				1.01	0.86[f]	1.60	1.49	1.73
Scotch Bonnet I.	0.12[a]		0.15[c]		1.10	1.01[f]			2.13
Mugg's Island					1.52[d]	1.47[d]	1.56		1.40
Lake Erie									
Pt. Colborne Bkw.		0.48[b]	0.65[b]	0.79[b]		1.45			
Middle Island						1.70	1.63	1.62	2.10
Lake Huron									
Chantry Island			1.48[c]		1.12	1.40	2.17	2.17	2.16
Double Island						1.57	2.17	2.25	2.23
Lake Superior									
Agawa Rocks						1.66	0.88	0.40	0.37
Granite Island			1.32[e]	1.55[e]		1.12	1.70	1.40	0.46

[a] Gilbertson, 1974.

[b] Morris and Haymes, 1977; uses 30 days to "fledging," these data are not strictly comparable but are not thought to differ greatly.

[c] Gilman et al., 1977.

[d] Fetterolf (personal communication).

[e] Ryder and Carroll, 1978.

[f] Minimum values, determined from one visit only. Values based on a single visit on unfenced colonies usually underestimate the fledging success by 25–40%.

in recent years although early values were still within normal limits. Results from Lake Superior are puzzling because reproductive levels there were excellent during the early 1970s, but have recently plummeted. Research is now underway to find a cause for this collapse. Preliminary information suggests that food may be in short supply but the effects of the probable contamination of this population by toxaphene (Section 3.3) have yet to be assessed.

It has become increasingly apparent that our current approach to monitoring contaminant-mediated effects on reproductive performance needs further modification because net chick production is probably not affected by current pollutant burdens. A measurement that is able to detect minor intrinsic problems with the least disturbance is being developed (Section 5).

4.5. The Incidence of Congenital Anomalies As An Indicator of the Presence of Teratogens

In the early 1970s, congenital anomalies (crossed bills, malformed eyes, extra limbs, etc.) were abnormally prevalent in chicks of some species of fish-eating birds of Lake Ontario (Gilbertson et al., 1976; Gilbertson and Fox, in press). They were not found in herring gulls, which may reflect their lower sensitivity to such agents. However, anomalies may have gone undetected since few herring gull chicks hatched. Embryos were grossly deformed in some eggs which failed to hatch in 1974 and 1975. Crossed bills and extra limbs have since been noted in herring gull chicks and an adult with a crossed bill was trapped. Congenital anomalies have most frequently been observed in terns (Gilbertson et al., 1976; Hays and Risebrough, 1972; Gochfeld, 1975) and therefore terns might be better field indicators for the presence of congenital anomalies. A registry of congenital anomalies in fish-eating colonial birds has been established at the National Wildlife Research Centre, Ottawa, in an attempt to better understand this phenomenon and perfect a monitoring technique.

4.6. Mutagenicity Studies

Studies of the rates of sister chromatid exchange in herring gull embryos collected throughout the Great Lakes in 1981 suggested that if the levels of contaminants present were causing genetic damage, it was not detectable with this sensitive technique (Ellenton and McPherson, 1983a). Extracts from the same clutches failed to produce point mutations in *Salmonella* bacteria. However, positive results were obtained in cytogenetic tests in mammalian cells (Ellenton et al., 1983b). There was no relationship between the activity of the extracts and any of the measured contaminant levels; the activity of extracts from "clean" eggs was similar to those from Great Lakes colonies.

4.7. Biochemical Monitoring

Results of AHH assays on 1981 embryos were reported in Section 4.2. They suggest that AHH activity may be a feasible screening technique for highly toxic microcontaminants such as TCDD which are potent AHH inducers. We are expanding our studies to measure other mixed-function oxidase activities.

5. WHERE DO WE GO FROM HERE?

We can summarize our knowledge of the herring gull as an indicator of organochlorine contamination as follows:

1. The herring gull accumulates very high levels of persistent organochlorine contaminants.
2. These contaminants are almost entirely obtained from the aquatic food chain of the Great Lakes and, therefore, the herring gull can be considered a general Great Lakes ecosystem monitor.
3. Several environmentally relevant compounds (e.g., the mirex derivatives) were first identified from herring gull tissues. In other cases (e.g., the contamination by 2,3,7,8-TCDD) the gull has supplied the only available information on the spatial and temporal scope of the contaminant problem.
4. The herring gull has proved useful in elucidating the fate of persistent organochlorines once control measures were established.
5. The herring gull can be useful in documenting contaminant levels in some problem areas of contamination, although it most probably integrates residue levels over wide areas.
6. Gull population in Lakes Ontario and Erie were suffering from severe reproductive problems in the early 1970s. Factors intrinsic to the egg as well as behavioral problems in the adults were implicated. Although a cause and effect relationship has not yet been proven, the evidence suggests that chick-edema active, teratogenic, porphyrinogenic compounds such as 2,3,7,8-dibenzo-*p*-dioxin (TCDD) were involved.
7. Current reproductive rates, as measured, are normal with the exception of Lake Superior.
8. On the biochemical level, induction of the enzyme aryl hydrocarbon hydroxylase (AHH) was correlated with current levels of 2,3,7,8-TCDD, indicating a systemic response to contaminants.
9. Efforts to find any evidence of genetic damage at current contaminant levels have not been successful.

More is probably known about the biology of the herring gull than about any other fish-eating bird species in the Great Lakes. However, there are definite gaps in our basic knowledge of contaminant dynamics in the species. Research is currently in progress to develop a lipid-oriented bioenergetics model for free-living herring gulls in the Great Lakes, which will allow *a priori* simulation of food intake throughout the year, and especially the pre-breeding season. This model will give an estimate of contaminant intake, provided the feeding ecology is well defined. Rates of clearance of contaminants from gulls are being determined and, when in hand, they will be fitted to a pharmacodynamics compartmental model. A combination of the bioenergetics and contaminant clearance models will put the interpretation of egg residue levels on a sounder quantitative footing, and give a means of testing the effects on these data of changes in feeding habits, rate of contaminant clearance and so on.

The current use of the herring gull as a monitor of deleterious biological effects requires critical examination. It was briefly mentioned previously that the herring gull is not very sensitive to DDE-induced shell thinning. For this reason, we also monitor contaminant levels and reproductive success in double-crested cormorants (*Phalacrocorax auritus*) in the Great Lakes. Cormorants and other members of the family Pelicaniformes are well known to be very sensitive to shell thinning. Likewise, it was mentioned that terns might be better indicators of congenital anomalies and more work is needed in that direction.

Present levels of contaminants in the lower Great Lakes do not appear to have any gross effect on the reproductive success of the herring gull. Historically, the successful use of the herring gull as a monitor of adverse biological effects has been strictly to document a crisis situation. Unfortunately, the gross parameters we have examined have not shown the graded response to contaminant levels which would be necessary for biological monitoring to be useful as contaminant levels decline. In the last few years, it has become clear that we must develop more subtle and easily quantifiable tests that can best mirror present levels of contamination. For example, by means of inspection of a limited number of eggs, opened part-way through incubation in 1977 and in 1981, it was ascertained that embryonic mortality is still highly variable between colonies even though fledging rates are fairly uniform. An electronic sensor of embryonic motility has been developed so that the viability of a *large* number of eggs can be determined nondestructively. This method has good potential for establishing correlations between contaminant levels and egg viability.

Our ability to measure impacts of contaminants at the subcellular level represents a new and exciting development. However, enzyme induction is only indicative of a chemically induced stress on a biological system, the implications of which are not fully known. Enzyme induction does have considerable potential as a screening tool for the *presence* of biologically significant toxic material and to indicate for which class of chemicals analysis will be required.

In summary, current developments in the herring gull monitoring program lean very strongly toward a reductionist approach. It is very crucial however, that we not lose the holistic view of the species. Although it seems unlikely that the major reproductive failures of the 1960s and 1970s will recur, it is a possibility we cannot altogether discount. Observations on basic biological parameters obtained by naturalists at that time were the first indication that something was very wrong. Any monitoring scheme will always be at least partially dependent on this type of data.

ACKNOWLEDGMENTS

It is impossible to acknowledge all of the individuals who have contributed to the Herring Gull Program over the years. We thank them all, wherever they now are. Mr. M. Gilbertson deserves special mention for getting things started, A. P. Gilman for much of the subsequent impetus and D. B. Peakall and then T. C. Dauphiné for the direction of the project. Earlier drafts were commented on by T. C. Dauphiné, D. B. Peakall, M. E. Gilbertson, M. R. Chadwick, W. K. Marshall, and P. Loshak. This study was funded by the Canadian Wildlife Service in support of the International Joint Commission's Great Lakes Water Quality Board Surveillance Program.

REFERENCES

Anderson, D. W. and J. J. Hickey. (1976). Dynamics of storage of organochlorine pollutants in herring gulls. *Environ. Pollut.* **10:** 183–200.

Armstrong, R. W. and R. J. Sloan. (1980). Trends in levels of several known chemical contaminants in fish from New York State waters. New York State Department of Environmental Conservation. Technical Report 80-2.

Bowes, C. W., B. R. Simoneit, A. L. Burlingame, B. W. de Lappe and R. W. Risebrough. (1973). The search for chlorinated dibenzofurans and chlorinated dibenzodioxins in wildlife populations showing elevated levels of embryonic death. *Environ. Health Persp.* **5:** 191–198.

Dauphiné, T. C. (1980). The surveillance of toxic substances in Great Lakes wildlife. In Proc. First Biological Surveillance Symposium, 22nd Conference on Great Lakes Research, Can. Tech. Rep. Fish. Aquat. Sci. 976, pp. 153–174.

Edwards, M. (1970). *Audubon field notes*, Dec. 1970. Item No. 43.

Ellenton, J. A. and M. F. McPherson. (1983a). Mutagenicity studies on herring gulls from different locations on the Great Lakes. I. Sister chromatid exchange rates in herring gull embryos. *J. Toxicol. Environ. Health.* **12:** (in press).

Ellenton, J. A., M. F. McPherson, and K. L. Maus. (1983b). Mutagenicity studies on herring gulls from different locations on the Great Lakes. II. Mutagenic evaluation of extracts of herring gull eggs in a battery of *in vitro* mammalian and microbial tests. *J. Toxicol. Environ. Health.* **12:** (in press).

Enderson, J. H. and D. D. Berger. (1970). Pesticides: Eggshell thinning and lowered production of young in Prairie Falcons. *BioScience* **20:** 355–356.

Firestone, D. (1973) Etiology of chick edema disease. *Environ. Health Perspect* **5:** 59–66.

Fox, G. A., A. P. Gilman, D. B. Peakall and F. W. Anderka. (1978). Behavioral abnormalities in nesting Lake Ontario herring gulls. *J. Wildl. Mgmt.* **42:** 477–483.

Frank, R., M., Holdrinet and H. E. Braun. (1977). Organochlorine insecticides and PCBs in sediments of Lake St. Clair (1970 and 1974) and Lake Erie (1971). *Sci. Total Environ.* **8:** 205–227.

Frank, R., R. L. Thomas, M. Holdrinet, A. L. W. Kemp, and H. E. Braun. (1979a). Organochlorine insecticides and PCB in surficial sediments (1968) and sediment cores (1976) from Lake Ontario. *J. Great Lakes Res.* **5**(1): 18–27.

Frank, R., R. L. Thomas, M. Holdrinet, A. L. W. Kemp, H. E. Braun, and R. Dawson. (1979b). Organochlorine insecticides and PCB in the sediments of Lake Huron (1969) and Georgian Bay and North Channel (1973). *Sci. Total Environ.* **13:** 101–117.

Gilbertson, M. E. (1974). Pollutants in breeding Herring Gulls in the Lower Great Lakes. *Can. Field-Nat.* **88:** 273–280.

Gilbertson, M. (1975) A Great Lakes tragedy. *Nat. Can.* **4**(1): 22–25.

Gilbertson, M. and G. A. Fox. (1977). Pollutant-associated embryonic mortality of Great Lakes herring gulls. *Environ. Pollut.* **12:** 211–216.

Gilbertson, M. (1983). Etiology of chick edema disease in herring gulls in the lower Great Lakes. *Chemosphere* **12**(3): 357–370.

Gilbertson, M. and R. Hale. (1974a). Early embryonic mortality in a herring gull colony in Lake Ontario. *The Can. Field-Nat.* **88:** 354–356.

Gilbertson, M. and R. Hale. (1974b). Characteristics of the breeding failure of a colony of herring gulls on Lake Ontario. *Can. Field-Nat.* **88:** 356–358.

Gilbertson, M. and L. Reynolds. (1974). DDE and PCB in Canadian birds, 1969–1972. Canadian Wildlife Service, Occasional Paper Number 19.

Gilbertson, M., R. D. Morris and R. A. Hunter. (1976). Abnormal chicks and PCB residue levels in eggs of colonial birds on the Lower Great Lakes (1971–73). *The Auk* **93:** 434–442.

Gilman, A. P., G. A. Fox, D. B. Peakall, S. M. Teeple, T. R. Carroll, and G. T. Haymes. (1977). Reproductive parameters and egg contaminant levels of Great Lakes herring gulls. *J. Wild. Mgmt.* **41**(3): 458–468.

Gilman, A. P., D. J. Hallett, G. A. Fox, L. J. Allan, W. J. Learning, and D. B. Peakall. (1978). Effects of injected organochlorines on naturally incubated herring gull eggs. *J. Wildl. Mgmt.* **42**(3): 484–493.

Gilman, A. P., D. B. Peakall, D. J. Hallett, G. A. Fox, and R. J. Norstrom. (1979). Herring Gulls (*Larus argentatus*) as monitors of contamination in the Great Lakes. In: *Animals as monitors of environmental pollutants.* National Academy of Sciences, Washington, D.C.

Gochfeld, M. (1975). Developmental defects in common terns of western Long Island, New York. *Auk* **92:** 58–65.

Hallett, D. J., R. J. Norstrom, F. I. Onuska, M. E. Comba, and R. Sampson. (1976). Mass spectral confirmation and analysis by the Hall detector of mirex and photomirex in herring gulls from Lake Ontario. *J. Agric. Food Chem.* **24**(6): 1189–1193.

Hallett, D. J., R. J. Norstrom, F. I. Onuska, and M. E. Comba. (1977a). Mirex, chlordane, dieldrin, DDT and PCBs: Metabolites and photoisomers in Lake Ontario herring gulls. In: G. W. Ivie and H. W. Dorough, Eds., *Fate of pesticides in large animals.* Academic Press, New York, pp. 183–192.

Hallett, D. J., R. J. Norstrom, F. I. Onuska, and M. E. Comba. (1977b). Analysis of polynuclear aromatic hydrocarbons and organochlorine pollutants in Great Lakes Herring Gulls by high resolution gas chromatography. In: Glass Capillary Chromatography, 2nd International Symposium, Hindelang, pp. 115–125.

Hallett, D. J., R. J. Norstrom, F. I. Onuska, and M. E. Comba. (1982). Incidence of chlorinated benzene and chlorinated ethylenes in Lake Ontario Herring Gulls. *Chemosphere* **11:** 277–285.

Hays, H. and R. W. Risebrough. (1972). Pollutant concentrations in abnormal young terns from Long Island sound. *Auk* **89:** 19–35.

Herman, S. C., R. L. Garrett, and R. L. Rudd. (1969). Pesticides and the western grebe. In: M. W. Miller and G. G. Berg, Eds., *Chemical fallout.* Current Research on Persistent Pesticides. Proceedings. Rochester Conference on Toxicity. Charles C. Thomas, USA. pp. 24–53.

Hickey, J. J. and D. W. Anderson. (1968). Chlorinated hydrocarbons and eggshell changes in raptorial and fish-eating birds. *Science* **162:** 271–273.

International Joint Commission. (1980). Great Lakes Water Quality Board Report to the International Joint Commission, Windsor, Ontario.

Kaiser, K. L. E. (1974). Mirex: An unrecognized contaminant of fishes from Lake Ontario. *Science* **185:** 523.

Keith, J. A. (1966). Reproduction in a population of herring gulls (*Larus argentatus*) contaminated by DDT. *J. Appl. Ecol.* **3:** 57–70.

Keith, J. A. and I. M. Gruchy. (1972). Residue levels of chemical pollutants in North American birdlife. In: K. H. Voous, Ed., Proceedings XV International Ornithol. Congress. E. J. Brill, Leiden. pp. 437–454.

Ludwig, J. P. and C. S. Tomoff. (1966). Reproductive success and insecticide residues in Lake Michigan herring gulls. *The Jack-Pine Warbler* **44**(2): 77–85.

McArthur, M. L. B., G. A. Fox, D. B. Peakall, and B. J. R. Philogène. (1983). Ecological significance of behavioral and hormonal abnormalities in breeding ring doves fed an organochlorine chemical mixture. *Arch. Environm. Contam. Toxicol.* **12:** 343–353.

Mineau, P. and D. V. Weseloh. (1981). Low disturbance monitoring of herring gull reproductive success on the Great Lakes. *Colonial Waterbirds* **4:** 138–142.

Mineau, P. (1982). Levels of major organochlorine contaminants in sequentially-laid herring gull eggs. *Chemosphere* **11:** 679–685.

Mineau, P., D. V. Weseloh, and D. J. Hallett. (1979). Contamination par les substances organochlorées et reproduction du goéland argenté, 1974 à 1978. In: Les contaminants dans l'environnement. Comptes rendus. Environnement Canada, Québec. pp. 24/1–24/17.

Mineau, P., G. E. J. Smith, R. Markel, and C. S. Lam. (1982). Aging herring gulls from hatching to fledging. *J. Field Ornithol.* **53**(4): 394–402.

Mirex Task Force. (1977). Mirex in Canada. A report of the task force on mirex, April 1, 1977 to the Enrivonmental Contaminants Committee of Fisheries and Environment Canada and Health and Welfare. *Canada. Tech Rept.* 77–1. 153p.

Moore, F. R. (1976). The dynamics of seasonal distribution of Great Lakes herring gulls. *Bird Banding* **47:** 141–159.

Morris, R. D. and G. T. Haymes. (1977). The breeding biology of two Lake Erie herring gull colonies. *Can. J. Zool.* **55**(5): 796–805.

Norstrom, R. J., D. J. Hallett, and R. A. Sonstegard. (1978). Coho salmon (*Oncorhynchus kisutch*) and herring gulls (*Larus argentatus*) as indicators of organochlorine contamination in Lake Ontario. *J. Fish. Res. Board Can.* **35**(11): 1401–1409.

Norstrom, R. J., D. J. Hallett, F. I. Onuska, and M. E. Comba. (1980a). Mirex and its degradation products in Great Lakes herring gulls. *Environ. Sci. Technol.* **14:** 860–866.

Norstrom, R. J., H. T. Won, M. Van Hove Holdrinet, P. G. Calway, and C. D. Naftel. (1980b). Gas-liquid chromatographic determination of mirex and photomirex in the presence of polychlorinated biphenyls: Interlaboratory study. *J. Assoc. Off. Anal. Chem.* **63:** 37–42.

Norstrom, R. J., A. P. Gilman, and D. J. Hallett. (1981). Total organically bound chlorine and bromine in Lake Ontario herring gull eggs, 1977, by instrumental neutron activation and chromatographic methods. *Sci. Total Environ.* **20:** 217–230.

Norstrom, R. J., D. J. Hallet, M. Simon, and M. J. Mulvihill. (1982). Analysis of Great Lakes herring gull eggs for tetrachlorodibenzo-*p*-dioxins. In: Hutzinger et al., Eds., *Chlorinated dioxins and related compounds; Impact on the environment.* Pergamon Press, New York, pp. 173–181.

Organization for Economic and Co-operative Development (1980) Chemical trends in wildlife: an international cooperative study. Publications de l'OCDE. France, 248 pp.

Peakall, D. B. and A. P. Gilman. (1979). Limitations of expressing organochlorine levels in eggs on a lipid-weight basis. *Bull. Environ. Contam. Toxicol.* **23:** 287–290.

Peakall, D. B., G. A. Fox, A. P. Gilman, D. J. Hallet, and R. J. Norstrom. (1978). The herring gull as a monitor of Great Lakes contamination. In: B. K. Afghan and D. Mackay, Eds., *Hydrocarbons and halogenated hydrocarbons in the aquatic environment.* Environ. Sci. Res. Vol. 16, Plenum Press, New York, pp. 337–344.

Poland, A., E. Glover, A. S. Kende, M. DeCamp, and C. M. Giandomenico. (1976). 3,4,3′,4′-Tetrachloroazoxybenzene and azobenzene: Potent inducers of aryl hydrocarbon hydroxylase. *Science* **194:** 627–630.

Poland, A., W. F. Greenlee, and A. S. Kende. (1979). Studies on the mechanism of action of the chlorinated dibenzo-*p*-dioxins and related compounds. *Ann. N.Y. Acad. Sci.* **320:** 214–230.

Reynolds, L. M. (1969) Polychlorobiphenyls (PCBs) and their interference with pesticide residue analysis. *Bull. Environ. Contam. Toxicol.* **4**(3): 128–143.

Reynolds, L. M. (1971). Pesticide residue analysis in the presence of polychlorobiphenyls (PCBs). *Residue Rev.* **34:** 27–57.

Reynolds, L. M. and T. Cooper. (1975). Analysis of organochlorine residues in fish. *Water Qual. Param.* ASTM STP. **573:** 196–205.

Ryder, J. P. (1974). Organochlorine and mercury residues in gulls' eggs from western Ontario. *Can. Field-Nat.* **88**(3): 349–352.

Ryder, J. P. and T. R. Carroll. (1978). Reproductive success of herring gulls on Granite Island, northern Lake Superior, 1975 and 1976. *Can. Field-Nat.* **92:** 51–54.

Shrankel, K. R., B. L. Kreamer, and M. T. Stephen Hsia. (1982). Embryotoxicity of 3,3′,4,4′-tetrachloroazobenzene and 3,3′,4,4′-tetrachloroazoxybenzene in the chick embryo. *Arch. Environ. Contam. Toxicol.* **11:** 195–202.

Strik, J. J. T, W. A., F. M. H. Debets, and G. Koss. (1980). Chemical porphyria. In: R. D. Kimbrough, Ed., *Halogenated biphenyls, terphenyls, naphtalenes, dibenzodioxins and related products.* Topics in environmental health, vol 4. Elsevier. pp. 191–239.

Swain, W. R., M. D. Mullin, and J. C. Filkins. (1982) Refined analysis of residue forming organic substances in lake trout from the vicinity of Isle Royale, Lake Superior. *Internat. Assoc. Great Lakes Res.* (abstract) May 4–6, 1982. Sault-Ste-Marie. p. 18.

Teeple, S. M. (1977) Reproductive success of herring gulls nesting on Brothers Island, Lake Ontario, in 1973. *Can. Field-Nat.* **91:** 148–157.

Thomas, R. L., J.-M. Jaquet, A. L. W. Kemp, and C. F. M. Lewis. (1976). Surficial sediments of Lake Erie. *J. Fish. Res. Board Can.* **33:** 385–403.

Vermeer, K. and L. M. Reynolds. (1970). Organochlorine residues in aquatic birds in the Canadian Prairie Provinces. *Can. Field-Nat.* **84:** 117–130.

Vos, J. G., J. H. Koeman, H. L. van der Maas, M. C. Ten Noever de Brauw, and R. H. de Vos. (1970). Identification and toxicological evaluation of chlorinated dibenzofuran and chlorinated naphtalene in two commercial polychlorinated biphenyls. *Food Cosmet. Toxicol.* **8:** 625–633.

Weseloh, D. V., P. Mineau, and D. J. Hallett. (1979). Organochlorine contaminants and trends in reproduction in Great Lakes herring gulls, 1974–78. In: Transactions of the 44th North American Wildlife and Natural Resources Conference. *Wildlife Mgmt. Inst.* pp. 543–557.

Won, H. T. and R. J. Norstrom. (1980). Analytical reference materials: Organochlorine residues in CWS-79-1, a Herring Gull Egg Pool from Lake Erie, 1979. Canadian Wildlife Service, Wildlife Toxicology Division, Manuscript Report No. 41.

APPENDIX

THE GREAT LAKES WATER QUALITY AGREEMENT OF 1978

This agreement, aimed at restoring and maintaining the chemical, physical, and biological integrity of the waters of the Great Lakes Basin Ecosystem, is unique and could serve as a model in dealing with the pollution of other transboundary waters. For this reason, the agreement is reproduced in its entirety.

The agreement stresses particularly the control strategy for the hazardous substances which have been defined rather broadly as any element or compound which, if discharged in quantity into or upon receiving waters or adjoining shorelines, "can cause death, disease, behavioral abnormalities, cancer, genetic mutations, physiological or reproductive malfunctions or physical deformities in any organism or its offspring, or can become poisonous after concentration in the food chain or in combination with other substances." It mandates the United States and Canada to develop research programs that deal effectively with water quality problems in the Great Lakes and to provide sufficient resources for implementing these programs. The 1978 Agreement thus provides the impetus and also acts as the backdrop for the recent research on toxic contaminants in the Great Lakes including most of those reported in this volume.

AGREEMENT BETWEEN CANADA AND THE UNITED STATES OF AMERICA ON GREAT LAKES WATER QUALITY, SIGNED AT OTTAWA ON NOVEMBER 22, 1978

Preamble

The Government of Canada and the Government of the United States of America, having in 1972 entered into an Agreement on Great Lakes Water Quality; reaffirming their determination to restore and enhance water quality in the Great Lakes System; continuing to be concerned about the impairment of water quality on each side of the boundary to an extent that is causing injury to health and property on the other side, as described by the International Joint Commission; reaffirming their intent to prevent further pollution of the Great Lakes Basin Ecosystem owing to continuing population growth, resource development and increasing use of water; reaffirming in a spirit of friendship and cooperation the rights and obligations of both countries under the Boundary Waters Treaty, signed on January 11, 1909, and in particular their obligation not to pollute boundary waters; continuing to recognize the rights of each country in the use of its Great Lakes waters; having decided that the Great Lakes Water Quality Agreement of April 15, 1972 and subsequent reports of the International Joint Commission provide a sound basis for new and more effective cooperative actions to restore and enhance water quality in the Great Lakes Basin Ecosystem; recognizing that restoration and enhancement of the boundary waters can not be achieved independently of other parts of the Great Lakes Basin Ecosystem with which these waters interact; concluding that the best means to preserve the aquatic ecosystem and achieve improved water quality throughout the Great Lakes System is by adopting common objectives, developing and implementing cooperative programs and other measures, and assigning special responsibilities and functions to the International Joint Commission; have agreed as follows:

ARTICLE I

Definitions

As used in this Agreement:

(a) "Agreement" means the present Agreement as distinguished from the Great Lakes Water Quality Agreement of April 15, 1972;

(b) "Annex" means any of the Annexes to this Agreement, each of which is attached to and forms an integral part of this Agreement;

(c) "Boundary waters of the Great Lakes System" or "boundary waters" means boundary waters, as defined in the Boundary Waters Treaty, that are within the Great Lakes System;

(d) "Boundary Waters Treaty" means the Treaty between the United States and Great Britain Relating to Boundary Waters, and Questions Arising Between the United States and Canada, signed at Washington on January 11, 1909;

(e) "Compatible regulations" means regulations no less restrictive than the agreed principles set out in this Agreement;

(f) "General Objectives" are broad descriptions of water quality conditions consistent with the protection of the beneficial uses and the level of environmental quality which the Parties desire to secure and which will provide overall water management guidance;

(g) "Great Lakes Basin Ecosystem" means the interacting components of air, land, water and living organisms, including man, within the drainage basin of the St. Lawrence River at or upstream from the point at which this river becomes the international boundary between Canada and the United States;

(h) "Great Lakes System" means all of the streams, rivers, lakes and other bodies of water that are within the drainage basin on the St. Lawrence River at or upstream from the point at which this river becomes the international boundary between Canada and the United States;

(i) "Harmful quantity" means any quantity of a substance that if discharged into receiving water would be inconsistent with the achievement of the General and Specific Objectives;

(j) "Hazardous polluting substance" means any element or compound identified by the Parties which, if discharged in any quantity into or upon receiving waters or adjoining shorelines, would present an imminent and substantial danger to public health or welfare; for this purpose, "public health or welfare" encompasses all factors affecting the health and welfare of man including but not limited to human health, and the conservation and protection of flora and fauna, public and private property, shorelines and beaches;

(k) "International Joint Commission" or "Commission" means the International Joint Commission established by the Boundary Waters Treaty;

(l) "Monitoring" means a scientifically designed system of continuing standardized measurements and observations and the evaluation thereof;

(m) "Objectives" means the General Objectives adopted pursuant to Article III and the Specific Objectives adopted pursuant to Article IV of this Agreement;

(n) "Parties" means the Government of Canada and the Government of the United States of America;

(o) "Phosphorus" means the element phosphorus present as a constituent of various organic and inorganic complexes and compounds;

(p) "Research" means development, demonstration and other research activities but does not include monitoring and surveillance of water or air quality;

(q) "Science Advisory Board" means the Great Lakes Science Advisory Board of the International Joint Commission established pursuant to Article VIII of this Agreement;

(r) "Specific Objectives" means the concentration or quantity of a substance or level of effect that the Parties agree, after investigation, to recognize as a maximum or minimum desired limit for a defined body of water or portion thereof, taking into account the beneficial uses or level of environmental quality which the Parties desire to secure and protect;

(s) "State and Provincial Governments" means the Governments of the States of Illinois, Indiana, Michigan, Minnesota, New York, Ohio, Wisconsin and the Commonwealth of Pennsylvania, and the Government of the Province of Ontario;

(t) "Surveillance" means specific observations and measurements relative to control or management;

(u) "Terms of Reference" means the Terms of Reference for the Joint Institutions and the Great Lakes Regional Office established pursuant to this Agreement, which are attached to and form an integral part of this Agreement;

(v) "Toxic substance" means a substance which can cause death, disease, behavioural abnormalities, cancer, genetic mutations, physiological or reproductive malfunctions or physical deformities in any organism or its offspring, or which can become poisonous after concentration in the food chain or in combination with other substances;

(w) "Tributary waters of the Great Lakes System" or "tributary waters" means all the waters within the Great Lakes System that are not boundary waters;

(x) "Water Quality Board" means the Great Lakes Water Quality Board of the International Joint Commission established pursuant to Article VIII of this Agreement.

ARTICLE II

Purpose

The purpose of the Parties is to restore and maintain the chemical, physical, and biological integrity of the waters of the Great Lakes Basin Ecosystem. In order to achieve this purpose, the Parties agree to make a maximum effort to develop programs, practices and technology necessary for a better understanding of the Great Lakes Basin Ecosystem and to eliminate or reduce to the maximum extent practicable the discharge of pollutants into the Great Lakes System.

Consistent with the provisions of this Agreement, it is the policy of the Parties that:

(a) The discharge of toxic substances in toxic amounts be prohibited and the discharge of any or all persistent toxic substances be virtually eliminated;
(b) Financial assistance to construct publicly owned waste treatment works be provided by a combination of local, state, provincial, and federal participation; and
(c) Coordinated planning processes and best management practices be developed and implemented by the respective jurisdictions to ensure adequate control of all sources of pollutants.

ARTICLE III

General Objectives

The Parties adopt the following General Objectives for the Great Lakes System. These waters should be:

(a) Free from substances that directly or indirectly enter the waters as a result of human activity and that will settle to form putrescent or otherwise objectionable sludge deposits, or that will adversely affect aquatic life or waterfowl;
(b) Free from floating materials such as debris, oil, scum, and other immiscible substances resulting from human activities in amounts that are unsightly or deleterious;
(c) Free from materials and heat directly or indirectly entering the water as a result of human activity that alone, or in combination with other materials, will produce colour, odour, taste, or other conditions in such a degree as to interfere with beneficial uses;
(d) Free from materials and heat directly or indirectly entering the water as a result of human activity that alone, or in combination with other materials, will produce conditions that are toxic or harmful to human, animal, or aquatic life; and
(e) Free from nutrients directly or indirectly entering the waters as a result of human activity in amounts that create growths of aquatic life that interfere with beneficial uses.

ARTICLE IV

Specific Objectives

1. The Parties adopt the Specific Objectives for the boundary waters of the Great Lakes System as set fourth in Annex 1, subject to the following:

   (a) The Specific Objectives adopted pursuant to this Article represent the minimum levels of water quality desired in the boundary waters of the Great Lakes System and are not intended to preclude the establishment of more stringent requirements.
   (b) The determination of the achievement of Specific Objectives shall be based on statistically valid sampling data.
   (c) Notwithstanding the adoption of Specific Objectives, all reasonable and practicable measures shall be taken to maintain or improve the existing water quality in those areas of the boundary waters of the Great Lakes System where such water quality is better than that prescribed by the Specific Objectives, and in those areas having outstanding natural resource value.
   (d) The responsible regulatory agencies shall not consider flow augmentation as a substitute for adequate treatment to meet the Specific Objectives.
   (e) The Parties recognize that in certain areas of inshore waters natural phenomena exist which, despite the best efforts of the Parties, will prevent the achievement of some of the Specific Objectives. As early as possible, these areas should be identified explicitly by the appropriate jurisdictions and reported to the International Joint Commission.
   (f) Limited use zones in the vicinity of present and future municipal, industrial and tributary point source discharges shall be designated by the responsible regulatory agencies within which some of the Specific Objectives may not apply. Establishment of these zones shall not be considered a substitute for adequate treatment or control of discharges at their source. The size shall be minimized to the greatest possible degree, being no larger than that attainable by all reasonable and practicable effluent treatment measures. The boundary of a limited use zone shall not transect the international boundary. Principles for the designation of limited use zones are set out in Annex 2.

2. The Specific Objectives for the boundary waters of the Great Lakes System or for particular portions thereof shall be kept under review by the Parties and by the International Joint Commission, which shall make appropriate recommendations.

3. The Parties shall consult on:

   (a) The establishment of Specific Objectives to protect beneficial uses from the combined effects of pollutants; and
   (b) The control of pollutant loading rates for each lake basin to protect the integrity of the ecosystem over the long term.

ARTICLE V

Standards, Other Regulatory Requirements, and Research

1. Water quality standards and other regulatory requirements of the Parties shall be consistent with the achievement of the General and Specific Objectives. The Parties shall use their best efforts to ensure that water quality standards and other regulatory requirements of the State and Provincial Governments shall similarly be consistent with the achievement of these Objectives. Flow augmentation shall not be considered as a substitute for adequate treatment to meet water quality standards or other regulatory requirements.
2. The Parties shall use their best efforts to ensure that:
   (a) The principal research funding agencies in both countries orient the research programs of their organizations in response to research priorities identified by the Science Advisory Board and recommended by the Commission; and
   (b) Mechanisms be developed for appropriate cost-effective international cooperation.

ARTICLE VI

Programs and Other Measures

1. The Parties shall continue to develop and implement programs and other measures to fulfil the purpose of this Agreement and to meet the General and Specific Objectives. Where present treatment is inadequate to meet the General and Specific Objectives, additional treatment shall be required. The programs and measures shall include the following:
   (a) *Pollution from Municipal Sources*. Programs for the abatement, control and prevention of municipal discharges and urban drainage into the Great Lakes System. These programs shall be completed and in operation as soon as practicable, and in the case of municipal sewage treatment facilities no later than December 31, 1982. These programs shall include:
      (i) Construction and operation of waste treatment facilities in all municipalities having sewer systems to provide levels of treatment consistent with the achievement of phosphorus requirements and the General and Specific Objectives, taking into account the effects of waste from other sources;
      (ii) Provision of financial resources to ensure prompt construction of needed facilities;
      (iii) Establishment of requirements for construction and operating standards for facilities;
      (iv) Establishment of pre-treatment requirements for all industrial plants discharging waste into publicly owned treatment works where such industrial wastes are not amenable to adequate treatment or removal using conventional municipal treatment processes;

(v) Development and implementation of practical programs for reducing pollution from storm, sanitary, and combined sewer discharges; and

(vi) Establishment of effective enforcement programs to ensure that the above pollution abatement requirements are fully met.

(b) *Pollution from Industrial Sources.* Programs for the abatement, control and prevention of pollution from industrial sources entering the Great Lakes System. These programs shall be completed and in operation as soon as practicable and in any case no later than December 31, 1983, and shall include:

(i) Establishment of waste treatment or control requirements expressed as effluent limitations (concentrations and/or loading limits for specific pollutants where possible) for all industrial plants, including power generating facilities, to provide levels of treatment or reduction or elimination of inputs of substances and effects consistent with the achievement of the General and Specific Objectives and other control requirements, taking into account the effects of waste from other sources;

(ii) Requirements for the substantial elimination of discharges into the Great Lakes System of persistent toxic substances;

(iii) Requirements for the control of thermal discharges;

(iv) Measures to control the discharge of radioactive materials into the Great Lakes System;

(v) Requirements to minimize adverse environmental impacts of water intakes;

(vi) Development and implementation of programs to meet industrial pre-treatment requirements as specified under sub-paragraph (a) (iv) above; and

(vii) Establishment of effective enforcement programs to ensure the above pollution abatement requirements are fully met.

(c) *Inventory of Pollution Abatement Requirements.* Preparation of an inventory of pollution abatement requirements for all municipal and industrial facilities discharging into the Great Lakes System in order to gauge progress toward the earliest practicable completion and operation of the programs listed in sub-paragraphs (a) and (b) above. This inventory, prepared and revised annually, shall include compliance schedules and status of compliance with monitoring and effluent restrictions, and shall be made available to the International Joint Commission and to the public. In the initial preparation of this inventory, priority shall be given to the problem areas previously identified by the Water Quality Board.

(d) *Eutrophication.* Programs and measures for the reduction and control of inputs of phosphorus and other nutrients, in accordance with the provisions of Annex 3.

(e) *Pollution from Agricultural, Forestry and Other Land Use Activities.* Measures for the abatement and control of pollution from agricultural, forestry and other land use activities including:

(i) Measures for the control of pest control products used in the Great

Lakes Basin to ensure that pest control products likely to have long-term deleterious effects on the quality of water or its biota be used only as authorized by the responsible regulatory agencies; that inventories of pest control products used in the Great Lakes Basin be established and maintained by appropriate agencies; and that research and educational programs be strengthened to facilitate integration of cultural, biological and chemical pest control techniques;

(ii) Measures for the abatement and control of pollution from animal husbandry operations, including encouragement to appropriate agencies to adopt policies and regulations regarding utilization of animal wastes, and site selection and disposal of liquid and solid wastes, and to strengthen educational and technical assistance programs to enable farmers to establish waste utilization, handling and disposal systems;

(iii) Measures governing the hauling and disposal of liquid and solid wastes, including encouragement to appropriate regulatory agencies to ensure proper location, design, and regulation governing land disposal, and to ensure sufficient, adequately trained technical and administrative capability to review plans and to supervise and monitor systems for application of wastes on land;

(iv) Measures to review and supervise road salting practices and salt storage to ensure optimum use of salt and all-weather protection of salt stores in consideration of long-term environmental impact;

(v) Measures to control soil losses from urban and suburban as well as rural areas;

(vi) Measures to encourage and facilitate improvements in land use planning and management programs to take account of impacts on Great Lakes water quality;

(vii) Other advisory programs and measures to abate and control inputs of nutrients, toxic substances and sediments from agricultural, forestry and other land use activities; and

(viii) Consideration of future recommendations from the International Joint Commission based on the Pollution from Land Use Activities Reference.

(f) *Pollution from Shipping Activities*. Measures for the abatement and control of pollution from shipping sources, including:

(i) Programs and compatible regulations to prevent discharges of harmful quantities of oil and hazardous polluting substances, in accordance with Annex 4;

(ii) Compatible regulations for the control of discharges of vessel wastes, in accordance with Annex 5;

(iii) Such compatible regulations to abate and control pollution from shipping sources as may be deemed desirable in the light of continuing reviews and studies to be undertaken in accordance with Annex 6;

(iv) Programs and any necessary compatible regulations in accordance with Annexes 4 and 5, for the safe and efficient handling of shipboard generated wastes, including oil, hazardous polluting substances, garbage, waste water and sewage, and for their subsequent disposal, including the type and quantity of reception facilities and, if applicable, treatment standards; and

(v) Establishment by the Canadian Coast Guard and the United States Coast Guard of a coordinated system for aerial and surface surveillance for the purpose of enforcement of regulations and the early identification, abatement and clean-up of spills of oil, hazardous polluting substances, or other pollution.

(g) *Pollution from Dredging Activities*. Measures for the abatement and control of pollution from all dredging activities, including the development of criteria for the identification of polluted sediments and compatible programs for disposal of polluted dredged material, in accordance with Annex 7. Pending the development of compatible criteria and programs, dredging operations shall be conducted in a manner that will minimize adverse effects on the environment.

(h) *Pollution from Onshore and Offshore Facilities*. Measures for the abatement and control of pollution from onshore and offshore facilities, including programs and compatible regulations for the prevention of discharges of harmful quantities of oil and hazardous polluting substances, in accordance with Annex 8.

(i) *Contingency Plan*. Maintenance of a joint contingency plan for use in the event of a discharge or the imminent threat of a discharge of oil or hazardous polluting substances, in accordance with Annex 9.

(j) *Hazardous Polluting Substances*. Implementation of Annex 10 concerning hazardous polluting substances. The Parties shall further consult from time to time for the purpose of revising the list of hazardous polluting substances and of identifying harmful quantities of these substances.

(k) *Persistent Toxic Substances*. Measures for the control of inputs of persistent toxic substances including control programs for their production, use, distribution and disposal, in accordance with Annex 12.

(l) *Airborne Pollutants*. Programs to identify pollutant sources and relative source contributions, including the more accurate definition of wet and dry deposition rates, for those substances which may have significant adverse effects on environmental quality including the indirect effects of impairment of tributary water quality through atmospheric deposition in drainage basins. In cases where significant contributions to Great Lakes pollution from atmospheric sources are identified, the Parties agree to consult on appropriate remedial programs.

(m) *Surveillance and Monitoring*. Implementation of a coordinated surveillance and monitoring program in the Great Lakes System, in accordance with Annex 11, to assess compliance with pollution control requirements and achievement of the Objectives, to provide information for measuring local and whole lake response to control measures, and to identify emerging problems.

2. The Parties shall develop and implement such additional programs as they jointly decide are necessary and desirable to fulfil the purpose of this Agreement and to meet the General and Specific Objectives.

ARTICLE VII

Powers, Responsibilities and Functions of the International Joint Commission

1. The International Joint Commission shall assist in the implementation of this Agreement. Accordingly, the Commission is hereby given, by a Reference pursuant to Article IX of the Boundary Waters Treaty, the following responsibilities:
   (a) Collation, analysis and dissemination of data and information supplied by the Parties and State and Provincial Governments relating to the quality of the boundary waters of the Great Lakes System and to pollution that enters the boundary waters from tributary waters and other sources;
   (b) Collection, analysis and dissemination of data and information concerning the General and Specific Objectives and the operation and effectiveness of the programs and other measures established pursuant to this Agreement;
   (c) Tendering of advice and recommendations to the Parties and to the State and Provincial Governments on problems of and matters related to the quality of the boundary waters of the Great Lakes System including specific recommendations concerning the General and Specific Objectives, legislation, standards and other regulatory requirements, programs and other measures, and intergovernmental agreements relating to the quality of these waters;
   (d) Tendering of advice and recommendations to the Parties in connection with matters covered under the Annexes to this Agreement;
   (e) Provision of assistance in the coordination of the joint activities envisaged by this Agreement;
   (f) Provision of assistance in and advice on matters related to research in the Great Lakes Basin Ecosystem, including identification of objectives for research activities, tendering of advice and recommendations concerning research to the Parties and to the State and Provincial Governments, and dissemination of information concerning research to interested persons and agencies;
   (g) Investigations of such subjects related to the Great Lakes Basin Ecosystem as the Parties may from time to time refer to it.
2. In the discharge of its responsibilities under this Reference, the Commission may exercise all of the powers conferred upon it by the Boundary Waters Treaty and by any legislation passed pursuant thereto including the power to conduct public hearings and to compel the testimony of witnesses and the production of documents.
3. The Commission shall make a full report to the Parties and to the State and Provincial Governments no less frequently than biennially concerning progress

toward the achievement of the General and Specific Objectives including, as appropriate, matters related to Annexes to this Agreement. This report shall include an assessment of the effectiveness of the programs and other measures undertaken pursuant to this Agreement, and advice and recommendations. In alternate years the Commission may submit a summary report. The Commission may at any time make special reports to the Parties, to the State and Provincial Governments and to the public concerning any problem of water quality in the Great Lakes System.

4. The Commission may in its discretion publish any report, statement or other document prepared by it in the discharge of its functions under this Reference.
5. The Commission shall have authority to verify independently the data and other information submitted by the Parties and by the State and Provincial Governments through such tests or other means as appear appropriate to it, consistent with the Boundary Waters Treaty and with applicable legislation.
6. The Commission shall carry out its responsibilities under this Reference utilizing principally the services of the Water Quality Board and the Science Advisory Board established under Article VIII of this Agreement. The Commission shall also ensure liaison and coordination between the institutions established under this Agreement and other institutions which may address concerns relevant to the Great Lakes Basin Ecosystem, including both those within its purview, such as those Boards related to Great Lakes levels and air pollution matters, and other international bodies, as appropriate.

ARTICLE VIII

Joint Institutions and Regional Office

1. To assist the International Joint Commission in the exercise of the powers and responsibilities assigned to it under this Agreement, there shall be two Boards:
   (a) A Great Lakes Water Quality Board which shall be the principal advisor to the Commission. The Board shall be composed of an equal number of members from Canada and the United States, including representatives from the Parties and each of the State and Provincial Governments; and
   (b) A Great Lakes Science Advisory Board which shall provide advice on research to the Commission and to the Water Quality Board. The Board shall further provide advice on scientific matters referred to it by the Commission, or by the Water Quality Board in consultation with the Commission. The Science Advisory Board shall consist of managers of Great Lakes research programs and recognized experts on Great Lakes water quality problems and related fields.
2. The members of the Water Quality Board and the Science Advisory Board shall be appointed by the Commission after consultation with the appropriate government or governments concerned. The functions of the Boards shall be as specified in the Terms of Reference appended to this Agreement.
3. To provide administrative support and technical assistance to the two Boards, and to provide a public information service for the programs, including public

hearings, undertaken by the International Joint Commission and by the Boards, there shall be a Great Lakes Regional Office of the International Joint Commission. Specific duties and organization of the Office shall be as specified in the Terms of Reference appended to this Agreement.

4. The Commission shall submit an annual budget of anticipated expenses to be incurred in carrying out its responsibilities under this Agreement to the Parties for approval. Each party shall seek funds to pay one-half of the annual budget so approved, but neither Party shall be under an obligation to pay a larger amount than the other toward this budget.

ARTICLE IX

Submission and Exchange of Information

1. The International Joint Commission shall be given at its request any data or other information relating to water quality in the Great Lakes System in accordance with procedures established by the Commission.
2. The Commission shall make available to the Parties and to the State and Provincial Governments upon request all data or other information furnished to it in accordance with this Article.
3. Each Party shall make available to the other at its request any data or other information in its control relating to water quality in the Great Lakes System.
4. Notwithstanding any other provision of this Agreement, the Commission shall not release without the consent of the owner any information identified as proprietary information under the law of the place where such information has been acquired.

ARTICLE X

Consultation and Review

1. Following the receipt of each report submitted to the Parties by the International Joint Commission in accordance with paragraph 3 of Article VII of this Agreement, the Parties shall consult on the recommendations contained in such report and shall consider such action as may be appropriate, including:
   (a) The modification of existing Objectives and the adoption of new Objectives;
   (b) The modification or improvement of programs and joint measures; and
   (c) The amendment of this Agreement or any Annex thereto.

   Additional consultations may be held at the request of either Party on any matter arising out of the implementation of this Agreement.
2. When a Party becomes aware of a special pollution problem that is of joint concern and requires an immediate response, it shall notify and consult the other Party forthwith about appropriate remedial action.
3. The Parties shall conduct a comprehensive review of the operation and effectiveness of this Agreement following the third biennial report of the Commission required under Article VII of this Agreement.

ARTICLE XI

Implementation

1. The obligations undertaken in this Agreement shall be subject to the appropriation of funds in accordance with the constitutional procedures of the Parties.
2. The Parties commit themselves to seek:
   (a) The appropriation of the funds required to implement this Agreement, including the funds needed to develop and implement the programs and other measures provided for in Article VI of this Agreement, and the funds required by the International Joint Commission to carry out its responsibilities effectively;
   (b) The enactment of any additional legislation that may be necessary in order to implement the programs and other measures provided for in Article VI of this Agreement; and
   (c) The cooperation of the State and Provincial Governments in all matters relating to this Agreement.

ARTICLE XII

Existing Rights and Obligations

Nothing in this Agreement shall be deemed to diminish the rights and obligations of the Parties as set forth in the Boundary Waters Treaty.

ARTICLE XIII

Amendment

1. This Agreement, the Annexes, and the Terms of Reference may be amended by agreement of the Parties. The Annexes may also be amended as provided therein, subject to the requirement that such amendments shall be within the scope of this Agreement. All such amendments to the Annexes shall be confirmed by an exchange of notes or letters between the Parties through diplomatic channels which shall specify the effective date or dates of such amendments.
2. All amendments to this Agreement, the Annexes, and the Terms of Reference shall be communicated promptly to the International Joint Commission.

ARTICLE XIV

Entry Into Force and Termination

This Agreement shall enter into force upon signature by the duly authorized representatives of the Parties, and shall remain in force for a period of five years and thereafter until terminated upon twelve months' notice given in writing by one of the Parties to the other.

ARTICLE XV

Supersession

This Agreement supersedes the Great Lakes Water Quality Agreement of April 15, 1972, and shall be referred to as the "Great Lakes Water Quality Agreement of 1978".

ANNEX 1

Specific Objectives

These Objectives are based on available information on cause/effect relationships between pollutants and receptors to protect the recognized most sensitive use in all waters. These Objectives may be amended, or new Objectives may be added, by mutual consent of the Parties.

I. CHEMICAL

A. *Persistent Toxic Substances*

1. Organic

(a) Pesticides

The sum of the concentrations of aldrin and dieldrin in water should not exceed 0.001 microgram per litre. The sum of concentrations of aldrin and dieldrin in the edible portion of fish should not exceed 0.3 microgram per gram (wet weight basis) for the protection of human consumers of fish.

Chlordane

The concentration of chlordane in water should not exceed 0.06 microgram per litre for the protection of aquatic life.

DDT and Metabolites

The sum of the concentrations of DDT and its metabolites in water should not exceed 0.003 microgram per litre. The sum of the concentrations of DDT and its metabolites in whole fish should not exceed 1.0 microgram per gram (wet weight basis) for the protection of fish-consuming aquatic birds.

Endrin

The concentration of endrin in water should not exceed 0.002 microgram per litre. The concentration of endrin in the edible portion of fish should not exceed 0.3 microgram per gram (wet weight basis) for the protection of human consumers of fish.

Heptachlor/Heptachlor Epoxide

The sum of the concentrations of heptachlor and heptachlor epoxide in water should not exceed 0.001 microgram per litre. The sum of the concentrations of heptachlor and heptachlor epoxide in edible portions of fish should not exceed 0.3 microgram per gram (wet weight basis) for the protection of human consumers of fish.

Lindane

The concentration of lindane in water should not exceed 0.01 microgram per litre for the protection of aquatic life. The concentration of lindane in edible portions of fish should not exceed 0.3 microgram per gram (wet weight basis) for the protection of human consumers of fish.

Methoxychlor

The concentration of methoxychlor in water should not exceed 0.04 microgram per litre for the protection of aquatic life.

Mirex

For the protection of aquatic organisms and fish-consuming birds and animals, mirex and its degradation products should be substantially absent from water and aquatic organisms. Substantially absent here means less than detection levels as determined by the best scientific methodology available.

Toxaphene

The concentration of toxaphene in water should not exceed 0.008 microgram per litre for the protection of aquatic life.

(b) Other Compounds

Phthalic Acid Esters

The concentration of dibutyl phthalate and di(2-ethylhexyl) phthalate in water should not exceed 4.0 micrograms per litre and 0.6 microgram per litre, respectively, for the protection of aquatic life. Other phthalic acid esters should not exceed 0.2 microgram per litre in waters for the protection of aquatic life.

Polychlorinated Biphenyls (PCBs)

The concentration of total polychlorinated biphenyls in fish tissues (whole fish, calculated on a wet weight basis), should not exceed 0.1 microgram per gram for the protection of birds and animals which consume fish.

Unspecified Organic Compounds

For other organic contaminants, for which Specific Objectives have not been defined, but which can be demonstrated to be persistent and are likely to be toxic, the concentrations of such compounds in water or aquatic organisms should be substantially absent, i.e., less than detection levels as determined by the best scientific methodology available.

2. Inorganic

(a) Metals

Arsenic

The concentrations of total arsenic in an unfiltered water sample should not exceed 50 micrograms per litre to protect raw waters for public water supplies.

Cadmium

The concentration of total cadmium in an unfiltered water sample should not exceed 0.2 microgram per litre to protect aquatic life.

Chromium

The concentration of total chromium in an unfiltered water sample should not exceed 50 micrograms per litre to protect raw waters for public water supplies.

Copper

The concentration of total copper in an unfiltered water sample should not exceed 5 micrograms per litre to protect aquatic life.

Iron

The concentration of total iron in an unfiltered water sample should not exceed 300 micrograms per litre to protect aquatic life.

Lead

The concentration of total lead in an unfiltered water sample should not exceed 10 micrograms per litre in Lake Superior, 20 micrograms per litre in Lake Huron and 25 micrograms per litre in all remaining Great Lakes to protect aquatic life.

Mercury

The concentration of total mercury in a filtered water sample should not exceed 0.2 microgram per litre nor should the concentration of total mercury in whole fish exceed 0.5 microgram per gram (wet weight basis) to protect aquatic life and fish-consuming birds.

Nickel

The concentration of total nickel in an unfiltered water sample should not exceed 25 micrograms per litre to protect aquatic life.

Selenium

The concentration of total selenium in an unfiltered water sample should not exceed 10 micrograms per litre to protect raw water for public water supplies.

Zinc

The concentration of total zinc in an unfiltered water sample should not exceed 30 micrograms per litre to protect aquatic life.

(b) Other Inorganic Substances

Fluoride

The concentration of total fluoride in an unfiltered water sample should not exceed 1200 micrograms per litre to protect raw water for public water supplies.

Total Dissolved Solids

In Lake Erie, Lake Ontario and the International Section of the St. Lawrence River, the level of total dissolved solids should not exceed 200 milligrams per litre. In the St. Clair River, Lake St. Clair, the Detroit River and the Niagara River, the level should be consistent with maintaining the levels of total dissolved solids in Lake Erie and Lake Ontario at not to exceed 200 milligrams per litre. In the remaining boundary waters, pending further study, the level of total dissolved solids should not exceed present levels.

B. *Non-Persistent Toxic Substances*

1. Organic Substances

(a) Pesticides

Diazinon

The concentration of diazinon in an unfiltered water sample should not exceed 0.08 microgram per litre for the protection of aquatic life.

Guthion

The concentration of guthion in an unfiltered water sample should not exceed 0.005 microgram per litre for the protection of aquatic life.

Parathion

The concentration of parathion in an unfiltered water sample should not exceed 0.008 microgram per litre for the protection of aquatic life.

Other Pesticides

The concentration of unspecified, non-persistent pesticides should not exceed 0.05 of the median lethal concentration on a 96-hour test for any sensitive local species.

(b) Other Substances

Unspecified Non-Persistent Toxic Substances and Complex Effluents

Unspecified non-persistent toxic substances and complex effluents of municipal, industrial or other origin should not be present in concentrations which exceed 0.05 of the median lethal concentration in a 96-hour test for any sensitive local species to protect aquatic life.

Oil and Petrochemicals

Oil and petrochemicals should not be present in concentrations that:

(i) can be detected as visible film, sheen or discolouration on the surface;

(ii) can be detected by odour;

(iii) can cause tainting of edible aquatic organisms; and

(iv) can form deposits on shorelines and bottom sediments that are detectable by sight or odour, or are deleterious to resident aquatic organisms.

2. Inorganic Substances

Ammonia

The concentration of un-ionized ammonia (NH_3) should not exceed 20 micrograms per litre for the protection of aquatic life. Concentrations of total ammonia should not exceed 500 micrograms per litre for the protection of public water supplies.

Hydrogen Sulfide

The concentration of undissociated hydrogen sulfide should not exceed 2.0 micrograms per litre to protect aquatic life.

C. *Other Substances*

1. Dissolved oxygen

In the connecting channels and in the upper waters of the Lakes, the dissolved oxygen level should not be less than 6.0 milligrams per litre at any time; in hypolimnetic waters, it should be not less than necessary for the support of fishlife, particularly cold water species.

2. pH

Values of pH should not be outside the range of 6.5 to 9.0, nor should discharge change the pH at the boundary of a limited use zone more than 0.5 units from that of the ambient waters.

3. Nutrients

Phosphorus

The concentration should be limited to the extent necessary to prevent nuisance growths of algae, weeds and slimes that are or may become injurious to any beneficial water use. (Specific phosphorus control requirements are set out in Annex 3.)

4. Tainting Substances

(a) Raw public water supply sources should be essentially free from objectionable taste and odour for aesthetic reasons.

(b) Levels of phenolic compounds should not exceed 1.0 microgram per litre in public water supplies to protect against taste and odor in domestic water.

(c) Substances entering the water as the result of human activity that cause tainting of edible aquatic organisms should not be present in concentrations which will lower the acceptability of these organisms as determined by organoleptic tests.

II. PHYSICAL

A. *Asbestos*

Asbestos should be kept at the lowest practical level and in any event should be controlled to the extent necessary to prevent harmful effects on human health.

B. *Temperature*

There should be no change in temperature that would adversely affect any local or general use of the waters.

C. *Settleable and Suspended Solids, and Light Transmission*

For the protection of aquatic life, waters should be free from substances attributable to municipal, industrial or other discharges resulting from human activity that will settle to form putrescent or otherwise objectionable sludge deposits or that will alter the value of Secchi disc depth by more than 10 per cent.

III. MICROBIOLOGICAL

Waters used for body contact recreation activities should be substantially free from bacteria, fungi, or viruses that may produce enteric disorders or eye, ear, nose, throat and skin infections or other human diseases and infections.

IV. RADIOLOGICAL

The level of radioactivity in waters outside of any defined source control area should not result in a TED_{50} (total equivalent dose integrated over 50 years as calculated in accordance with the methodology established by the International Commission on Radiological Protection) greater than 1 millirem to the whole body from a daily ingestion of 2.2 litres of lake water for one year. For dose commitments between 1 and 5 millirem at the periphery of the source control area, source investigation and corrective action are recommended if releases are not as low as reasonably achievable. For dose commitments greater than 5 millirem, the responsible regulatory authorities shall determine appropriate corrective action.

ANNEX 2

Limited Use Zones

1. The Parties, in consultation with the State and Provincial Governments, shall take measures to define and describe all existing and future limited use zones, and shall prepare an annual report on these measures. The measures shall include:
   (a) Identification and quantitative and qualitative description of all point source waste discharges (including tributaries) to boundary waters;
   (b) Delineation of boundaries for limited use zones assigned to identified discharges;
   (c) Assessment of the impact of the proposed limited use zones on existing and potential beneficial uses; and
   (d) Continuing review and revision of the extent of limited use zones to achieve maximum possible reduction in size and effect of such zones in accordance with improvements in waste treatment technology.
2. Limited use zones within the boundary waters of the Great Lakes System shall be designated for industrial discharges, and for municipal discharges in excess of 1 million gallons per day before January 1, 1980, in accordance with the following principles:
   (a) The boundary of a limited use zone shall not transect the international boundary.
   (b) The size, shape and exact location of a limited use zone shall be specified on a case-by-case basis by the responsible regulatory agency. The size shall be minimized to the greatest possible degree, being no larger than that attainable by all reasonable and practicable effluent treatment measures.
   (c) Specific Objectives and conditions applicable to the receiving water body shall be met at the boundary of limited use zones.

(d) Existing biological, chemical, physical and hydrological conditions shall be defined before considering the location of a new limited use zone or restricting an existing one.

(e) Areas of extraordinary natural resource value shall not be designated as limited use zones.

(f) Limited use zones shall not form barriers to migratory routes of aquatic species or interfere with biological communities or populations of important species to a degree which damages the ecosystem, or diminishes other beneficial uses disproportionately. Routes of passage for specific organisms which require protection and which would normally inhabit or pass through limited use zones shall be assured either by location of the zones, or by design of conditions within the zones.

(g) Conditions shall not be permitted within the limited use zones which:
   (i) are rapidly lethal to important aquatic life;
   (ii) cause irreversible responses which could result in detrimental post-exposure effects; or
   (iii) result in bioconcentration of toxic substances which are harmful to the organism or its consumers.

(h) Concentrations of toxic substances at any point in the limited use zone where important species are physically capable of residing shall not exceed the 24-hour LC_{50}.

(i) Every attempt shall be made to insure that the zones are free from:
   (i) objectionable deposits;
   (ii) unsightly or deleterious amounts of flotsam, debris, oil, scum and other floating matter;
   (iii) substances producing objectionable colour, odour, taste or turbidity; and
   (iv) substances and conditions or combinations thereof at levels which produce aquatic life in nuisance quantities that interfere with other uses.

(j) Limited use zones may overlap unless the combined effects exceed the conditions set forth in other guidelines.

(k) As a general condition, limited use zones should not overlap with municipal and other water intakes and recreational areas. However, knowledge of local effluent characteristics and effects could allow such a combination of uses.

3. Candidate areas for designation as limited use zones shall be reported, in all available detail, by the responsible regulatory agencies to the International Joint Commission. Within 60 days, the Commission may comment upon the extent of the area proposed for designation as a limited use zone, or any other aspect or measure to promote the attainment of the General and Specific Objectives of this Agreement. The responsible regulatory agency will take the comments of the Commission into account prior to making a formal designation of the area as a limited use zone. If no comment is received from the Commission within 60 days, it may be assumed that the Commission agrees with the proposed designation.

4. The Parties shall consult to develop more definitive procedures to delineate the extent of individual limited use zones and to develop scientific guidelines for determining the maximum portions of the boundary waters of each of the Great Lakes and connecting channels which may be occupied by limited use zones.

ANNEX 3

Control of Phosphorus

1. The purpose of the following programs is to minimize eutrophication problems and to prevent degradation with regard to phosphorus in the boundary waters of the Great Lakes System. The goals of phosphorus control are:
   (a) Restoration of year-round aerobic conditions in the bottom waters of the Central Basin of Lake Erie;
   (b) Substantial reduction in the present levels of algal biomass to a level below that of a nuisance condition in Lake Erie;
   (c) Reduction in present levels of algal biomass to below that of a nuisance condition in Lake Ontario including the International Section of the St. Lawrence River;
   (d) Maintenance of the oligotrophic state and relative algal biomass of Lakes Superior and Huron;
   (e) Substantial elimination of algal nuisance growths in Lake Michigan to restore it to an oligotrophic state; and
   (f) The elimination of algal nuisance in bays and in other areas wherever they occur.
2. The following programs shall be developed and implemented to reduce input of phosphorus to the Great Lakes:
   (a) Construction and operation of municipal waste treatment facilities in all plants discharging more than one million gallons per day to achieve, where necessary to meet the loading allocations to be developed pursuant to paragraph 3 below, or to meet local conditions, whichever are more stringent, effluent concentrations of 1.0 milligram per litre total phosphorus maximum for plants in the basins of Lakes Superior, Michigan, and Huron, and of 0.5 milligram per litre total phosphorus maximum for plants in the basins of Lakes Ontario and Erie.
   (b) Regulation of phosphorus introduction from industrial discharges to the maximum practicable extent.
   (c) Reduction to the maximum extent practicable of phosphorus introduced from diffuse sources into Lakes Superior, Michigan, and Huron; and reduction by 30 per cent of phosphorus introduced from diffuse sources into Lakes Ontario and Erie, where necessary to meet the loading allocations to be developed pursuant to paragraph 3 below, or to meet local conditions, whichever are more stringent.
   (d) Reduction of phosphorus in household detergents to 0.5 per cent by weight where necessary to meet the loading allocations to be developed pursuant to paragraph 3 below, or to meet local conditions, whichever are more stringent.

(e) Maintenance of a viable research program to seek maximum efficiency and effectiveness in the control of phosphorus introductions into the Great Lakes.

3. The following table establishes phosphorus loads for the base year (1976) and future phosphorus loads. The Parties, in cooperation with the State and Provincial Governments, shall within eighteen months after the date of entry into force of this Agreement confirm the future phosphorus loads, and based on these establish load allocations and compliance schedules, taking into account the recommendations of the International Joint Commission arising from the Pollution from Land Use Activities Reference. Until such loading allocations and compliance schedules are established, the Parties agree to maintain the programs and other measures specified in Annex 2 of the Great Lakes Water Quality Agreement of 1972.

Basin	1976 Phosphorus Load in Metric Tonnes Per Year	Future Phosphorus Load in Metric Tonnes Per Year
Lake Superior	3600	3400*
Lake Michigan	6700	5600*
Main Lake Huron	3000	2800*
Georgian Bay	630	600*
North Channel	550	520*
Saginaw Bay	870	440**
Lake Erie	20000	11000**
Lake Ontario	11000	7000**

* These loadings would result if all municipal plants over one million gallons per day achieved an effluent of 1 milligram per litre of phosphorus.

** These loadings are required to meet the goals stated in paragraph 1 above.

ANNEX 4

Discharges of Oil and Hazardous Polluting Substances From Vessels

1. *Definitions*. As used in this Annex:
   (a) "Discharge" includes, but is not limited to, any spilling, leaking, pumping, pouring, emitting or dumping; it does not include unavoidable direct discharges of oil from a properly functioning vessel engine;
   (b) "Harmful quantity of oil" means any quantity of oil that, if discharged from a ship that is stationary into clear calm water on a clear day, would produce a film or a sheen upon, or discolouration of, the surface of the water or adjoining shoreline, or that would cause a sludge or emulsion to be deposited beneath the surface of the water or upon the adjoining shoreline;
   (c) "Oil" means oil of any kind or in any form, including, but not limited to, petroleum, fuel oil, oil sludge, oil refuse, oil mixed with ballast or bilge water, and oil mixed with wastes other than dredged material;
   (d) "Tanker" means any vessel designed for the carriage of liquid cargo in bulk; and

(e) "Vessel" means any ship, barge or other floating craft, whether or not self-propelled.

2. *General Principles*. Compatible regulations shall be adopted for the prevention of discharges into the Great Lakes System of harmful quantities of oil and hazardous polluting substances from vessels in accordance with the following principles:
   (a) The discharge of a harmful quantity of oil or hazardous polluting substance shall be prohibited and made subject to appropriate penalties; and
   (b) As soon as any person in charge has knowledge of any discharge of harmful quantities of oil or hazardous polluting substances, immediate notice of such discharge shall be given to the appropriate agency in the jurisdiction where the discharge occurs; failure to give this notice shall be made subject to appropriate penalties.
3. *Oil*. The programs and measures to be adopted for the prevention of discharges of harmful quantities of oil shall include:
   (a) Compatible regulations for design, construction, and operation of vessels based on the following principles:
      (i) Each vessel shall have a suitable means of containing on board cargo oil spills caused by loading or unloading operations;
      (ii) Each vessel shall have a suitable means of containing on board fuel oil spills caused by loading or unloading operations, including those from tank vents and overflow pipes;
      (iii) Each vessel shall have the capability of retaining on board oily wastes accumulated during vessel operation;
      (iv) Each vessel shall be capable of off-loading retained oily wastes to a reception facility;
      (v) Each vessel shall be provided with a means for rapidly and safely stopping the flow of cargo or fuel oil during loading, unloading or bunkering operations in the event of an emergency;
      (vi) Each vessel shall be provided with suitable lighting to adequately illuminate all cargo and fuel oil handling areas if the loading, unloading or bunkering operations occur at night;
      (vii) Hose assemblies used on board vessels for oil loading, unloading, or bunkering shall be suitably designed, identified, and inspected to minimize the possibility of failure; and
      (viii) Oil loading, unloading, and bunkering systems shall be suitably designed, identified, and inspected to minimize the possibility of failure; and
   (b) Programs to ensure that merchant vessel personnel are trained in all functions involved in the use, handling, and stowage of oil and in procedures for abatement of oil pollution.
4. *Hazardous Polluting Substances*. The programs and measures to be adopted for the prevention of discharges of harmful quantities of hazardous polluting substances carried as cargo shall include:
   (a) Compatible regulations for the design, construction, and operation of vessels using as a guide the Code for the Construction and Equipment of Ships

Carrying Dangerous Chemicals in Bulk as established through the Inter-Governmental Maritime Consultative Organization (IMCO), including the following requirements:

(i) Each vessel shall have a suitable means of containing on board spills caused by loading or unloading operations;

(ii) Each vessel shall have a capability of retaining on board wastes accumulated during vessel operation;

(iii) Each vessel shall be capable of off-loading wastes retained to a reception facility;

(iv) Each vessel shall be provided with a means for rapidly and safely stopping the flow during loading or unloading operations in the event of an emergency; and

(v) Each vessel shall be provided with suitable lighting to adequately illuminate all cargo handling areas if the loading or unloading operations occur at night;

(b) Identification of vessels carrying cargoes of hazardous polluting substances in bulk, containers, and package form, and of all such cargoes;

(c) Identification in vessel manifests of all hazardous polluting substances;

(d) Procedures for notification to the appropriate agency by the owner, master or agent of a vessel of all hazardous polluting substances; and

(e) Programs to ensure that merchant vessel personnel are trained in all functions involving the use, handling, and stowage of hazardous polluting substances; the abatement of pollution from such substances; and the hazards associated with the handling of such substances.

5. *Additional Measures.* Both Parties shall take, as appropriate, action to ensure the provision of adequate facilities for the reception, treatment, and subsequent disposal of oil and hazardous polluting substances wastes from all vessels.

ANNEX 5

Discharges of Vessel Wastes

1. *Definitions.* As used in this Annex:

(a) "Discharge" includes, but is not limited to, any spilling, leaking, pumping, pouring, emitting, and dumping;

(b) "Garbage" means all kinds of victual, domestic, and operational wastes, excluding fresh fish and parts thereof generated during the normal operation of the ship and liable to be disposed of continually or periodically;

(c) "Sewage" means human or animal waste generated on board ship and includes wastes from water closets, urinals, or a hospital facility;

(d) "Vessel" means any ship, barge or other floating craft, whether or not self-propelled; and

(e) "Waste water" means water in combination with other substances, including ballast water and water used for washing cargo holds, but excluding water in combination with oil, hazardous polluting substances, or sewage.

2. *General Principles*. Compatible regulations shall be adopted governing the discharge into the Great Lakes System of garbage, sewage, and waste water from vessels in accordance with the following principles:
   (a) The discharge of garbage shall be prohibited and made subject to appropriate penalties;
   (b) The discharge of waste water in amounts or in concentrations that will be deleterious shall be prohibited and made subject to appropriate penalties; and
   (c) Every vessel operating in these waters that is provided with toilet facilities shall be equipped with a device or devices to contain, incinerate, or treat sewage to an adequate degree; appropriate penalties shall be provided for failure to comply with the regulations.
3. *Critical Use Areas*. Critical use areas of the Great Lakes System may be designated where the discharge of waste water or sewage shall be limited or prohibited.
4. *Containment Devices*. Regulations may be established requiring a device or devices to contain the sewage of pleasure craft or other classes of vessels operating in the Great Lakes System or designated areas thereof.
5. *Additional Measures*. The Parties shall take, as appropriate, action to ensure the provision of adequate facilities for the reception, treatment, and subsequent disposal of garbage, waste water, and sewage from all vessels.

ANNEX 6

Review of Pollution From Shipping Sources

1. *Review*. The Canadian Coast Guard and the United States Coast Guard shall continue to review services, systems, programs, recommendations, standards, and regulations relating to shipping activities for the purpose of maintaining or improving Great Lakes water quality. The reviews shall include:
   (a) Review of vessel equipment, vessel manning, and navigation practices or procedures, and of aids to navigation and vessel traffic management, for the purpose of precluding casualties which may be deleterious to water quality;
   (b) Review of practices and procedures regarding waste water and their deleterious effect on water quality;
   (c) Review of practices and procedures, as well as current technology for the treatment of vessel sewage; and
   (d) Review of current practices and procedures regarding the prevention of pollution from the loading, unloading, or on board transfer of cargo.
2. *Consultation*. Representatives of the Canadian Coast Guard and the United States Coast Guard, and other interested agencies, shall meet at least annually to consider this Annex. A report of this annual consultation shall be furnished to the International Joint Commission prior to its annual meeting on Great Lakes water quality. The purpose of the consultation shall be to:

(a) Provide an interchange of information with respect to continuing reviews, ongoing studies, and areas of concern;
(b) Identify and determine the relative importance of problems requiring further study; and
(c) Apportion responsibility, as between the Canadian Coast Guard and the United States Coast Guard, for the studies, or portions thereof, which were identified in subparagraph 2(b) above.

3. *Studies*. Where a review identifies additional areas for improvement, the Canadian Coast Guard and the United States Coast Guard, and other interested agencies, will undertake a study to establish improved procedures for the abatement and control of pollution from shipping sources, and will:
(a) Develop a brief study description which will include the nature of the perceived problem, procedures to quantify the problem, alternative solutions to the problem, procedures to determine the best alternative, and an estimated completion date;
(b) Transmit study descriptions to the International Joint Commission and other interested agencies;
(c) Transmit the study, or a brief summary of its conclusions, to the International Joint Commission and other interested agencies; and
(d) Transmit a brief status report to the International Joint Commission and other interested agencies if the study is not completed by the estimated completion date.

4. *Responsibility*. Responsibility for the coordination of the review, consultation, and studies is assigned to the Canadian Coast Guard and the United States Coast Guard.

ANNEX 7

Dredging

1. There shall be established, under the auspices of the Water Quality Board, a Subcommittee on Dredging. The Subcommittee shall:
(a) Review the existing practices in both countries relating to dredging activities, as well as the previous work done by the International Working Group on Dredging, with the objective of developing, within one year of the date of entry into force of this Agreement, compatible guidelines and criteria for dredging activities in the boundary waters of the Great Lakes System;
(b) Maintain a register of significant dredging projects being undertaken in the Great Lakes System with information to allow for the assessment of the environmental effects of the projects. The register shall include pertinent statistics to allow for the assessment of pollution loadings from dredged materials to the Great Lakes System;
(c) Encourage the exchange of information relating to developments of dredging technology and environmental research.

2. The Subcommittee shall identify specific criteria for the classification of polluted sediments of designated areas of intensive and continuing dredging activities

within the Great Lakes System. Pending development of criteria and guidelines by the Subcommittee, and their acceptance by the Parties, the Parties shall continue to apply the criteria now in use by the regulatory authorities; however, neither Party shall be precluded from applying standards more stringent than those now in use.

3. The Parties shall continue to direct particular attention to the identification and preservation of significant wetland areas in the Great Lakes Basin Ecosystem which are threatened by dredging and disposal activities.
4. The Parties shall encourage research to investigate advances in dredging technology and the pathways, fate and effects of nutrients and contaminants of dredged materials.
5. The Subcommittee shall undertake any other activities as the Water Quality Board may direct.

ANNEX 8

Discharges From Onshore and Offshore Facilities

1. *Definitions.* As used in this Annex:

   (a) ''Discharge'' means the introduction of polluting substances into receiving waters and includes, but is not limited to, any spilling, leaking, pumping, pouring, emitting or dumping; it does not include continuous effluent discharges from municipal or industrial treatment facilities;
   (b) ''Harmful quantity of oil'' means any quantity of oil that, if discharged into clear calm waters on a clear day, would produce a film or sheen upon, or discolouration of the surface of the water or adjoining shoreline, or that would cause a sludge or emulsion to be deposited beneath the surface of the water or upon the adjoining shoreline;
   (c) ''Facility'' includes motor vehicles, rolling stock, pipelines, and any other facility that is used or capable of being used for the purpose of processing, producing, storing, disposing, transferring or transporting oil or hazardous polluting substances, but excludes vessels;
   (d) ''Offshore facility'' means any facility of any kind located in, on or under any water;
   (e) ''Onshore facility'' means any facility of any kind located in, on or under, any land other than submerged land;
   (f) ''Oil'' means oil of any kind or in any form, including, but not limited to petroleum, fuel oil, oil sludge, oil refuse, and oil mixed with wastes, but does not include constituents of dredged spoil.

2. *Principles.* Regulations shall be adopted for the prevention of discharges into the Great Lakes System of harmful quantities of oil and hazardous polluting substances from onshore and offshore facilities in accordance with the following principles:
   (a) Discharges of harmful quantities of oil or hazardous polluting substances shall be prohibited and made subject to appropriate penalties;

(b) As soon as any person in charge has knowledge of any discharge of harmful quantities of oil or hazardous polluting substances, immediate notice of such discharge shall be given to the appropriate agency in the jurisdiction where the discharge occurs; failure to give this notice shall be made subject to appropriate penalties.

3. *Programs and Measures.* The programs and measures to be adopted shall include the following:

(a) Review of the design, construction, and location of both existing and new facilities for their adequacy to prevent the discharge of oil or hazardous polluting substances;

(b) Review of the operation, maintenance and inspection procedures of facilities for their adequacy to prevent the discharge of oil or hazardous polluting substances;

(c) Development and implementation of regulations and personnel training programs to ensure the safe use and handling of oil or hazardous polluting substances;

(d) Programs to ensure that at each facility plans and provisions are made and equipment provided to stop rapidly and safely, contain, and clean up discharges of oil or hazardous polluting substances; and

(e) Compatible regulations and other programs for the identification and placarding of containers, vehicles and other facilities containing, carrying or handling oil or hazardous polluting substances; and, where appropriate, notification to appropriate agencies of vehicle movements, maintenance of a registry, and identification in manifests of such substances to be carried.

4. *Implementation.*

(a) Each Party shall submit a report to the International Joint Commission outlining its programs and measures, existing or proposed, for the implementation of this Annex within six months of the date of entry into force of this Agreement.

(b) The report shall outline programs and measures, existing or proposed, for each of the following types of onshore and offshore facilities:

(i) land transportation including rail and road modes;

(ii) pipelines on land and submerged under water;

(iii) offshore drilling rigs and wells;

(iv) storage facilities both onshore and offshore; and

(v) wharves and terminals with trestle or underwater pipeway connections to land and offshore island type structures and buoys used for the handling of oil or hazardous polluting substances.

(c) The report shall outline programs and measures, existing or proposed, for any other type of onshore or offshore facility.

(d) Upon receipt of the reports, the Commission, in consultation with the Parties, shall review the programs and measures outlined for adequacy and compatibility and, if necessary, make recommendations to rectify any such inadequacy or incompatibility it finds.

ANNEX 9

Joint Contingency Plan

1. *The Plan.* The "Joint Canada-United States Marine Pollution Contingency Plan for the Great Lakes (CANUSLAK)" adopted on June 20, 1974, shall be maintained in force, as amended from time to time. The Canadian Coast Guard and the United States Coast Guard shall, in cooperation with other affected parties, identify and provide detailed Supplements for areas of high risk and of particular concern in augmentation of CANUSLAK. It shall be the responsibility of the United States Coast Guard and the Canadian Coast Guard to coordinate and to maintain the Plan and the Supplements appended to the Plan.
2. *Purpose.* The purpose of the Plan is to provide for coordinated and integrated response to pollution incidents in the Great Lakes System by responsible federal, state, provincial and local agencies. The Plan supplements the national, provincial and regional plans of the Parties.
3. *Pollution Incidents.*
   (a) A pollution incident is a discharge, or an imminent threat of discharge of oil, hazardous polluting substance or other substance of such magnitude or significance as to require immediate response to contain, clean up, and dispose of the material.
   (b) The objectives of the Plan in pollution incidents are:
      (i) To develop appropriate preparedness measures and effective systems for discovery and reporting the existence of a pollution incident within the area covered by the Plan;
      (ii) To institute prompt measures to restrict the further spread of the pollutant; and
      (iii) To provide adequate cleanup response to pollution incidents.
4. *Funding.* The costs of operations of both Parties under the Plan shall be borne by the Party in whose waters the pollution incident occurred, unless otherwise agreed.
5. *Amendment.* The Canadian Coast Guard and the United States Coast Guard are empowered to amend the Plan subject to the requirement that such amendments shall be consistent with the purpose and objectives of this Annex.

ANNEX 10

Hazardous Polluting Substances

1. The Parties shall:
   (a) Maintain a list, to be known as Appendix 1 of this Annex (hereinafter referred to as Appendix 1), of substances known to have toxic effects on aquatic and animal life and a risk of being discharged to the Great Lakes System;

(b) Maintain a list, to be known as Appendix 2 of this Annex (hereinafter referred to as Appendix 2), of substances potentially having such effects and such a risk of discharge, and to give priority to the examination of these substances for possible transfer to Appendix 1;

(c) Ensure that these lists are continually revised in the light of growing scientific knowledge; and

(d) Develop and implement programs and measures to minimize or eliminate the risk of release of hazardous polluting substances to the Great Lakes System.

2. Hazardous polluting substances to be listed in Appendix 1 shall be determined in accordance with the following procedures:

   (a) Selection of all hazardous substances for listing in Appendix 1 shall be based upon documented toxicological and discharge potential data which have been evaluated by the Parties and deemed to be mutually acceptable.

   (b) Revisions to Appendix 1 may be made by mutual consent of the Parties and shall be treated as amendments to this Annex for the purposes of Article XIII of this Agreement.

   (c) Using the agreed selection criteria, either Party may recommend at any time a substance to be added to the list in Appendix 1. Such substance need not previously have been listed in Appendix 2. The Party receiving the recommendation will have 60 days to review the associated documentation and either reject the proposed substance or accept the substance pending completion of appropriate procedural or domestic regulatory requirements. Cause for rejection must be documented and submitted to the initiating Party and may be the basis for any further negotiations.

3. The criteria to be applied to the selection of substances as candidates for listing in Appendix 1 are:

   (a) Acute toxicological effects, as determined by whether the substance is lethal to:

      (i) One-half of a test population of aquatic animals in 96 hours or less at a concentration of 500 milligrams per litre or less; or

      (ii) One-half of a test population of animals in 14 days or less when administered in a single oral dose equal to or less than 50 milligrams per kilogram of body weight; or

      (iii) One-half of a test population of animals in 14 days or less when dermally exposed to an amount equal to or less than 200 milligrams per kilogram body weight for 24 hours; or

      (iv) One-half of a test population of animals in 14 days or less when exposed to a vapour concentration equal to or less than 20 cubic centimeters per cubic meter in air for one hour; or

      (v) Aquatic flora as measured by a maximum specific growth rate or total yield of biomass which is 50 per cent lower than a control culture over 14 days in a medium at concentrations equal to or less than 100 milligrams per litre.

   (b) Risk of discharge into the Great Lakes System, as determined by:

      (i) Gathering information on the history of discharges or accidents;

(ii) Assessing the modal risks during transport and determining the use and distribution patterns;

(iii) Identifying quantities manufactured or imported.

4. Potentially hazardous polluting substances to be listed in Appendix 2 of this Annex shall be determined in accordance with the following procedures:

(a) Either Party may add new substances to Appendix 2 by notifying the other in writing that the substance is considered to be a potential hazard because of documented information concerning aquatic toxicity, mammalian and other vertebrate toxicity, phytotoxicity, persistence, bio-accumulation, mutagenicity, teratogenicity, carcinogenicity, environmental translocation or because of documented information on risk of discharge to the environment. The documentation of the potential hazard and the selected criteria upon which it is based will also be submitted.

(b) Removal of substances from Appendix 2 shall be by mutual consent of the Parties.

(c) The Parties shall give priority to the examination of substances listed in Appendix 2 for possible transfer to Appendix 1.

5. Programs and measures to control the risk of pollution from transport, storage, handling and disposal of hazardous polluting substances are contained in Annexes 4 and 8.

APPENDIX 1

Hazardous Polluting Substances

Acetaldehyde
Acetic Acid
Acetic Anhydride
Acetone Cyanohydrin
Acetyl Bromide
Acetyl Chloride
Acrolein
Acrylonitrile
Aldrin
Allyl Alcohol
Allyl Chloride
Aluminum Sulfate
Ammonia
Ammonium Acetate
Ammonium Benzoate
Ammonium Bicarbonate
Ammonium Bichromate
Ammonium Bifluoride
Ammonium Bisulfite
Ammonium Carbamate
Ammonium Carbonate
Ammonium Chloride

Ammonium Chromate
Ammonium Citrate, Dibasic
Ammonium Fluoborate
Ammonium Fluoride
Ammonium Hydroxide
Ammonium Oxalate
Ammonium Silicofluoride
Ammonium Sulfamate
Ammonium Sulfide
Ammonium Sulfite
Ammonium Tartrate
Ammonium Thiocyanate
Ammonium Thiosulfate
Amyl Acetate
Aniline
Antimony Pentachloride
Antimony Potassium Tartrate
Antimony Tribromide
Antimony Trichloride
Antimony Trifluoride
Antimony Trioxide
Arsenic Disulfide
Arsenic Pentoxide
Arsenic Trichloride
Arsenic Trioxide
Arsenic Trisulfide
Barium Cyanide
Benzene
Benzoic Acid
Benzonitrile
Benzoyl Chloride
Benzyl Chloride
Beryllium Chloride
Beryllium Fluoride
Beryllium Nitrate
Butyl Acetate
Butylamine
Butyric Acid
Cadmium Acetate
Cadmium Bromide
Cadmium Chloride
Calcium Arsenate
Calcium Arsenite
Calcium Carbide
Calcium Chromate
Calcium Cyanide
Calcium Dodecylbenzenesulfonate
Calcium Hydroxide
Calcium Hypochlorite

Calcium Oxide
Captan
Carbaryl
Carbon Disulfide
Chlordane
Chlorine
Chlorobenzene
Chloroform
Chlorosulfonic Acid
Chlorpyrifos
Chromic Acetate
Chromic Acid
Chromic Sulfate
Chromous Chloride
Cobaltous Bromide
Cobaltous Formate
Cobaltous Sulfamate
Coumaphos
Cresol
Cupric Acetate
Cupric Acetoarsenite
Cupric Chloride
Cupric Nitrate
Cupric Oxalate
Cupric Sulfate
Cupric Sulfate, Ammoniated
Cupric Tartrate
Cyanogen Chloride
Cyclohexane
2,4-D Acid
2,4-D Esters
Dalapon
DDT
Diazinon
Dicamba
Dichlobenil
Dichlone
Dichlorvos
Dieldrin
Diethylamine
Dimethylamine
Dinitrobenzene (mixed)
Dinitrophenol
Diquat
Disulfoton
Diuron
Dodecylbenzenesulfonic Acid
Endosulfan
Endrin

Ethion
Ethylbenzene
Ethylenediamine
EDTA
Ferric Ammonium Citrate
Ferric Ammonium Oxalate
Ferric Chloride
Ferric Fluoride
Ferric Nitrate
Ferric Sulfate
Ferrous Ammonium Sulfate
Ferrous Chloride
Ferrous Sulfate
Formaldehyde
Formic Acid
Fumaric Acid
Furfural
Guthion
Heptachlor
Hydrochloric Acid
Hydrofluoric Acid
Hydrogen Cyanide
Isoprene
Isopropanolamine Dodecylbenzenesulfonate
Kelthane
Lead Acetate
Lead Arsenate
Lead Chloride
Lead Fluoborate
Lead Fluoride
Lead Iodide
Lead Nitrate
Lead Stearate
Lead Sulfate
Lead Sulfide
Lead Thiocyanate
Lindane
Lithium Chromate
Malathion
Maleic Acid
Maleic Anhydride
Mercuric Cyanide
Mercuric Nitrate
Mercuric Sulfate
Mercuric Thiocyanate
Mercurous Nitrate
Methoxychlor
Methyl Mercaptan
Methyl Methacrylate

Methyl Parathion
Mevinphos
Mexacarbate
Monoethylamine
Monomethylamine
Naled
Naphthalene
Naphthenic Acid
Nickel Ammonium Sulfate
Nickel Chloride
Nickel Hydroxide
Nickel Nitrate
Nickel Sulfate
Nitric Acid
Nitrobenzene
Nitrogen Dioxide
Nitrophenol (mixed)
Paraformaldehyde
Parathion
Pentachlorophenol
Phenol
Phosgene
Phosphoric Acid
Phosphorus
Phosphorus Oxychloride
Phosphorus Pentasulfide
Phosphorus Trichloride
Polychlorinated Biphenyls
Potassium Arsenate
Potassium Arsenite
Potassium Bichromate
Potassium Chromate
Potassium Cyanide
Potassium Hydroxide
Potassium Permanganate
Propionic Acid
Propionic Anhydride
Pyrethrins
Quinoline
Resorcinol
Selenium Oxide
Sodium
Sodium Arsenate
Sodium Arsenite
Sodium Bichromate
Sodium Bifluoride
Sodium Bisulfite
Sodium Chromate
Sodium Cyanide

Sodium Dodecylbenzenesulfonate
Sodium Fluoride
Sodium Hydrosulfide
Sodium Hydroxide
Sodium Hypochlorite
Sodium Methylate
Sodium Nitrite
Sodium Phosphate, Dibasic
Sodium Phosphate, Tribasic
Sodium Selenite
Strontium Chromate
Strychnine
Styrene
Sulfuric Acid
Sulfur Monochloride
2,4,5-T Acid
2,4,5-T Esters
TDE
Tetraethyl Lead
Tetraethyl Pyrophosphate
Toluene
Toxaphene
Trichlorfon
Trichlorophenol
Triethanolamine Dodecylbenzenesulfonate
Triethylamine
Trimethylamine
Uranyl Acetate
Uranyl Nitrate
Vanadium Pentoxide
Vanadyl Sulfate
Vinyl Acetate
Xylene (mixed)
Xylenol
Zinc Acetate
Zinc Ammonium Chloride
Zinc Borate
Zinc Bromide
Zinc Carbonate
Zinc Chloride
Zinc Cyanide
Zinc Fluoride
Zinc Formate
Zinc Hydrosulfite
Zinc Nitrate
Zinc Phenolsulfonate
Zinc Phosphide
Zinc Silicofluoride
Zinc Sulfate

Zirconium Nitrate
Zirconium Potassium Fluoride
Zirconium Sulfate
Zirconium Tetrachloride

APPENDIX 2

Potential Hazardous Polluting Substances

Acridine
Allethrin
Aluminum Fluoride
Aluminum Nitrate
Ammonium Bromide
Ammonium Hypophosphite
Ammonium Iodide
Ammonium Pentaborate
Ammonium Persulfate
Antimony Pentafluoride
Antimycin A
Arsenic Acid
Barhan
Benfluralin
Bensulide
Benzene Hexachloride
Beryllium Sulfate
Butifos
Cadmium
Cadmium Cyanide
Cadmium Nitrate
Captafol
Carbophenothion
Chlorflurazole
Chlorothion
Chlorpropham
Chromic Chloride
Chromium
Chromyl Chloride
Cobaltous Fluoride
Copper
Crotoxyphos
Cupric Carbonate
Cupric Citrate
Cupric Formate
Cupric Glycinate
Cupric Lactate
Cupric Paraamino Benzoate
Cupric Salicylate

Cupric Subacetate
Cuprous Bromide
Demeton
Dibutyl Phthalate
Dicapthon
2,4-Dinitrochlorobenzene
p-Dinitrocresol
Dinocap
Dinoseb
Dioxathion
Dodine
EPN
Gold Trichloride
Hexachlorophene
Hydrogen Sulfide
m-Hydroxybenzoic Acid
p-Hydroxybenzoic Acid
Hydroxylamine
2-Hydroxyphenazine-1-Carboxylic Acid
Lactonitrile
Lead Tetraacetate
Lead Thiosulfate
Lead Tungstate
Lithium Bichromate
Malachite Green
Manganese Chloride, Anhydrous
MCPA
Mercuric Acetate
Mercuric Chloride
Mercury
Metam-Sodium
p-Methylamino-Phenol
2-Methyl-Napthoquinone
Neburon
Nickel Formate
Phenylmercuric Acetate
n-Phenyl Naphthylamine
Phorate
Phosphamidon
Picloram
Potassium Azide
Potassium Cuprocyanide
Potassium Ferricyanide
Propyl Alcohol
Pyridyl Mercuric Acetate
Rotenone
Silver
Silver Nitrate
Silver Sulfate

Sodium Azide
Sodium 2-Chlorotoluene-5-Sulfonate
Sodium Pentachlorophenate
Sodium Phosphate, Monobasic
Sodium Sulfide
Stannous Fluoride
Strontium Nitrate
Sulfoxide
Temephos
Thallium
Thionazin
1,2,4-Trichlorobenzene
Uranium Peroxide
Uranyl Sulfate
Zinc Bichromate
Zinc Potassium Chromate
Zirconium Acetate
Zirconium Oxychloride

ANNEX 11

Surveillance and Monitoring

1. Surveillance and monitoring activities shall be undertaken for the following purposes:
   - (a) *Compliance*. To assess the degree to which jurisdictional control requirements are being met.
   - (b) *Achievement of General and Specific Objectives*. To provide definitive information on the location, severity, areal or volume extent, frequency and duration of non-achievement of the Objectives, as a basis for determining the need for more stringent control requirements.
   - (c) *Evaluation of Water Quality Trends*. To provide information for measuring local and whole lake response to control measures using trend analyses and cause/effect relationships, and to provide information which will assist in the development and application of predictive techniques for assessing impact of new developments and pollution sources. The results of water quality evaluations will be used for:
     - (i) assessing the effectiveness of remedial and preventative measures and identifying the need for improved pollution control;
     - (ii) assessing enforcement and management strategies; and
     - (iii) identifying the need for further technology development and research activities.
   - (d) *Identification of Emerging Problems*. To determine the presence of new or hitherto undetected problems in the Great Lakes Basin Ecosystem, leading to the development and implementation of appropriate pollution control measures.
2. A joint surveillance and monitoring program necessary to ensure the attainment

of the foregoing purposes shall be developed and implemented among the Parties and the State and Provincial Governments. The Great Lakes International Surveillance Plan contained in the Water Quality Board Annual Report of 1975 and revised in subsequent reports shall serve as a model for the development of the joint surveillance and monitoring program.

3. The program shall include baseline data collection, sample analysis, evaluation and quality assurance programs (including standard sampling and analytical methodology, inter-laboratory comparisons, and compatible data management) to allow assessments of the following:
   (a) Inputs from tributaries, point source discharges, atmosphere, and connecting channels;
   (b) Whole lake data including that for nearshore areas (such as harbours and embayments, general shoreline and cladophora growth areas), open waters of the Lakes, fish contaminants, and wildlife contaminants; and
   (c) Outflows including connecting channels, water intakes and outlets.

ANNEX 12

Persistent Toxic Substances

1. *Definitions*. As used in this Annex:
   (a) "Persistent toxic substance" means any toxic substance with a half-life in water of greater than eight weeks;
   (b) "Half-life" means the time required for the concentration of a substance to diminish to one-half of its original value in a lake or water body;
   (c) "Early warning system" means a procedure to anticipate future environmental contaminants (i.e., substances having an adverse effect on human health or the environment) and to set priorities for environmental research, monitoring and regulatory action.
2. *General Principles*.
   (a) Regulatory strategies for controlling or preventing the input of persistent toxic substances to the Great Lakes System shall be adopted in accordance with the following principles:
      (i) The intent of programs specified in this Annex is to virtually eliminate the input of persistent toxic substances in order to protect human health and to ensure the continued health and productivity of living aquatic resources and man's use thereof;
      (ii) The philosophy adopted for control of inputs of persistent toxic substances shall be zero discharge.
   (b) The Parties shall take all reasonable and practical measures to rehabilitate those portions of the Great Lakes System adversely affected by persistent toxic substances.
3. *Programs*. The Parties, in cooperation with the State and Provincial Governments, shall develop and adopt the following programs and measures for the elimination of discharges of persistent toxic substances:

(a) Identification of raw materials, processes, products, by-products, waste sources and emissions involving persistent toxic substances, and quantitative data on the substances, together with recommendations on handling, use and disposition. Every effort shall be made to complete this inventory by January, 1982;

(b) Establishment of close coordination between air, water and solid waste programs in order to assess the total input of toxic substances to the Great Lakes System and to define comprehensive, integrated controls;

(c) Joint programs for disposal of hazardous materials to ensure that these materials such as pesticides, contaminated petroleum products, contaminated sludge and dredge spoils and industrial wastes are properly transported and disposed of. Every effort shall be made to implement these programs by 1980.

4. *Monitoring*. Monitoring and research programs in support of the Great Lakes International Surveillance Plan should be established at a level sufficient to identify:

   (a) Temporal and spatial trends in concentration of persistent toxic substances such as PCB, mirex, DDT, mercury and dieldrin, and of other substances known to be present in biota and sediment of the Great Lakes System;

   (b) The impact of persistent toxic substances on the health of humans and the quality and health of living aquatic systems;

   (c) The sources of input of persistent toxic substances; and

   (d) The presence of previously unidentified persistent toxic substances.

5. *Early Warning System*. An early warning system consisting of, but not restricted to, the following elements shall be established to anticipate future toxic substances problems:

   (a) Development and use of structure-activity correlations to predict environmental characteristics of chemicals;

   (b) Compilation and review of trends in the production, import, and use of chemicals;

   (c) Review of the results of environmental testing on new chemicals;

   (d) Toxicological research on chemicals, and review of research conducted in other countries;

   (e) Maintenance of a biological tissue bank and sediment bank to permit retroactive analysis to establish trends over time;

   (f) Monitoring to characterize the presence and significance of chemical residues in the environment;

   (g) Development and use of mathematical models to predict consequences of various loading rates of different chemicals;

   (h) Development of a data bank for storage of information on physical/chemical properties, toxicology, use and quantities in commerce of known and suspected persistent toxic substances.

6. *Human Health*. The Parties shall establish action levels to protect human health from the individual and interactive effects of toxic substances.

7. *Research*. Research should be intensified to determine the pathways, fate and effects of toxic substances aimed at the protection of human health, fishery

resources and wildlife of the Great Lakes Basin Ecosystem. In particular, research should be conducted to determine:

(a) The significance of effects of persistent toxic substances on human health and aquatic life;

(b) Interactive effects of residues of toxic substances on aquatic life, wildlife, and human health; and

(c) Approaches to calculation of acceptable loading rates for persistent toxic substances, especially those which, in part, are naturally occurring.

TERMS OF REFERENCE

For the Joint Institutions and the Great Lakes Regional Office

1. *Great Lakes Water Quality Board*

   (a) This Board shall be the principal advisor to the International Joint Commission with regard to the exercise of all the functions, powers and responsibilities (other than those functions and responsibilities of the Science Advisory Board pursuant to paragraph 2 of these Terms of Reference) assigned to the Commission under this Agreement. In addition, the Board shall carry out such other functions, related to the water quality of the boundary waters of the Great Lakes System, as the Commission may request from time to time.

   (B) The Water Quality Board, at the direction of the Commission, shall:

      (i) Make recommendations on the development and implementation of programs to achieve the purpose of this Agreement;

      (ii) Assemble and evaluate information evolving from such programs;

      (iii) Identify deficiencies in the scope and funding of such programs and evaluate the adequacy and compatibility of results;

      (iv) Examine the appropriateness of such programs in the light of present and future socio-economic imperatives; and

      (v) Advise the Commission on the progress and effectiveness of such programs and submit appropriate recommendations.

   (c) The Water Quality Board, on behalf of the Commission, shall undertake liaison and coordination between the institutions established under this Agreement and other institutions and jurisdictions which may address concerns relevant to the Great Lakes Basin Ecosystem so as to ensure a comprehensive and coordinated approach to planning and to the resolution of problems, both current and anticipated.

   (d) The Water Quality Board shall report to the Commission periodically as appropriate, or as required by the Commission, on all aspects relating to the operation and effectiveness of this Agreement.

2. *Great Lakes Science Advisory Board*

   (a) This Board shall be the scientific advisor to the Commission and the Water Quality Board.

   (b) The Science Advisory Board shall be responsible for developing recommendations on all matters related to research and the development of scientific knowledge pertinent to the identification, evaluation and resolution of current and anticipated problems related to Great Lakes water quality.

(c) To effect these responsibilities the Science Advisory Board shall:

(i) Review scientific information in order to:

a. examine the impact and adequacy of research and the reliability of research results, and ensure the dissemination of such results;

b. identify additional research requirements;

c. identify specific research programs for which international cooperation is desirable; and

(ii) Advise jurisdictions of relevant research needs, solicit their involvement and promote coordination.

(d) The Science Advisory Board shall seek analyses, assessments and recommendations from other scientific, professional, academic, governmental or intergovernmental groups relevant to Great Lakes Basin Ecosystem research.

(e) The Science Advisory Board shall report to the Commission and the Water Quality Board periodically as appropriate, or as required by the Commission, on all matters of a scientific or research nature relating to the operation and effectiveness of this Agreement.

3. *The Great Lakes Regional Office*

(a) This Office, located at Windsor, Ontario, shall assist the Commission and the two Boards in the discharge of the functions specified in subparagraph (b) below.

(b) The Office shall perform the following functions:

(i) Provide administrative support and technical assistance for the Water Quality Board and the Science Advisory Board and their suborganizations, to assist the Boards in discharging effectively the responsibilities, duties and functions assigned to them.

(ii) Provide a public information service for the programs, including public hearings, undertaken by the Commission and its Boards.

(c) The Office shall be headed by a Director who shall be appointed by the Commission in consultation with the Parties and with the Co-Chairmen of the Boards. The position of Director shall alternate between a Canadian citizen and a United States citizen. The term of office for the Director shall be determined in the review referred to in subparagraph (d) below.

(d) The Parties, mindful of the need to staff the Great Lakes Regional Office to carry out the functions assigned the Commission by this Agreement, shall, within six months from the date of entry into force of this Agreement, complete a review of the staffing of the Office. This review shall be conducted by the Parties based upon recommendations of the Commission after consultation with the Co-Chairmen of the Boards. Subsequent reviews may be requested by either Party, or recommended by the Commission, in order to ensure that the staffing of the Regional Office is maintained at a level and character commensurate with its assigned functions.

(e) Consistent with the responsibilities assigned to the Commission, and under the general supervision of the Water Quality Board, the Director shall be responsible for the management of the Regional Office and its staff in carrying out the functions described herein.

(f) The Co-Chairmen of the Boards, in consultation with the Director, will determine the activities which they wish the Office to carry out on behalf of, or in support of the Boards, within the current capability of the Office and its staff. The Director is responsible to the Co-Chairmen of each Board for activities carried out on behalf of, or in support of such Board, by the Office or individual staff members.

(g) The Commission, in consultation with the Director, will determine the public information activities to be carried out on behalf of the Commission by the Regional Office.

(h) The Director shall be responsible for preparing an annual budget to carry out the functions of the Boards and the Regional Office for submission jointly by the two Boards to the Commission for approval and procurement of resources.

AUTHOR INDEX

SUBJECT INDEX

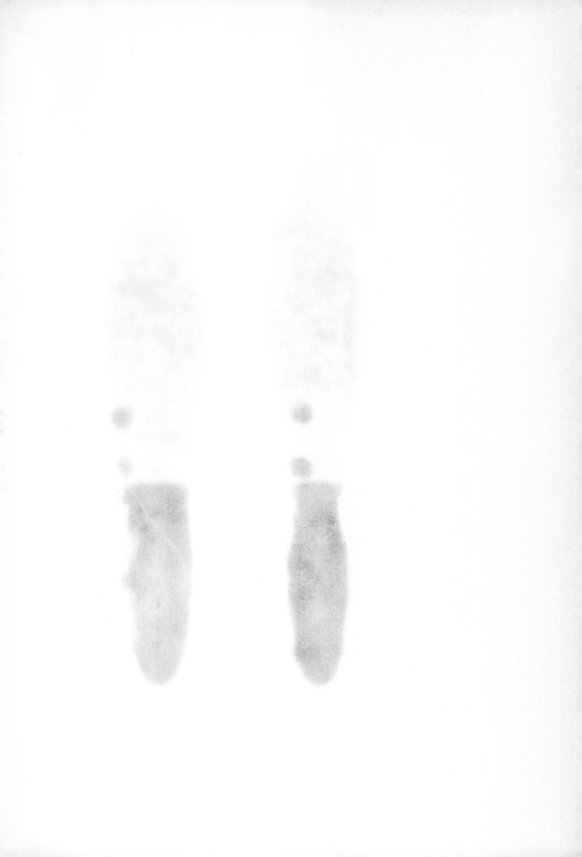

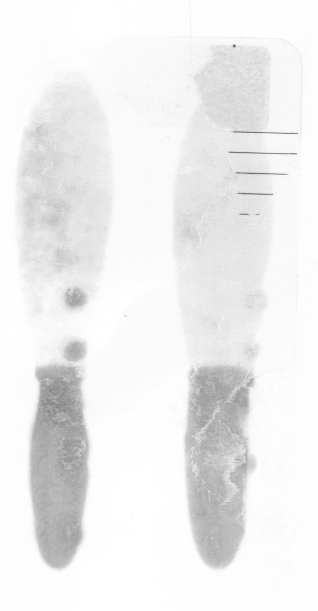